AF411435

Titanium Alloys and Titanium-Based Matrix Composites

Contents

About the Editor

Maciej Motyka is Associate Professor in the Department of Materials Science at the Faculty of Mechanical Engineering and Aeronautics and a researcher in the R&D Laboratory for Aerospace Materials, with both positions at the Rzeszow University of Technology. He graduated from the Faculty of Non-Ferrous Metals in AGH University of Science and Technology in Krakow in 1998 and earned his PhD degree in 2004 and DSc degree in 2015.

The scientific interests of Dr. Motyka cover relationships between processing, microstructure and mechanical properties of advanced structural materials, mainly titanium alloys. His activity is focused on hot plasticity and fine-grained superplasticity phenomena. He is also working on the characterization of ultrafine-grained materials obtained by severe plastic deformation and plastic consolidation methods and directionally solidified nickel-based superalloys.

Dr. Motyka is an author and co-author of over 100 papers in scientific journals and conference proceedings.

Editorial

Titanium Alloys and Titanium-Based Matrix Composites

Maciej Motyka

Department of Materials Science/R&D Laboratory for Aerospace Materials, Rzeszow University of Technology, 35-959 Rzeszow, Poland; motyka@prz.edu.pl; Tel.: +48-17-7432416

Citation: Motyka, M. Titanium Alloys and Titanium-Based Matrix Composites. *Metals* **2021**, *11*, 1463. https://doi.org/10.3390/met11091463

Received: 7 September 2021
Accepted: 15 September 2021
Published: 15 September 2021

Publisher's Note: MDPI stays neutral with regard to jurisdictional claims in published maps and institutional affiliations.

1. Introduction and Scope

Titanium alloys have been considered unique materials for many years. Their development has led to the design of several groups of structural alloys, including single-phase α or β alloys, two-phase α + β alloys—the most popular ones—and TiAl intermetallic alloys. The main application areas of titanium alloys include transportation (mainly aerospace structures), machine building, the fuel–energy industry and medicine. Titanium alloys are also good materials for metal matrix composites (MMCs). Their main attractions are high strength and stiffness—which depend on the type of reinforcement. An important feature of titanium is its remarkable biocompatibility. The low Young's modulus of titanium alloys—especially β alloys—makes them valuable biomaterials used for bone implants.

The constant development of titanium-based materials is also related to new technologies introduced—e.g., 3D printing or friction welding. It has restored the need for further research on titanium components produced using modern manufacturing and processing methods. This Special Issue covers various aspects of research on titanium alloys and Ti-based metal matrix composites—related to the development of their microstructure and operational properties by chemical composition modification, heat treatment and plastic deformation, but also by modern manufacturing and processing methods such as powder metallurgy, additive manufacturing or friction stir welding.

2. Contributions

Twenty articles—18 research ones, 1 review and 1 technical note—have been published in the present Special Issue of *Metals*. Their subject matter is quite extensive, encompassing the fields of deformation behavior, microstructure and properties development, as well as special applications (biomedical and functional ones) and technologies. For the ease of the reader, papers have been divided into three groups, although among them thematic links can be established.

2.1. Deformation Behavior

Four papers in this section concern the hot deformation behavior of metallic materials, including the evaluation of microstructure evolution and processing maps elaboration. Liu et al. [1] investigated the hot deformation process of metastable β Ti-6Cr-5Mo-5V-4Al alloy (Ti-6554), which is considered as a new type of high-strength and high-toughness titanium alloy for manufacturing large-scale aircraft structural parts. It is of great significance to improve the properties of the alloy. The results of hot compression tests at a temperature range of 715–840 °C and strain rate range of 0.001–1 s^{-1} were the basis of a processing maps calculation, enabling the determination of the instability regions in the process and improvement of forging effectiveness. Zherebtsov et al. [2] analyzed the deformation behavior of Ti-15Mo/TiB composite produced by spark plasma sintering. They found its gradual softening with an increase in deformation temperature in the range of 500–1000 °C to be associated with the dynamic recovery and recrystallization processes, in addition to the reorientation and shortening of TiB whiskers. The effect of pre-deformation on the hot plasticity of Ti-6Al-4V alloy was researched by Wu et al. [3]. It was found that pre-strain of 0.05 enhanced flow stress and elongation in tensile tests at a temperature of 850 °C. Moreover, pre-strain promoted dynamic recrystallization by increasing the deformation substructure. Zhang et al. [4] analyzed the asymmetry of tension–compression behavior of an extruded Ti-6.5Al-2Zr-1Mo-1V alloy (TA15) at a temperature range of 650–750 °C. The corresponding yield stress and asymmetric strain hardening behavior were studied.

2.2. Development of Microstructure and Operational Properties

The largest section contains ten papers related to the development of the microstructure and operational properties of titanium-based materials—alloys and composites—by heat treatment, plastic deformation methods and modification of chemical composition.

The first six papers concern titanium alloys manufactured by conventional metallurgical methods, though with a reference to 3D printing techniques. Guo et al. [5] investigated the precipitation behavior of the ω phase and $\omega \rightarrow \alpha$ transformation in Ti-5Al-5Mo-5V-1Cr-1Fe near-β alloy (Ti-55511) during isothermal ageing at 450 °C. It was observed that the size of α precipitates increase with increasing ageing time, resulting from the $\beta \rightarrow \alpha$ and $\omega \rightarrow \alpha$ transformations. Phase transformations upon ageing in the other β alloy were the subject of research conducted by Bartha et al. [6]. Ti-15Mo alloy was subjected to two techniques of intensive plastic deformation—high-pressure torsion and rotary swaging at room temperature—leading to the formation of an ultrafine-grained microstructure. It was found that, after isothermal ageing in both deformed conditions, precipitations of the α phase remain small and equiaxed even after ageing at 500 °C for 16 h. The other β alloy—Ti-Nb-Zr-Fe-O—was investigated by Cojocaru et al. [7]. The influence of complex thermo-mechanical processing on its mechanical properties was analyzed. The advantages and disadvantages of such treatment were analyzed for expanding the database of possible β-Ti bio-alloys that could be used, depending on the specific requirements of different biomedical implant applications. Ding et al. [8] evaluated the solidification microstructure of Ti–6Al–4V–xFe (x = 0.1, 0.3, 0.5, 0.7 and 0.9) alloys fabricated by levitation melting. The growth of grains in the function of Fe content, as well as the composition distribution mechanisms during the solidification process of the alloy, were discussed. Qiao et al. [9] researched the microstructure, corrosion behavior and tensile properties of a newly developed low-cost titanium alloy, Ti-4Al-2V-1Mo-1Fe, and compared it to the widespread Ti-6Al-4V alloy. It was revealed that the corrosion resistance and ductility of the Ti-4Al-2V-1Mo-1Fe alloy were higher than that of the Ti-6Al-4V alloy, with a slight reduction in strength. Motyka [10], in his review paper, pointed to another aspect. He noticed that the introduction of new manufacturing methods—e.g., 3D printing—requires further research on titanium alloys already considered as well-known and that have been described in detail in the case of conventional technologies. In his article, he focused on martensite formation and decomposition processes in two-phase titanium alloys and emphasized their important role in microstructure development during conventional and additive manufacturing processes.

The next four papers are devoted to titanium-based alloys and composites produced by powder metallurgy. The influence of the hot isostatic pressing (HIP) post-processing step on the microstructure, porosity and mechanical properties of Ti-35Nb-2Sn alloy was studied by Lario et al. [11]. They confirmed that field-assisted consolidation processes, such as HIP, can be employed to reduce residual porosity and to increase the chemical and phase homogeneity of sintered β titanium alloys intended for biomedical applications. Markovsky et al. [12] investigated the influence of a strain rate on the mechanical response and microstructure evolution of the selected titanium-based materials, i.e., Ti-6Al-4V alloy produced via a conventional cast and wrought technology, as well as fabricated using blended elemental powder metallurgy. It was found out, among others, that Ti-6Al-4V with a globular microstructure is characterized by high strength and high plasticity in comparison to this alloy with a lamellar microstructure, whereas Ti-6Al-4V obtained via the powder metallurgy method reveals the highest plastic flow stress, with good plasticity at the same time. Hou et al. [13] successfully manufactured TiB-whiskers-reinforced Ti-15Mo-3Al-

2.7Nb-0.2Si alloys by adding TiB_2 powder and pre-sintering followed by canned hot extrusion. It was observed that the extrusion led to grain refinement and the strengthening of the composite, especially at the process temperature of 1000 °C. Powder-metallurgy-titanium-based composites were also researched by Montealegre-Meléndez et al. [14]. Their study was aimed at the analysis of the reaction layer between the titanium matrix and reinforcement: B_4C particles and/or intermetallic Ti_xAl_y. The authors revealed that composites with ceramic reinforcement showed excellent hardness and good wear resistance.

2.3. Special Applications and Technologies

The last section comprises six papers and concerns special applications of titanium-based materials—such as the biomedical one—as well as modern technologies—such as additive manufacturing, plastic consolidation or friction stir processing. Regarding biomedicine, titanium alloys belong to the few biomaterials that naturally match the requirement of bone implants or bone tissue replacement in the human body. The mechanical and bio-functional behavior of a Ti-30Nb-13Ta alloy in the form of foam, fabricated by mechanical alloying and subsequent spark plasma sintering, was studied by Giner et al. [15]. The material was evaluated as a potential prosthetic biomaterial used for cortical bone replacement and compared with commercial, pure Ti used for bone replacement implants. Moiduddin et al. [16] proposed an integrated system methodology for the reconstruction of complex zygomatic bone defects, using Ti-6Al-4V ELI alloy, comprising several steps, right from the patient scan to implant fabrication, while maintaining proper aesthetic and facial symmetry. Finally, it was concluded that a mirror-designed titanium implant (fabricated by the electron beam melting method) satisfies the aesthetic, functional, and mechanical properties for efficient zygomatic bone reconstruction. Moreover, the authors stated that the proposed design methodology could also be applied for other bone reconstruction surgeries.

The next three papers describe the effects of unconventional manufacturing and processing methods. Topolski et al. [17] consolidated titanium chips into solid, bulk material by extrusion of the briquettes into the form of solid rods using the KOBO method. Structural aspects of the solid-state processing of various titanium chips were analyzed. It was found that the manufactured rods were consolidated and near fully dense. Kim et al. [18] applied the step rolling process for the production of ultrafine-grained Ti-6Al-4V sheets. This study clarified the effect of subsequent annealing on the tensile properties of step-rolled Ti-6Al-4V at room and elevated temperature. Mironov et al. [19] used friction stir processing for the development of a globular α microstructure in Ti-6Al-4V alloy. Characterization of the crystallographic aspects of such a microstructure was accomplished by the electron backscatter diffraction technique.

The last paper in this section covers issues related to micro-actuators based on shape memory alloys. The most popular among them is nitinol (NiTi), which in fact is a nickel-based alloy, but is quite often described together with functional titanium alloys. Shimoga et al., in their technical report [20], reflected on the characteristics of the NiTi coil spring structure with its phase transformations and thermal transformation properties. It was postulated that the micro-actuators based on NiTi could be used for advanced high-tech applications.

3. Conclusions and Outlook

A variety of interrelated topics have been raised in the present Special Issue of *Metals*, providing a wide overview of recent research developments on different aspects of titanium-based materials. The number of articles and the wide range of topics prove the continued interest in this group of materials. It is worth mentioning that the authors represent scientific institutions from 13 countries: Austria, China, the Czech Republic, Japan, Korea, Pakistan, Poland, Romania, Russia, Saudi Arabia, Spain, Ukraine and the USA.

As a Guest Editor of this Special Issue, I am very happy with the final result, and hope that the published papers will be useful to researchers working on titanium alloys and titanium-based composites. I would like to warmly thank all the authors for their contributions, and all of the reviewers for their efforts in ensuring a high-quality publication. Thanks are also due to the *Metals* Editorial Office, especially to Toliver Guo, Assistant Editor, who managed and facilitated the publication process.

Funding: This research received no external funding.

Conflicts of Interest: The author declares no conflict of interest.

References

1. Liu, Q.; Wang, Z.; Yang, H.; Ning, Y. Hot Deformation Behavior and Processing Maps of Ti-6554 Alloy for Aviation Key Structural Parts. *Metals* **2020**, *10*, 828. [CrossRef]
2. Zherebtsov, S.; Ozerov, M.; Klimova, M.; Moskovskikh, D.; Stepanov, N.; Salishchev, G. Mechanical Behavior and Microstructure Evolution of a Ti-15Mo/TiB Titanium–Matrix Composite during Hot Deformation. *Metals* **2019**, *9*, 1175. [CrossRef]
3. Wu, T.; Wang, N.; Chen, M.; Zuo, D.; Xie, L.; Shi, W. Effect of Pre-Strain on Microstructure and Tensile Properties of Ti-6Al-4V at Elevated Temperature. *Metals* **2021**, *11*, 1321. [CrossRef]
4. Zhang, C.; Li, D.; Li, X.; Li, Y. An Experimental Study of the Tension-Compression Asymmetry of Extruded Ti-6.5Al-2Zr-1Mo-1V under Quasi-Static Conditions at High Temperature. *Metals* **2021**, *11*, 1299. [CrossRef]
5. Guo, Y.; Wei, S.; Yang, S.; Ke, Y.; Zhang, X.; Zhou, K. Precipitation Behavior of ω Phase and $\omega \rightarrow \alpha$ Transformation in Near β Ti-5Al-5Mo-5V-1Cr-1Fe Alloy during Aging Process. *Metals* **2021**, *11*, 273. [CrossRef]
6. Bartha, K.; Stráský, J.; Veverková, A.; Veselý, J.; Čížek, J.; Málek, J.; Polyakova, V.; Semenova, I.; Janeček, M. Phase Transformations upon Ageing in Ti15Mo Alloy Subjected to Two Different Deformation Methods. *Metals* **2021**, *11*, 1230. [CrossRef]
7. Cojocaru, V.D.; Nocivin, A.; Trisca-Rusu, C.; Dan, A.; Irimescu, R.; Raducanu, D.; Galbinasu, B.M. Improving the Mechanical Properties of a β-type Ti-Nb-Zr-Fe-O Alloy. *Metals* **2020**, *10*, 1491. [CrossRef]
8. Ding, L.; Hu, R.; Gu, Y.; Zhou, D.; Chen, F.; Zhou, L.; Chang, H. Effect of Fe Content on the As-Cast Microstructures of Ti–6Al–4V–xFe Alloys. *Metals* **2020**, *10*, 989. [CrossRef]
9. Qiao, Y.; Xu, D.; Wang, S.; Ma, Y.; Chen, J.; Wang, Y.; Zhou, H. Corrosion and Tensile Behaviors of Ti-4Al-2V-1Mo-1Fe and Ti-6Al-4V Titanium Alloys. *Metals* **2019**, *9*, 1213. [CrossRef]
10. Motyka, M. Martensite Formation and Decomposition during Traditional and AM Processing of Two-Phase Titanium Alloys—An Overview. *Metals* **2021**, *11*, 481. [CrossRef]
11. Lario, J.; Vicente, Á.; Amigó, V. Evolution of the Microstructure and Mechanical Properties of a Ti35Nb2Sn Alloy Post-Processed by Hot Isostatic Pressing for Biomedical Applications. *Metals* **2021**, *11*, 1027. [CrossRef]
12. Markovsky, P.E.; Janiszewski, J.; Bondarchuk, V.I.; Stasyuk, O.O.; Savvakin, D.G.; Skoryk, M.A.; Cieplak, K.; Dziewit, P.; Prikhodko, S.V. Effect of Strain Rate on Microstructure Evolution and Mechanical Behavior of Titanium-Based Materials. *Metals* **2020**, *10*, 1404. [CrossRef]
13. Hou, J.; Gao, L.; Cui, G.; Chen, W.; Zhang, W.; Tian, W. Grain Refinement of Ti-15Mo-3Al-2.7Nb-0.2Si Alloy with the Rotation of TiB Whiskers by Powder Metallurgy and Canned Hot Extrusion. *Metals* **2020**, *10*, 126. [CrossRef]
14. Montealegre-Meléndez, I.; Arévalo, C.; Beltrán, A.M.; Kitzmantel, M.; Neubauer, E.; Pérez Soriano, E.M. Reaction Layer Analysis of In Situ Reinforced Titanium Composites: Influence of the Starting Material Composition on the Mechanical Properties. *Metals* **2020**, *10*, 265. [CrossRef]
15. Giner, M.; Chicardi, E.; Costa, A.d.F.; Santana, L.; Vázquez-Gámez, M.Á.; García-Garrido, C.; Colmenero, M.A.; Olmo-Montes, F.J.; Torres, Y.; Montoya-García, M.J. Biocompatibility and Cellular Behavior of TiNbTa Alloy with Adapted Rigidity for the Replacement of Bone Tissue. *Metals* **2021**, *11*, 130. [CrossRef]
16. Moiduddin, K.; Hammad Mian, S.; Umer, U.; Ahmed, N.; Alkhalefah, H.; Ameen, W. Reconstruction of Complex Zygomatic Bone Defects Using Mirroring Coupled with EBM Fabrication of Titanium Implant. *Metals* **2019**, *9*, 1250. [CrossRef]
17. Topolski, K.; Jaroszewicz, J.; Garbacz, H. Structural Aspects and Characterization of Structure in the Processing of Titanium Grade4 Different Chips. *Metals* **2021**, *11*, 101. [CrossRef]
18. Kim, G.; Lee, T.; Lee, Y.; Kim, J.N.; Choi, S.W.; Hong, J.K.; Lee, C.S. Ambivalent Role of Annealing in Tensile Properties of Step-Rolled Ti-6Al-4V with Ultrafine-Grained Structure. *Metals* **2020**, *10*, 684. [CrossRef]
19. Mironov, S.; Sato, Y.S.; Kokawa, H.; Hirano, S.; Pilchak, A.L.; Semiatin, S.L. Microstructural Characterization of Friction-Stir Processed Ti-6Al-4V. *Metals* **2020**, *10*, 976. [CrossRef]
20. Shimoga, G.; Kim, T.-H.; Kim, S.-Y. An Intermetallic NiTi-Based Shape Memory Coil Spring for Actuator Technologies. *Metals* **2021**, *11*, 1212. [CrossRef]

Article

Hot Deformation Behavior and Processing Maps of Ti-6554 Alloy for Aviation Key Structural Parts

Qi Liu, Zhaotian Wang, Hao Yang and Yongquan Ning *

School of Materials Science and Engineering, Northwestern Polytechnical University, Xi'an 710072, China; liuqi2018@mail.nwpu.edu.cn (Q.L.); wzt199604043138@163.com (Z.W.); yryanghao@126.com (H.Y.)
* Correspondence: luckyning@nwpu.edu.cn; Tel.: +86-29-8849-2642

Received: 2 June 2020; Accepted: 17 June 2020; Published: 21 June 2020

Abstract: With the development of the aviation industry, the performance requirements of materials for aviation large-scale structural parts are getting higher and higher. Ti-6554 alloy is the material of choice for aviation large-scale structural parts, but its forming process window is narrow and its microstructure is sensitive to process parameters, which affects the performance of the alloy. By adjusting the existing hot deformation process, it is of great significance to improve the properties of the alloy. Hot compression tests of Ti-6554 alloy were carried out at temperatures of 715–840 °C and strain rates of 0.001–1 s^{-1}. The results show that the flow stress and peak stress increased significantly with the increase of strain rate. At the same strain rate, the strain required for the stress to reach the peak point is smaller with the temperature increases. When the deformation temperature is below the phase transition point, the volume fraction and size of primary α phase gradually decrease with the increase of deformation temperature, while when the temperature is above the phase transition point, with the increase of deformation temperature, β grains grow up gradually, and the grain boundary bending effect is more obvious. The hyperbolic-sine Arrhenius constitutive equation was established. The correlation coefficient between experimental data and model calculated data reached 0.994. It indicates that the stress constitutive model proposed in this study can accurately reflect the stress characteristics of Ti-6554 alloy. Based on the dynamic material model, the processing maps of the alloy were established. The optimum hot deformation parameters range of the alloy was determined by analyzing the processing maps: the deformation temperature range of 800–830 °C, the strain rate range of 0.001–0.01 s^{-1}. Through the analysis of the processing maps, the instability regions in the process of cross-phase forging can be effectively avoided, and the performance of the forging can be effectively improved.

Keywords: Ti-6554 alloy; hot deformation; microstructure evolution; constitutive model; processing map

---

1. Introduction

Metastable β titanium alloy has high specific strength, deep hardenability, good corrosion resistance, and excellent processing performance, it is the material of choice for manufacturing large structural parts of aircraft [1–4]. Ti-6554 (Ti-6Cr-5Mo-5V-4Al) alloy is a new type of metastable beta high strength and high toughness titanium alloy, which is mainly used for manufacturing large-scale aviation structural parts, such as aircraft frames and landing gear. Its [Mo]$_{eq}$ is 13.3 and [Al]$_{eq}$ is 4 [5,6]. This alloy is a titanium alloy with high structural efficiency and excellent comprehensive performance. When the tensile strength of Ti-6554 alloy reaches 1270 MPa, its fracture toughness can reach more than 80 MPa.m$^{1/2}$. Its strength and toughness are better than BT-22 and Ti-1023 [7–9]. However, the alloy has large deformation resistance and poor thermal conductivity during thermal deformation. The flow stress and the microstructure are sensitive to the thermal deformation parameters [10,11]. In the actual production, various macroscopic and microscopic defects are easily generated, and the mechanical properties of the forgings are reduced. At present, there are few studies on the flow behavior and microstructure evolution of this alloy during thermal deformation. Hence, in order to obtain good performance of the alloy, it is necessary to study the flow behavior of the alloy.

During the thermal deformation of materials, the constitutive model can be used to describe the effects of deformation temperature, strain rate, and strain on the flow behavior, which is helpful to reduce the unnecessary forming defects and costs. In recent years, a large number of constitutive equations have been proposed or improved, which can be divided into three categories [12–14]: phenomenological, physically-based and artificial neural network (ANN). The phenomenological constitutive model is a model developed in the past to simulate the forming process of metals, alloys or composites at high temperature and high strain rate, Arrhenius constitutive relation is a typical representative of this constitutive model [15–17]. The model is widely used to describe the flow behavior of alloy materials such as titanium alloy [18,19], magnesium alloy [20], aluminum alloy [21], and nickel-base superalloy [22]. The physically-based constitutive model is based on the microscopic mechanism within the material to establish a model related to the deformation mechanism [23]. Due to the fact that the actual forming often involves multiple deformation mechanisms at the same time, the solution of the model parameters is relatively complicated, and the theory may deviate greatly from the reality, which leads to the difficulty in the test and application of the model. ANN model is composed of many simple and widely connected units. These units are arranged in layers, which can dynamically respond to input information and process information, and can describe complex nonlinear relationships [24]. However, ANN model has poor generality and requires long learning time. Therefore, for finite element simulation of engineering applications and macroscopic deformation, the phenomenological constitutive model is more practical.

The processing maps can be used as the basis to evaluate the machining performance of materials, and it is a powerful auxiliary tool for the process design of metal materials. At present, there are about two types of processing maps that are highly recognized. One is the Raj [25] processing maps based on the atomic model, and the other is the processing maps based on the dynamic material model (DMM). The Raj processing maps is only effective under steady-state conditions, and general complex alloys cannot be used directly, so it has great limitations in practical applications. The processing maps developed on the basis of DMM not only reflects the region where the specific deformation mechanism of specific microstructure is located, but also describes the unstable state that should be avoided during the hot processing [26]. Therefore, this paper chooses the processing maps based on DMM. The plastic deformation mechanism under different deformation conditions can be predicted by using the processing maps, and the unstable deformation parameters during the hot deformation process can be avoided. The cross-phase forging process can be guided by the processing maps.

In this study, the thermal deformation behavior of Ti-6554 alloy was studied by hot simulation compression test within the deformation temperatures of 715~840 °C and strain rates of 0.001 s^{-1}~1 s^{-1}. Through the analysis of the test results, the effects of deformation temperature and strain rates on the flow stress and microstructure evolution of the alloy were studied. The hyperbolic sinusoidal Arrhenius constitutive equation was established and the accuracy of the model was tested, which provides reference for the numerical simulation of the forming process of Ti-6554 alloy. Finally, the processing maps of the alloy was established, and the optimum range of hot processing parameters were determined. Through the analysis of the processing maps, the better forging process parameters of the cross-phase region were determined, which provides a reference for improving the service performance of forgings.

2. Materials and Methods

The test material was Ti-6554 alloy hot forging bar. It was prepared by three times of vacuum arc remelting, and the chemical composition is shown in Table 1. The phase transition point was 790 (±5) °C as measured by metallographic method. A sample having a diameter of 8 mm and a height of 12 mm was taken out from the original bar blank. Figure 1 shows the original microstructure of Ti-6554 alloy bar. From Figure 1, we can see that the microstructure of the sample is equiaxed α phase and β matrix, and its specific morphology is: along the grain boundary, there are slightly larger equiaxed α phase, while in the β matrix within the crystal, there are even smaller equiaxed α phases, and the number of equiaxed α phases in the crystal is obviously more. The volume fraction of the equiaxed α phase is about 13%, and the average diameter is about 5 μm.

The compression test was carried on a Gleebe-1500D thermal simulation tester (DSI, America) with six sets of deformation temperatures (715, 735, 755, 775, 810, 840 °C) and four sets of strain rates (0.001, 0.01, 0.1, 1 s^{-1}), the height was reduced by 60% to 4 mm. Before the experiment, tantalum pieces coated with glass lubricant were placed on both ends of the sample to reduce the friction between the table and the sample. During the experiment, the heating rate of the sample was set to 10 °C/s. When the temperature reached the set value, the sample was kept warm for 5 min in order to make the temperature of the center of the sample consistent with that of the surface. After the experiment, the samples were quenched immediately to retain the high temperature microstructure of the deformed samples. The system can automatically collect the relevant data in the process of compression. The sample was mechanically polished and then etched with a solution of 2 mL HF:10 mL HNO_3:88 mL H_2O. Observe and take metallographic photographs under the OLYMPUS GX-71 optical microscope (Olympus, Tokyo, Japan).

Table 1. Chemical composition of Ti-6554 alloy (mass fraction, %).

Cr	Mo	V	Al	Fe	Si	C	Ti
5.7	4.7	4.81	3.93	0.080	0.028	0.025	Bal.

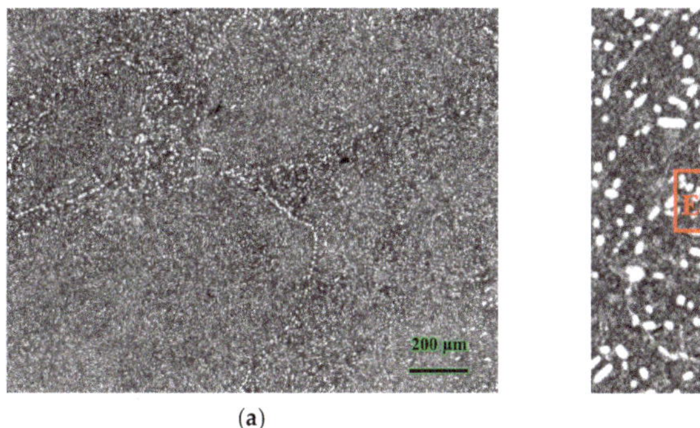
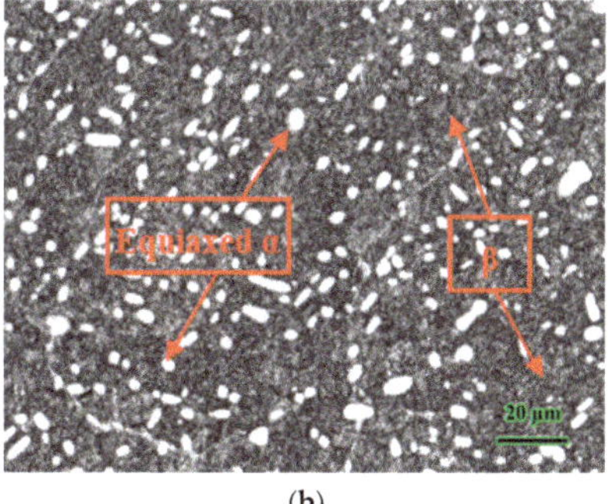

(a) (b)

Figure 1. The initial microstructure of Ti-6554 alloy bar. (**a**) Microstructure at 50× magnification; (**b**) Microstructure at 500× magnification.

3. Results and Discussion

3.1. Flow Stress Characteristics

The true stress–strain curves of Ti-6554 titanium alloy obtained through the thermal compression test are depicted in Figure 2. It can be seen from the figure that Ti-6554 titanium alloy has similar true stress–strain curve variation rule under different strain rates. In the early deformation stage, the flow stress rises rapidly with increasing strain. The reason is that at the early stage of deformation, dislocations proliferate and accumulate at the grain boundary through slip. The increase of dislocation density makes the critical shear stress of slip increase rapidly [27]. Therefore, in the early deformation stage, the overall work hardening rate of the deformed alloy is relatively high. When the strain reaches the critical strain of dynamic recrystallization (generally less than 0.1 [28]). With the beginning of dynamic recrystallization (DRX), the dynamic softening effect gradually increases, and the work hardening rate of the alloy decreases. The dynamic softening and work hardening reached the first dynamic equilibrium at the peak strain [29]. After that, it enters the dynamic softening stage, and the dislocation network generated in the previous stage evolves to form a large number of new crystal nuclei, which absorb the surrounding dislocation and gradually grow into dynamic recrystallized grains, reducing the dislocation density in the alloy and weakening the work hardening effect. At last, it entered the steady state flow stage. The dislocation growth rate in the alloy was basically equal to the consumption rate of the growth of new grain nucleation with the further increase of strain, and the work hardening and dynamic softening reached the second dynamic equilibrium.

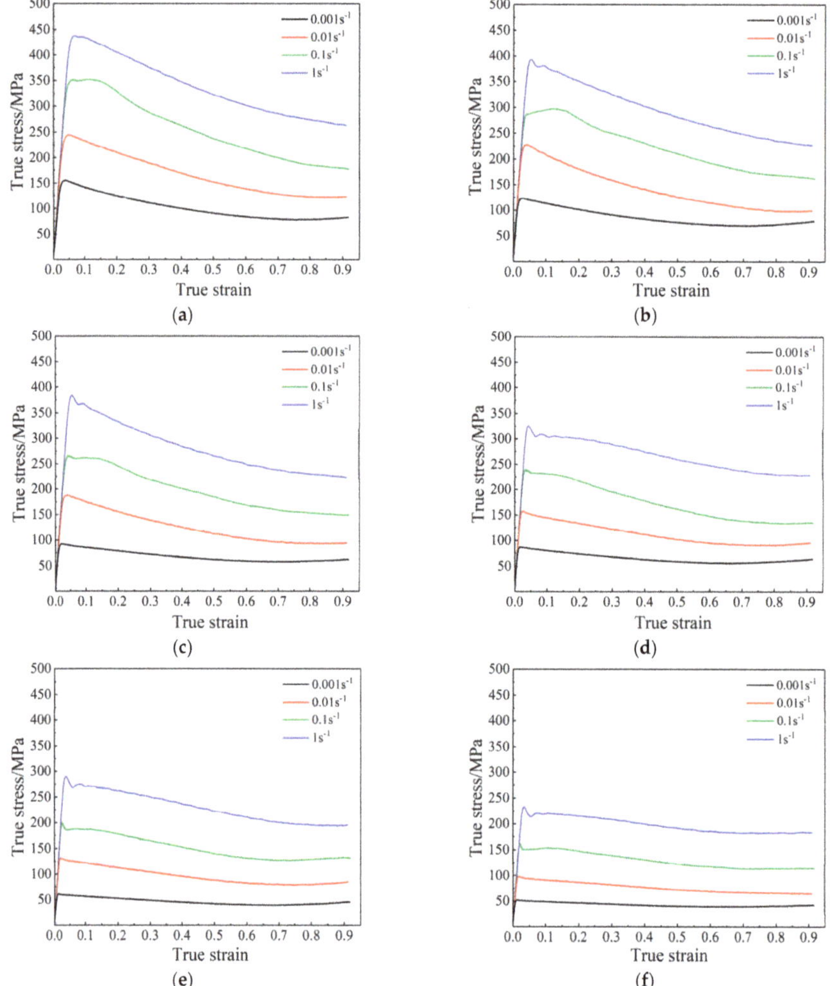

Figure 2. The true stress–strain curves of Ti-6554 alloy under deformation temperatures of (**a**) 715 °C; (**b**) 735 °C; (**c**) 755 °C; (**d**) 775 °C; (**e**) 810 °C; (**f**) 840 °C.

Thermal deformation process parameters have an important effect on the microstructure and properties of the material after hot forming. It can be seen from Figure 2 that the flow stress of Ti6554 alloy has high strain rate sensitivity, and its main characteristics are as follows:

(1) Under the same temperature conditions, the greater the strain rate, the greater the stress corresponding to the same strain. This is because when the strain rate is large, the time of unit strain is short, a large number of dislocations move simultaneously in a short time, which will increase the distortion degree of Ti-6554 titanium alloy, leading to the increase of critical shear strain force, the decrease of grain boundary slip momentum, and leading to the increase of work-hardening rate.

(2) Under the condition of high strain rate, the stress–strain curves have a sharp initial stress peak, discontinuous yield occurs, and then continues to soften. This is mainly because: in the elastic deformation stage, the dislocation density in Ti-6554 alloy is low, but when the plastic deformation occurs, the dislocations starts to move and multiply, when the dislocation movement to the grain boundary will be blocked and produce plug accumulation, so that the stress increases rapidly. As the dislocation accumulates until the critical value is exceeded, dynamic recovery (DRV) will occur in the β phase, and the dislocation of the plugging product will move through climbing, cross slip, and other means, and a large number of dissimilar dislocations will annihilate each other, so that the dislocation density in the grain will decrease, which shows that the stress will drop significantly in a short time.

(3) At the same strain rate, the strain required for the stress to reach the peak point is smaller with the temperature increases. This is mainly because: when the alloy is deformed at a higher temperature, a small amount of recrystallization occurs under a smaller strain, which causes the alloy to enter the softening stage in advance and presents the characteristics of an earlier peak point in the curve.

3.2. Establishment of Constitutive Model

Arrhenius equation is widely used to describe the relationship between material flow stress and deformation parameters [30]. Its specific equation is:

$$\dot{\varepsilon} = AF(\sigma)\exp\left(-\frac{Q}{RT}\right) \tag{1}$$

$F(\sigma)$ is the stress function, which has the following three forms under different stress states:

$$F(\sigma) = \begin{cases} \sigma^{n'} & \alpha\sigma < 0.8 \\ \exp(\beta\sigma) & \alpha\sigma > 1.2 \\ [\sinh(\alpha\sigma)]^{n} & \text{for all } \sigma \end{cases} \tag{2}$$

where $\dot{\varepsilon}$ is deformation strain rate (s^{-1}), T is deformation temperature (K), σ is the flow stress (MPa), R is gas constant (8.31 J·mol^{-1}·K^{-1}), Q is activation energy (KJ·mol^{-1}), $\alpha = \beta/n'$, A, β, n, and n′ are material constants. The power exponent-type is usually the preferred method for describing creep, but it cannot be used when the stress is large. Under the conditions of relatively low temperature or high strain rate, the exponential type is applicable. The hyperbolic-sine type can be applied to various temperatures and strain rates [31]. Therefore, the hyperbolic sine is more general expression. In order to obtain the above material constants, substitute Equation (2) into Equation (1), and then take natural logarithm for both sides at the same time to get the following formula:

$$\ln\dot{\varepsilon} = n'\ln\sigma + \left(\ln A_1 - \frac{Q}{RT}\right) \tag{3}$$

$$\ln\dot{\varepsilon} = \beta\sigma + \left(\ln A_2 - \frac{Q}{RT}\right) \tag{4}$$

$$\ln\dot{\varepsilon} = n\ln[\sin h(\alpha\sigma)] + \left(\ln A - \frac{Q}{RT}\right) \tag{5}$$

In addition, the thermal deformation condition can be expressed by the Zener-Hollomon [32] parameter as follows:

$$Z = \dot{\varepsilon}\exp\left(\frac{Q}{RT}\right) = AF(\sigma) = A[\sin h(\alpha\sigma)]^{n} \tag{6}$$

Through the transformation of Equation (6), the expression of stress can be obtained:

$$\sigma = \frac{1}{\alpha} \ln \left\{ \left(\frac{Z}{A} \right)^{\frac{1}{n}} + \left[\left(\frac{Z}{A} \right)^{\frac{2}{n}} + 1 \right]^{\frac{1}{2}} \right\} \tag{7}$$

The results of thermal simulation compression test show that the stress value of Ti-6554 alloy varies greatly under different thermal deformation parameters. However, according to Figure 2, most of the stress values tend to be stable or only change in a small range as the strain increases when the true strain reaches 0.6. It can be considered that the Ti-6554 alloy has entered the steady state flow stage. Therefore, stress $\sigma_{0.6}$ is selected to establish the constitutive equation of Ti-6554 alloy. Since the large difference between the deformation microstructure of the titanium alloy in different phase region, in order to further understand its phase region characteristics, the constitutive equations of the $\alpha + \beta$ phase region and the β region are solved separately.

According to Equations (3) and (4), the constants $n' = \partial \ln \dot{\varepsilon} / \partial \ln \sigma$ and $\beta = \partial \ln \dot{\varepsilon} / \partial \sigma$ are the coefficients of $\ln \sigma - \ln \dot{\varepsilon}$ and $\sigma - \ln \dot{\varepsilon}$ at a certain strain, respectively. When the true strain is 0.6, the relationship diagrams of $\ln \sigma - \ln \dot{\varepsilon}$ and $\sigma - \ln \dot{\varepsilon}$ at different temperatures are drawn respectively, as depicted in Figure 3. The least square method is used to fit the data points linearly, and the values of n' and β in different phase regions can be obtained, the specific values are shown in Table 2. Hence, we can derive the value of α in different phase regions according to formula $\alpha = \beta / n'$, and the specific α value is shown in Table 2.

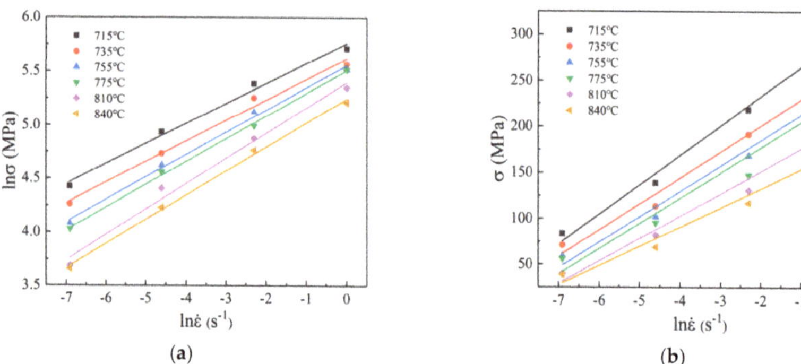

Figure 3. The relationships of (**a**) $\ln \sigma$ and $\ln \dot{\varepsilon}$; (**b**) σ and $\ln \dot{\varepsilon}$.

Table 2. Constitutive model parameters for Ti-6554 alloy.

Model Parameters	Parameters Values	
	A + β Phase	β Phase
n'	5.003402	4.331211
β	0.034779	0.04428
α	0.006951	0.010223
n	3.706985	3.197382
Q	244.8959	183.815
$\ln A$	24.82687	16.08101

Find the partial differential of Equation (5) about $\ln [\sinh(\alpha \sigma)]$, and sort it out as follows:

$$n = \frac{\partial \ln \dot{\varepsilon}}{\partial \ln [\sinh(\alpha \sigma)]} \tag{8}$$

The variation of $\ln[\sinh(\alpha\sigma)]$ with $\ln\dot{\varepsilon}$ as shown in Figure 4a. The data points at different temperatures are fitted into a straight line, and the inverse of the slope is the approximate value of n at that temperature. Finding the average of the inverse of the slope of the straight line at different phase region temperature is the value of n, as shown in Table 2.

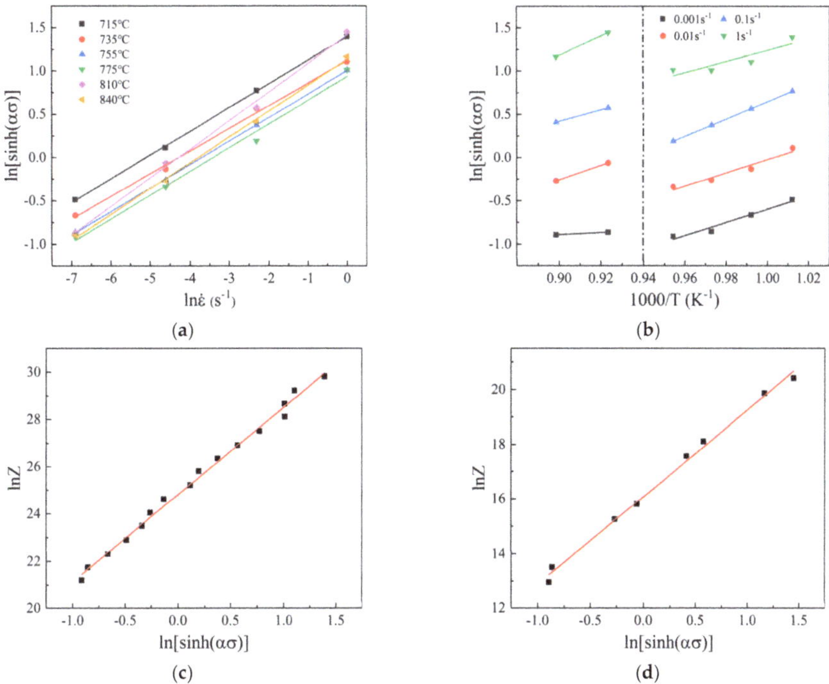

Figure 4. The relationships of (**a**); (**b**) $\ln[\sinh(\alpha\sigma)]$ and $1000/T$; (**c**) $\ln Z$ and $\ln[\sinh(\alpha\sigma)]$ in $\alpha + \beta$ phase; (**d**) $\ln Z$ and $\ln[\sinh(\alpha\sigma)]$ in β phase.

At a given strain rate, solve both sides of Equation (6) for partial differentials about T^{-1} at the same time, and sort out the following equation:

$$Q = nR \left. \frac{\partial \ln[\sinh(\alpha\sigma)]}{\partial\left(\frac{1}{T}\right)} \right|_{\varepsilon,\dot{\varepsilon}} \tag{9}$$

The theory of plastic deformation believes that the deformation activation energy is an energy barrier that must be overcome in the process of atomic transition of the material during thermal deformation, and it can reflect the difficulty of material deformation. According to Figure 4b, Q-value is further determined by the linear fitting of $\ln[\sinh(\alpha\sigma)]$ and $1000/T$. Table 3 lists the Q values of several typical high strength-toughness titanium alloys during processing in different phase regions. During the β-phase region deformation, the Q values of all the listed alloys are small, which may be related to the difference in the crystal structure of the α-phase and the β-phase. The Q of the pure titanium β phase is 153 KJ/mol, and the Q of the α phase is 242 KJ/mol [33]. Under different deformation conditions, the thermal deformation activation energy obtained reflects the thermal deformation mechanism of the material to a certain extent. The Q value of the Ti-6554 alloy when deformed in the β phase region is 184 KJ/mol, which is between the Q values of α phase and β phase, indicating that the main deformation mechanism may be the DRV caused by dislocation slip and climbing. The Q value during deformation in the $\alpha + \beta$ phase region is 245 KJ/mol, which is slightly higher than the pure titanium α phase, indicating that the main deformation mechanism may be dynamic and metastable recrystallization.

Table 3. Activation energy for several typical high strength-toughness titanium alloys.

Alloy	Transformation Point $T_\beta/°C$	Deformation Temperature $T/°C$	Strain Rate $\dot{\varepsilon}/s^{-1}$	$Q/kJ \cdot mol^{-1}$
Ti-1300 [34]	875	860~890 800~860	10^{-2}~10	178 216
Ti-55511 [35]	845	850~950 700~800	10^{-3}~10	137 288
Ti-55531 [36]	803	823~843 763~783	10^{-2}~1	148 275
Ti-6554	790	810~840 715~775	10^{-3}~10	184 245

Substituting Q, n, α, T, and $\dot{\varepsilon}$ into Equation (6), the Z values of different thermal deformation parameters can be obtained. As shown in Figure 4c,d, draw the relation diagram of $\ln Z - \ln[\sin h(\alpha\sigma)]$ and perform linear fitting with the least square method. The intercept of the line is the value of lnA.

Substitute all the above data into Equation (5), the flow stress model of Ti-6554 alloy under different process parameters can be obtained as follows:

$\alpha + \beta$ two-phase region:

$$\dot{\varepsilon} = 6.06 \times 10^{10}[\sinh(0.006951\sigma)]^{3.70} \exp\left(-\frac{2.45 \times 10^5}{RT}\right)$$

β single phase region:

$$\dot{\varepsilon} = 9.64 \times 10^6[\sinh(0.010223\sigma)]^{3.20} \exp\left(-\frac{1.84 \times 10^5}{RT}\right)$$

3.3. Verification of Different Strain Constitutive Models

So far, the constitutive equation of a certain true strain of 0.6 has been obtained. However, during thermal deformation of metallic materials, the effect of strain on flow stress is very obvious, so in order to describe the flow stress of materials more accurately, it is necessary to verify the constitutive model under different strains. Hence, the material constants (a, lnA, n, Q) corresponding to different strains can be obtained by the above method. Then select Equations (10)–(13) to perform fifth-order polynomial regression on the four material constants. The polynomial fitting curves of the material constants to the true-stress as depicted in Figure 5. The polynomial fitting results are shown in Tables 4 and 5.

$$\alpha = X_0 + X_1\varepsilon + X_2\varepsilon^2 + X_3\varepsilon^3 + X_4\varepsilon^4 + X_5\varepsilon^5 \tag{10}$$

$$n = N_0 + N_1\varepsilon + N_2\varepsilon^2 + N_3\varepsilon^3 + N_4\varepsilon^4 + N_5\varepsilon^5 \tag{11}$$

$$Q = Q_0 + Q_1\varepsilon + Q_2\varepsilon^2 + Q_3\varepsilon^3 + Q_4\varepsilon^4 + Q_5\varepsilon^5 \tag{12}$$

$$\ln A = Y_0 + Y_1\varepsilon + Y_2\varepsilon^2 + Y_3\varepsilon^3 + Y_4\varepsilon^4 + Y_5\varepsilon^5 \tag{13}$$

Through the above method, we can build constitutive models under different strains. In order to verify the reliability of the established Arrhenius constitutive model, as shown in Figure 6, the model calculated stress value is compared with the experimentally stress value. The observation shows that the deviation between the theoretical value and the measured value is not large, which indicates that the established Arrhenius model is reliable. To further quantitatively test the accuracy of the proposed stress constitutive model of Ti-6554 alloy, the correlation coefficient R and the average absolute relative error (AARE) are used to estimate the model accuracy. The mathematical expression is as follows:

$$R = \frac{\sum_{i=1}^{N}\left(E_i - \overline{E}\right)\left(P_i - \overline{P}\right)}{\sqrt{\sum_{i=1}^{N}\left(E_i - \overline{E}\right)^2 \sum_{i=1}^{N}\left(P_i - \overline{P}\right)^2}} \tag{14}$$

$$AARE(\%) = \frac{1}{N}\sum_{i=1}^{N}\left|\frac{E_i - P_i}{E_i}\right| \times 100 \tag{15}$$

where E_i is the measured stress, P_i is the model calculated stress, $\overline{E}$ is the measured stress average. As shown in Figure 7, the values of R and AARE are 0.994 and 3.80%, respectively. This indicated that the established constitutive model has good precision. From the above qualitative and quantitative analysis, it can be seen that the constitutive model proposed in this study can accurately reflect the stress characteristics of Ti-6554 alloy, and has good precision, thus verifying the reliability of the model.

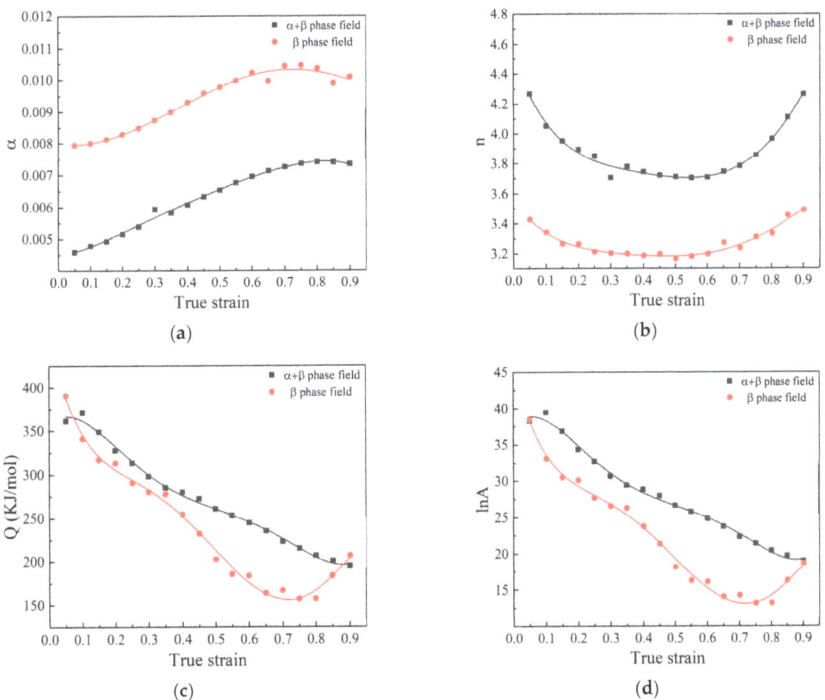

Figure 5. The fitted curves of (**a**) α; (**b**) n; (**c**) Q; (**d**) lnA.

Table 4. Polynomial fitting results of a, lnA, n, Q in α + β phase region.

	α		n		Q		lnA
X_0	0.00449	N_0	4.53506	Q_0	356.55303	Y_0	37.73193
X_1	0.00117	N_1	−6.57274	Q_1	390.72639	Y_1	46.276
X_2	0.01945	N_2	23.81874	Q_2	−4473.36504	Y_2	−531.88745
X_3	−0.04781	N_3	−45.73538	Q_3	12113.23612	Y_3	1449.71709
X_4	0.05363	N_4	42.39728	Q_4	−13845.55598	Y_4	−1663.14237
X_5	−0.02422	N_5	−13.74889	Q_5	5700.25554	Y_5	685.98485

Table 5. Polynomial fitting results of a, lnA, n, Q in β phase region.

	α		n		Q		lnA
X_0	0.00794	N_0	3.56092	Q_0	457.62985	Y_0	42.82754
X_1	−0.000679	N_1	−3.21573	Q_1	−1759.56953	Y_1	−191.89453
X_2	0.01243	N_2	12.24894	Q_2	8148.57959	Y_2	890.00338
X_3	0.00287	N_3	−25.02692	Q_3	−19516.71413	Y_3	−2131.70623
X_4	−0.2847	N_4	25.43849	Q_4	20285.7594	Y_4	2216.20461
X_5	0.01556	N_5	−9.37857	Q_5	−7368.92828	Y_5	−805.47528

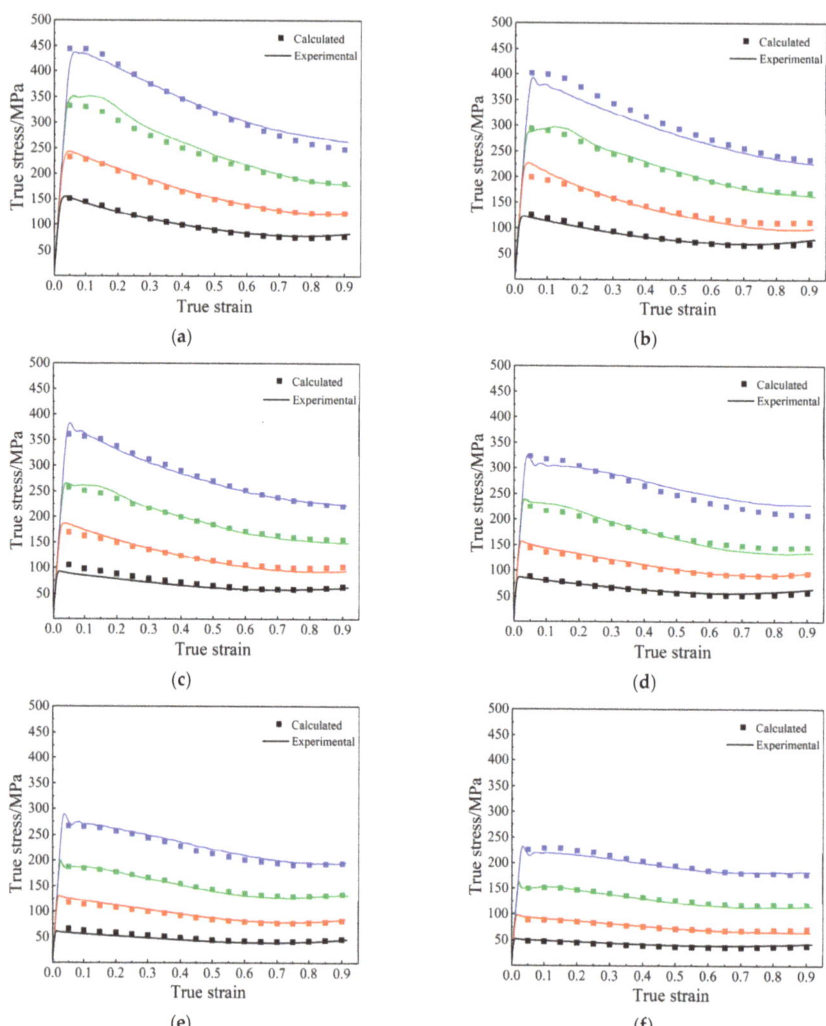

Figure 6. Comparison between experimental and model calculated flow stress at deformation temperatures of (**a**) 715 °C; (**b**) 735 °C; (**c**) 755 °C; (**d**) 775 °C; (**e**) 810 °C; (**f**) 840 °C.

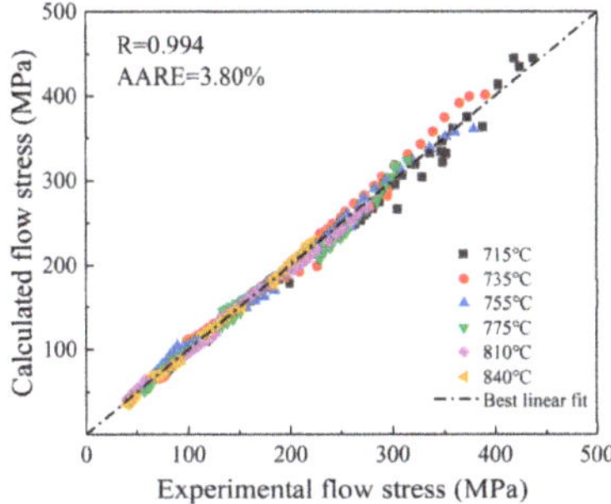

Figure 7. Correlations between experimental and model calculated flow stress.

3.4. Establishment of Processing Map

The processing map is mainly established through the DMM proposed by Prasad and Gegel [37]. It can evaluate the processing performance of metal materials and optimize the process parameters, and can achieve the role of controlling product performance and microstructure evolution. According to the DMM, the workpiece undergoing thermal deformation is a non-linear power dissipation unit, and the power input P of the workpiece during the hot forming process can be divided into two parts: dissipation content (G) and dissipation co-content (J), as follows:

$$P = \sigma\dot{\varepsilon} = G + J = \int_0^{\dot{\varepsilon}} \sigma d\dot{\varepsilon} + \int_0^{\sigma} \dot{\varepsilon}\, d\sigma \tag{16}$$

Dissipation content (G) refers to the power dissipated during thermal deformation, and most of the dissipation is converted into thermal energy, and a small part is stored in the workpiece; dissipation co-content (J) refers to the power consumed in the process of microstructure evolution during thermal deformation of materials, which is finally released as thermal energy. The distribution relationship of the total input power P between the dissipation content G and dissipation co-content J is determined by strain rate sensitivity index (m), as follows:

$$m = \frac{\partial J}{\partial G} = \frac{\dot{\varepsilon}\partial\sigma}{\sigma\partial\dot{\varepsilon}} = \frac{\partial\ln\sigma}{\partial\ln\dot{\varepsilon}} \tag{17}$$

For an ideal linear dissipative unit, the m value is 1, and J has a maximum value $J = J_{max} = \frac{\sigma\dot{\varepsilon}}{2}$. For the nonlinear dissipation process, the power dissipation efficiency η can be used to describe the efficiency of the energy consumed by microstructural changes during plastic deformation, which is shown as the following:

$$\eta = \frac{J}{J_{max}} = \frac{2m}{m+1} \tag{18}$$

By calculating the value of η, the power dissipation maps under different hot deformation conditions can be obtained. Generally, the power dissipation efficiency is proportional to the deformation mechanism, the higher the value of η, the better the deformation mechanism. In order to deal with the deformation instability of materials during hot working, Prasad et al. [38] proposed an instability criterion that can distinguish the thermal deformation instability region from the stability region of materials. The conditions of deformation instability of materials are as follows:

$$\xi(\dot{\varepsilon}) = \frac{\partial\ln\left(\frac{m}{m+1}\right)}{\partial\ln\dot{\varepsilon}} + m < 0 \tag{19}$$

15

The region with a negative parameter $\xi(\dot{\varepsilon})$ is the instability region, which is the processing risk region, and may produce adiabatic shear band, crack, mechanical twinning, and so on. By changing the instability criterion, the instability maps of the material can be drawn, and the processing maps of the material can be obtained by superposing the instability maps with the power dissipation maps. Based on the above method, the power dissipation maps under different strains are shown in Figure 8. It can be seen from Figure 8 that the peak value of power dissipation efficiency of the alloy varies greatly under different strains, and the peak value of power dissipation efficiency decreases with the increase of strain. When the strain is 0.3, at the deformation temperature of 790–840 °C and the strain is less than 0.01 s^{-1}, the peak power dissipation efficiency can reach 0.56, which indicates that the better microstructure can be obtained by hot working in this range. With the increase of strain, the peak area of power dissipation efficiency is concentrated in the deformation temperature of 800–830 °C, and the strain rate is 0.01–0.001 s^{-1}, which indicates that this range is the better processing parameter range of Ti-6554 alloy. At this time, the material's DRX is more sufficient, the microstructure is fine, and it has good mechanical properties. Conversely, the shaded area in the processing maps (as shown in Figure 9) shows the characteristic of instability. The instability region is basically the same under different strains, mainly concentrated in the low temperature (715–750 °C) and high strain rate (0.1–1 s^{-1}) area. However, when the strain reached 0.9, a small part of the instability region was added near the medium temperature (770–795 °C) and high strain rate (0.001–0.003 s^{-1}) region. Therefore, in the process of hot working, processing in these regions should be avoided, otherwise it will lead to uneven distribution of material structure, crack, and mechanical property decline, which will seriously deteriorate the mechanical properties of the material.

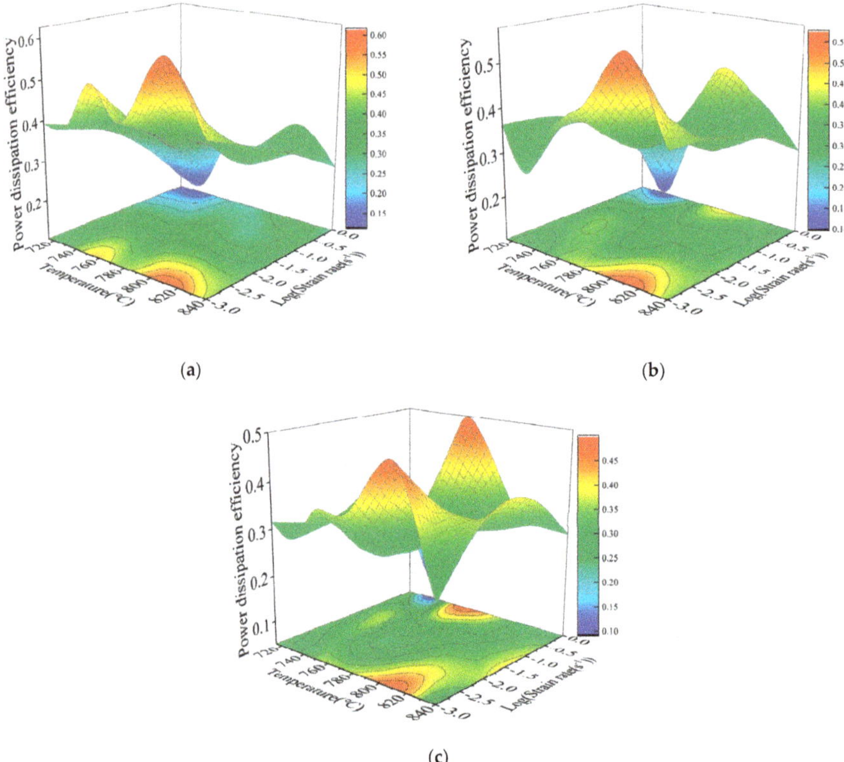

Figure 8. The power dissipation map of Ti-6554 alloy at the true strain of: (**a**) 0.3; (**b**) 0.6; (**c**) 0.9.

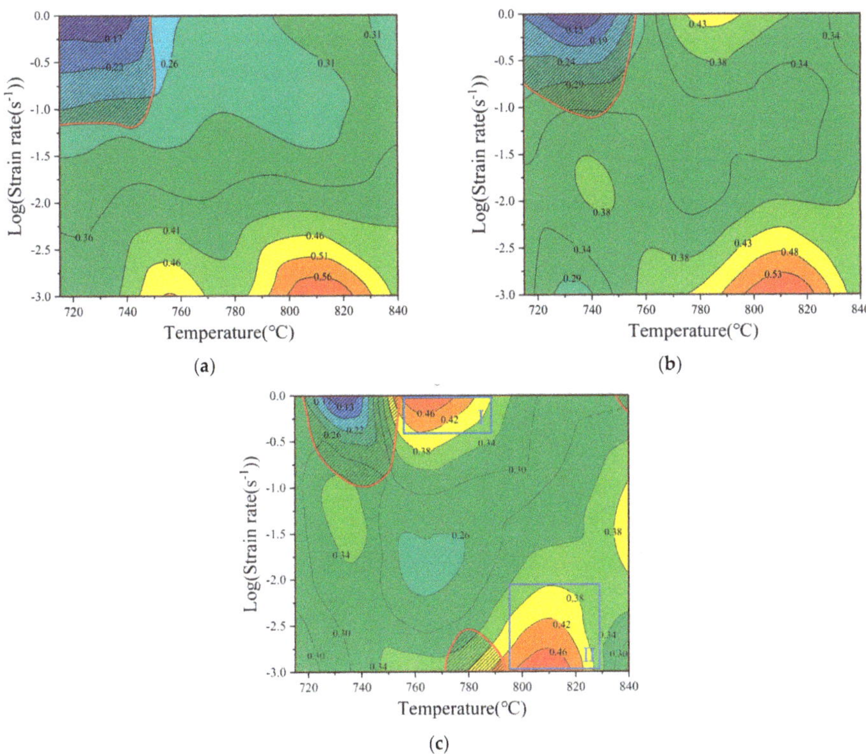

Figure 9. Processing maps of Ti-6554 alloy at the true strain of: (**a**) 0.3; (**b**) 0.6; (**c**) 0.9.

As shown in Figure 9c, when the true strain is 0.9, the stable deformation region is composed of two regions: domain I with the deformation temperatures of 760–790 °C and strain rates of 0.178–1 s^{-1}, the η-value is 0.34–0.46; domain II with the deformation temperatures of 800–830 °C and strain rates of 0.001–0.01 s^{-1}, the η-value is 0.38–0.46. However, the range of effective deformation temperature and strain rate in domain I is narrow, which is not conducive to large-scale processing of the alloy. Therefore, the optimal hot processing area determined by the processing map are the deformation temperatures of 800–830 °C and the strain rates of 0.001–0.01 s^{-1}. The following shows a typical microstructure of instability region and safety region, as shown in Figure 10. Among them, Figure 10a shows the microstructure in the instability area. It can be seen from the figure that the α phase is elongated along the direction perpendicular to the compression direction, and the aspect ratio of some α phases reaches 10:1, which is easy to be broken into fine α phase. Moreover, due to the high strain rate, the center area of the alloy is prone to a large temperature rise effect, and the thermal conductivity of Ti6554 alloy is poor, which is easy to cause unstable tissue properties. Figure 10b shows the microstructure in the safety area, mainly composed of β grains, and contains a small amount of equiaxed β grains, indicating that DRX has occurred under this condition, and the alloy has better overall performance.

Through the above analysis, we have determined the hot workability of the alloy for an area of temperatures and strain rates. During the forging process of the titanium alloy in the cross-phase region, since forging needs to be performed in different phase regions, it is particularly important to ensure that the hot processing process parameters selected in the different phase regions are not in the instability region. According to the processing maps, we can determine the better cross-phase forging process parameters: in the β single-phase region, the forging temperatures of 800–830 °C and the strain rates of 0.001–0.01 s^{-1}; in the α + β phase region, the forging temperatures of 740–765 °C and the strain rates of 0.001–0.01 s^{-1}. The distribution of the deformation of different phase regions in the cross-phase forging can also be determined by reference to the processing maps. Therefore, the processing maps can be used to guide the cross-phase forging and provide reference for determining the best cross-phase forging process parameters.

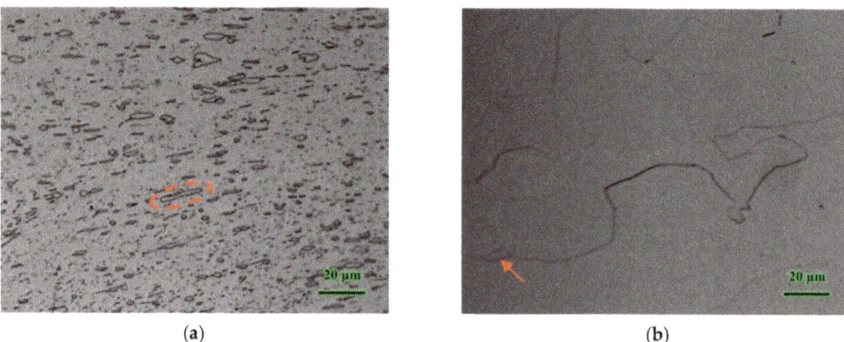

<table>
<tr><td align="center">(a)</td><td align="center">(b)</td></tr>
</table>

Figure 10. Typical microstructure of instability region and safety region. (a) 735 °C/1 s^{-1}; (b) 810 °C/0.001 s^{-1}.

3.5. Microstructure Evolution

Figure 11 shows the microstructure of Ti-6554 alloy after a 60% thermal deformation at different deformation temperatures under the condition of strain rate 0.001 s^{-1}. During thermal deformation below the phase transition point (790 °C), the deformed structure is mainly composed of equiaxed α phases evenly distributed on the β matrix. As shown in Figure 11a,b, when the deformation temperature rises from 715 °C to 735 °C, the shape of the primary α phase is equiaxed, and the volume fraction and size of the primary α phase does not change much. When the deformation temperature rises to 755 °C, as shown in Figure 11c, the shape of the primary α phase is basically unchanged, but its volume fraction and size decrease. When the deformation temperature reaches 775 °C, as shown in Figure 11d, the equiaxed α phase gradually dissolves, and its volume fraction and size decrease amplitude increases. This is because the higher the deformation temperature, the greater the degree of transformation of the α phase into the β phase within the alloy. On the other hand, as the deformation temperature increases, the α phase will swallow the surrounding fine α phase, thereby increasing the size of the α phase. During the thermal deformation of the alloy, these two mechanisms coexist, but when the temperature rises to a certain level, the former mechanism dominates, resulting in a rapid decline in the volume fraction and size of the primary α phase. Below the phase change point, the general change trend is as the deformation temperature increases, the volume fraction and size of the primary α phase in the microstructure of Ti-6554 alloy decrease simultaneously.

During thermal deformation above the phase transition point, the α phase in the microstructure is completely dissolved and the microstructure of the alloy consists mainly of flattened β grains, as depicted in Figure 11e,f. At this point, the β grain is severely deformed, elongated, and forms distinct streamlines, the grain boundary is large serrated, and a small amount of relatively small equiaxed β grains are observed around the flattened β grain. Among them, the flattened β grains are the remains of the original microstructure, while the equiaxed β grains are the new grains produced by dynamic recrystallization.

Above the phase transition point, with the deformation temperature increases, the β grain grows gradually, and the bending effect of grain boundary is more obvious. This is because: on the one hand, at higher temperatures, the ability of atoms to move is enhanced due to thermal activation, the faster the atom diffusion rate, the faster the migration rate, the smaller the hindrance effect of defects such as vacancy on dislocation, and the enhanced grain boundary slip ability. On the other hand, as the temperature increases, the large β grains are more likely to swallow the small β grains, resulting in the β grain coarsening. At the same time, although with the deformation temperature increases, the ability of atom migration and diffusion is stronger, the occurrence of DRV in Ti-6554 alloy reduces the distortion energy, thus reducing the driving force of DRX nucleation, but the high temperature promotes the growth of recrystallized grains, which makes the size of DRX grains larger and the volume fraction increased.

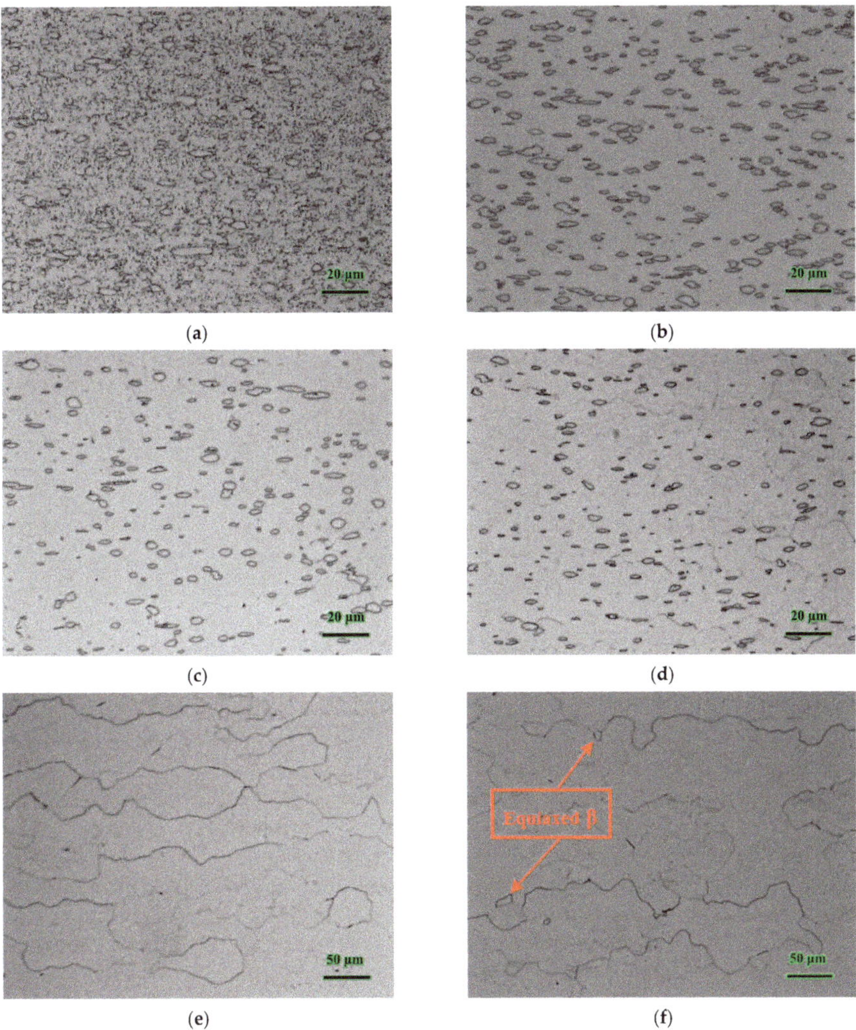

Figure 11. The microstructure of different deformation temperatures at 0.001 s^{-1}: (**a**) 715 °C; (**b**) 735 °C; (**c**) 755 °C; (**d**) 775 °C; (**e**) 810 °C; (**f**) 840 °C.

Figure 12 shows the microstructures of Ti-6554 titanium alloy in single phase region after hot deformation. According to Figure 12a,b, when the strain rates were 0.001 s^{-1} and 0.1 s^{-1}, the β grain boundary is relatively curved, and some of the β grains present an equiaxed shape. This is because with the acceleration of strain, the deformation duration is shortened, the mechanisms of grain boundary sliding and diffusion creep are gradually replaced by dislocation movement, and the shape of the grain changes into a band. At a low strain rate, the dynamic recovery consumes only a small amount of dislocations in the alloy grain, resulting in a large distortion energy. The driving force of dynamic recrystallization is large, so there are many nucleation points and a relatively large amount of recrystallization. The softening mechanism inside the alloy is mainly DRX.

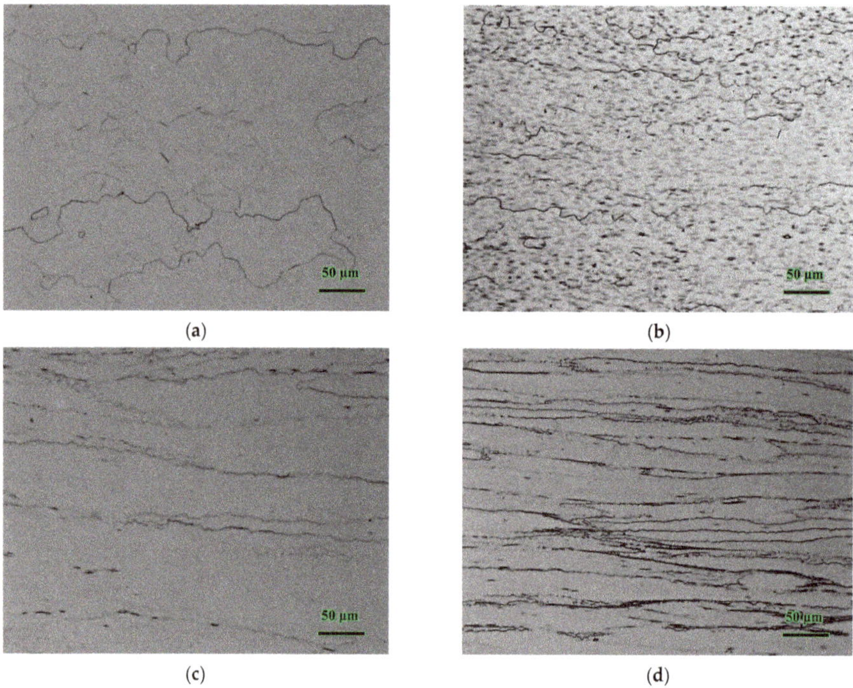

Figure 12. The microstructure of different strain rate at 840 °C: (**a**) 0.001 s^{-1}; (**b**) 0.01 s^{-1}; (**c**) 0.1 s^{-1}; (**d**) 1 s^{-1}.

However, under the higher strain rate (0.1 s^{-1}, 1 s^{-1}), it can be observed from Figure 12c,d that β grains are seriously elongated and strip like. The β grain boundary is relatively smooth, has a fine zigzag shape, and no recrystallized grains appear. When the strain rate is high, the deformation time is shortened, the atom has no time to diffuse, the grain boundary slip and diffusion creep are difficult, and the cross slip has no time to complete, the dislocation density increases rapidly in a short time, resulting in dislocation entanglement, and the softening mechanism in the alloy is mainly dynamic recovery. By means of dynamic recovery and local migration of grain boundaries, the surface tension of grain boundaries can reach a dynamic balance with the ever-changing dislocation density, thus forming a fine jagged grain boundary [39].

4. Conclusions

In this study, the thermal deformation behavior of Ti-6554 alloy under the deformation temperatures of 715–840 °C and the strain rates of 0.001 s^{-1}–1 s^{-1} was studied by thermal simulation compression test. Some important conclusions are summarized as follows:

(1) The flow stress is sensitive to the technological parameters of thermal deformation. The flow stress and peak stress increased with the increase of strain rate. At the same strain rates, the strain required for the stress to reach the peak point is smaller with the temperature increases.

(2) The constitutive equations of Ti-6554 alloy in different phase regions were established, respectively. $\alpha + \beta$ two-phase region:

$$\dot{\varepsilon} = 6.06 \times 10^{10} [\sinh(0.006951\sigma)]^{3.70} \exp\left(-\frac{2.45 \times 10^5}{RT}\right)$$

β single-phase region:

$$\dot{\varepsilon} = 9.64 \times 10^{6} [\sinh(0.010223\sigma)]^{3.20} \exp\left(-\frac{1.84 \times 10^5}{RT}\right)$$

(3) According to the processing maps, the best range of hot working process parameters were determined as follows: the deformation temperature range of 800–830 °C, the strain rate range of 0.001–0.01 s^{-1}.

(4) Below the phase transition point, with the deformation temperature increases, the volume fraction and size of primary α phase gradually decrease due to the gradual isomeric transformation of primary α phase.

(5) Above the phase transition point, with the deformation temperature increases, β grains grow up gradually, and the grain boundary bending effect is more obvious. With the increase of strain rate, the β grains deformation becomes more serious, and grain boundary changes from big ripple shape to fine zigzag shape. Dynamic recrystallization mechanism is gradually replaced by dynamic recovery mechanism.

Author Contributions: Methodology, Q.L.; investigation, Q.L. and H.Y.; validation, Y.N.; writing—original draft preparation, Q.L. and Z.W.; writing—reviewing and editing, Y.N.; funding acquisition, Y.N. All authors have read and agreed to the published version of the manuscript.

Funding: This work was supported by the National Natural Science Foundation of China [Grant No. 51775440] and the National Key Research and Development Program of China.

References

1. Xiao, J.F.; Nie, Z.H.; Tan, C.W.; Zhou, G.; Chen, R.; Li, M.R.; Yu, X.D.; Zhao, X.C.; Hui, S.X.; Ye, W.J.; et al. The dynamic response of the metastable β titanium alloy Ti-2Al-9.2Mo-2Fe at ambient temperature. *Mater. Sci. Eng. A* **2019**, *751*, 191–200. [CrossRef]
2. Zhao, Q.Y.; Yang, F.; Torrens, R.; Bolzoni, L. Evaluation of the hot workability and deformation mechanisms for a metastable beta titanium alloy prepared from powder. *Mater. Charact.* **2019**, *146*, 226–238. [CrossRef]
3. Fan, J.; Li, J.; Li, J.S.; Zhang, Y.D.; Kou, H.C.; Germain, L.; Esling, C. Formation and crystallography of nano/ultrafine-trimorphic structure in metastable β titanium alloy Ti-5Al-5Mo-5V-3Cr-0.5Fe processed by dynamic deformation at low temperature. *Mater. Charact.* **2017**, *130*, 149–155. [CrossRef]
4. Banerjee, D.; Williams, J.C. Perspective on Titanium science and technology. *Acta. Mater.* **2013**, *61*, 844–879. [CrossRef]
5. Long, S.; Xia, Y.F.; Wang, P.; Zhou, Y.T.; Gong, F.J.; Zhou, J.; Zhang, J.S.; Cui, M.L. Constitutive modelling, dynamic globularization behavior and processing map for Ti-6Cr-5Mo-5V-4Al alloy during hot deformation. *J. Alloys Compd.* **2015**, *796*, 65–76. [CrossRef]
6. Li, C.L.; Mi, X.J.; Ye, W.J.; Hui, S.X.; Yu, Y.; Wang, W.Q. Effect of solution temperature on microstructures and tensile properties of high strength Ti-6Cr-5Mo-5V-4Al alloy. *Mater. Sci. Eng. A* **2013**, *578*, 103–109. [CrossRef]
7. Kumar, J.; Singh, V.; Ghosal, P.; Kumar, V. Characterization of fracture and deformation mechanism in a high strength beta titanium alloy Ti-10-2-3 using EBSD technique. *Mater. Sci. Eng. A* **2015**, *623*, 49–58. [CrossRef]
8. Wang, W.Q.; Yang, Y.L.; Zhang, Y.Q.; Li, F.L.; Yang, H.L.; Zhang, P.H. The microstructure and mechanical properties of high-strength and high-toughness titanium alloy BTi-6554 bar. *Mater. Sci. Forum* **2009**, *618–619*, 173–176. [CrossRef]
9. Boyer, R.R.; Briggs, R.D. The use of β titanium alloys in the aerospace industry. *J. Mater. Eng. Perform.* **2005**, *14*, 681–685. [CrossRef]

10. Zhan, H.; Kent, D.; Wang, G.; Wang, G.; Dargusch, M. The dynamic response of a β titanium alloy to high strain rates and elevated temperatures. *Mater. Sci. Eng. A* **2014**, *607*, 417–426. [CrossRef]

11. Zhan, H.; Zeng, W.; Wang, G.; Kent, D.; Dargusch, M. Microstructural characteristics of adiabatic shear localization in a metastable beta titanium alloy deformed at high strain rate and elevated temperatures. *Mater. Charact.* **2015**, *102*, 103–113. [CrossRef]

12. Trimble, D.; O'Donnell, G.E. Constitutive modelling for elevated temperature flow behaviour of AA7075. *Mater. Des.* **2015**, *76*, 150–168. [CrossRef]

13. Lin, Y.C.; Chen, X.M. A critical review of experimental results and constitutive descriptions for metals and alloys in hot working. *Mater. Des.* **2011**, *32*, 1733–1759. [CrossRef]

14. Rusinek, A.; Rodríguez-Martínez, J.A.; Arias, A. A thermo-viscoplastic constitutive model for FCC metals with application to OFHC copper. *Int. J. Mech. Sci.* **2010**, *52*, 120–135. [CrossRef]

15. He, A.; Xie, G.; Zhang, H.; Wang, X.T. A comparative study on Johnson-Cook, modified Johnson-Cook and Arrhenius-type constitutive models to predict the high temperature flow stress in 20CrMo alloy steel. *Mater. Des.* **2013**, *52*, 677–685. [CrossRef]

16. Senthilkumar, V.; Balaji, A.; Narayanasamy, R. Analysis of hot deformation behavior of Al 5083-TiC nanocomposite using constitutive and dynamic material models. *Mater. Des.* **2012**, *37*, 102–110. [CrossRef]

17. Liu, Y.H.; Ning, Y.Q.; Yao, Z.K.; Guo, H.Z. Hot deformation behavior of Ti-6.0Al-7.0Nb biomedical alloy by using processing map. *J. Alloys Compd.* **2014**, *587*, 183–189. [CrossRef]

18. Pilehva, F.; Zarei-Hanzaki, A.; Ghambari, M.; Abedi, H.R. Flow behavior modeling of a Ti-6Al-7Nb biomedical alloy during manufacturing at elevated temperatures. *Mater. Des.* **2013**, *51*, 457–465. [CrossRef]

19. Qin, C.; Yao, Z.K.; Ning, Y.Q.; Shi, Z.F.; Guo, H.Z. Hot deformation behavior of TC11/Ti-22Al-25Nb dual-alloy in isothermal compression. *Trans. Nonferrous Met. Soc. China* **2015**, *25*, 2195–2205. [CrossRef]

20. Jia, W.T.; Xu, S.; Le, Q.C.; Fu, L.; Ma, L.F.; Tang, Y. Modified Fields-Backofen model for constitutive behavior of as-cast AZ31B magnesium alloy during hot deformation. *Mater. Des.* **2016**, *106*, 120–132. [CrossRef]

21. Haghdadi, N.; Zarei-Hanzaki, A.; Abedi, H.R. The flow behavior modeling of cast A356 aluminum alloy at elevated temperatures considering the effect of strain. *Mater. Sci. Eng. A* **2012**, *535*, 252–257. [CrossRef]

22. Ning, Y.Q.; Fu, M.W.; Chen, X. Hot deformation behavior of GH4169 superalloy associated with stick δ phase dissolution during isothermal compression process. *Mater. Sci. Eng. A* **2012**, *540*, 164–173. [CrossRef]

23. Zhang, H.M.; Chen, G.; Chen, Q.; Han, F.; Zhao, Z.D. A physically-based constitutive modelling of a high strength aluminum alloy at hot working conditions. *J. Alloys Compd.* **2018**, *743*, 283–293. [CrossRef]

24. Sun, Y.; Zeng, W.D.; Zhao, Y.Q.; Zhang, X.M.; Shu, Y.; Zhou, Y.G. Modeling constitutive relationship of Ti40 alloy using artificial neural network. *Mater. Des.* **2011**, *32*, 1537–1541. [CrossRef]

25. Raj, R. Development of a processing map for use in warm-forming and hot-forming processes. *Metall. Trans. A* **1981**, *12*, 1089–1097. [CrossRef]

26. Zhang, B.Y.; Liu, X.M.; Yang, H.; Ning, Y.Q.; Wen, S.F.; Wang, Q.D. The deformation behavior, microstructural mechanism, and process optimization of PM/Wrought dual superalloys for manufacturing the dual-property turbine disc. *Metals* **2019**, *9*, 1127. [CrossRef]

27. Chen, X.M.; Lin, Y.C.; Wen, D.X.; Zhang, J.L.; He, M. Dynamic recrystallization behavior of a typical nickel-based superalloy during hot deformation. *Mater. Des.* **2014**, *57*, 568–577. [CrossRef]

28. Wu, K.; Liu, G.Q.; Hu, B.F.; Wang, C.Y.; Zhang, Y.W.; Tao, Y.; Liu, J.T. Effect of processing parameters on hot compressive deformation behavior of a new Ni-Cr-Co based P/M superalloy. *Mater. Sci. Eng. A* **2011**, *528*, 4620–4629. [CrossRef]

29. Chen, F.; Liu, J.; Ou, H.G.; Liu, B.; Gui, Z.S.; Long, H. Flow characteristics and intrinsic workability of IN718 superalloy. *Mater. Sci. Eng. A* **2015**, *642*, 279–287. [CrossRef]

30. Sellars, C.M.; Mctegart, W.J. On the mechanism of deformation. *Acta Metall.* **1966**, *14*, 1136–1138. [CrossRef]

31. Mcqueen, H.J.; Ryan, N.D. Constitutive analysis in hot working. *Mater. Sci. Eng. A* **2002**, *322*, 43–63. [CrossRef]

32. Zener, C.; Hollomon, J.H. Effect of strain rate upon plastic flow of steel. *J. Appl. Phys.* **1944**, *15*, 22–32. [CrossRef]

33. Sargent, P.M.; Ashby, M.F. Deformation maps for titanium and zirconium. *Scr. Metall.* **1982**, *16*, 1415–1422. [CrossRef]

34. Zhao, H.Z.; Xiao, L.; Ge, P.; Sun, J.; Xi, Z.P. Hot deformation behavior and processing maps of Ti-1300 alloy. *Mater. Sci. Eng. A* **2014**, *604*, 111–116. [CrossRef]

35. Nie, X.A.; Hu, Z.; Liu, H.Q.; Yi, D.Q.; Chen, T.X.; Wang, B.F.; Gao, Q.; Wang, D.C. High temperature deformation and creep behavior of Ti-5Al-5Mo-5V-1Fe-1Cr alloy. *Mater. Sci. Eng. A* **2014**, *613*, 306–316. [CrossRef]
36. Warchomicka, F.; Poletti, C.; Stockinger, M. Study of the hot deformation behaviour in Ti-5Al-5Mo-5V-3Cr-1Zr. *Mater. Sci. Eng. A* **2011**, *528*, 8277–8285. [CrossRef]
37. Prasad, Y.V.R.K.; Gegel, H.L.; Doraivelu, S.M.; Malas, J.C.; Morgan, J.T.; Lark, K.A.; Barker, D.R. Modeling of dynamic material behavior in hot deformation: Forging of Ti-6242. *Metall. Trans. A* **1984**, *15*, 1883–1892. [CrossRef]
38. Prasad, Y.V.R.K.; Rao, K.P. Processing maps and rate controlling mechanisms of hot deformation of electrolytic tough pitch copper in the temperature range 300–950 °C. *Mater. Sci. Eng. A* **2005**, *391*, 141–150. [CrossRef]
39. Dikovits, M.; Poletti, C.; Warchomicka, F. Deformation mechanisms in the near-β titanium alloy Ti-55531. *Metall. Mater. Trans. A* **2014**, *45*, 1586–1596. [CrossRef]

Article

Mechanical Behavior and Microstructure Evolution of a Ti-15Mo/TiB Titanium–Matrix Composite during Hot Deformation

Sergey Zherebtsov [1], Maxim Ozerov [1,*], Margarita Klimova [1], Dmitry Moskovskikh [2], Nikita Stepanov [1] and Gennady Salishchev [1]

[1] Laboratory of Bulk Nanostructured Materials, Belgorod State University, Belgorod 308015, Russia; zherebtsov@bsu.edu.ru (S.Z.); klimova_mv@bsu.edu.ru (M.K.); stepanov@bsu.edu.ru (N.S.); salishchev@bsu.edu.ru (G.S.)

[2] Centre of Functional Nanoceramics, National University of Science and Technology, Moscow 119049, Russia; mos@misis.ru

* Correspondence: ozerov@bsu.edu.ru; Tel.: +7-919-223-8528

Received: 26 September 2019; Accepted: 29 October 2019; Published: 31 October 2019

Abstract: A Ti-15Mo/TiB titanium–matrix composite (TMC) was produced by spark plasma sintering at 1400 °C under a load of 40 MPa for 15 min using a Ti-14.25(wt.)%Mo-5(wt.)%TiB$_2$ powder mixture. Microstructure evolution and mechanical behavior of the composite were studied during uniaxial compression at room temperature and in a temperature range of 500–1000 °C. At room temperature, the composite showed a combination of high strength (the yield strength was ~1500 MPa) and good ductility (~22%). The microstructure evolution of the Ti-15Mo matrix was associated with the development of dynamic recovery at 500–700 °C and dynamic recrystallization at higher temperatures (≥800 °C). The apparent activation energy of the plastic deformation was calculated and a processing map for the TMC was constructed using the obtained results.

Keywords: titanium–matrix composite; deformation; microstructure evolution; mechanical properties

---

1. Introduction

Due to a combination of high specific strength, excellent corrosion properties, and remarkable biocompatibility, titanium alloys are widely used in industry (e.g., shipbuilding, aircraft building, chemistry, food industry, etc.) and medicine (e.g., orthopedic and dental implants, surgery instruments) [1,2]. However, the absolute strength and hardness of titanium alloys are rather low and that limits their utilization for certain applications, especially in the case of most biocompatible pure titanium and some low-alloyed titanium alloys. A considerable increase in strength can be attained by complex alloying; however, the most accepted alloying elements (e.g., Al, V) can substantially deteriorate biocompatibility [3]. Searching for new titanium alloy compositions with satisfactory properties is still a challenging problem attracting a great deal of interest from materials scientists.

Another promising way to attain high strength and hardness without loss of biocompatibility and corrosion properties is associated with the production of titanium-based composites [4,5]. Among a variety of reinforcements, TiB seems to be the most suitable option due to very similar properties (density, thermal expansion coefficient, good crystallographic matching) with the Ti matrix [5,6]. Titanium–matrix composites (TMCs) can be produced using traditional metallurgical methods through the addition of B into melted Ti [6]. However, powder metallurgy is more preferable in order to obtain a finer microstructure. In this case, TiB reinforcements form in the Ti matrix during the in-situ $3Ti + TiB_2 = 2Ti + 2TiB$ reaction [6–8]. Due to high heating rate and high pressures, the spark plasma sintering (SPS) process can be used for powder consolidation at relatively low temperatures and for short time intervals, thereby preserving a very fine microstructure in the specimen [8]. Hard fibers of TiB increase the strength and hardness of TMC, however the ductility of the composite can decrease to nearly zero [9,10].

Some improvement in both ductility and technological properties was attained in thermomechanically treated TMCs [11–17] due to the shortening and redistribution of TiB whiskers, along with the development of dynamic recrystallization in the titanium matrix. Another promising way to improve ductility of the composite can be associated with an increase in plasticity of the titanium matrix by adding a β-stabilizing element. In this case, a hard-to-deform hexagonal close-packed (hcp) lattice should become more ductile due to the greater number of slip systems in a body-centered cubic (bcc) lattice [1]. Meanwhile, there is a lack of information on both the mechanical properties of β-Ti-based TMCs and an influence of thermomechanical treatment on structure and properties of these materials.

In this work a bcc Ti-15 (wt.%)Mo matrix (which is widely used for bio-medical applications) was reinforced by 8.5 vol.% of TiB. This amount of TiB was selected using literature data for hcp Ti/TiB composites to attain a combination of high strength and satisfactory ductility [6]. It was selected to investigate the mechanical properties of the TBC at room temperature, to study the effect of hot deformation at 500–1000 °C and strain rates in the interval 5×10^{-4}–10^{-2} s^{-1} on microstructure evolution and mechanical behavior. The obtained results will be used for the development of a proper thermomechanical treatment of the TMC.

2. Materials and Procedure

Powders of Ti (99.1% purity), Mo (99.95% purity), and TiB$_2$ (99.9% purity) were used for the sintering of the Ti-15Mo/TiB composites with the bcc Ti matrix. The average sizes of the Ti, Mo, and TiB$_2$ particles were 25, 3, and 4 μm, respectively (Figure 1). A mixture of the powders containing 80.75 wt.% Ti, 14.25 wt.% of Mo, and 5 wt.% of TiB$_2$ (to obtain a Ti-15wt.% Mo alloy with 8.5 vol.% of TiB) were produced using a Retsch RS200 vibrating cup mill (RETSCH, Haan, Germany) in ethanol at a milling rotation speed of 700 rpm. The mixing duration was 1 h.

Figure 1. SEM images of (**a**) Ti, (**b**) Mo, and (**c**) TiB$_2$ powders.

Cylindrical Ti-15Mo/TiB TMC specimens measured 15 mm in height and 19 mm in diameter and were produced using the SPS process under vacuum on a Thermal Technology SPS 10-3 machine (Thermal Technology, LLC, Santa Rosa, CA, USA) at 1400 °C and 40 MPa for 15 min. Then, the sintered samples of the composite were homogenized at 1200 °C for 24 h and cooled in air. To prevent oxidation during the homogenization annealing, the specimens were sealed in vacuumed quartz tubes filled with a titanium getter. The homogenized condition of the TMC is referred to as the initial one hereafter.

Specimens Ø7 × 10 mm in high were cut from the homogenized composite using a Sodick AQ300L electro-discharge machine (Sodick Inc., Schaumburg, IL, USA). Then, the specimens were isothermally strained by compression in air at 20, 500, 600, 700, 800, 900, or 1000 °C on an Instron mechanical testing machine (INSTRON, Horwood, MA, USA). This was done at a nominal strain rate of 10^{-3} s^{-1} to 70% height reduction (a corresponding true strain was $\varepsilon \approx 1.2$) or to the specimens' fracture. The holding time at the required temperatures before the onset of deformation was 15 min. After deformation, the specimens were air cooled. Strain-rate jump compressive tests were conducted at strain rates of 10^{-2}, 5×10^{-3}, 10^{-3}, and 5×10^{-4} s^{-1} in the temperature interval 400–1000 °C. The obtained data were used to determine the activation energy of deformation for the Ti-15Mo/TiB TMC.

Microstructure of the TMC in the initial condition and after deformation was studied using an FEI Quanta 600 FEG scanning electron microscope (Thermo Fisher Scientific, Hillsboro, OR, USA) and a JEOL JEM 2100 transmission electron microscope (JEOL, Tokyo, Japan). Mechanically polished specimens for SEM were etched with Kroll's reagent (95% H_2O, 3% HNO_3, 2% HF). Samples for TEM analysis were prepared by twin-jet electro-polishing in a mixture of 60 mL perchloric acid, 600 mL methanol, and 360 mL butanol at -35 °C and 29 V. The spacing between the TiB whiskers was evaluated based on seven TEM images (more than 100 measurements in total) using the linear-intercept method. Phase composition of the TMC was evaluated using X-ray diffraction (XRD) on an ARL-Xtra diffractometer (Thermo Fisher Scientific, Portland, OR, USA) with Cu Kα radiation. The volume fraction of the phases was determined using the Rietveld method [18].

To determine the optimal processing window, a processing map was constructed using the obtained results for a true strain of 1.2. Toward this end, power dissipation efficiency (η) was evaluated in comparison with an ideal linear dissipator (i.e., with $m = 1$): $\eta = 2m/(m+1)$ [19], with the strain rate sensitivity of flow stress $m = \frac{\Delta \log \sigma}{\Delta \log \dot{\varepsilon}}$ (here σ and $\dot{\varepsilon}$ are the flow stress and strain rate, respectively). The processing map shows a three-dimensional area projection describing the η isolines, depending on the strain rate and temperature on the $T - \dot{\varepsilon}$ surface.

3. Results

Microstructure of the Ti-15Mo/TiB composite in the initial condition (i.e., after the homogenization annealing at 1200 °C for 24 h) consisted of the β-Ti matrix reinforced by TiB whiskers (Figure 2a–c). Due to rather high sintering and/or homogenization temperatures, the cross-section of some TiB whiskers attained ~3–4 μm (Figure 2b,c). Nevertheless, the average diameter was almost an order of magnitude smaller (~200 nm). Dislocation density in the titanium matrix was higher nearby or between the TiB whiskers (Figure 2d). Grain boundaries in the titanium matrix were not detected on TEM images and the spacing between the TiB whiskers (which can be accepted in a first approximation as a free dislocation path) was ~0.5–0.7 μm.

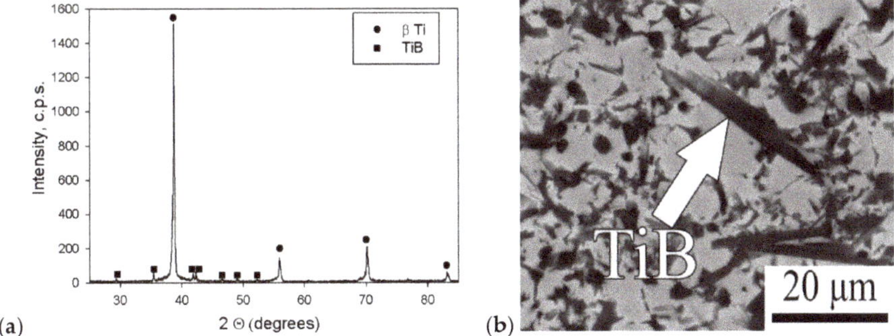

(a)   (b)

Figure 2. *Cont.*

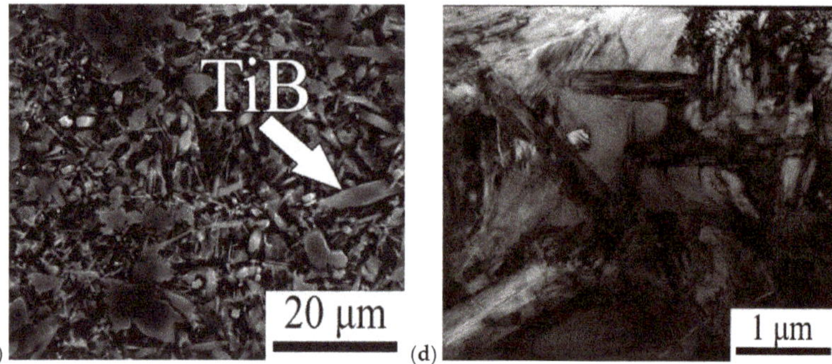

Figure 2. Microstructure of the Ti-15Mo/TiB titanium–matrix composite (TMC) in the initial condition: (**a**) XRD pattern, (**b**) SEM image of unetched surface; (**c**) SEM image of etched surface; (**d**) TEM bright field image.

During compression at room temperature, the composite showed an attractive combination of strength (the yield strength was ~1480 MPa and the ultimate strength was almost 2 GPa) and ductility (~22% height reduction) (Figure 3a). A short plateau just before the fracture can most likely be associated with strain localization.

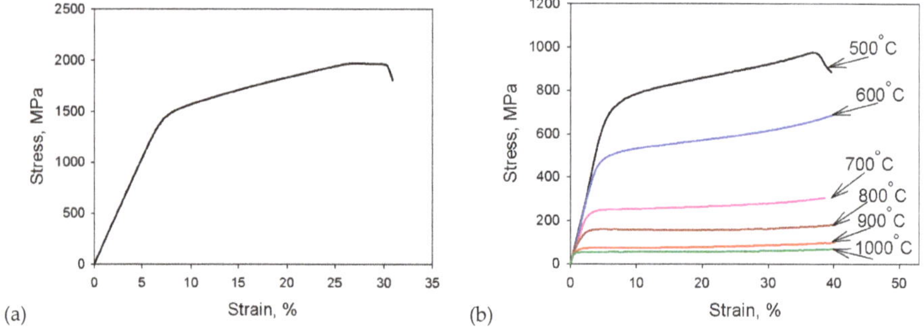

Figure 3. Flow curves obtained during compression at (**a**) 20 °C or at (**b**) 500–1000 °C of the Ti-15Mo/TiB TMC at a nominal strain rate 10^{-3} s^{-1}.

Deformation of the Ti-15Mo/TiB TMC in compression resulted—after initial hardening transient—in continuous strengthening at 500–700 °C (this interval is below the $\alpha + \beta \leftrightarrow \beta$ transition, which occurs at 727 °C for the Ti-15Mo alloy [20]) or steady-state flow at higher temperatures, that is, at 800–1000 °C (Figure 3b). The latter can be ascribed to dynamic recrystallization/recovery processes typical of hot deformation. The yield strength gradually decreased from 630 MPa at 500 °C to 45 MPa at 1000 °C. At 500 °C, the TMC fractured at ~30% reduction.

The microstructure evolution during deformation was associated with some shortening of TiB whiskers (from ~6 μm in the initial condition to ~3.5 μm in all specimens deformed at 500–1000 °C) and their preferred orientation along the metal flow direction (shown by arrows in Figure 4). The redistribution process appeared to be more pronounced with an increase in temperature.

XRD analysis indicated the formation of the α phase in the β matrix during deformation. The volume fraction of the α phase was found to be ~17%–22% in the interval 500–800 °C (this interval corresponds to the minimal stability of the β phase [21]) and decreased to ~7% while increasing in deformation temperature to 1000 °C (Figure 5). The volume fraction of TiB was found to be ~9% in the whole temperature interval.

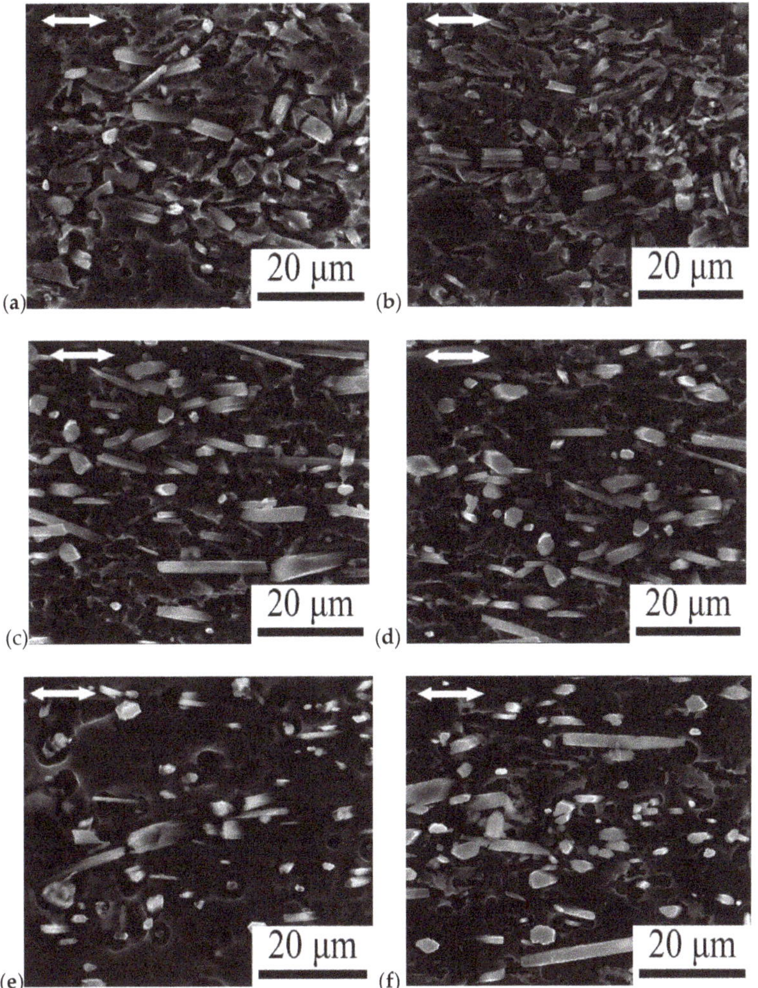

Figure 4. SEM microstructure of the Ti-15Mo/TiB TMC produced using spark plasma sintering (SPS) at 1400 °C; uniaxial compression at a nominal strain rate of 10^{-3} s^{-1} and temperatures (**a**) 500 °C, (**b**) 600 °C, (**c**) 700 °C, (**d**) 800 °C, (**e**) 900 °C, and (**f**) 1000 °C. The compression axis is vertical in all cases and the metal flow direction is shown by arrows.

The TEM investigation showed the formation of a typical deformed microstructure with a rather high dislocation density and dislocation cells/pile-ups after deformation at low temperatures (500–600 °C) (Figure 6a,b). However, areas with a decreased dislocation density were already observed at 600 °C (Figure 6b). In addition, lens-shaped α-phase lamellae (up to 0.1 μm width) were found (Figure 6b). An increase in deformation temperature resulted in the development of dynamic recrystallization, associated with an overall decrease in dislocation density, formation of new recrystallized grains, and straightening of grain boundaries (Figure 6c,d). The size of the recrystallized grains increased with temperature from ~1 μm at 800 °C (Figure 6c) to 2–3 μm at 1000 °C (Figure 6d). At all temperatures, the Ti/TiB interfaces were quite clean and did not contain any cracks or pores.

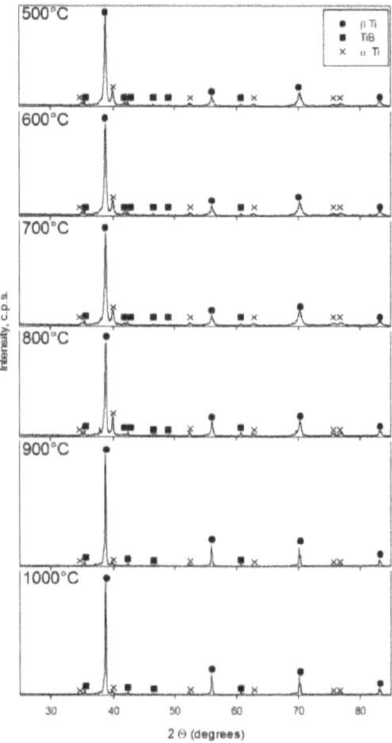

Figure 5. XRD patterns of the Ti-15Mo/TiB MMC after compression at different temperatures.

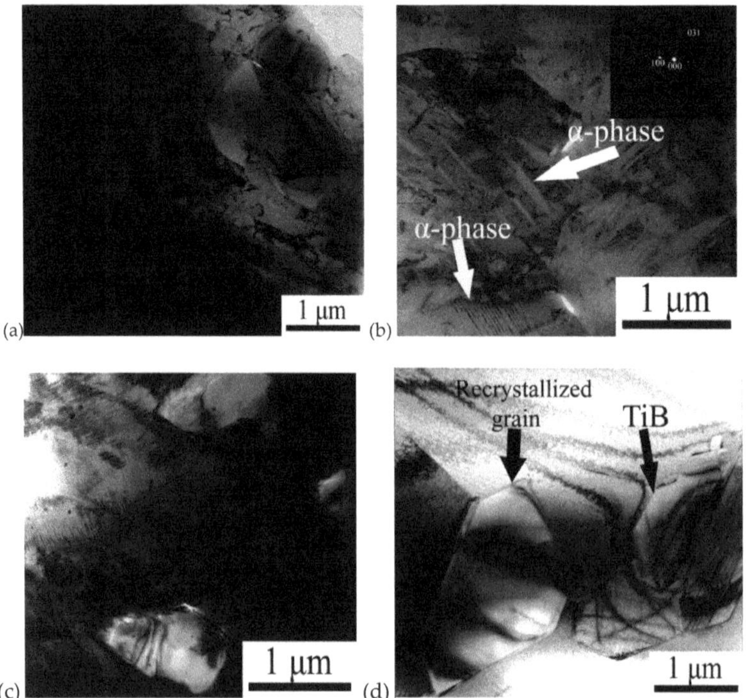

Figure 6. TEM microstructure of the Ti-15Mo/TiB TMC after uniaxial compression at (**a,b**) 600 °C, (**c**) 800 °C and (**d**) 1000 °C. Selected area diffraction pattern for the α-phase is inserted in (**b**).

4. Discussion

The obtained results demonstrate a definite beneficial effect of the matrix transition from the hcp alpha titanium to the bcc beta titanium. The latter was obtained through the addition of 15 wt.% of Mo into Ti. The composite produced by SPS showed a very attractive combination of strength and ductility (1.5–2 GPa and ~22%, respectively) (Figure 3a) at room temperature, which is superior to most of the β-rich alloys in the heat-strengthened condition [22] or metal–matrix composites with the hcp α-Ti matrix [4,23,24]. The observed increase in ductility is obviously associated with a greater number of slip systems in the bcc lattice in comparison with that of the hcp one [25]. Lower compression ductility of the hcp Ti/TiB (4%–17%, depending on sintering methods) was also reported in [23,24]. The very high strength of the Ti-15Mo/TiB composite can be ascribed to the superposition of several strengthening mechanisms with Orowan strengthening providing the main contribution [26,27]. Some additional strengthening effects in comparison with the hcp Ti/TiB composites [28] is most likely the result of solid solution strengthening.

An increase in testing temperature led to a decrease in both yield and flow stresses and an improvement in ductility (Figure 3b) due to the activation of additional deformation mechanisms. For example, the pronounced softening at T $\geq$ 700 °C can be associated with the development of dynamic recrystallization (Figure 6c,d). Meanwhile, precipitations of the α-phase particles (Figures 5 and 6b) should result in some increase in the overall strength. Deformation-induced precipitation of the α-phase particles could be the reason for the hardening transient at the stress–strain curves obtained at 500–700 °C (Figure 3b). Note that the α phase was found in the composite after compression in the full temperature range of 500–1000 °C (Figure 5). However, the α phase formation at temperatures above the α + β $\leftrightarrow$ β transition (727 °C for the Ti-15Mo alloy [18]) obviously occurred during cooling from the deformation temperatures. That is why the fraction of the α phase was lower after deformation at the high temperatures.

Since some improvement in mechanical properties can be attained in the composites through hot/warm working [14,29], the obtained results on mechanical behavior during hot deformation were used to determine the optimal processing window. The temperature–strain rate map in Figure 7 shows domains where deformation capacity of the composite was high enough for hot/warm working (i.e., areas associated, for example, with superplasticity or dynamic recrystallization). The map also predicts domains of unstable plastic flow where deformation of the composite can be associated with early necking and/or crack formation [30]. The maximum values of dissipation are expectably related to the highest temperature and the lowest strain rate (the bottom-right corner in Figure 7). However, even at relatively low temperatures (700–800 °C) and strain rates 5×10^{-4}–10^{-3} s^{-1} the value of η fell in the interval 0.35–04, thereby suggesting sufficiently favorable hot working conditions. The obtained results (Figure 7) suggest the occurrence of dynamic recrystallization/recovery in these domains, warranting satisfactory deformability. At higher temperatures, good ductility of the composite was accompanied by intensive grain growth, which is not advantageous for the performance properties.

To gain insight into the operative deformation mechanisms, the apparent activation energy of plastic deformation was calculated using the obtained results (Figure 8). The calculated values of the apparent activation energy were found to be Q = 241 kJ/mol for the high-temperature interval (700–1000 °C) with $n \approx 5$ and Q = 135 kJ/mol with n = 10 for lower temperatures. The former value of Q is very similar to that observed in [20] for the Ti/TiB composite with the hcp α-Ti matrix during deformation in the β phase field region (i.e., at temperatures above 900 °C), where Q was found to be 250 kJ/mol. This value being higher than that reported for self-diffusion in β titanium (153 kJ/mol) [31] can be ascribed to dislocation slip inhibition due to the presence of the TiB whiskers [32]. In addition, similar values of Q = 250–330 kJ/mol were reported in [33] for glide along the prism planes with the thermally activated overcoming of solute atoms as the main rate-controlling mechanism. An increase in the value of Q can also be related to the occurrence of discontinuous dynamic recrystallization during hot deformation [34].

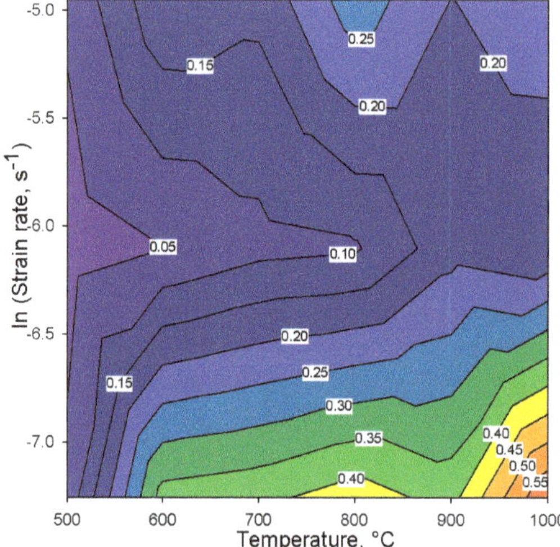

Figure 7. Processing map for the Ti-15Mo/TiB composite.

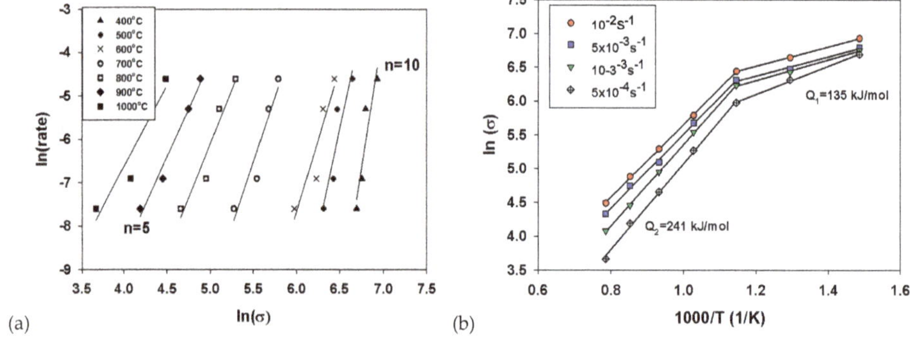

(a)

(b)

Figure 8. (**a**) Log–log dependence of strain rate on stress and (**b**) Arrhenius semi-log plot of steady-state flow stress vs. inversed temperature for the Ti-15Mo/TiB TMC strained in the interval T = 400–1000 °C.

In the lower temperature range (400–600 °C) the value of Q = 135 kJ/mol suggests that deformation is controlled by volume or pipe diffusion [31,35] associated with the development of dynamic recovery or continuous dynamic recrystallization. Microstructural observations confirmed the development of dynamic recovery in the TMC during compression at 600 °C (Figure 6b).

Quite low values of strain rate sensitivity $m = 1/n \approx 0.1$–0.2 (Figure 8a) along the obtained values of activation energy suggest the operation of dislocation-related mechanisms, mainly (dislocation glide/climb) during deformation in the studied temperature interval [33,36].

5. Summary

In this work, the structure and compression mechanical behavior of a Ti-15Mo/TiB titanium–matrix composite (TMC) produced by spark plasma sintering were studied. The following conclusions were drawn:

(1) The Ti-15Mo/TiB TMC fabricated by spark plasma sintering at 1400 °C under a load of 40 MPa for 15 min using a Ti-14.25(wt.)%Mo-5(wt.)%TiB$_2$ powder mixture was composed of β-Ti matrix reinforced with TiB whiskers with the average diameter of 200 nm. The as-sintered TMC had

attractive compression mechanical properties at room temperature: yield strength ~1480 MPa and ductility ~22%.

(2) The uniaxial compression in the temperature range of 500–1000 °C demonstrated gradual softening with an increase in deformation temperature. Microstructure evolution of the Ti-15Mo matrix was associated with the development of dynamic recovery at 500–700 °C and dynamic recrystallization at temperatures ≥800 °C. Reorientation of TiB whiskers toward the metal flow direction and some shortening of the whiskers also occurred. In addition, the precipitation of the α particles was found after deformation at 500–1000 °C.

(3) The optimal parameters of thermomechanical processing for the Ti-15Mo/TiB TMC were associated with deformation temperatures of 700–800 °C and strain rates of 5×10^{-4}–10^{-3} s^{-1}.

Author Contributions: S.Z. conceived and designed the experiments. M.O. performed the experiments. D.M. prepared the TEM specimens. M.K. carried out microstructure analysis. S.Z., N.S., M.O., and G.S. analyzed the data and wrote the paper.

Funding: This research was funded by Russian Science Foundation, grant number 15-19-00165.

Acknowledgments: The authors gratefully acknowledge the financial support from the Russian Science Foundation (Grant Number 15-19-00165). The authors are grateful to the personnel of the Joint Research Centre at Belgorod State University for their assistance with the instrumental analysis.

Conflicts of Interest: The authors declare no conflict of interest.

References

1. Leyens, C.; Peters, M. *Titanium and Titanium Alloys: Fundamentals and Applications*; Wiley-VCH: Weinheim, Germany, 2003; pp. 1–499.

2. Khorasani, A.M.; Goldberg, M.; Doeven, E.H.; Littlefair, G. Titanium in biomedical applications—Properties and fabrication: A review. *J. Biomater. Tissue Eng.* **2015**, *5*, 593–619. [CrossRef]

3. Chen, Q.; Thouas, G.A. Metallic implant biomaterials. *Mater. Sci. Eng. R Rep.* **2015**, *87*, 1–57. [CrossRef]

4. Saito, T.; Furuta, T.; Yamaguchi, T. Development of low cost titanium matrix composite. In *Advances in Titanium Metal Matrix Composites, the Minerals, Metals and Materials Society*; Froes, F.H., Storer, J., Eds.; TMS: Warrendale, PA, USA, 1995; pp. 33–44.

5. Godfrey, T.M.T.; Goodwin, P.S.; Ward-Close, C.M. Titanium Particulate Metal Matrix Composites—Reinforcement, Production Methods, and Mechanical Properties. *Adv. Eng. Mater.* **2000**, *2*, 85–91. [CrossRef]

6. Morsi, K.; Patel, V.V. Processing and properties of titanium-titanium boride (TiBw) matrix composites—A review. *J. Mater. Sci.* **2007**, *42*, 2037–2047. [CrossRef]

7. Ravi Chandran, K.S.; Panda, K.B.; Sahay, S.S. TiBw-reinforced Ti composites: Processing, properties, application, prospects, and research needs. *JOM* **2004**, *56*, 42–48. [CrossRef]

8. Feng, H.; Zhou, Y.; Jia, D.; Meng, Q.; Rao, J. Growth mechanism of in situ TiB whiskers in spark plasma sintered TiB/Ti metal matrix composites. *Cryst. Growth Des.* **2006**, *6*, 1626–1630. [CrossRef]

9. Ozerov, M.; Stepanov, N.; Kolesnikov, A.; Sokolovsky, V.; Zherebtsov, S. Brittle-to-ductile transition in a Ti–TiB metal-matrix composite. *Mater. Lett.* **2017**, *187*, 28–31. [CrossRef]

10. Ozerov, M.; Klimova, M.; Vyazmin, A.; Stepanov, N.; Zherebtsov, S. Orientation relationship in a Ti/TiB metal-matrix composite. *Mater. Lett.* **2017**, *186*, 168–170. [CrossRef]

11. Gaisin, R.A.; Imayev, V.M.; Imayev, R.M. Effect of hot forging on microstructure and mechanical properties of near α titanium alloy/TiB composites produced by casting. *J. Alloys Compd.* **2017**, *723*, 385–394. [CrossRef]

12. Ozerov, M.S.; Gazizova, M.Y.; Klimova, M.V.; Stepanov, N.D.; Zherebtsov, S.V. Effect of Plastic Deformation on the Structure and Properties of the Ti/TiB Composite Produced by Spark Plasma Sintering. *Russ. Metall. (Met.)* **2018**, *7*, 638–644. [CrossRef]

13. Zherebtsov, S.; Ozerov, M.; Stepanov, N.; Klimova, M.; Ivanisenko, Y. Effect of high-pressure torsion on structure and microhardness of Ti/TiB metal-matrix composite. *Metals* **2017**, *7*, 507. [CrossRef]

14. Ozerov, M.; Klimova, M.; Sokolovsky, V.; Stepanov, N.; Popov, A.; Boldin, M.; Zherebtsov, S. Evolution of microstructure and mechanical properties of Ti/TiB metal-matrix composite during isothermal multiaxial forging. *J. Alloys Compd.* **2019**, *770*, 840–848. [CrossRef]

15. Imayev, V.; Gaisin, R.; Gaisina, E.; Imayev, R.; Fecht, H.-J.; Pyczak, F. Effect of hot forging on microstructure and tensile properties of Ti–TiB. *Mater. Sci. Eng. A* **2014**, *609*, 34–41. [CrossRef]

16. Ozerov, M.S.; Klimova, M.V.; Stepanov, N.D.; Zherebtsov, S.V. Microstructure evolution of a Ti/TiB metal-matrix composite during high-temperature deformation. *Mater. Phys. Mech.* **2018**, *38*, 54–63. [CrossRef]

17. Zherebtsov, S.; Ozerov, M.; Stepanov, N.; Klimova, M. Structure and properties of Ti/TiB metal–matrix composite after isothermal multiaxial forging. *Acta Phys. Pol. A* **2018**, *134*, 695–698. [CrossRef]

18. Will, G. *Powder Diffraction: The Rietveld Method and the Two-Stage Method to Determine and Refine Crystal Structures from Powder Diffraction Data*; Springer: Berlin, Germany, 2005.

19. Prasad, Y.V.R.K.; Rao, K.P.; Sasidhara, S. *Hot Working Guide: A Compendium of Processing Maps*; ASM International: Materials Park, OH, USA, 2015; pp. 1–625.

20. Weiss, I.; Semiatin, S.L. Thermomechanical processing of beta titanium alloys—An overview. *Mater. Sci. Eng. A* **1998**, *243*, 46–65. [CrossRef]

21. Jiang, B.; Tsuchiya, K.; Emura, S.; Min, X. Effect of high-pressure torsion process on precipitation behavior of α phase in β-type Ti-15Mo alloy. *Mater. Trans.* **2014**, *55*, 877–884. [CrossRef]

22. Ilyin, A.A.; Kolachev, B.A.; Polkin, I.S. *Titanium Alloys. Composition, Structure, Properties*; VILS-MATI Publishing: Moscow, Russia, 2009.

23. Jeong, H.W.; Kim, S.J.; Hyun, Y.T.; Lee, Y.T. Densification and Compressive Strength of In-situ Processed Ti/TiB Composites by Powder Metallurgy. *Metals Mater. Int.* **2002**, *8*, 25–35. [CrossRef]

24. Attar, H.; Bönisch, M.; Calin, M.; Zhang, L.-C.; Scudino, S.; Eckert, J. Selective laser melting of in situ titanium–titanium boride composites: Processing, microstructure and mechanical properties. *Acta Mater.* **2014**, *76*, 13–22. [CrossRef]

25. Meyers, M.A.; Chawla, K.K. *Mechanical Behavior of Materials*; Cambridge University Press: New York, NY, USA, 2009.

26. Casati, R.; Vedani, M. Metal Matrix Composites Reinforced by Nano-Particles—A review. *Metals* **2004**, *4*, 65–83. [CrossRef]

27. Zherebtsov, S.; Ozerov, M.; Klimova, M.; Stepanov, N.; Vershinina, T.; Ivanisenko, Y.; Salishchev, G. Effect of High-Pressure Torsion on Structure and Properties of Ti-15Mo/TiB Metal-Matrix Composite. *Materials* **2018**, *11*, 2426. [CrossRef] [PubMed]

28. Morsi, K. Review: Titanium–titanium boride composites. *J. Mater. Sci.* **2019**, *54*, 6753–6771. [CrossRef]

29. Ozerov, M.; Klimova, M.; Kolesnikov, A.; Stepanov, N.; Zherebtsov, S. Deformation behavior and microstructure evolution of a Ti/TiB metal-matrix composite during high-temperature compression tests. *Mater. Des.* **2016**, *112*, 17–26. [CrossRef]

30. Rao, K.P.; Prasad, Y.V.R.K. Advanced Techniques to Evaluate Hot Workability of Materials. *Compr. Mater. Process.* **2014**, *3*, 397–426. [CrossRef]

31. Frost, H.J.; Ashby, M.F. *Deformation-Mechanism Maps*; Pergamon Press: Oxford, UK, 1982; pp. 1–166.

32. Zhang, Y.; Huang, L.; Liu, B.; Geng, L. Hot deformation behavior of in-situ TiBw/Ti6Al4V composite with novel network reinforcement distribution, Trans. *Nonferrous Metals Soc.* **2012**, *22*, 465–471. [CrossRef]

33. Conrad, H. Effect of interstitial solutes on the strength and ductility of titanium. *Prog. Mater. Sci.* **1981**, *26*, 123–403. [CrossRef]

34. Raj, S.V.; Langdon, T.G. Creep behavior of copper at intermediate temperatures—I. Mechanical characteristics. *Acta Metall.* **1989**, *37*, 843–852. [CrossRef]

35. Walsöe De Reca, N.E.; Libanati, C.M. Autodifusion de titanio beta y hafnio beta. *Acta Metall.* **1968**, *16*, 1297–1305. [CrossRef]

36. Kumari, S.; Prasad, N.E.; Chandran, K.S.R.; Malakondaiah, G. High-temperature deformation behavior of Ti-TiBw in-situ metal-matrix composites. *JOM* **2004**, *56*, 51–55. [CrossRef]

 *metals*

Article

Effect of Pre-Strain on Microstructure and Tensile Properties of Ti-6Al-4V at Elevated Temperature

Taowen Wu , Ning Wang *, Minghe Chen, Dunwen Zuo, Lansheng Xie and Wenxiang Shi

Nanjing University of Aeronautics and Astronautics, Nanjing 210016, China; meewtw0116@nuaa.edu.cn (T.W.); meemhchen@nuaa.edu.cn (M.C.); imit505@nuaa.edu.cn (D.Z.); meelsxie@nuaa.edu.cn (L.X.); swx1996@nuaa.edu.cn (W.S.)
* Correspondence: meewn1987@nuaa.edu.cn

Abstract: Research on pre-deformation influences on material properties in multistep hot forming is of important scientific interest. In this paper, hot tensile tests at 850 °C and a strain rate of 0.001 s^{-1} were performed to study the microstructural evolution and mechanical properties of Ti-6Al-4V with pre-strains at 0.05, 0.1 and 0.15. The tensile test results showed that the specimen with 0.05 pre-strain exhibited higher flow stress and larger elongation. Additionally, increasing the pre-strain resulted in a decrease in ultimate tensile strength (UTS) and elongation (*EL*). The EBSD results showed that the main deformation mechanism of Ti-6Al-4V was high-angle grain boundary sliding. Pre-strain promoted dynamic recrystallization (DRX) by increasing the deformation substructure. The refinement of grains and the eradication of dislocations enhanced the deformability, resulting in an increase in flow stress.

Keywords: Ti-6Al-4V titanium alloy; pre-deformation; dynamic recrystallization; microstructure evolution; mechanical properties

check for
updates

Citation: Wu, T.; Wang, N.; Chen, M.; Zuo, D.; Xie, L.; Shi, W. Effect of Pre-Strain on Microstructure and Tensile Properties of Ti-6Al-4V at Elevated Temperature. *Metals* **2021**, *11*, 1321. https://doi.org/10.3390/met11081321

Academic Editors: Maciej Motyka and Marcello Cabibbo

Received: 25 June 2021
Accepted: 18 August 2021
Published: 20 August 2021

Publisher's Note: MDPI stays neutral with regard to jurisdictional claims in published maps and institutional affiliations.

1. Introduction

Titanium alloys have a wide range of applications in aerospace and other industries due their high specific strength, corrosion resistance, high temperature resistance, etc. [1,2]. Ti-6Al-4V was the first titanium alloy put into structural production of which the long-term service temperature can reach 400 °C. In the aviation industry, Ti-6Al-4V is mainly used to manufacture aero-engine compressor discs and blades [3]. The alloy is composed of the hcp α phase, which plays an important role in hindering plastic deformation, and the bcc β phase [4]. However, Ti-6Al-4V has poor forming performance and strong work hardening at room temperature. Hot-forming techniques are widely used to form complicated shape parts because the formability of Ti-6Al-4V is largely improved at elevated temperature. Ti-6Al-4V has been selected to manufacture hollow aero-engine blades for weight reduction. The hollow blade forming process consists of diffusion bonding, hot twisting, hot stamping, and gas bulging. To investigate the hot twisting procedure influence on Ti-6Al-4V deformation behavior in the subsequent hot stamping process, it is necessary to study the microstructural evolution and deformation mechanisms of Ti-6Al-4V with pre-strain [5–7].

Research has been performed to reveal the relationship between the rheological properties and microstructure of titanium alloys in high-temperature deformation [8–11]. Cui et al. [8] presented a constitutive model of TC11 based on the Arrhenius-type hyperbolic sine method by the cubic piecewise function of strain which ensured the high precision of the model. Liu et al. [9] considered that the deformation and softening mechanism of the Ti55 titanium alloy under temperatures ranging from 885 °C to 935 °C were grain boundary sliding and discontinuous dynamic recrystallization, respectively. Davis et al. [10] analyzed how the process parameters affect the β-phase recrystallization level and the final grain size by controlling the specimen deformation and heating rate. In addition, some researchers have achieved excellent microstructure and mechanical properties through multiple deformations [11,12] and proposed effective step-by-step forming methods to refine the grains of titanium alloys and increase the strength of titanium alloys [13–15].

Pre-deformation effects on microstructural and mechanical properties have also been studied. Wang et al. [16] found that pre-deformation can promote the precipitation of a stripe wall-like phase, which improves the material strength and elongation. Sachtleber et al. [17] established a relationship between pre-deformation and the grain size of 6022-T4 aluminum alloys through uniaxial tensile tests. However, little research has been conducted on the effects of pre-deformation on the microstructure and properties of titanium alloys during the hot-forming process.

The aim of this research was to study the microstructural evolution and mechanical properties of Ti-6Al-4V titanium alloys with different pre-strains during hot tensile deformation. Ti-6Al-4V specimens were first stretched by hot tensile tests at a temperature of 850 °C and strain rate of 0.001 s^{-1} to obtain various pre-strains and microstructures. The grain size, dislocation density and dynamic recrystallization ratio were quantitatively analyzed by the electron backscatter diffraction (EBSD) technique. The deformation mechanism of Ti-6Al-4V alloy was analyzed to explain the tensile behaviors. The influence of pre-strain on the properties of the Ti-6Al-4V titanium alloy was analyzed, combined with dislocation density and dynamic recrystallization ratio, which provided a theoretical basis for the stepwise hot torsion process.

2. Materials and Methods

The chemical composition of Ti-6Al-4V titanium alloy used in this study was Ti-5.92Al-3.92V-0.108O-0.1Fe-0.02N. Before the experiment, we used Wire Electrical Discharge Machining (WEDM) to cut it into tensile specimens as shown in Figure 1a. The gauge length of the specimen was ground to reduce burrs and scratches on the surface of specimens by sandpaper. A boron nitride solder stopper was sprayed on the surface of specimens to prevent the specimen from oxidizing. Schematic diagram of hot tensile test heating path is shown in Figure 2.

All the hot tensile tests were carried out on a UTM 5504X electronic universal testing machine (SUNSTEST Inc., Shenzhen, China) equipped with an environment chamber. The experimental scheme is given in Table 1.

Table 1. Experimental schemes of hot tensile test.

Test No.	Pre-Strain	Tensile Test Parameter	Final Strains
Test-0	0	850 °C, 0.001 s^{-1}	0.3 0.5 0.7 fracture
Test-1	0.05	850 °C, 0.001 s^{-1}	0.3 0.5 0.7 fracture
Test-2	0.1	850 °C, 0.001 s^{-1}	0.3 0.5 0.7 fracture
Test-3	0.15	850 °C, 0.001 s^{-1}	0.3 0.5 0.7 fracture

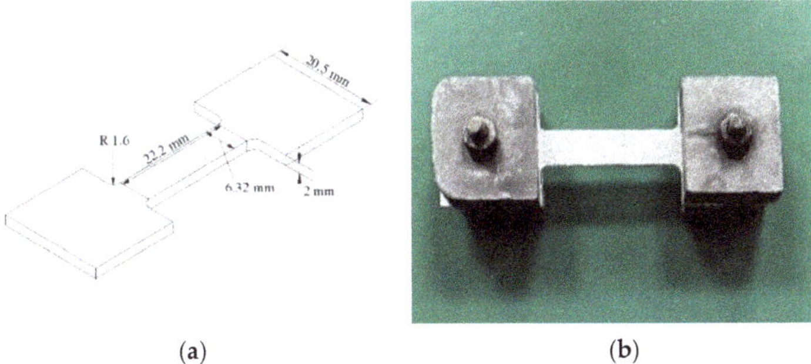

(a)   (b)

Figure 1. (a) Design of the hot tensile test specimen; (b) hot tensile fixtures.

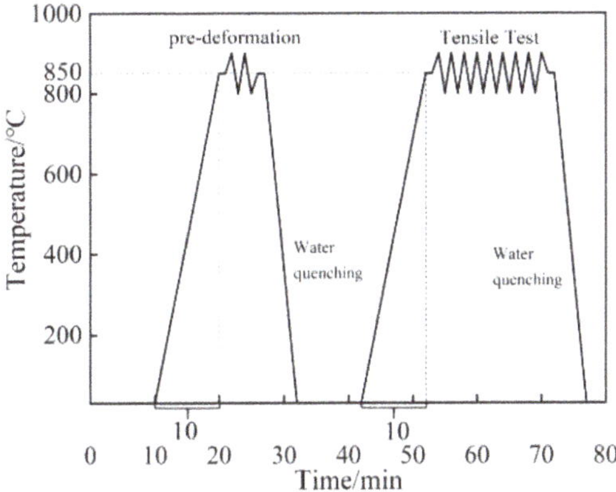

Figure 2. Schematic diagram of the Ti-6Al-4V hot tensile test heating path.

The specimens were first installed in the grips and soaked at 850 °C for 10 min. Then, they were stretched to the true strains of 0, 0.05, 0.1, and 0.15 at the strain rate of 0.001 s^{-1}, held for 10 min, and cooled in water, which represented the hot twisting process. Tensile specimens with pre-deformation were placed in the grips and soaked at 850 °C for 10 min. Then, they were stretched to the true strains of 0.3, 0.5, and 0.7 and fractured. Specimens were immediately quenched in water to retain the microstructures. For each condition, the test was repeated three times. Metallographic observations were carried out through EBSD testing. EBSD data were obtained by SIGMA 500 (ZEISS Inc., Oberkochen, Germany) and analyzed through HKL-Channel 5 software (Version 5, Oxford Instruments Inc., London, UK). XRD data were obtained by Bruker D8 advance (Bruker Inc., Karlsruhe, Germany) and analyzed through Jade (Version 6.5, Materials Data, Livemore, CA, USA).

3. Results and Discussion

3.1. Hot Tensile Properties of Specimens with Pre-Strain

Figure 3a shows the true stress-strain curve of specimens under uniaxial tension at 850 °C and 0.001 s^{-1} with different pre-deformation strains. It is obvious that the tensile behavior was affected by the pre-deformation. Tensile properties such as yield strength (*YS*), ultimate tensile strength (*UTS*), flow stress σ (ε = 0.5) and elongation (*EL*) are given in Figure 3b. It was found that the *EL* was more than 200%, regardless of the pre-strain, indicating that the material had superplasticity under this tensile condition [18,19]. The results show that the specimen with a pre-strain of 0.05 had the highest ultimate tensile strength and maximum elongation, which are 77.3 MPa and 234.5%, respectively. As the pre-deformation strain increased from 0.05 to 0.1, the *YS* decreased gradually from

63.4 MPa to 55.5 MPa, the *UTS* decreased from 77.3 MPa to 66.8 MPa, and the *EL* decreased from 234.5% to 205.2%. When the pre-strain was raised from 0.1 to 0.15, the *UTS* further declined to 61.4 MPa and the change in the *EL* was small. Additionally, Figure 3c shows the samples stretched to fracture.

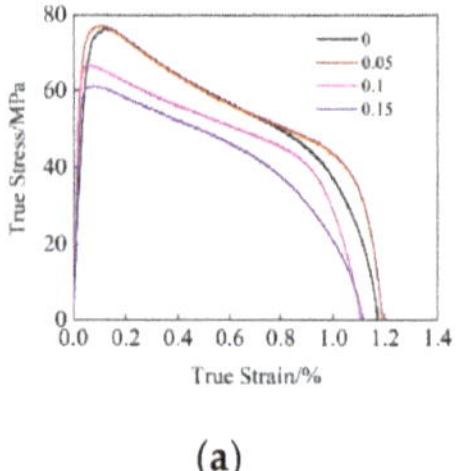

(a)

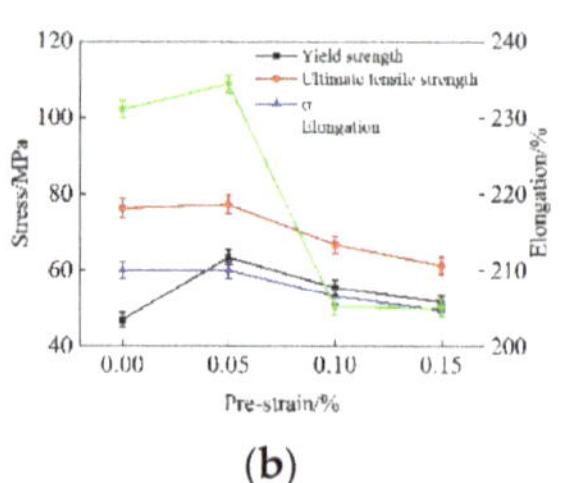

(b)

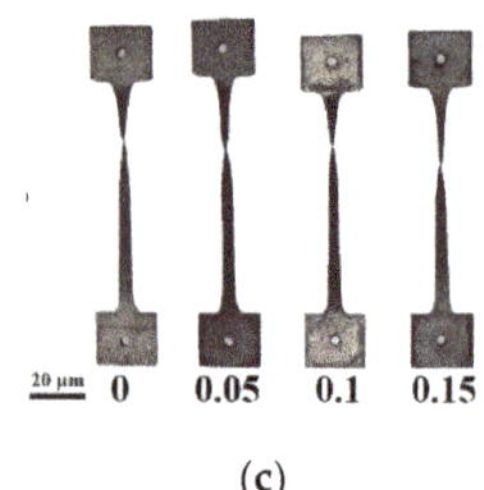

(c)

Figure 3. Hot tensile behavior of different Ti-6Al-4V specimens stretched at 850 °C with 0.001 s^{-1}: (a) True stress-strain curves; (b) Tensile properties; (c) Tested samples with different pre-strain.

3.2. Microstructure Evolution during Hot Tensile Test

Microstructures of the tensile specimens with true strains of 0.3, 0.5 and 0.7 were compared to investigate the microstructure evolution and DRX during hot deformation, which could explain the changes in the tensile properties.

3.2.1. Initial Microstructure

The microstructure and the inverse pole figure (IPF) are given in Figure 4a,b, respectively, in which the color code represents the orientations of α grains: black lines represent the high-angle grain boundaries (HAGBs, $\theta > 15°$), and blue lines represent the low angle grain boundaries (LAGBs, $2° < \theta < 1\,5°$). The initial structure consisted of an equiaxed α phase and intergranular β phase. Figure 4c shows the average grain size distribution, which indicates that the initial structure was mostly composed of equiaxed grains with size ≤2 μm. The average grain size was 2.77 μm. Figure 4d shows the misorientation distribution. The fraction of LAGBs was 32.6%, and LAGBs were concentrated within 5°. HAGBs were concentrated between 55° and 65°. The abundant HAGBs indicate that the as-received material had a few deformation substructures. As shown in Figure 4e, the initial samples were composed of α and β phases.

3.2.2. Microstructures of Specimens without Pre-Strain

Figure 5 shows the microstructures and corresponding IPFs of the tensile specimens without pre-strain at different deformation stages. The average sizes of grains in specimens with different strains were 2.57 μm, 2.55 μm and 2.44 μm, respectively. The phase volume fraction changed slightly with the increasing strain because the stretching temperature was much lower than the phase transition point, which was 998 °C. The grain size decreased continuously with the increase in strain, whereas the DRX fraction increased significantly. It can be deduced that with the increase in strain, LAGBs transformed into HAGBs, and finally formed small crystal grains [20]. In the early tensile stage, the dislocation density increased rapidly with strain. Additionally, the dislocations accumulated to form LAGBs. Many small equiaxed grains appeared on the grain boundaries when the strain reached 0.5, marked by the white tip in Figure 5b. This may have been the result of dynamic recrystallization (DRX) as the subgrains with LAGBs evolved into grains with HAGBs. The abundant internal LAGBs promoted the occurrence of DRX. The merge of the subgrains and the formation of HAGBs during the hot tensile procedure caused the decrease in dislocation density, which resulted in a decrease in the flow stress. DRX plays a softening role in the deformation process. DRX is mainly divided into continuous dynamic recrystallization (CDRX), discontinuous dynamic recrystallization (DDRX), and geometric dynamic recrystallization (GDRX), and is distinguished by the mechanism of recrystallization grain formation [21,22].

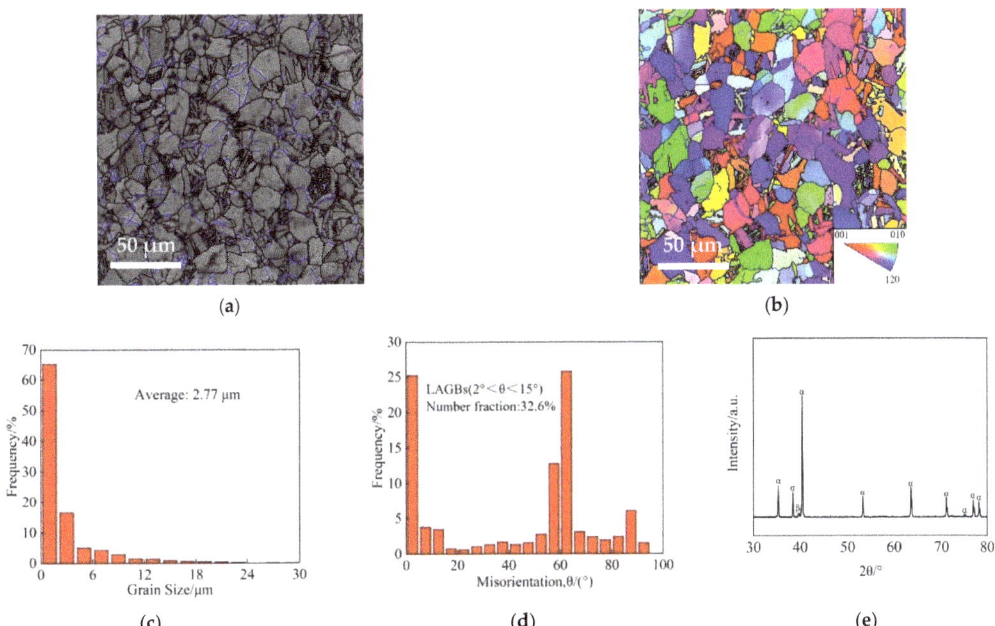

Figure 4. Characterization of initial Ti-6Al-4V: (**a**) SEM; (**b**) IPF; (**c**) Grain size distribution; (**d**) Misorientation distribution; and (**e**) XRD.

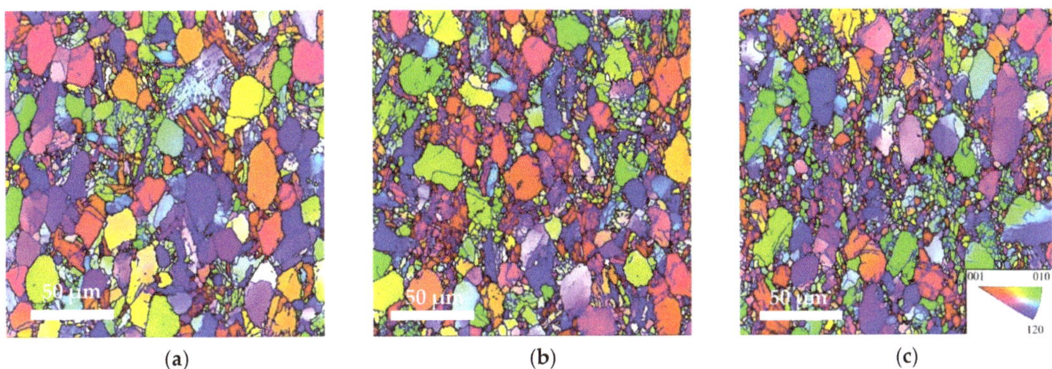

Figure 5. IPFs of specimens without pre-deformation at different true strains: (**a**) 0.3; (**b**) 0.5; and (**c**) 0.7.

The proportion of LAGBs first increased from 61.1% to 61.4% and then decreased from 61.4% to 58.3% when the strain increased from 0.3 0.7. The change in LAGBs was related to the change in dislocation density. Additionally, the evolution of dislocation density could be reflected by the geometric dislocation density (GND) [23,24]. Figure 6 shows the GND density maps of specimens without pre-deformation at true strains of 0.3, 0.5, and 0.7. The GND density first increased and then decreased. This indicates that the DRX was not obvious when the true strains were low [25]. However, a large number of deformation substructures accumulated during the initial deformation stage. The proportion increased rapidly with the true strain beyond the critical strain, which accelerated the aggregation and growth of sub-grains, eradicated dislocations in the grains, and reduced the content of LAGBs. In addition, recrystallized grains promoted by the DRX were concentrated in the HAGBs, and only a few grains were inside the elongated grains. This suggests that high-temperature deformation can promote the migration of HAGBs and nucleation with the expansion of HAGBs. The growth in grain number depends on the grain boundary migration, which indicates that DRX is dominated by DDRX [26].

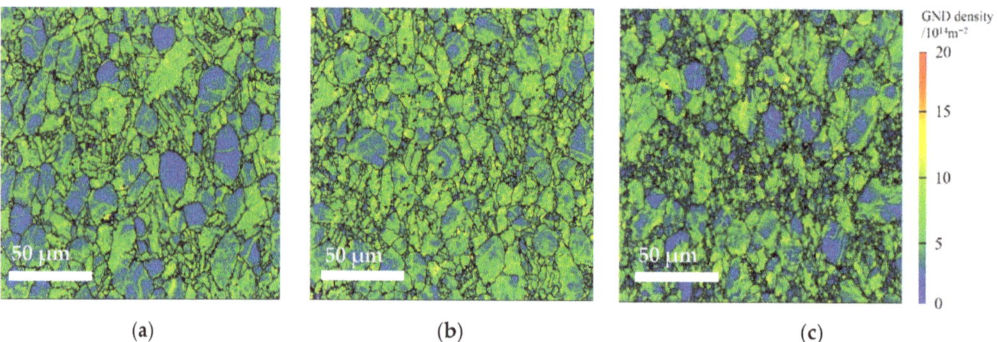

Figure 6. GND density maps of specimens without pre-deformation: (**a**) 0.3; (**b**) 0.5; and (**c**) 0.7.

3.2.3. Microstructure of Specimens with Pre-Strain

Figure 7 shows the IPFs of specimens with pre-strain of 0.05 at true strains of 0.3, 0.5, and 0.7. The increasing deformation promoted the DRX, leading to the formation of a large number of small grains [27,28]. Images show that the structure is mainly composed of elongated equiaxed grains and small equiaxed grains. This indicated that DRX is the main mode of microstructural evolution [29]. After the deformation, the initial grains were obviously elongated along the tensile direction, which indicated that grain boundary sliding was the main hot-deformation mechanism of Ti-6Al-4V. When the true strain rose from 0.3 to 0.5, some lamellar α grains broke and formed small equiaxed grains. When the true strain reached 0.7, the lamellar α grains were transformed into fine equiaxed grains.

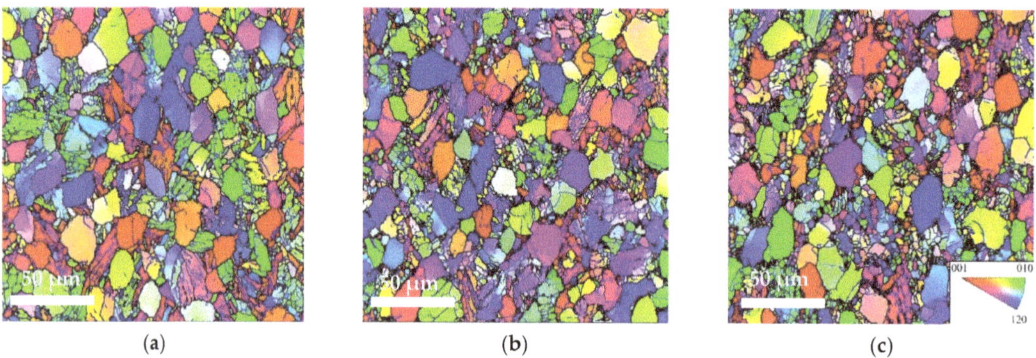

Figure 7. IPFs of specimens with pre-deformation strain of 0.05 at different true strains: (**a**) 0.3; (**b**) 0.5; and (**c**) 0.7.

Figure 8a shows the average grain size and dynamic recrystallization fraction of the specimens with different pre-strains. At a true strain of 0.3, the average grain sizes of the specimens with pre-strains of 0.05, 0.1 and 0.15 were 2.42 μm, 2.85 μm, and 2.79 μm, and the dynamic recrystallization fractions were 8.03%, 11%, and 12.6%, respectively. At strains larger than 0.3, the DRX fraction increased with the decrease in pre-strain, which promoted the formation of small grains and reduced the average grain size. This caused a decrease in HAGBs, which hindered the movement of dislocations, leading to a decrease in flow stress. Figure 8b shows that the DRX fraction increased with the pre-deformation when the strain was 0.3. This indicates that the smaller pre-deformation failed to promote the DRX. However, a large number of substructures were accumulated in the material at the early stages of deformation. When the deformation continued, the substructure accelerated the annihilation and rearrangement of dislocations, leading to increased DRX [30]. Pre-deformation can promote the DRX, which accelerated the transformation of sub-grains and LAGBs into small equiaxed grains and HAGBs. With the decrease in pre-deformation, the DRX fraction increased, leading to the generation of more small-sized grains. Additionally, the increase in HAGBs, which hindered the movement of dislocations, led to the increase in strength. At the same time, the DRX fraction increased, which further intensified the softening. Compared with the specimen without pre-deformation, the pre-deformed specimens have small equiaxed grains in the largely elongated grains. It may be because the pre-deformation increased the dislocation and promoted the CDRX. In addition, more compact equiaxed grains appeared at the HAGBs, indicating that pre-deformation can effectively promote the migration of HAGBs. As the true strain promoted the DRX, during which the small grains gradually formed, the deformed substructure reduced, resulting in a gradual decrease in the flow stress. More deformed substructures were accumulated for the less pre-deformed specimens. Additionally, the increase in deformation provides the driving force for DRX.

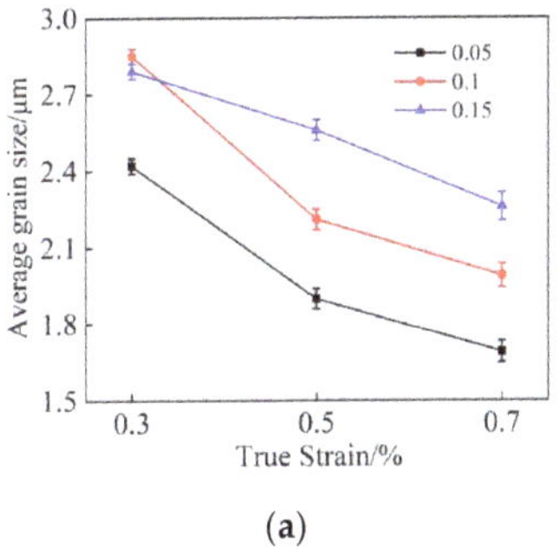
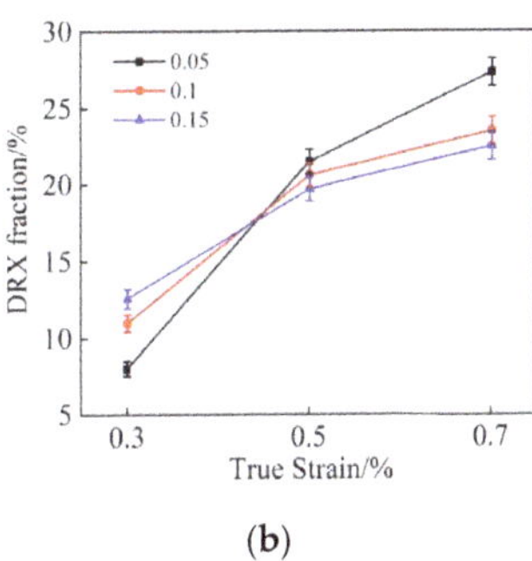

Figure 8. The (**a**) average grain size and (**b**) DRX fraction of the specimens with pre-strains.

Pre-strain can increase the deformation substructures inside the material, promote the DRX, and then reduce the dislocations in the material, so that softening occurs in the deformation process of the material and the flow stress decreases. In addition, the deformation substructure decreases with the increase in pre-strain. When the strain increases, the DRX degree of samples with smaller pre-strain is higher, and more small-sized equiaxed grains are generated. The average grain size decreases, and the proportion of HAGBs per unit area is higher, which hindrances the dislocation movement and leads to the increase in flow stress.

4. Conclusions

- The deformation behavior of Ti-6Al-4V titanium alloys with different pre-deformation was studied. When the pre-strain was 0.05, the elongation of the specimen was the highest: 234.5%. When the pre-strain reached 0.15, the ultimate tensile strength was the lowest: 66.8 MPa. The hot-forming performance of Ti-6Al-4V increased first and then decreased with the increase in pre-strain;

- During the hot tensile deformation, the deformation mechanism of Ti-6Al-4V was dominated by high-angle grain boundaries sliding. Dislocation movement also played an important role in the hot deformation process, which could be considered as the adjustment process of grain boundary sliding. In the non-pre-deformed specimen, the recrystallized grains all appeared at the grain boundary, indicating that the dynamic recrystallization was dominated by discontinuous dynamic recrystallization;
- Pre-deformation provides more deformation substructures for subsequent deformation and promotes dynamic recrystallization of the material in the subsequent deformation process. Recrystallized grains appeared at the grain boundaries and inside the grains during the hot tensile process. The pre-deformation promoted both continuous dynamic recrystallization and discontinuous dynamic recrystallization, which caused major changes in mechanical properties during the hot deformation.

Author Contributions: Conceptualization, T.W., M.C., D.Z. and L.X.; methodology, N.W., D.Z. and L.X.; software, T.W. and N.W.; validation, T.W., N.W. and W.S.; formal analysis, T.W.; investigation, W.S.; resources, M.C.; data curation, T.W.; writing—original draft preparation, T.W.; writing—review and editing, T.W. and N.W.; visualization, T.W.; supervision, M.C., D.Z. and L.X.; project administration, M.C.; funding acquisition, M.C. All authors have read and agreed to the published version of the manuscript.

Funding: This research received no external funding.

Data Availability Statement: Data presented in this article are available at request from the corresponding author.

Conflicts of Interest: The authors declare no conflict of interest.

References

1. Zhao, Z.B.; Wang, Q.J.; Liu, J.R.; Yang, R. Effect of heat treatment on the crystallographic orientation evolution in a near-α titanium alloy Ti60. *Acta Mater.* **2017**, *131*, 305–314. [CrossRef]
2. Ma, L.X.; Wan, M.; Li, W.D.; Shao, J.; Bai, X.P. Constitutive modeling and processing map for hot deformation of Ti-15Mo-3Al-2.7Nb-0.2Si. *J. Alloys Compd.* **2019**, *808*, 151759. [CrossRef]
3. Cui, C.X.; Hu, B.M.; Zhao, L.C.; Liu, S.J. Titanium alloy production technology, market prospects and industry development. *Mater. Des.* **2011**, *32*, 1684–1691. [CrossRef]
4. Wang, Q.Q.; Liu, Z.Q. Plastic deformation induced nano-scale twins in Ti-6Al-4V machined surface with high speed machining. *Mater. Sci. Eng. A* **2016**, *675*, 271–279. [CrossRef]
5. Momeni, A.; Abbasi, S.M.; Sadeghpour, S. A comparative study on the hot deformation behavior of Ti-5Al-5Mo-5V-3Cr and newly developed Ti-4Al-7Mo-3V-3Cr alloys. *Vaccum* **2019**, *161*, 401–408. [CrossRef]
6. Li, W.S.; Yamasaki, S.; Mitsuhara, M.; Hakashima, H. In situ EBSD study of deformation behavior of primary α phase in a bimodal Ti-6Al-4V alloy during uniaxial tensile tests. *Mater. Charact.* **2020**, *163*, 110282. [CrossRef]
7. Bodunrin, M.O.; Chown, L.H.; Merwe, J.W.; Alaneme, K.K. Hot working of Ti-6Al-4V with a complex initial microstructure. *Int. J. Mater. Form.* **2018**, *12*, 857–874. [CrossRef]
8. Cui, J.H.; Yang, H.; Sun, Z.C.; Li, H.W.; Li, Z.J.; Shen, W.C. Flow behavior and constitutive model using piecewise function of strain for TC11 alloy. *Rare Met. Mater. Eng.* **2012**, *41*, 397–401.
9. Liu, Z.G.; Li, P.J.; Xiong, L.T.; Liu, T.Y.; He, L.J. High-temperature deformation behavior and microstructure evolution of Ti55 titanium alloy. *Mater. Sci. Eng: A* **2016**, *680*, 259–269. [CrossRef]
10. Ma, T.F.; Zhou, X.; Du, Y.; Li, L.; Zhang, L.C.; Zhang, Y.S.; Zhang, P.X. High temperature deformation and microstructure evolution of core-shell structured titanium alloy. *J. Alloys Compd.* **2019**, *775*, 316–321. [CrossRef]
11. Wang, H.F.; Ban, C.Y.; Zhao, N.N.; Kang, Y.Y.; Qin, T.P.; Liu, S.T.; Cui, J.Z. Enhance strength and ductility of nano-grain titanium processed by two-step severe plastic deformation. *Mater. Lett.* **2020**, *266*, 127485. [CrossRef]
12. Hajizadeh, K.; Eghbali, B. Effect of two-step severe plastic deformation on the microstructure and mechanical properties of commercial purity titanium. *Met. Mater. Int.* **2014**, *20*, 343–350. [CrossRef]
13. Stolyarov, V.V.; Zeipper, L.; Mingler, B.; Zehetbauer, M. Influence of post-deformation on CP-Ti processed by equal channel angular pressing. *Mater. Sci. Eng. A* **2008**, *476*, 98–105. [CrossRef]
14. Stolyarov, V.V.; Zhu, Y.T.; Raab, G.I.; Zharikov, A.I.; Valiev, R.Z. Effect of initial microstructure on the microstructural evolution and mechanical properties of Ti during cold rolling. *Mater. Sci. Eng. A* **2004**, *385*, 309–313. [CrossRef]
15. Zhu, T.; Huang, J.Y.; Gubicza, J.; Ungar, T.; Wang, Y.M.; Ma, E.; Valiev, R.Z. Nanostructures in Ti processed by severe plastic deformation. *J. Mater. Res.* **2003**, *18*, 1908–1917. [CrossRef]

16. Wang, S.H.; Liu, C.H.; Chen, J.H.; Li, X.L.; Zhu, D.H.; Tao, G.H. Hierarchical nanostructures strengthen Al–Mg–Si alloys processed by deformation and aging. *Mater. Sci. Eng. A* **2013**, *585*, 233–242. [CrossRef]
17. Sachtleber, M.; Raabe, D.; Weiland, H. Surface roughening and color changes of coated aluminum sheets during plastic straining. *J. Mater. Process. Technol.* **2004**, *148*, 68–76. [CrossRef]
18. Seshacharyulu, T.; Medeiros, S.C.; Frazier, W.G.; Prasad, Y.V.R.K. Hot Working of Commercial Ti–6Al–4V with an Equiaxed α–β Microstructure: Materials Modeling Considerations. *Mater. Sci. Eng. A* **2000**, *284*, 184–194. [CrossRef]
19. Zhang, W.; Liu, H.; Ding, H.; Fujii, H. Superplastic Deformation Mechanism of the Friction Stir Processed Fully Lamellar Ti-6Al-4V Alloy. *Mater. Sci. Eng. A* **2020**, *785*, 139390. [CrossRef]
20. Belyakov, A.; Sakai, T.; Miura, H.; Kaibyshev, R.; Tsuzaki, K. Continuous recrystallization in austenitic stainless steel after large strain deformation. *Acta Mater.* **2002**, *50*, 1547–1557. [CrossRef]
21. Sheppard, T.; Norley, J. Deformation Characteristics of Ti–6Al–4V. *Mater. Sci. Technol.* **1988**, *4*, 903–908. [CrossRef]
22. Park, N.-K.; Yeom, J.-T.; Na, Y.-S. Characterization of Deformation Stability in Hot Forging of Conventional Ti–6Al–4V Using Processing Maps. *J. Mater. Process. Technol.* **2002**, *130–131*, 540–545. [CrossRef]
23. Kumar, S.; Pavithra, B.; Singh, V.; Ghosal, P.; Raghu, T. Tensile anisotropy associated microstructural and microtextural evolution in a metastable beta titanium alloy. *Mater. Sci. Eng. A* **2019**, *747*, 1–16. [CrossRef]
24. Moussa, C.; Bernacki, M.; Besnard, R.; Bozzolo, N. Statistical analysis of dislocations and dislocation boundaries from EBSD data. *Ultramicroscopy* **2017**, *179*, 63–72. [CrossRef] [PubMed]
25. Yi, Z.X.; Li, X.Q.; Pan, C. Dynamic Recrystallization Behavior and Model Study of Equiaxed Fine Grain Structure TC4. *J. Phys. Conf. Ser.* **2020**, *1676*, 012159. [CrossRef]
26. Sun, Y.G.; Zhang, C.J.; Feng, H.; Zhang, S.Z.; Han, J.C.; Zhang, W.G.; Zhao, E.; Wang, H.W. Dynamic recrystallization mechanism and improved mechanical properties of a near α high temperature titanium alloy processed by severe plastic deformation. *Mater. Charact.* **2020**, *163*, 110281. [CrossRef]
27. Weiss, Y.; Semiatin, S.L. Thermomechanical processing of alpha titanium alloys—An overview. *Mater. Sci. Eng. A* **1999**, *263*, 243–256. [CrossRef]
28. Zherebtsov, S.; Murzinova, M.; Salishchev, G.; Semiatin, S.L. Spheroidization of the lamellar microstructure in Ti–6Al–4V alloy during warm deformation and annealing. *Acta Mater.* **2011**, *59*, 4138–4150. [CrossRef]
29. Matsumoto, H.; Velay, V. Mesoscale modeling of dynamic recrystallization behavior, grain size evolution, dislocation density, processing map characteristic, and room temperature strength of Ti-6Al-4V alloy forged in the (α+β) region. *J. Alloys Compd.* **2017**, *708*, 404–413. [CrossRef]
30. Zhao, J.; Wang, K.H.; Huang, K.; Liu, G. Recrystallization behavior during hot tensile deformation of TA15 titanium alloy sheet with substantial prior deformed substructures. *Mater. Charact.* **2019**, *151*, 429–435. [CrossRef]

Article

An Experimental Study of the Tension-Compression Asymmetry of Extruded Ti-6.5Al-2Zr-1Mo-1V under Quasi-Static Conditions at High Temperature

Chen Zhang, Dongsheng Li, Xiaoqiang Li * and Yong Li

School of Mechanical Engineering and Automation, Beihang University, Beijing 100191, China; zhangchen1993@buaa.edu.cn (C.Z.); lidongs@buaa.edu.cn (D.L.); liyong19@buaa.edu.cn (Y.L.)
* Correspondence: lixiaoqiang@buaa.edu.cn; Tel.: +86-010-82316584

Abstract: The tension-compression asymmetry (TCA) behavior of an extruded titanium alloy at high temperatures has been investigated experimentally in this study. Uniaxial tensile and compressive tests were conducted from 923 to 1023 K with various strain rates under quasi-static conditions. The corresponding yield stress and asymmetric strain hardening behavior were obtained and analyzed. In addition, the microstructure at different temperatures and stress states indicates that the extruded TA15 profile exhibits a significant yield stress asymmetry at different testing temperatures. The flow stress and yield stress during tension are greater than compression. The yield stress asymmetry decreases with the increase in temperature. The alloy also exhibits TCA behavior on the strain hardening rate. Its mechanical response during compression is more sensitive than tension. A dynamic recrystallization phenomenon is observed instead of twin generated in tension and compression under high-temperature quasi-static conditions. The grains are elongated along the tensile direction and deformed by about 45° along the compressive load axis. Finally, the TCA of Ti-6.5Al-2Zr-1Mo-1V (TA15) alloy is due to slip displacement. The tensile deformation activates basal <a>, prismatic <a> and pyramidal <c + a> slip modes, while the compressive deformation activates only prismatic <a> and pyramidal <c + a> slip modes.

Keywords: extruded titanium alloy profile; tension–compression asymmetry (TCA); hot stretch bending (HSB); high temperature; microstructure

Citation: Zhang, C.; Li, D.; Li, X.; Li, Y. An Experimental Study of the Tension-Compression Asymmetry of Extruded Ti-6.5Al-2Zr-1Mo-1V under Quasi-Static Conditions at High Temperature. *Metals* **2021**, *11*, 1299. https://doi.org/10.3390/met11081299

Academic Editors: Maciej Motyka and Marcello Cabibbo

Received: 13 July 2021
Accepted: 12 August 2021
Published: 17 August 2021

Publisher's Note: MDPI stays neutral with regard to jurisdictional claims in published maps and institutional affiliations.

1. Introduction

Titanium and its alloys for their high corrosion resistance, excellent mechanical properties and relatively light weight are widely used in aerospace materials [1–4]. Additionally, unlike aluminum alloys, they have similar thermal expansion coefficients and outstanding electrochemical compatibility with carbon fiber reinforced plastics (CFRP) [5,6]. Titanium alloys are highly desirable for use in contoured airframe structures with complex curvatures, such as the frame and reinforced frames of a cabin door, see in Figure 1 [7].

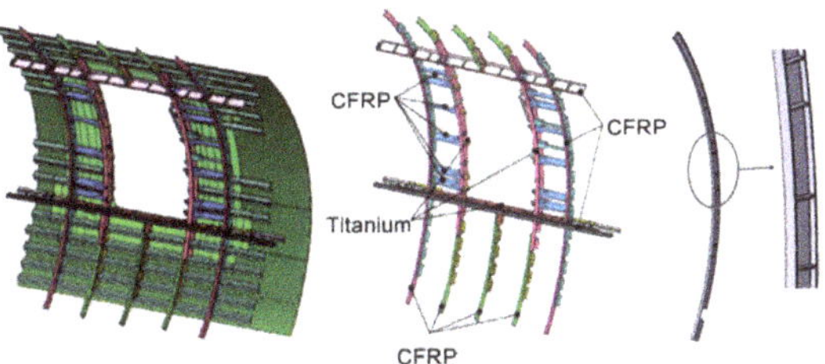

Figure 1. Reinforced frames of cabin door in Boeing 787 [7]. Reprinted with permission from Ref. [7]. Copyright 2021 Springer Nature.

To form such contoured titanium components, Hot Stretch Bending (HSB) is developed by the Cyril Bath Company (Monroe, NC, USA) to fabricate such components in a new generation of commercial aircraft. Additionally, its process procedure includes the following four stages: heating, pre-stretch, bending and cooling, as seen in Figure 2 [6,8]. In the HSB, the bending moment produces tensile stress above the neutral axis, and compressive stress below the neutral axis. There are significantly TCA stress states in the alloy profile during forming; therefore, it is necessary to research the mechanical response under different stress states (tension and compression) to achieve accurate control of the geometry of the formed parts.

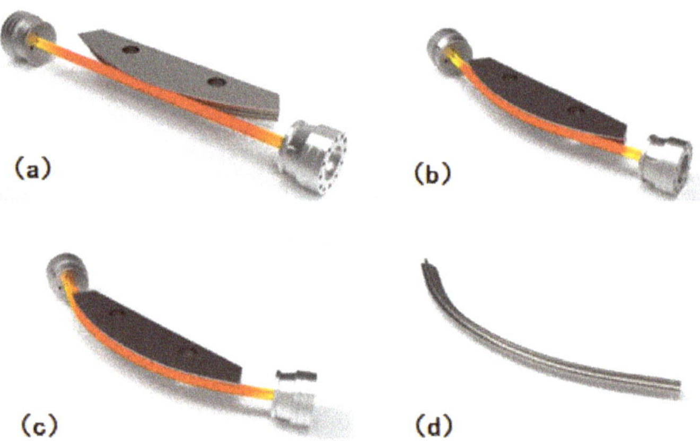

Figure 2. Process procedure of HSB: (**a**) heating, (**b**) pre-stretch, (**c**) bending and (**d**) cooling [6].

Many scholars have found that there is TCA in pure titanium and its alloys at room temperature and medium-high temperatures ($\leq$873 K). Lin et al. [9] studied the TCA of pure titanium (CP-Ti) at room temperature and stated that the difference of the stress values activating the secondary twins is the main cause of TCA at room temperature. Tuninetti et al. [10] investigated the mechanical response of Ti-6Al-4V based on the target strain rate and the results indicated there is obvious TCA at ambient temperature. Hao et al. [11] revealed that the strain hardening and yield asymmetry of CP-Ti are prominent due to the large number of twinning grains generated during the compression process. Neeraj et al. [12] suggested that the <c + a> dislocation phenomenon occurred in the compression, but not in the tensile process. Additionally, the different critical shear stress (CRSS) of <a> slip was also one reason leading to the TCA of titanium alloy. Sarsfield et al. [13] founded that the tensile fracture of the material was in the shape of 'V', while the compression specimen was in the shape of 45° shear failure.

Additionally, it is also widely observed that TCA is significantly related to deformation temperatures and strain rates. Akhtar et al. [14] have carried out compression tests on three Ti-6Al-4V materials at 233–755 K with strain rates from 10^{-6} to 3378 s^{-1}. The results supported that the yield stress and strain hardening of the three materials are more sensitive to temperature than strain rate. Adharapurapu et al. [15] have conducted dynamic compression and tension tests on a Ni-Ti shape memory alloy at temperatures from 469 to 673 K. The observations illustrated that temperature variation was one of the main causes of TCA.

Previous studies on the TCA of titanium and its alloys mainly focused on room temperature or medium-high temperatures ($\leq$873 K). Compared to titanium plates and hot forging rods, HSB generally uses extruded profiles, and the forming temperature can achieve to the range of 868 to 1088 K [6]. However, there are few reports on the TCA of the extruded titanium profile in this forming temperature range. The hexagonal lattice metals have fewer sliding planes at ambient temperature, which mainly depends on the stress modes, grain orientation, twinning, etc. [16,17]. It is not clear whether alternative mechanisms such as twinning, grain boundary sliding are activated at such high temperatures. In addition, the extruded titanium alloy profile has a lamellar structure with uniform and coarse grains, and its plasticity and strength are lower than those of bimodal and tri-modal microstructures [18]. It remains unknown whether this microstructure pattern affects the TCA and, thus, influences the forming behavior of the alloys.

Hence, this paper is to investigate the tensile and compressive responses of the extruded titanium alloy profiles at high temperatures with HSB forming conditions. The stress-strain behavior and microstructure evolutions under tension and compression tests of the extrusion at 923 to 1023 K were studied. Based on the testing results, the effects of deformation temperature and strain rate on TCA were discussed.

2. Materials and Methods

2.1. Material

The TA15 extrusion profiles, provided by Baoji Titanium Industry Co., Ltd. (Baoji, China) [19] according to AMST9046B [20] was used in this research. The alloys were annealed at 823 K for 2 h and then cooled to room temperature to eliminate the initial residual stress. The chemical composition (in wt.%) and cross-sectional dimensions of TA15 are listed in Table 1 and Figure 3. The tensile and compressive specimens are processed along the extrusion direction of the titanium profile, see in Figure 4.

Table 1. Main chemical composition (in wt.%) of Ti-6.5Al-2Zr-1Mo-1V.

Al	Zr	Mn	V	Ti
5.5~7.0	1.5~2.5	0.5~2.0	0.8~2.5	allowance

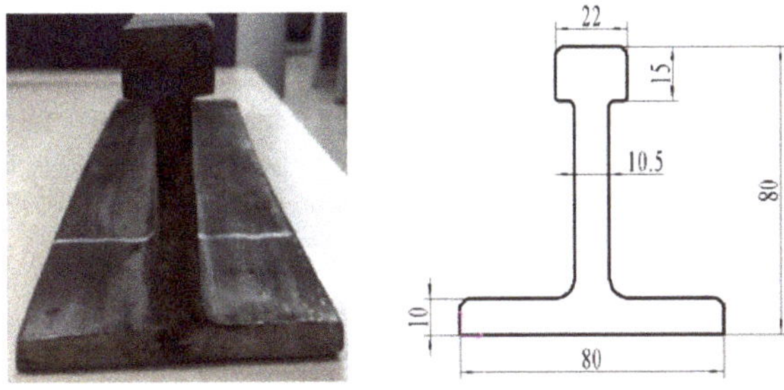

Figure 3. Illustration and detailed dimensions of the section of extruded I profile (unit: mm).

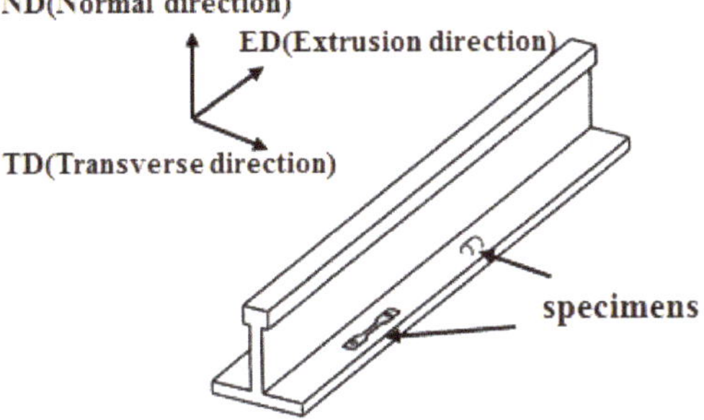

Figure 4. Cutting specimens on extruded titanium alloy profile.

2.2. Mechanical Tests

The tensile tests were carried out on DDL50 testing machine at high temperature (923–1023 K). The dimension of tensile samples (refer to ISO 6892-2 [21]) is shown in Figure 5a. For tension tests, the samples are heated at a rate of 50 K/min and then held for 5 min to maintain temperature uniformity [19].

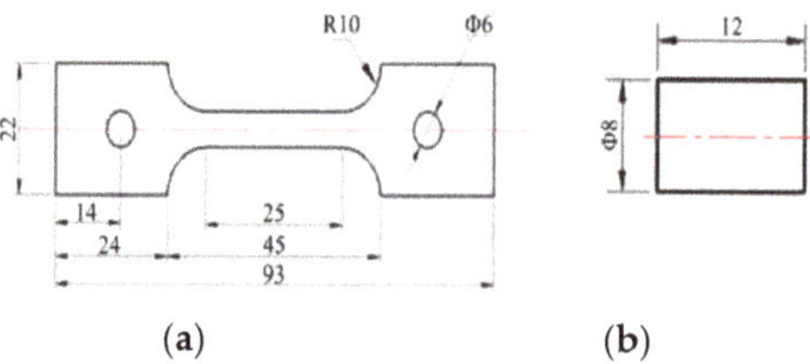

Figure 5. High temperature sample (unit: mm): (**a**) uniaxial tensile specimen and (**b**) uniaxial compressive specimen.

The compressive specimens are solid cylinders with a diameter of 8.0 mm and a length of 12 mm, as shown in Figure 5b. Additionally, the test was conducted on the Gleeble-1500 thermal/mechanical simulator (Gleebel Heat Simulator, Dynamic Systems Inc., Austin, TX USA). These specimens are heated to a specified temperature by 5 K/s and then kept for 5 min. It should be noted that although different heating rates have been adopted in the tensile and compressive tests due to the limitations of testing equipment, few effects are believed to be raised in the mechanical properties, as stated in a previous study by Shen et al. [22]. Considering the heat dissipation capacity of the equipment, the strain rate of the compression test is between 0.001–0.05 s^{-1}.

2.3. OM Observations

To identify the grain sizes of TA15, some tested samples for metallographic observations were prepared according to ISO 4499-1 [23]. The tensile specimens (the true strain is 0.25–0.4) were processed from the gauge length, while the compressive samples (the true stain is 0.7) were conducted along parallel compression direction by using a wire-electrode cutting machine. Then, the microstructure of TA15 alloy was observed under a Leica DM4000M metallographic microscope.

In this paper, the test scheme of TA15 titanium alloy with different temperature and strain rates is shown in Table 2.

Table 2. The experiments of uniaxial tension and compression of TA15.

Groups	Temperature/K	Strain Rate/s^{-1}	Microstructure Test
1	923 973 102	0.001	Optical microscope
2	973	0.0001 0.0005	-
		0.001 0.005	-

3. Results

3.1. Stress–Strain Behavior at Different Temperatures

The true stress–strain curves (σ-ε) were derived from the engineering stress–strain data based on the assumption of incompressibility plastic deformation. Meanwhile, the corresponding strain hardening–true strain curves (θ-ε) were also derived from the true stress–strain curve (σ-ε) to analyze the evolution of mechanical behavior [16]. The strain hardening rate (θ) is the slope of the stress–strain curves (σ-ε) and it can be defined as follows:

$$\theta = \left[\frac{d\sigma}{d\sigma}\right]_{\dot\varepsilon} \tag{1}$$

where σ is the true flow stress (MPa), ε is the true strain and $\dot\varepsilon$ is the strain rate (s^{-1}).

Figure 6 shows the σ-ε curves and corresponding work hardening rate (θ-ε) curves of the TA15 profiles at different temperatures. The 0.2% offset yield stress $\sigma_{0.2}$ was defined as the experimental Young's modulus fitted to the initial part of the stress-strain curves [24]. We can conclude that the flow stress of tension and compression decreases when the temperature rises from 923 to 1023 K. Figure 6a,c show that the yield stress $\sigma_{0.2}$ decreases about 150 MPa from 923 to 973 K, and the stress difference between 973 and 1023 K is about 100 MPa. Due to the high temperature softening effect [25], the stress curve shows a downward trend after passing the yield stress point, and the flow stress of titanium profiles is sensitive to the temperature. Figure 6b,d reveal that the different temperature also affects the work hardening rate (θ). The θ drops sharply at the beginning of the deformation and then becomes flat. In the uniaxial tension tests (Figure 6b), when the strain is larger than 0.08, it has a significant decrease and becomes negative at 923 K, which may be related to the local instability deformation [25]; when the temperature is 973 K and the strain exceed 0.2, θ starts to turn negative; it maintains stable when the temperature is 1023 K. In contrast, in uniaxial compression tests (Figure 6d), the difference of θ at different deformation temperature is small, and the curve is stable.

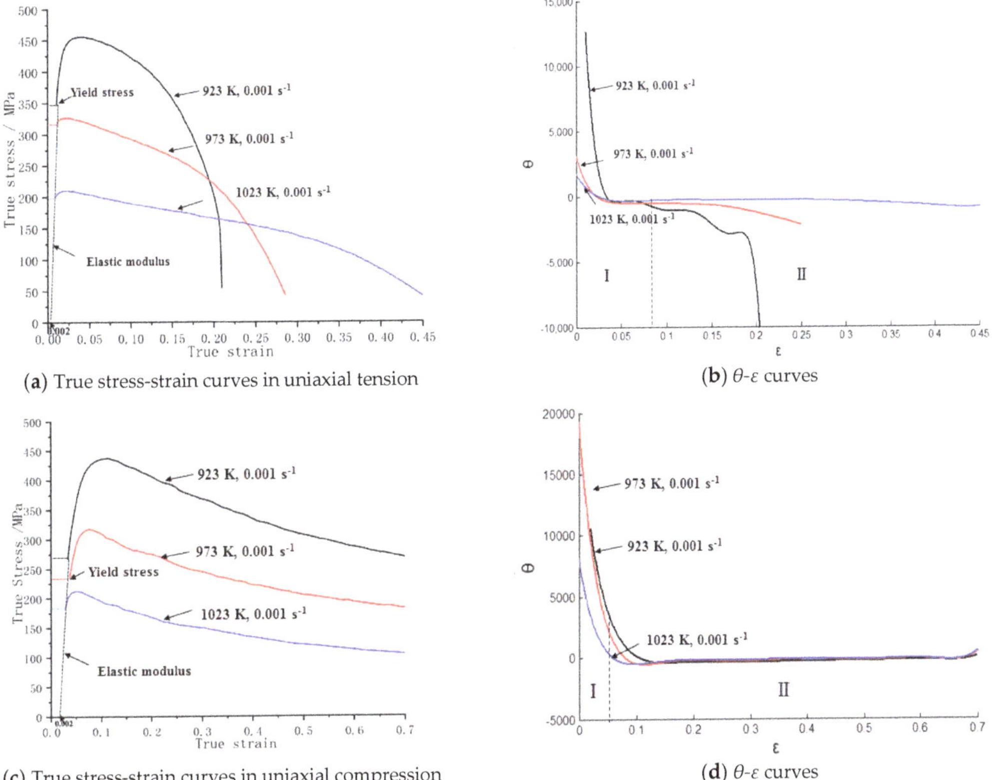

(**a**) True stress-strain curves in uniaxial tension

(**b**) θ-ε curves

(**c**) True stress-strain curves in uniaxial compression

(**d**) θ-ε curves

Figure 6. The σ-ε (**a**,**c**) and θ-ε (**b**,**d**) curves of tension and compression of TA15 at different temperatures.

3.2. Stress-Strain Behavior at Different Strain Rates

Figure 7 shows the σ-ε and θ-ε curves at 973 K with different strain rates. It can be seen from Figure 7a,c that the tension and compression responses of TA15 profiles have a positive strain rate correlation within the tested rate range. As the strain rate increases from 0.0001 to 0.005 s^{-1}, the yield stress rises from 150 to 368 MPa. According to Figure 7b,d, the θ changes little both in tensile and compressive deformation, and there is no three-stage hardening phenomenon [9] in (θ-ε) curves at 973 K.

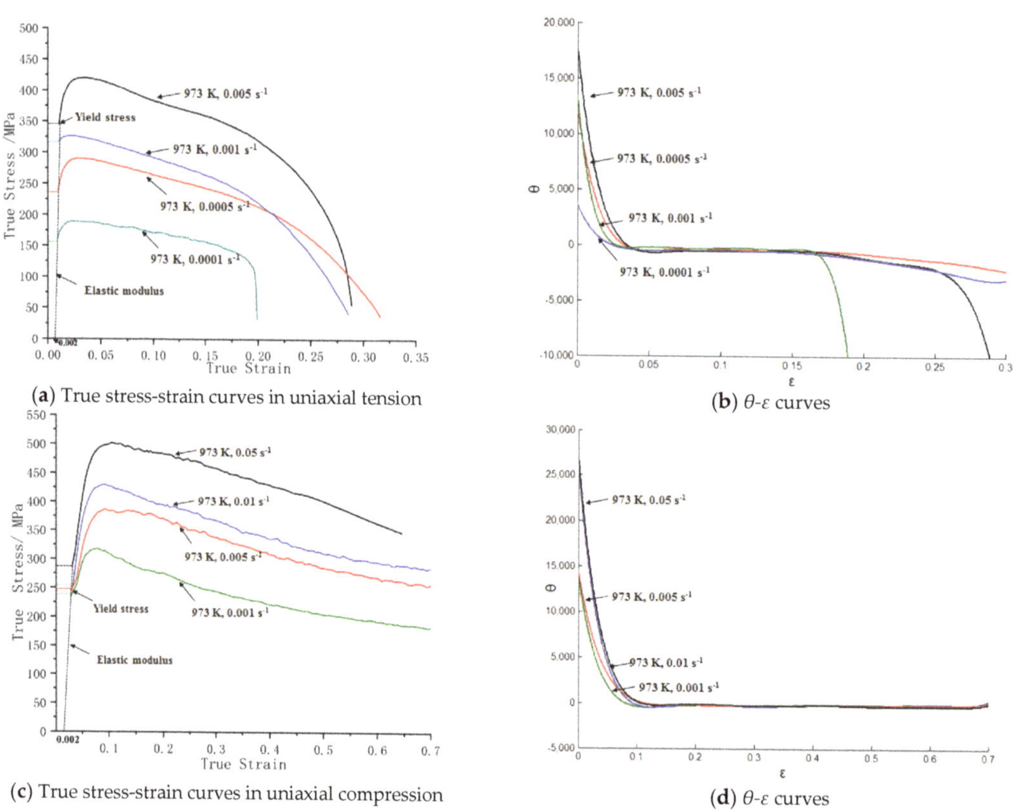

(**a**) True stress-strain curves in uniaxial tension   (**b**) θ-ε curves

(**c**) True stress-strain curves in uniaxial compression   (**d**) θ-ε curves

Figure 7. The σ-ε (**a**,**c**) and θ-ε (**b**,**d**) curves of tension and compression of TA15 with different strain rates.

3.3. OM Microstructure

TA15 is a near-alpha titanium alloy with excellent thermal stability, outstanding room and medium temperature strength as well as weldability. It can be used for aircraft structural components such as bulkheads and ribbed wall panels [26]. The original microstructure of TA15 titanium alloy is presented in Figure 8. TA15 was heated to above the phase transition temperature (about 1263 K) and kept for one hour [19]. Then, the profile was extruded in the beta range using glass lubrication and flat extrusion dies. Since the heat and deformation temperature is higher than its phase transition temperature (about 30–50 K), the microstructure displays a typical lamellar structure, which is characterized by complete β grain boundaries, a well-developed lamellar α structure within the grains and an obvious grain boundary. Hence, the tested alloy with a lamellar structure has excellent fracture toughness and creep properties [27].

The specimens are water quenched immediately after deformation to retain the deformed structure and then observe the microstructure. The microstructures of the tested alloy at different temperatures are presented in Figure 9. Figure 9a is the metallographic structure of the tensile deformation sample at 923 K with a true stain of 0.2. We can see that there are clear β grain boundaries within lamellar α, and there are dynamically recrystallized grains distributed at the β grain boundaries. As the temperature increases, we can observe from Figure 9c that the β grains tend to elongate and grow at 973 K, the thickness of lamellar α increases and small fine grains also appeared in the grain boundaries. When the deformation temperature rises to 1023 K, we can conclude from Figure 9e that the β grains are further elongated and the true strain is about 0.45, the average grain size has increased significantly. During the tensile tests, we can tell from the OM results that the strain and temperature have little effect on dynamic recrystallization under the quasi-static conditions (0.001 s^{-1}).

Figure 8. Microstructure of the unformed alloy.

Compared with tensile tests, we also observed that the microstructure of the compressed samples with true strain is 0.7 at different temperatures. We can see from Figure 9b that there is dynamic recrystallization, which is characterized by a large deal of fine grains at the grain boundaries at 923 K, resulting in the average grain size being smaller than that in tension. When the temperature rises to 973 K, as shown in Figure 9d, more fine grains appear at the grain boundaries, and the processing flow is approximately along the 45° to the axis of compression loading. According to Figure 9f, the grown β grains are easily to observe due to the external force at a high temperature and the dynamic recrystallization of 1023 K is higher than that of 923 K. The average grain size is the smallest and the processing flow is more evident and clearer.

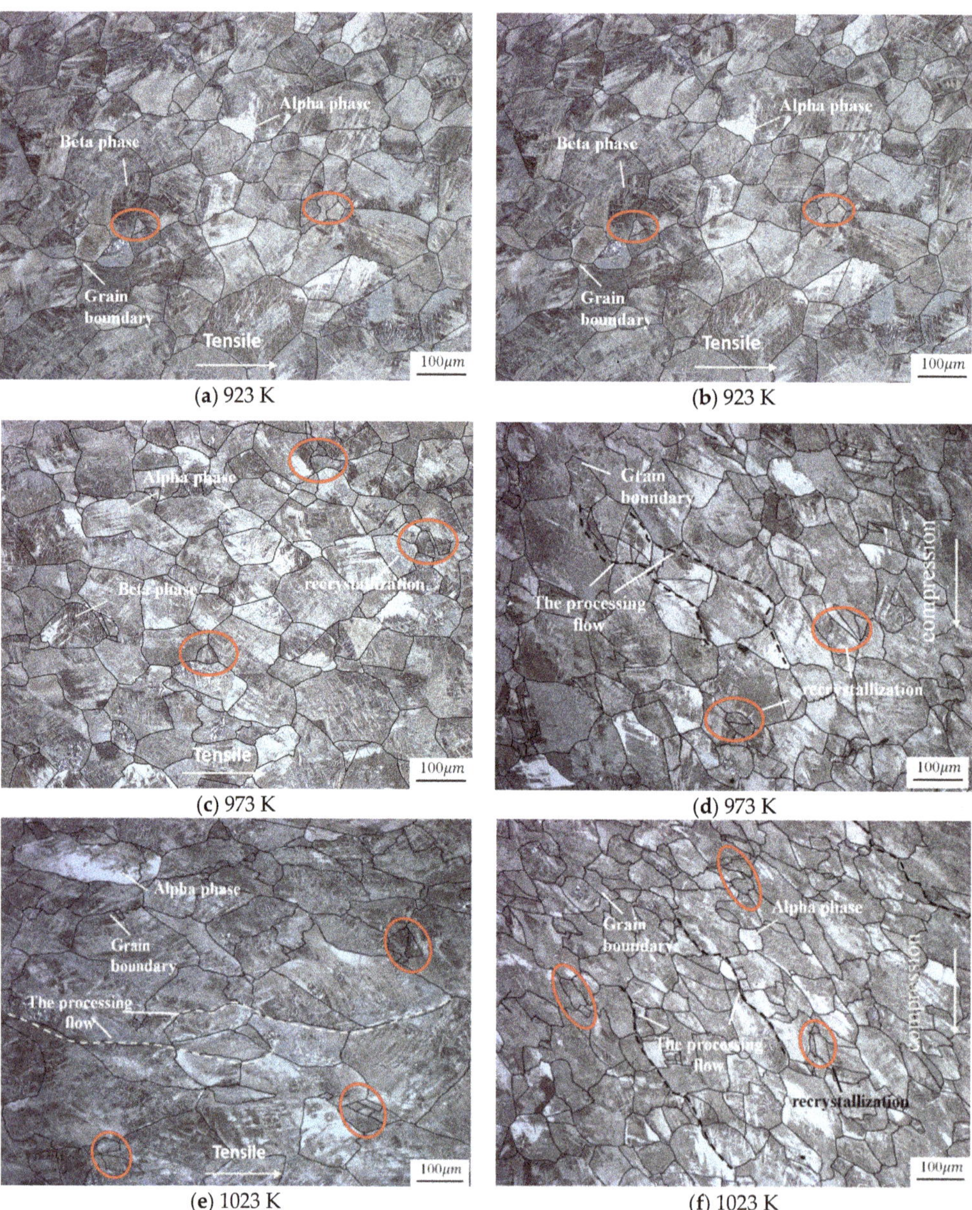

Figure 9. Metallographic structure of TA15 in tension and compression at different temperatures.

3.3.1. The Average Grain Size

The intercept point method [11] was used to analyze the average size of β grains from the OM results in Figure 9, and the results are shown in Figure 10. Figure 10 indicates that the average grain size in tension grows from 79.3 to 95.6 μm when the temperature increases from 923 to 1023 K, while that in compression is from 70.6 to 61.5 μm mostly due to recrystallization.

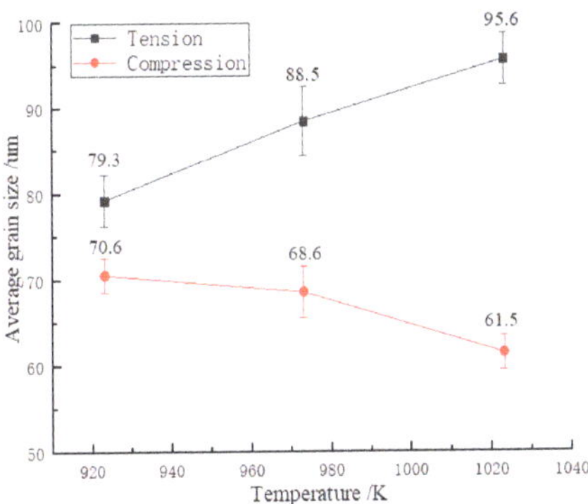

Figure 10. Average grain size vs. temperature of TA15.

3.3.2. The Recrystallization Fraction

Due to the low strain rate and long deformation time in this experiment, the dynamically recrystallized grains may be swallowed up or grow up, the accuracy of the data obtained using metallographic methods have some fluctuations [28]. Ou [29] combined the metal-graphic observation method and data extrapolation method to quantify and simplified the Avrami recrystallization kinetic equation [30,31]. The dynamic recrystallization kinetic model generally adopts the following Johnson-Mehl-Avrami (JMA) Equation:

$$X_{dRX} = 1 - \exp\left[-k \times \left(\frac{\varepsilon - \varepsilon_c}{\varepsilon_{0.5}}\right)^{n_d}\right], \ \varepsilon \geq \varepsilon_c \tag{2}$$

where k and n_d are the material parameters, and their values are 0.6245 and 1.0848 [29], respectively; X_{dRX} is the dynamic recrystallization volume fraction; ε_c is the critical strain of dynamic recrystallization; ε is the true strain and $\varepsilon_{0.5}$ is the strain when the dynamic recrystallization reaches 50%. According to Yue [32], when the strain rate maintains a constant, $\varepsilon_{0.5}$ and the temperature T (K) are approximately linear in relation, as shown in Equation (3), the value of $\varepsilon_{0.5}$ at a strain rate of 0.001 s^{-1} at 1323 K is 0.42, for ease of calculation, we take the $\varepsilon_{0.5,T}$ of 1023 K as the calculation basis and the value of $\varepsilon_{0.5}$ in this paper is 0.6.

$$\varepsilon_{0.5,T} = -0.0006T + 1.2138 \tag{3}$$

The dynamic recrystallization critical strain ε_c is represented by the Sellars model [33] as follows:

$$\varepsilon_c = k_1 \cdot \varepsilon_p \tag{4}$$

$$\varepsilon_p = a_1 \times Z^{a_2} \tag{5}$$

$$Z = \dot{\varepsilon} \cdot exp(Q/RT) \tag{6}$$

where ε_p is the peak strain; k_1 and a_1 are constants, and their values are 0.504 and 0.0341, respectively; a_2 is 0.0810 and 0.0605 in tension and compression, respectively [29]. Z is the Zener–Hollomon [34] parameter; $\dot{\varepsilon}$ is the strain rate (s^{-1}), 0.001 s^{-1}; Q is the deformation activation energy, 228,000 J/mol; R is the gas constant, 8.314 J (mol·K) and T is the absolute temperature (K) [35].

Combining Equation (5) with Equation (6), Equation (4) can be simplified as follows:

$$\varepsilon_c = 1.72 \times 10^{-2} \cdot Z^{a_2} \tag{7}$$

The calculation results of the parameters are shown in Table 3.

Table 3. The calculation results of σ_c, X_{dRX}.

Temperature (K)	Strain Rate (s⁻¹)	Z	ε_c		X_{dRX}	
			Tension	Compression	Tension	Compression
923	0.001	8.0075×10^9	0.109	0.068	0.0704	0.4390
973	0.001	1.7402×10^9	0.096	0.062	0.1681	0.4690
1023	0.001	4.3866×10^8	0.085	0.057	0.2668	0.4898

The above-described calculation results into Equation (2); it can be simplified as follows:

$$X_{dRX} = 1 - \exp\left[-0.6245 \times \left(\frac{\varepsilon - \varepsilon_c}{\varepsilon_{0.5,T}} \right)^{1.0848} \right], \varepsilon \geq \varepsilon_c \tag{8}$$

where ε_c is the dynamic recrystallization critical strain at 923–1023 K, ε takes the average strain value 0.2, 0.3 and 0.4 in tension and takes 0.7 in compression tests, respectively.

The calculated results were compared with the results obtained from the microscopic observation [9] to verify the validity of the JMA parameters. Additionally, on this basis, the X_{dRX} of compression test under small strain was calculated by using the JMA equation, and the result was compared with that of tensile test, as shown in Figure 11.

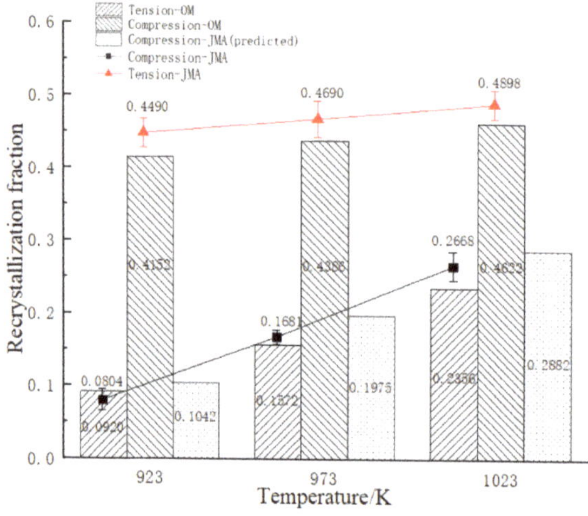

Figure 11. The experimental and calculated recrystallization fraction of TA15 at different temperatures.

We can conclude from Figure 11 that the experimental X_{dRX} in tension at 923 K, 973 K and 1023 K is 0.0920, 0.1572 and 0.2358, while the calculated X_{dRX} in tension at the same temperature is 0.0804, 0.1681 and 0.2668, respectively. Additionally, the experimental X_{dRX} in compression at 923 K, 973 K and 1023 K is 0.4152, 0.4386 and 0.4622, while the calculated X_{dRX} in compression at the same temperature is 0.4490, 0.4690 and 0.4898, respectively. Due to the low strain rate, part of the recrystallized grains grew, causing the calculated X_{dRX} to be higher than that of the OM results, the calculation result error is less than 10%. Additionally, the X_{dRX} of the compression test was predicted by using the JMA equation to be 0.1042, 0.1975 and 0.2882, respectively. We can tell that the recrystallization fraction in compression is higher than that in tension under the same strain condition, and the recrystallization fraction increase as the temperature rises.

4. Discussion

4.1. Effect of Deformation Temperature on TCA

Various studies have [9,16] found that the work hardening rate (θ-ε) curves of hexagonal close-packed (HCP) materials usually consist of a three-stage character of deformation curves at room temperature, which is classified by its slope [9], as shown in Figure 12. The sudden drop of θ in stage I is mainly due to the dynamic recovery [25]. Additionally, the twin phenomenon generates in stage II and the slope presents a visible increase; this phenomenon is easy to observe in the study of TCA at room temperature [11,25]. In stage III, there is a decrease, mainly because of the saturation of the twin volume fraction.

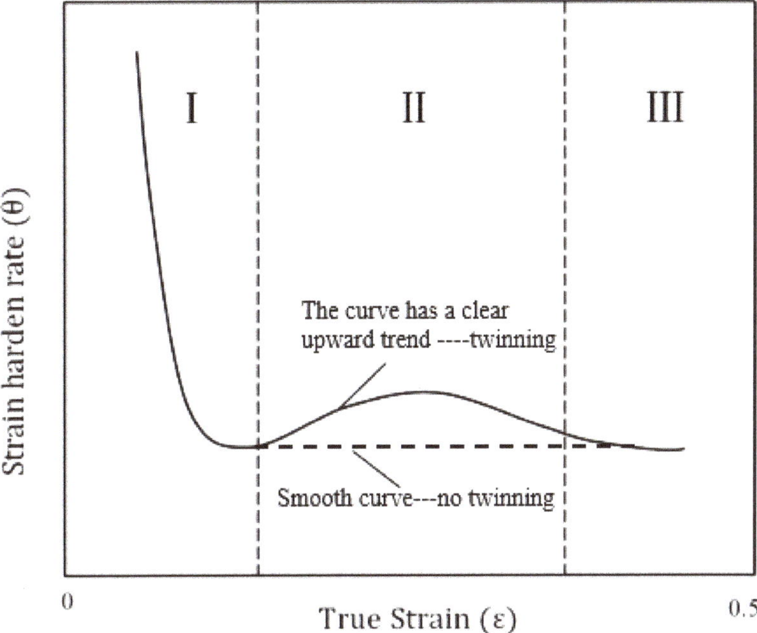

Figure 12. Curves of strain hardening rate θ versus true strain ε [9].

As the deformation temperature increases, the strength of the grains and grain boundaries decreases, and the dislocation resistance decreases [22,27], leading to the strain hardening rate reducing sharply in stage I. As the true strain increases, as shown in Figure 10, the dynamic recrystallized grains aggregate near the grain boundaries, which can not only activate more slip systems but also can increase the number of slip systems and change the grain orientation like twins [16], reducing the force that hinders plastic deformation. Therefore, we can see from Figure 6b,d and Figure 7b,d that there is no uptrend in stage II and the σ-ε curves (Figure 6a,c and Figure 7a,c) display a significant softening effect at high temperatures. Hao [11] studied the asymmetry of the strain hardening of CP-Ti at 298–873 K and found that the θ increased significantly and peaked in stage II due to the occurrence of twins in a low temperature, but there are no twins generated and the θ-ε curves of stage II become flat at 873 K. Similarly, we can judge from the (θ-ε) curves (Figure 7b,d) that the θ drops dramatically in the initial deformation and then flattens at 923–1023 K, and there is a recrystallization phenomenon instead of twins in Figure 9. Consequently, we can preliminarily tell from Figure 12 that there is no twin generated in the two stress states under the high-temperature conditions.

To study the effect of temperature on TCA clearly, the asymmetry coefficient q is defined as follows [36]:

$$q = \frac{\sigma_{y-t}^{0.2} - \sigma_{y-c}^{0.2}}{\sigma_{y-t}^{0.2}} \tag{9}$$

where $\sigma_{y-t}^{0.2}$ represents the tensile yield stress and $\sigma_{y-c}^{0.2}$ is the compressive yield stress. The deformed yield stress and the asymmetry coefficient q of uniaxial tension and compression at different temperatures and strain rates are shown in Table 4.

Table 4. Tensile and compressive yield stress, asymmetry coefficient at different temperatures.

Temperature/K	Tension/MPa	Compression/MPa	Asymmetry Coefficient q	Strain Rates/s^{-1}
923	355.25	272	0.23	
973	308.03	248	0.20	0.001
1023	212.93	185	0.13	
973	348.25	251	0.28	0.005

It can be concluded that the asymmetry coefficient q gradually decreased from 0.23 to 0.13 when the temperature rose from 923 to 1023 K. Additionally, the asymmetry coefficient q increased from 0.20 to 0.28 as the strain rate rose from 0.001 to 0.005 s^{-1}. Thus, the TCA decreases with the increasing temperature. Therefore, TCA is negatively correlated with temperature and positively correlated with strain rate at high temperatures and quasi-static conditions.

In the initial stage of deformation, the initial grain size increases as the deformation temperature rises, thereby reducing the grain strength and grain boundary strength, and the plastic flow of the material becomes easier. As the strain increases, the high-density dislocation produced by the dynamic recrystallization effect is significant, resulting in a decrease in flow stress. This is consistent with the σ-ε curves of TA15 in Figure 6. Under high-temperature conditions, the dynamic recrystallization phenomenon causes recrystallized grains to appear near the grain boundaries, increasing the grain boundary slip system. Although the reduction in the grain size has a strengthening effect on the fine grains, the dynamic recrystallization softening effect is significant, thereby the flow stress decreases. As shown in Figure 9, the dynamic recrystallization in compression is higher than that in tension, resulting in the tensile flow stress being higher than the compressive flow stress. Additionally, this asymmetry behavior is consistent with the results of titanium and its alloys reported in other studies [11,14]. Finally, the strain rate also affects the TCA when the temperature is constant. As the thermal deformation time will decrease with the increase in the strain rate, leading to a small grain size and obvious strengthening effect, which causes the yield strength and the TCA to increase.

The activation criterion during the yield stage for the deformation modes based on the Schmid law [37] is used to research the influence of the load direction on the deformation modes; the critical resolved shear stress (CRSS) and the Schmidt factor (SF) of HCP structural metals slip and twin systems at high temperature are shown in Table 5 [38].

$$\sigma_s = \frac{\tau_c}{m^{SF}} \tag{10}$$

where σ_s, τ_c and m^{SF} are the yield stress, the CRSS and the SF, respectively.

Due to the HCP structure, the deformation is coordinated by dislocation slip and twinning at room temperature [39,40]. However, the TCA has been generally attributed to activating additional sliding systems at high temperatures. From Table 4, it can be concluded that the prismatic $< a >$ slip, the basal $< a >$ slip and the pyramidal $< c + a >$ slip with the theoretical starting stress are 91.5–162.0 MPa, 114.9–152.1 MPa and 297.5 MPa, respectively. The tensile yield stress is 355.25 MPa, and the compressive stress is 272 MPa at 923 K; therefore, the tensile deformation activates the basal $< a >$, prismatic $< a >$ and pyramidal $< c + a >$ slip modes. In contrast, the compressive deformation can only activate the prismatic $< a >$ and pyramidal $< c + a >$ slip modes, which could be one of the reasons for the TCA behavior observed by TA15 at high temperatures. The yield strength difference between tension and compression decreases with the increase in temperature and the amount of activated slip system gradually becomes equal, resulting in a decrease in tension and compression asymmetry. Therefore, under high temperatures (923–1023 K) and low strain rate conditions (<0.05 s^{-1}), the TCA of TA15 is mainly caused by slip deformation.

Table 5. SF and CRSS for different slip systems of HCP under high temperature (1088 K) [9,38].

Slip/Twin System	Burgers Vector	Slip/Twin Plane and Direction	CRSS τ_c	Initial SF(m^{SF})	Stress /MPa
Prismatic	$< a >$	$\{10-10\}\langle 11-20\rangle$	43	0.47	91.5
			61	0.49	124.5
			81	0.50	162.0
Basal	$< a >$	$\{0002\}\langle 11-20\rangle$	73	0.48	152.1
			61	0.50	122.0
			54	0.47	114.9
Pyramidal	$< c + a >$	$\{10-11\}\langle 11-23\rangle$	119	0.40	297.5

4.2. Effect of Strain Rates on TCA

To assess the effect of strain rate on the TCA of the TA15 titanium alloy profile, the general relationship between strain rate and flow stress at a constant temperature is shown in Equation (11) [41].

$$\sigma = C\dot{\varepsilon}^m\big|_{\varepsilon,T} \tag{11}$$

where C is a constant; the strain rate sensitivity m is calculated as follows [41]:

$$m = \frac{\ln\left(\dfrac{\sigma_y^{0.2}}{\sigma_y^{0.2^0}}\right)}{\ln\left(\dfrac{\dot{\varepsilon}}{\dot{\varepsilon}^0}\right)} \tag{12}$$

where $\sigma_y^{0.2}$ and $\sigma_y^{0.2^0}$ are the yield stress under specific strain rate $\dot{\varepsilon}$ and $\dot{\varepsilon}^0$, respectively. $\dot{\varepsilon}^0$ is the reference strain rate 0.001 s^{-1}. The value of m in tension is 0.214 within the rate range of 0.0001–0.005 s^{-1}, and the value of m in compression is 0.318 at 0.001–0.05 s^{-1}. The results indicate that the mechanical response in compression has more obvious strain rate sensitivity compared with those in tension.

The following least square method used by Zhang [36] is used to indicate the relationship between the yield strength and strain rate:

$$\begin{aligned}
\sigma_{y-t}^{0.2} &= 674.9 + 117.35\log\dot{\varepsilon} \\
\sigma_{y-c}^{0.2} &= 568.5 + 106.8\log\dot{\varepsilon}
\end{aligned} \tag{13}$$

The linear correlation coefficients (Adj. R-square) of Equation (13) are 0.940 and 0.925, showing a good linear correlation. Additionally, the $\sigma_{0.2}$ in tension and compression against the strain rate from experimental data are shown in Figure 13.

We can see from Figure 13 that the yield stress during tension is higher than that during compression. Moreover, the asymmetry parameter q increases from 0.172 to 0.186 as the strain rate grows from 0.001 to 0.005 s^{-1} at 973 K, and the TCA shows strain rate sensitivity.

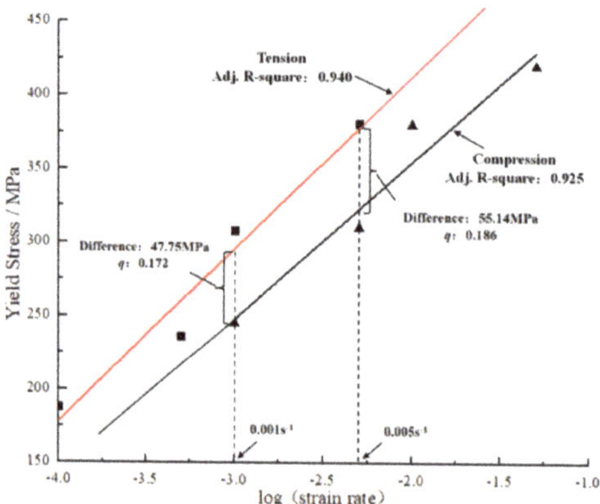

Figure 13. Effect of strain rate on the 0.2% yield stress of the alloy.

5. Conclusions

In this paper, the TCA in the yield stress and strain hardening behavior of TA15 have been investigated with both tension and compression tests and selected optical observations at high temperatures under quasi-static conditions. The results are discussed as follows:

1. The extruded TA15 profile exhibits a significant TCA at high temperatures, and the flow stress and yield stress during tension are larger than compression. Additionally, the asymmetry coefficient q decreases from 0.23 to 0.13 when the temperature increases from 923 to 1023 K.

2. The alloy also exhibits TCA on the strain hardening rate. Its mechanical response during compression is more sensitive than during tension. The asymmetry parameter q increases from 0.172 to 0.186 when the strain rate grows from 0.001 to 0.005 s^{-1} at 973 K. There is no three-stage hardening phenomenon due to the dynamic recrystallization-induced softening effect at high temperatures.

3. The dynamic recrystallization phenomenon is observed under quasi-static conditions at high temperatures. The β grains are easy to fracture with the boundaries clear. Moreover, the grains are elongated along the tensile direction and deformed by about 45° along the compressive load axis. Finally, the recrystallization fractions in compression at 923 to 1023 K with a strain rate of 0.001 s^{-1} are 0.4152, 0.4386 0.4622, respectively.

4. The TCA is negatively correlated with temperature and positively correlated with strain rate at high temperatures under quasi-static conditions. The TCA of the TA15 alloy is due to slip displacement. Additionally, the tensile deformation activates the basal $< a >$, prismatic $< a >$ and pyramidal $< c + a >$ slip modes, while the compressive deformation activates the prismatic $< a >$, pyramidal $< c + a >$ slip modes.

Author Contributions: This article is the result of the joint efforts of several authors. Conceptualization, D.L. and X.L.; formal analysis, C.Z. and X.L.; methodology, C.Z. and Y.L.; resources, D.L. and X.L.; writing-original draft, C.Z.; writing-review and editing, Y.L. All authors have read and agreed to the published version of the manuscript.

Funding: This research was funded by the National Key Research and Development Program of China, grant number 2017YFB0306200; National Natural Science Foundation of China, grant number 51775023 and 51975032.

Data Availability Statement: The data presented in this study are available on request from the corresponding author.

Conflicts of Interest: This article does not contain any studies with human participants or animals performed by any of the authors. Therefore, ethics approval is not applicable. The authors declare no competing interests.

References

1. Boyer, R. An overview on the use of titanium in the aerospace industry. *Mater. Sci. Eng. A* **1996**, *213*, 103–114. [CrossRef]
2. Li, C.-L.; Mi, X.-J.; Ye, W.-J.; Hui, S.-X.; Yu, Y.; Wang, W.-Q. A study on the microstructures and tensile properties of new beta high strength titanium alloy. *J. Alloy. Compd.* **2012**, *550*, 23–30. [CrossRef]
3. Li, Y.; Loretto, M.; Rugg, D.; Voice, W. Effect of heat treatment and exposure on microstructure and mechanical properties of Ti–25V–15Cr–2Al–0.2C (wt%). *Acta Mater.* **2001**, *49*, 3011–3017. [CrossRef]
4. Odenberger, E.-L.; Pederson, R.; Oldenburg, M. Finite element modeling and validation of springback and stress relaxation in the thermo-mechanical forming of thin Ti-6Al-4V sheets. *Int. J. Adv. Manuf. Technol.* **2019**, *104*, 3439–3455. [CrossRef]
5. Xiao, J.; Li, D.; Li, X.; Deng, T. Constitutive modeling and microstructure change of Ti–6Al–4V during the hot tensile deformation. *J. Alloy. Compd.* **2012**, *541*, 346–352. [CrossRef]
6. Astarita, A.; Armentani, E.; Ceretti, E.; Giorleo, L.; Mastrilli, P.; Paradiso, V.; Scherillo, F.; Squillace, A.; Velotti, C. Hot Stretch Forming of a Titanium Alloy Component for Aeronautic: Mechanical and Modeling. *Key Eng. Mater.* **2013**, *554–557*, 647–656. [CrossRef]
7. Guo, G.; Li, D.; Li, X.; Deng, T.; Wang, S. Finite element simulation and process optimization for hot stretch bending of Ti-6Al-4V thin-walled extrusion. *Int. J. Adv. Manuf. Technol.* **2017**, *92*, 1707–1719. [CrossRef]
8. Astarita, A.; Giorleo, L.; Scherillo, F.; Squillace, A.; Ceretti, E.; Carrino, L. Titanium Hot Stretch Forming: Experimental and Modeling Residual Stress Analysis. *Key Eng. Mater.* **2014**, *611–612*, 149–161. [CrossRef]
9. Lin, P.; Hao, Y.; Zhang, B.; Zhang, S.; Chi, C.; Shen, J. Tension-compression asymmetry in yielding and strain hardening behavior of CP-Ti at room temperature. *Mater. Sci. Eng. A* **2017**, *707*, 172–180. [CrossRef]
10. Tuninetti, V.; Gilles, G.; Milis, O.; Pardoen, T.; Habraken, A. Anisotropy and tension–compression asymmetry modeling of the room temperature plastic response of Ti–6Al–4V. *Int. J. Plast.* **2015**, *67*, 53–68. [CrossRef]
11. Hao, Y.G.; Lin, P.; Zhang, B.Y.; Cui, X.L.; Zhang, C.J.; Chi, C.Z. Effect of Planar Anisotropy and Tension-Compression Asymmetry of CP-Ti on its Deep Drawing Behavior at Room Temperature. *Mater. Sci. Forum* **2018**, *913*, 190–195. [CrossRef]
12. Neeraj, T.; Savage, M.; Tatalovich, J.; Kovarik, L.; Hayes, R.; Mills, M. Observation of tension–compression asymmetry in α and titanium alloys. *Philos. Mag.* **2005**, *85*, 279–295. [CrossRef]
13. Sarsfield, H.; Wang, L.; Petrinic, N. An experimental investigation of rate-dependent deformation and failure of three titanium alloys. *J. Mater. Sci.* **2007**, *42*, 5085–5093. [CrossRef]
14. Khan, A.S.; Kazmi, R.; Farrokh, B.; Zupan, M. Effect of oxygen content and microstructure on the thermo-mechanical response of three Ti–6Al–4V alloys: Experiments and modeling over a wide range of strain-rates and temperatures. *Int. J. Plast.* **2007**, *23*, 1105–1125. [CrossRef]
15. Adharapurapu, R.R.; Jiang, F.; Vecchio, K.S.; Gray, G.T. Response of NiTi shape memory alloy at high strain rate: A systematic investigation of temperature effects on tension–compression asymmetry. *Acta Mater.* **2006**, *54*, 4609–4620. [CrossRef]
16. Qin, H.; Jonas, J.J.; Yu, H.; Brodusch, N.; Gauvin, R.; Zhang, X. Initiation and accommodation of primary twins in high-purity titanium. *Acta Mater.* **2014**, *71*, 293–305. [CrossRef]
17. Haertel, S.; Graf, M.; Lehmann, T.; Ullman, M. Influence of Tension-Compression Anomaly during Bending of Magnesium Alloy AZ31. *Mater. Sci. Eng. A* **2017**, *705*, 62–71. [CrossRef]
18. Lei, Z.; Gao, P.; Li, H.; Cai, Y.; Li, Y.; Zhan, M. Comparative analyses of the tensile and damage tolerance properties of tri-modal microstructure to widmanstätten and bimodal microstructures of TA15 titanium alloy. *J. Alloy. Compd.* **2019**, *788*, 831–841. [CrossRef]
19. Zhang, C.; Li, D.; Li, X.; Xia, Q. Numerical simulation and process optimization for hot stretch bending of Ti-6.5Al-2Zr-1Mo-1V large-section extrusion. *Procedia Manuf.* **2020**, *50*, 483–487. [CrossRef]
20. AMST9046B. *Titanium and Titanium Alloy, Sheet, Strip, and Plate*; Society of Automotive Engineers (SAE): Washington, DC, USA, 2006.
21. ISO 6892-2. *Metallic Materials-Tensile Testing-Part2: Method of Test at Elevated Temperature*; Case Postale; International Organization for Standardization: Geneva, Switzerland, 2011.
22. Shen, F.J.; Chen, M.H.; Feng, J.C. Stress Relaxation and Flow Stress of TA15 Alloy at Elevated Temperature. *Aerosp. Mater. Technol.* **2013**, *3*, 114–118.
23. ISO 4499-1. *Hardmetals-Metallographic Determination of Microstructure-Part 1: Photomicrographs and Description*; International Organization for Standardization: Geneva, Switzerland, 2020.
24. Park, S.H.; Lee, J.H.; Moon, B.G.; You, B.-S. Tension–compression yield asymmetry in as-cast magnesium alloy. *J. Alloy. Compd.* **2014**, *617*, 277–280. [CrossRef]
25. Xu, W.C.; Shan, B.D.; Lv, Y.; Li, C.F. Effects of hot deformation parameters on flow stress and establishment of constitutive relationship system of BT20 titanium alloy. *Trans. Nonferrous Met. Soc. China* **2005**, *15*, 167–172. [CrossRef]
26. Shi, Y.; Zhong, F.; Li, X.; Gong, S.; Chen, L. Effect of laser beam welding on fracture toughness of a Ti-6.5Al-2Zr-1Mo-1V alloy sheet. *J. Mater. Sci.* **2007**, *42*, 6651–6657. [CrossRef]
27. Wang, T.; Li, B.; Wang, Z.; Nie, Z. Hot deformation behavior and microstructure evolution of a high-temperature titanium alloy modified by erbium. *J. Mater. Res.* **2017**, *32*, 1517–1527. [CrossRef]
28. Meng, Q.G.; Zhao, G.D.; Wang, L.N.; Wang, Y.F.; Guo, J.; Han, D. A Method for Determining the Volume Fraction of Dynamic Recrystallization of Metals Av-rami's Method of Mathematical Model Coefficient. CN 201910387211.8, 17 September 2019.

29. Ouyang, D.L.; Lu, S.; Cui, X.; Wang, K.L.; Wu, C. Dynamic Recrystallization of Titanium Alloy TA15 during β Hot Process at Different Strain Rates. *Rare. Metal. Mat. Eng.* **2011**, *40*, 325–330.
30. Avrami, M. Kinetics of Phase Change. *I General Theory. J. Chem. Phys.* **1939**, *7*, 1103–1112.
31. Avrami, M. Kinetics of Phase Change. II Transformation-Time Relations for Random Distribution of Nuclei. *J. Chem. Phys.* **1939**, *8*, 212–224. [CrossRef]
32. Yue, L.Y.; Huang, Z.F.; He, Z.L.; Jiang, M.G.; Chen, W. Microstructure Evolution and Numerical Simulation of DP Steel TWBs During Warm Tensile Deformation. *Adv. Mater. Res.* **2012**, *418–420*, 1222–1227. [CrossRef]
33. Nagarajan, V.; Palmiere, E.J.; Sellars, C.M. New approach for modelling strain induced precipitation of Nb(C,N) in HSLA steels during multipass hot deformation in austenite. *Mater. Sci. Technol.* **2009**, *25*, 1168–1174. [CrossRef]
34. Medina, S.F.; Hernandez, C.A. General expression of the Zener-Hollomon parameter as a function of the chemical composition of low alloy and microalloyed steels. *Acta Mater.* **1996**, *44*, 137–148. [CrossRef]
35. Jia, Y.; Xu, L.; Ma, P.; Gokuldoss Prashanth, K.; Yao, C.; Wang, G. Microstructure evolution and hot deformation behavior of spray-deposited Ti-Al alloy. *J. Mater. Res.* **2018**, *33*, 1–9. [CrossRef]
36. Zhang, Q.; Zhang, J.; Wang, Y. Effect of strain rate on the tension–compression asymmetric responses of Ti–6.6Al–3.3Mo–1.8Zr–0.29Si. *Mater. Des.* **2014**, *61*, 281–285. [CrossRef]
37. Yin, D.L.; Wang, J.T.; Liu, J.Q.; Zhao, X. On tension–compression yield asymmetry in an extruded Mg–3Al–1Zn alloy. *J. Alloy. Compd.* **2009**, *478*, 789–795. [CrossRef]
38. Salem, A.; Semiatin, S. Anisotropy of the hot plastic deformation of Ti–6Al–4V single-colony samples. *Mater. Sci. Eng. A* **2009**, *508*, 114–120. [CrossRef]
39. Mann, G.; Griffiths, J.R.; Cáceres, C.H. Hall-Petch parameters in tension and compression in cast Mg-2Zn alloys. *J. Alloys Compd.* **2004**, *378*, 188–191. [CrossRef]
40. Duc-Toan, N.; Seung-Han, Y.; Dong-Won, J.; Tien-Long, B.; Young-Suk, K. A study on material modeling to predict spring-back in V-bending of AZ31 magnesium alloy sheet at various temperatures. *Int. J. Adv. Manuf. Technol.* **2011**, *62*, 551–562. [CrossRef]
41. Dieter, G.E. *Mechanical Metallurgy*, 3rd ed.; McGraw-Hill Book Company: London, UK, 1988.

metals

Article

Precipitation Behavior of ω Phase and ω→α Transformation in Near β Ti-5Al-5Mo-5V-1Cr-1Fe Alloy during Aging Process

Yi Guo [1], Shaohong Wei [2,3], Sheng Yang [4], Yubin Ke [2,3], Xiaoyong Zhang [1,*] and Kechao Zhou [1]

[1] State Key Laboratory of Powder Metallurgy, Central South University, Lu Mountain South Road, Changsha 410083, China; guo-yi@csu.edu.cn (Y.G.); zhoukechao@csu.edu.cn (K.Z.)

[2] Spallation Neutron Source Science Center (SNSC), Dongguan 523808, China; weish@ihep.ac.cn (S.W.); keyb@ihep.ac.cn (Y.K.)

[3] Dongguan Research Department, Institute of High Energy Physics (IHEP), Chinese Academy of Sciences (CAS), Dongguan 523808, China

[4] Hunan Goldsky Titanium Industry Technology Co., Ltd., No. 97 Qianming Road, Deshan Town, Changde ECO-TECH Development Zone, Changde 415001, China; 13874850716@163.com

* Correspondence: zhangxiaoyong@csu.edu.cn

Abstract: In this work, the precipitation behavior of the ω phase and ω→α transformation in Ti-5Al-5Mo-5V-1Cr-1Fe (Ti-55511) alloy was investigated during isothermal aging at 450 °C. The results show that the α precipitates increase with the increasing of aging time, resulting from the β→α and ω→α transformations. The ω→→α transformation involves the formation and evolution of the isothermal ω phase. The formation of the isothermal ω phase occurs after 30 min and ends at 120 min, which is caused by the embryonic ω phase to isothermal ω phase transformation. Small angle neutron scattering (SANS) results indicates that the evolution of the isothermal ω phase goes through the increasing average size and aspect ratio from 24.7 to 47.0 nm and from 2.1 to 2.7 respectively, and the morphology evolution of the ω particle from ellipsoid to spindle-like. Moreover, the isothermal ω phase assists the α phase to nucleate at the ω/β interface, which involves the changes in elemental composition. The α phase is enriched in Al. Compared to the α phase, the element of Mo, V and Cr in the isothermal ω phase is lower. The Fe element is uniformly distributed in the isothermal ω phase and β matrix but lean in the α phase.

Keywords: precipitation behavior; Ti-55511; aging process; isothermal ω phase; SANS

Citation: Guo, Y.; Wei, S.; Yang, S.; Ke, Y.; Zhang, X.; Zhou, K. Precipitation Behavior of ω Phase and ω→α Transformation in Near β Ti-5Al-5Mo-5V-1Cr-1Fe Alloy during Aging Process. *Metals* **2021**, *11*, 273. https://doi.org/10.3390/met11020273

Academic Editors: Maciej Motyka and Irina P. Semenova

Received: 19 January 2021

Accepted: 2 February 2021

Published: 5 February 2021

Publisher's Note: MDPI stays neutral with regard to jurisdictional claims in published maps and institutional affiliations.

1. Introduction

Near β-titanium alloys have received great attention as the main materials for aerospace application in section forgings and fasteners because of their outstanding properties, including ultra-high specific strength, excellent fracture toughness and yield strength [1–5]. For example, Ti-10V-2Fe-3Al (Ti-1023) and Ti-5Al-5Mo-5V-3Cr (Ti-5553) alloy have been widely used in the landing gear of Airbus and Boeing passenger aircraft, respectively [6,7]. In addition, Ti–5Al–5Mo–5V–1Cr–1Fe (Ti-55511) near β titanium alloy (also called TC18 alloy) has good comprehensive properties, including the light weight, high strength and good fatigue performance. It has excellent plasticity and toughness at room temperature, and its tensile strength can reach more than 1150 MPa [8–10].

The excellent performances of near β-titanium alloys are mainly ascribed to its improved internal microstructures. Precipitation of the small dispersed α phase in the β matrix by appropriate processing methods and heat treatment is an effective method to achieve improved microstructures to obtain ultra-high performances [11–13]. However, the necessary nucleation sites for α precipitating are usually absent in near β-titanium alloys [14]. In addition, near β-titanium alloy has a high content of β stabilizers, such as Mo, V and others. Due to the low diffusion rate, the diffusion distance of Mo is short during rapid cooling or low temperature aging, which means that the β matrix can be retained [15]. Generally, when quenched from high temperature (above the β transus), the β matrix would precipitate the athermal ω phase as a transient phase [16]. After isothermal aging treatment at a lower temperature (<500 °C), athermal β matrix would convert into the isothermal ω phase with a hexagonal structure [17,18]. In general, the ω phase is considered as an effective nucleation assistant to promote the dispersed precipitation of the α phase [19]. However, not all types of ω phases can be used as α nucleation assistant. It was reported that the athermal ω phase had not shown the ability to assist the nucleation of the α phase [20,21]. The embryonic ω phase (also called the incommensurate ω phase, and is considered as the precursor of isothermal ω phase) has also no direct effect on the formation of α phase [22]. Whereas, the commensurate isothermal ω phase promotes the nucleation of the α phase zealously [23].

The dispersed precipitation of isothermal ω phase has a positive effect on precipitation and refinement of intragranular α phase, which provides an idea for improving the properties of alloys. For this purpose, the nature of the ω→α transformation has been discussed for many years. For now, two transformation mechanisms are widely recognized. The first is that α phase nucleates near the ω/β interface, which is attributed to the formation of Al-rich region during formation of ω phase [24]. The second mechanism demonstrates that ω→α transformation is a mixed-mode including reconstruction of ω/β interface and diffusion of Al element [12,23]. To further elaborate the mechanism of ω-assisted α nucleation, the impact of ω/β interface on α phase nucleation in the metastable β alloy was revealed, which was related to the size and morphology of isothermal ω phase [25,26]. In other words, the size and morphology of ω phase has a certain effect on the nucleation of α phase. In general, the isothermal ω is shown as ellipsoidal shape in low β-ω misfit or cubic shape in high β-ω misfit [19,27]. The mechanism of the ω phase with an ellipsoidal or cubic shape on the α nucleation has been proposed [28,29]. However, the research on the statistical particles size measurement of the ω phase during the early stage of the aging process is still limited. The size of the ω particle was generally measured by multiple inverse Fourier filtered transform (IFFT) images or dark field TEM images [23,30,31]. Nevertheless, the amount of ω particles measured by this method is usually low, and the data may not be representative since the distribution of ω phase may not uniform in some areas. Therefore, conducting a large number of statistical sizes of ω particles is needed.

In present study, the precipitation behavior of the ω phase and ω→α transformation in near β Ti-55511 alloy was studied during the aging process, including particle morphology, size, chemical distribution and formation of α assisted by ω phase. The phase composition was analyzed by X-ray diffraction (XRD) during the aging process. Transmission electron microscopy (TEM) and high-angle annular dark field scanning transmission electron microscopy (HAADF-STEM) tests were conducted to observe and confirm the nano-scale particles precipitated during aging process. In addition, the small angle neutron scattering (SANS) technique was applied to analyze the morphology and statistical size of the dispersed nano-scale precipitates.

2. Materials and Methods

Ti-55511 ingot for this work was provided by Hunan Goldsky Titanium Company (Changde, China). The raw ingot was prepared by vacuum arc-melting firstly, and forged to the bars with a diameter of 500 mm subsequently. The samples were carried out at 920 °C (above 875 °C transus) for 2 h firstly, followed by water quenching. Subsequently, a series of aging treatments were performed on the β-quenched specimens. Generally, the ω phase precipitates during isothermal aging at low temperature (below 500 °C) [18]. In order to better obtain the ω phase to study its precipitation behavior, aging temperature of 450 °C was designed. Aging times of 15, 30, 60 and 120 min were chosen. All aged samples were water-quenched to retain their microstructures.

The aged specimens were tested for via XRD, SANS and TEM. The XRD test was examined using the instrument (Lausanne, Switzerland) made in The Brock Company in Switzerland. The instrument uses Cu-Kα radiation with 40 KV voltage and 40 mA current.

Specimens for TEM and HAADF-STEM tests were prepared by mechanical thinning firstly, and then electropolished using twin-jet thinning technology with the condition parameter of $-25\,°C$ and 30 V, in which the solution was composed of $CH_3(CH_2)_3OH$ (35%), $HClO_4$ (5%) and CH_3OH. TEM images were observed using Titan G2 60–300 high resolution spherical aberration correction TEM (FEI, Hillsboro, OR, USA) at an accelerating voltage of 300 KV. Microstructures of HAADF-STEM specimens were examined using the Talos F200X instrument (FEI, Brno, Czech) at 200 KV.

In this study, the SANS measurement was conducted at China Spallation Neutron Source (CSNS, Dongguan, China) [32]. During the experiment, the sample to detector distance was set to 4 m and the diameter of sample aperture was 6 mm. The selected incident neutrons wavelength is 1–10 Å, which can cover a wide Q range from $0.005\,Å^{-1}$ to $0.60\,Å^{-1}$. To obtain high quality data, we collected approximate 90 min scattering information for each sample. The scattering data were set to absolute unit after normalization, background subtraction and correction with transmission and standard sample calibration.

3. Results and Discussion

3.1. Phase Transformation during Aging Process

Figure 1 shows the XRD patterns of different aged samples at 450 °C. An extremely weak peak appears in XRD pattern for solution treatment (ST) samples. It's supposed to be α'' phase, produced by non-diffusion shear of β matrix [33]. The α phase peak at 35.2° (labelled (010)α) appears at 15 min. It is evident that, α phase precipitates at early aging time. With the increasing aging time at 450 °C, the peak intensity of α phase arise clearly, which indicates the enhancement of α precipitates. Compared with the intensity of the other α peaks, the higher relative intensity of the (010)α peak resulted from the preferential precipitation of the α phase during the aging process, which was in accordance with the previously reported Burgers relationships between β and α, ω and α: $\{110\}_\beta//\{0001\}_\alpha$, $(11\bar{2}0)_\omega//(0001)_\alpha$ [34,35]. This may be the result of $\beta\rightarrow\alpha$ and/or $\omega\rightarrow\alpha$ transformations.

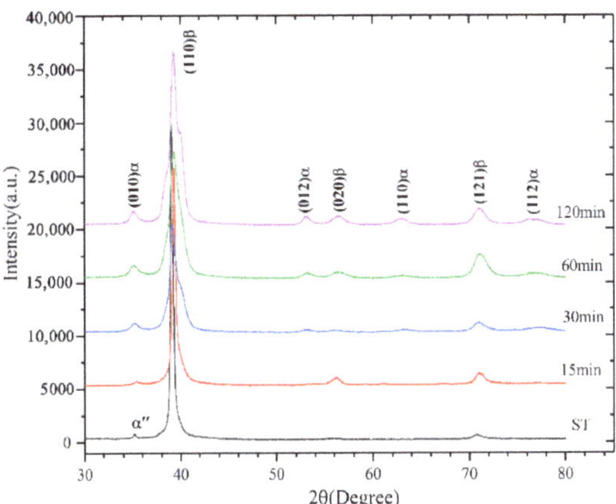

Figure 1. XRD patterns of samples solution treatment (ST) at 920 °C/2 h and aged at 450 °C for different times.

However, the peaks of the ω phase are not observed in the XRD patterns. The $\omega\rightarrow\alpha$ process still needs further direct evidences. For this purpose, HAADF-STEM analysis was performed, as shown in Figure 2. From Figure 2a, a large amount of short rod-shaped α precipitates after aging for 15 min. The triangle-like distribution of some α precipitates suggests that these α phases are directly precipitated from β matrix and satisfy the Burgers relationship. Meanwhile, the long needle-shaped α phase near the ellipsoidal ω phase can also observed in Figure 2b. It provides evidence that the α is formed by the ellipsoidal ω phase and β matrix together at 60 min. Only the isothermal ω would directly participate in the formation of α [23]. Therefore, it can be seen that the isothermal ω precipitates and participates in the formation of α at 60 min. Combined with the XRD results, the formation of α is the result of the β→α and ω→α transformations. In the early stage of aging, the β→α transformation occurs. With the aging time increases, the precipitated the ω phase participates in the precipitation of α phase. The aging time required for the isothermal ω phase to start to assist the precipitation of α phase was 4–8 h at a low aging temperature [23,36]. In this work, the aging temperature of 450 °C significantly accelerates the formation of α phase. This may be caused by two reasons. On the one hand, the high aging temperature can provide the sufficient driving force for β→α transformation. On the other hand, a high elemental diffusion rate given by a high aging temperature can promote the formation of isothermal ω phase which assists the precipitates of α phase.

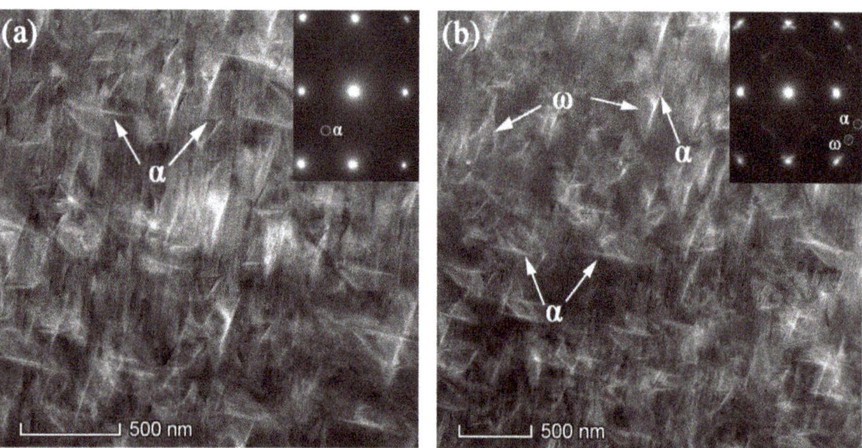

Figure 2. High-angle annular dark field scanning transmission electron microscopy (HAADF-STEM) images showing the microstructure of Ti-55511 aged for (**a**) 15 min at 450 °C and (**b**) 60 min at 450 °C.

3.2. Formation of Isothermal ω Phase

Although the isothermal ω→α transformation is confirmed at the later aging process (Figure 2), the question arises of how the isothermal ω phase assisted in the formation of α phase was formed. For this, the selected-area electron diffraction (SAED) images observed along the [110]β zone axis for different aging times at 450 °C is shown in Figure 3. After aging for 15 min, the existence of diffuse scattering streaking is observed. The reflections at 1/3 and 2/3 <112>β appear as lines. The diffuse scattering streaking is attributed to the embryonic ω phase [7,37]. After increasing aging time to 30 min (Figure 3b), the diffuse scattering streaking begins to weaken, and the spots at 1/3 and 2/3 <112>β appear. This provides the evidence showing that the remaining embryonic ω phase further transforms to the isothermal ω phase. After aging for 60 min (Figure 3c), the diffuse scattering streaking disappears and distinct spots appear, indicating that the embryonic ω phase transforms into the isothermal ω phase completely and the amount of isothermal ω precipitates increase. The spots in Figure 3c, d clearly show that the relationships between ω, α and β satisfy the following: $(111)_β // (0001)_ω // (11\bar{2}0)_α$ [38]. In other words, the formation of isothermal ω is a continuous process involving the transformation of embryonic ω phase to isothermal ω phase. The transformation process starts beginning with 15 min and ends at 120 min.

In order to visually observe the information of isothermal ω phase, dark field images were performed to assist SAED analysis. Figure 4a shows sparsely close ellipsoidal ω particles, which is typical for the near β titanium alloys, such as in Ti-5Al-5Mo-5V and Ti-7Mo-3Al-3Cr-3Nb [16,39]. From the analysis of Figure 3b–d, it suggests that the embryonic ω phase and isothermal ω phase already coexist after 30 min of aging. With the increasing aging time, the formation step-by-step of isothermal ω phase is obvious. Moreover, the ω phase distribution becomes denser and the size is larger, reaching the maximum after aging for 120 min (Figure 4d). It is clear that the formation of isothermal ω involves the size change of ω particles. However, the distribution of ω particles is not uniform. There are many smaller ω particles distributing around the larger ω phase, and the unique phenomenon of small ω agglomeration appear in some areas. Therefore, the SANS technique is used in this work, which can probe the dispersed ω phase in the entire neutron penetration volume of the bulk sample, which usually can provide a reliable statistical information of the nano-scale inhomogeneity within ~0.1 cm^3.

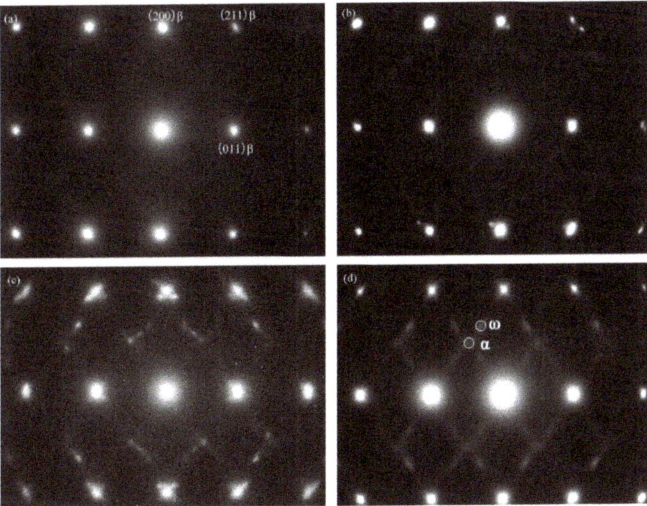

Figure 3. Selected area diffraction patterns of Ti-55511 after aging for (**a**) 15 min, (**b**) 30 min, (**c**) 60 min, (**d**) 120 min taken from the [110]$_\beta$ zone axis.

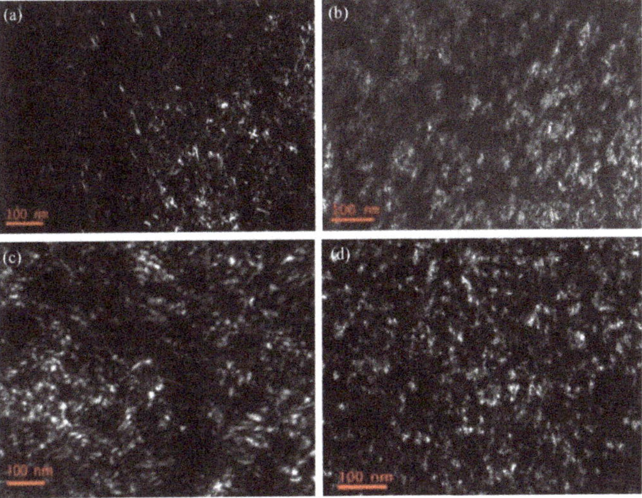

Figure 4. Dark field TEM images of Ti-55511 samples aged for (**a**) 15 min, (**b**) 30 min, (**c**) 60 min, (**d**) 120 min taken from 1/3 <112> reflection spot.

3.3. Evolution of Isothermal ω Phase

Figure 5 shows the result of the SANS test conducted at room temperature with different aged samples. By using an incident neutron wavelength of 1~10 Å, the scattering intensity as a function of scattering vector Q was collected in the range from 0.005 Å^{-1} to 0.6 Å^{-1}. As shown in Figure 5, the scattering intensity arises gradually as the aging time increases and changes sharply after aging for 30 min. With aging time increasing, the embryonic ω transforms to isothermal ω and causes the growth of scattering intensity, which is proportional to the volume fraction of nano-scale α precipitates. After aging for 60 min, the intensity difference was expanded due to the large number of nanoparticles α precipitates assisted by isothermal ω phase.

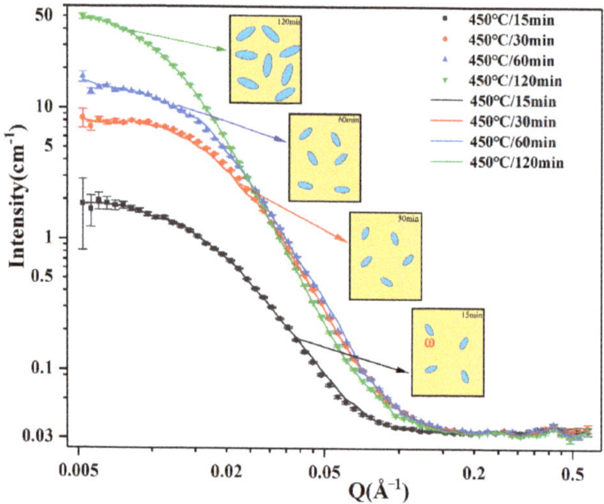

Figure 5. The SANS curves measured at room temperature of the aged Ti-55511 samples. The figure is plotted on a log-log scale. A schematic diagram of the evolution of ω particles is also shown.

To deduce the ω phase evolution, an elliptical particle model is employed to fit the observed SANS data by using the Sasview software [40]. According to this model the scattering intensity can be described by using the following function [41]:

$$I(q) = \frac{\text{scale}}{V} F^2(Q) + \text{background}, \tag{1}$$

where

$$F(q) = \Delta\rho V \frac{3(\sin qr - qr \cos qr)}{(qr)^3}, \tag{2}$$

for

$$r = [R_e^2 + R_p^2]^{1/2}, \tag{3}$$

where, scale is source intensity, $V = (4/3)\pi R_p R_e^2$ is the volume of one ellipsoid particle. R_p and R_e are the polar radius and the equatorial radius of the ellipsoid particle model, respectively. $\Delta\rho$ (contrast) is the difference of the scattering length density between the scatter particle and matrix. The background is used to avoid the influence of incoherent scattering. In the fitting process, the hard sphere structure factor is used as the structure factor which is defined as 1 due to the low volume fraction of precipitated phase. Equatorial radius and polar radius of the ω particles are obtained after fitting. The length (twice the equatorial radius) and width (twice the polar radius) of ω phase are obtained from the fitting parameters and plotted as a function of aging time in Figure 6. As it is shown, average size of the ω particle reaches a length of 24.7 nm and a width of 11.8 nm at 15 min. Subsequently, the size of ω particle changes slightly at 30 min, but grows significantly after 60 min. After aging for 120 min, the ω precipitates grow to a maximum with length of 47.0 nm and width of 16.8 nm. Furthermore, the ratio of the equatorial radius axis and polar radius is also shown in Figure 6, which demonstrates a noteworthy change. According to the results (Figures 4 and 6), the schematic diagram of the ω particles is shown in Figure 5. With increasing aging time, the changes that occur in the ω particles are clearly displayed. The aspect ratio increases from 2.1 to 2.7, which suggests the transition from ellipsoidal-shaped embryonic ω phase to spindle-shaped isothermal ω phase.

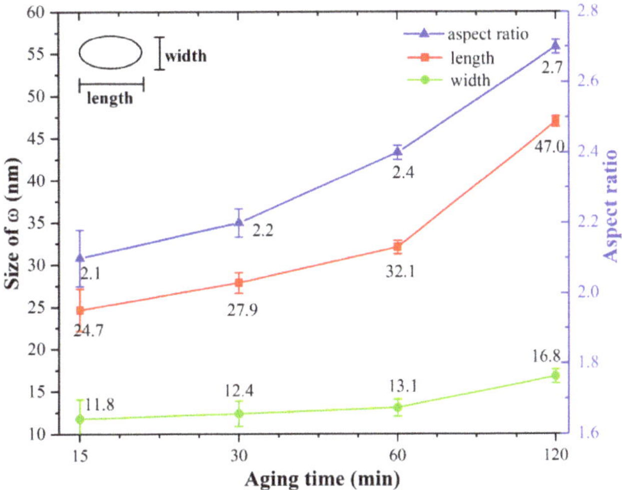

Figure 6. The average size parameters of ω particles obtained by small angle neutron scattering SANS) data. The size parameters are obtained by doubling the value of equatorial radius and polar radius to intuitively reflect the size of ω particles. The aspect ratios of ω particles are also shown.

Generally speaking, the aging process promotes the continuous increase of the size of ω phase, and the aspect ratio increases significantly after aging for 60 min. Combining the previous results (Figures 3 and 4), it is that the evolution of ω phase involves changes in the size and morphology. The embryonic ω phase to isothermal ω phase transformation causes an increasing average size. The size evolution of the ω particles involves a noteworthy change in the aspect ratio, which may result from the structural difference between embryonic ω phase and isothermal ω phase. The structural difference between embryonic ω and isothermal ω is attribute to constrain by coherent elastic strain [23]. Because of this elastic strain, the isothermal ω precipitates was observed to be slightly elongated along its length axis, compared to the embryonic ω precipitates. In the previous analysis (Figures 2 and 3), it has been known that the embryonic ω phase and isothermal ω phase coexist at 30 min of aging. In the SANS results, the sizes of the embryonic ω and isothermal ω are counted indistinguishably at 15 min and 30 min aging, and the average size is obtained in the fitting. The ratio of ω changes proves that the proportion of embryonic ω has little change at 15 and 30 min. After 60 min, the embryonic ω phase transforms into the isothermal ω phase completely, which causes the aspect ratio significantly different than before. In summary, the evolution of ω particles goes through the increase in size and particle density, and the transformation of the morphology from ellipsoid to like spindle. Among them, the conversion of embryonic ω phase → isothermal ω phase causes the aspect ratio change.

3.4. ω Assisted α Nucleation

The direct observation of isothermal ω assisting α formation is shown in Figure 2b. For more detailed information, the HRTEM image of sample aged at 450 °C/120 min recorded along the $[113]_\beta$ axis is shown in Figure 7a, and the FFT results at three regions corresponding to β, α and ω phase are shown in Figure 7b–d, respectively. The result proves that the α nucleates at surface of isothermal ω phase and grows into the β matrix. This is similar with what was observed in Ref. [23]. Ledges are visible at β/ω and ω/α interface. The β/ω interface plays an effective role in formation of α phase. The reasons for the existence of ledges at β/ω interface is that ledges lower the interfacial energy and relax the coherency strain energy [42]. The α phase grows near isothermal ω phase and satisfies the reported orientation relationship: $(10\bar{1}0)_\alpha//(1\bar{1}00)_\omega//(211)_\beta$ [43].

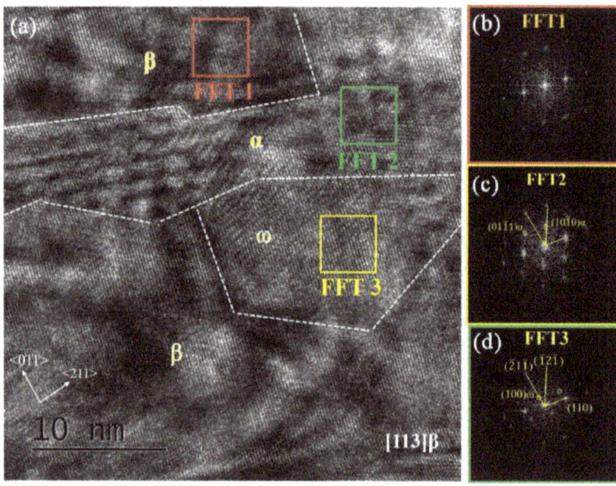

Figure 7. (**a**) Aberration corrected HRTEM image of the sample aged for 120 min at 450 °C taken along $[113]_\beta$ zone axis; (**b**) Fourier filtered transform (FFT) image of area 1; (**c**) FFT image of area 2; (**d**) FFT image of area 3.

In order to investigate the chemical distribution of the ω phase and α phase, HAADF-STEM mapping images of the sample treated at 450 °C/120 min were shown in Figure 8. The α grows in the side of isothermal ω is observed. Due to the long aging time (120 min), the α grows intensively and part of the α has been slatted. The growth of α is constrained after touching other α, so that the growth of α presents a dendritic cross growth, as shown in Figure 8a. Figure 8c–g shows the distribution of different elements in ω phase and α phase. It can be observed from the mapping images that Mo, V and Cr are almost not enrich in α phase, while Al is enriched in α phase with a larger content. For the ω phase, the content of Mo, V, and Cr is lower than that of the β matrix. The Fe element is relatively low in the Ti-55511 alloy, and it is less distributed in α phase, but is evenly distributed in β matrix and ω phase. The element distribution differences are clear between the ω and α phase. The main reason is that these elements play different roles in the formation of the different phases. Mo, V and Cr are β-stabilizing elements, and Al is α-stabilizing element. When isothermal ω is formed, ω discharges Al which is a α-stabilizer, and the content of Mo, V, Cr discharges to the β matrix is also reduced. When the isothermal ω phase assists the formation of α phase, the Al element accumulates in α as stable component.

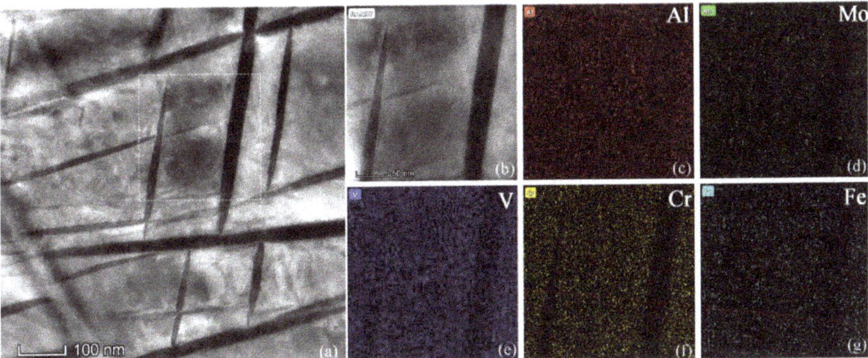

Figure 8. (**a**) HAADF-STEM image of the Ti-55511 sample aged for 120 min at 450 °C; (**b**) the image large magnification image of corresponding area in (a); (**c–g**) the element distribution mapping images of Al, Mo, V, Cr and Fe, respectively.

4. Conclusions

The precipitation behavior of the ω phase and $\omega \rightarrow \alpha$ transformation in the early aging process of Ti-55511 alloy was investigated. For this purpose, XRD, TEM and SANS tests were used. The formation of the α phase, the formation and evolution of the ω phase during the aging process and the formation of the α assisted by ω phase have been studied carefully. The specific conclusions obtained in this work are as follows:

1. The increased α precipitates are the result of the joint action of $\beta \rightarrow \alpha$ and $\omega \rightarrow \alpha$ transformation processes. When the aging condition is 450 °C/30 min, the isothermal ω phase has coexisted with the embryonic ω phase. This may be due to the aging temperature of 450 °C accelerating the element diffusion process, which can promote the transformation process from the embryonic ω phase to the isothermal ω phase.
2. The embryonic ω phase precipitates at 15 min is not obvious. As the aging time continues to increase, the embryonic ω phase precipitates and transforms to isothermal ω phase after 30 min of aging. The transformation process completes after 120 min. The evolution of the isothermal ω phase goes through the increasing average size and aspect ratio from 24.7 to 47.0 nm and from 2.1 to 2.7 respectively, and the change of the ω particle morphology goes from ellipsoid to spindle-like with the increasing aging time.
3. The α phase nucleates at the ω/β interface and satisfy the orientation relationship: $(10\bar{1}0)_{\alpha}//(1\bar{1}00)_{\omega}//(211)_{\beta}$. The element of Mo, V and Cr in the isothermal ω phase is low. Additionally, Al, as an α stabilizing element, is enriched in the α phase. The Fe elements are evenly distributed in the isothermal ω phase and β matrix but lean in the α phase.

Author Contributions: Data curation, S.W. and Y.K.; writing—original draft preparation, Y.G.; writing—review and editing, Y.G. and X.Z.; supervision, K.Z.; project administration, K.Z. and S.Y.; funding acquisition, X.Z. and K.Z. All authors have read and agreed to the published version of the manuscript.

Funding: The authors are grateful to thanks to funding supports from the National Natural Science Foundation of China (No. 51871242), the National Key R&D Program of China (2018YFB0704100) and Scientific and Technological Innovation Projects of Hunan Province, China (No. 2017GK2292).

Institutional Review Board Statement: Studies not involving humans or animals.

Informed Consent Statement: Studies not involving humans.

Data Availability Statement: All raw data supporting the conclusion of this paper are provided by the authors.

Acknowledgments: China Spallation Neutron Source at Dongguan is thanked for providing beam time and help with performing the SANS experiment and data-reduction. Besides, the authors would like to gratitude to Wei Chen for performing the HRTEM examination.

Conflicts of Interest: The authors declare no conflict of interest.

References

1. Banerjee, D.; Williams, J. Perspectives on titanium science and technology. *Acta Mater.* **2013**, *61*, 844–879. [CrossRef]
2. Boyer, R.R.; Briggs, R.D. The use of beta titanium alloys in the aerospace industry. *J. Mater. Eng. Perform* **2005**, *14*, 681–685. [CrossRef]
3. Ankem, S.; Greene, C.A. Recent developments in microstructure/property relationships of beta titanium alloys. *Mater. Sci. Eng. A* **1999**, *263*, 127–131. [CrossRef]
4. Okulov, I.V.; Wendrock, H.; Volegov, A.S.; Attar, H.; Kühn, U.; Skrotzki, W.; Eckert, J. High strength beta titanium alloys: New design approach. *Mater. Sci. Eng. A* **2015**, *628*, 297–302. [CrossRef]
5. Okulov, I.V.; Bönisch, M.; Volegov, A.S.; Shahabi, H.S.; Wendrock, H.; Gemming, T.; Eckert, J. Micro-to-nano-scale deformation mechanism of a Ti-based dendritic-ultrafine eutectic alloy exhibiting large tensile ductility. *Mater. Sci. Eng. A* **2017**, *682*, 673–678. [CrossRef]
6. Franco, L.A.L.; Lourenco, N.J.; Graca, M.L.A.; Silva, O.M.M.; De Campos, P.P.; Von Dollinger, C.F.A. Fatigue fracture of a nose landing gear in a military transport aircraft. *Eng. Fail. Anal.* **2006**, *13*, 474–479. [CrossRef]
7. Warchomicka, F.; Poletti, C.; Stockinger, M. Study of the hot deformation behavior in Ti-5Al-5Mo-5V-3Cr. *Mater. Sci. Eng. A* **2008**, *490*, 369–377.
8. Li, C.; Zhang, X.Y.; Zhou, K.C.; Peng, C.Q. Relationship between lamellar α evolution and flow behavior during isothermal deformation of Ti–5Al–5Mo–5V–1Cr–1Fe near β titanium alloy. *Mater. Sci. Eng. A* **2012**, *558*, 668–674. [CrossRef]
9. Tsybanev, G.V.; Ageev, M.A.; Titarenko, R.V. Analysis of the specific features of loading of the elements of landing gears of aircrafts aimed at the evaluation of the load-bearing capacity of the structure. *Strength Mater.* **2008**, *40*, 458–462. [CrossRef]
10. Lan, C.; Wu, Y.; Guo, L.; Chen, H.; Chen, F. Microstructure texture evolution and mechanical properties of cold rolled Ti-32.5Nb-6.8Zr-2.7Sn biomedical beta titanium alloy. *J. Mater. Sci. Technol.* **2018**, *34*, 788–792. [CrossRef]
11. Cotton, J.D.; Briggs, R.D.; Boyer, R.R.; Tamirisakandala, S.; Russo, P.; Shchetnikov, N.; Fanning, J.C. State of the art in beta titanium alloys for airframe applications. *JOM* **2015**, *67*, 1281–1303. [CrossRef]
12. Haghighi, S.E.; Attar, H.; Okulov, I.V.; Dargusch, M.S.; Kent, D. Microstructural evolution and mechanical properties of bulk and porous low-cost Ti-Mo-Fe alloys produced by powder metallurgy. *J. Alloys Compd.* **2021**, *853*, 156768. [CrossRef]
13. Chen, W.; Wang, H.D.; Lin, Y.C.; Zhang, X.Y.; Chen, C.; Zhou, Y.P.L.K.C. The dynamic responses of lamellar and equiaxed near β-Ti alloys subjected to multi-pass cross rolling. *J. Mater. Sci. Technol.* **2020**, *43*, 220–229. [CrossRef]
14. Jones, N.G.; Dashwood, R.J.; Jackson, M.; Dye, D. Beta phase decomposition in Ti-5Al-5Mo-5V-3Cr. *Acta Mater.* **2009**, *57*, 3830–3839. [CrossRef]
15. Lutjering, G.; Williams, J.C. *Titanium*; Springer: Berlin, Germany, 2007; pp. 30–33.
16. Nag, S.; Banerjee, R.; Srinivasan, R.; Hwang, J.Y.; Harper, M.; Fraser, H.L. Omega-Assisted nucleation and growth of alpha precipitates in the Ti-5Al-5Mo-5V-3Cr-0.5Fe beta titanium alloy. *Acta Mater.* **2009**, *57*, 2136–2147. [CrossRef]
17. Williams, J.C.; Jaffee, R.I.; Burte, H.M. *Ti1973 Science and Technology*; Plenum Press: New York, NY, USA, 1973; pp. 1433–1494.
18. Williams, J.C.; Fontaine, D.D.; Paton, N.E. The ω-phase as an example of an unusual shear transformation. *Metall. Transform.* **1973**, *4*, 2701–2708. [CrossRef]
19. Blackburn, M.J.T.; Williams, J.C. *Phase Transformation in Ti-Mo and Ti-V Alloys*; Boeing Scientific Research Labs: Seattle, WA, USA, 1968.
20. Zheng, Y.; Williams, R.E.; Sosa, J.M.; Alam, T.; Wang, Y.; Banerjee, R.; Fraser, H.L. The indirect influence of the omega phase on the degree of refinement of distributions of the alpha phase in metastable beta-Titanium alloys. *Acta Mater.* **2016**, *103*, 850–858. [CrossRef]
21. Zheng, Y.F.; Choudhuri, D.; Alam, T.; Williams, R.E.; Benerjee, R.; Fraser, H.L. The role of cuboidal omega precipitates on alpha precipitation in a Ti-20V alloy. *Scr. Mater.* **2016**, *123*, 81. [CrossRef]
22. Li, T.; Kent, D.; Sha, G.; Cairney, J.M.; Dargusch, M.S. The role of ω in the precipitation of α in near-β Ti alloys. *Scr. Mater.* **2016**, *117*, 92–95. [CrossRef]
23. Li, T.; Kent, D.; Sha, G.; Stephenson, L.T.; Ceguerra, A.V.; Ringer, S.P.; Dargusch, M.S.; Cairney, J.M. New insights into the phase transformations to isothermal ω and ω-assisted α in near β-Ti alloys. *Acta Mater.* **2016**, *106*, 353–366. [CrossRef]
24. Azimzaden, S.; Rack, H. Phase transformations in Ti-6.8Mo-4.5Fe-1.5Al. *Met. Mater. Trans. A* **1998**, *29*, 2455–2467. [CrossRef]
25. Shi, R.; Zheng, Y.; Banerjee, R.; Fraser, H.L.; Wang, Y. ω-Assisted α nucleation in a metastable β titanium alloy. *Scr. Mater.* **2019**, *171*, 62–66. [CrossRef]
26. Cao, S.; Jiang, Y.; Yang, R.; Hu, Q.M. Properties of β/ω phase interfaces in Ti and their implications on mechanical properties and ω morphology. *Comput. Mater. Sci.* **2019**, *158*, 49–57. [CrossRef]
27. Sass, S.L. The structure and decomposition of Zr and Ti b.c.c. solid solutions. *J. Less Common. Met.* **1972**, *28*, 157–173. [CrossRef]
28. Ivasishin, P.E.; Markovsky, S.L.; Semiatin, C.H. Ward, Aging response of coarse and fine-grained β titanium alloys. *Mater. Sci. Eng. A* **2005**, *405*, 296–305. [CrossRef]

29. Prima, F.; Vermaut, P.; Texier, G.; Ansel, D.; Gloriant, T. Evidence of α-nanophase heterogeneous nucleation from ω particles in a β-metastable Ti-based alloy by high-resolution electron microscopy. *Scr. Mater.* **2006**, *54*, 645–648. [CrossRef]
30. Mantri, S.A.; Choudhuri, D.; Behera, A.; Hendrickson, M.; Alam, T.; Banerjee, R. Role of isothermal omega phase precipitation on the mechanical behavior of a Ti-Mo-Al-Nb alloy. *Mater. Sci. Eng. A* **2019**, *767*, 138397. [CrossRef]
31. Xu, Y.F.; Yi, D.Q.; Liu, H.Q.; Wang, B.; Yang, F.L. Age-hardening behavior, microstructural evolution and grain growth kinetics of isothermal ω phase of Ti–Nb–Ta–Zr–Fe alloy for biomedical applications. *Mater. Sci. Eng. A* **2011**, *529*, 326–334. [CrossRef]
32. Ke, Y.; He, C.; Zheng, H.; Geng, Y.; Fu, J.; Zhang, S.; Hu, H.; Wang, S.; Zhou, B.; Wang, F.; et al. The time-of-flight Small-Angle Neutron Spectrometer at China Spallation Neutron Source. *Neutron News* **2018**, *29*, 14–17. [CrossRef]
33. Coakley, J.; Vorontsov, V.A.; Jones, N.G.; Radecka, A.; Bagot, P.A.; Littrell, K.C.; Heenan, R.K.; Hu, F.; Magyar, A.P.; Bell, D.C.; et al. Precipitation processes in the Beta-Titanium alloy Ti-5Al-5Mo-5V-3Cr. *J. Alloys Compd.* **2015**, *646*, 946–953. [CrossRef]
34. Sikka, S.K.; Vohra, Y.K.; Chidambaram, R. Omega phase in materials. *Prog. Mater. Sci.* **1982**, *27*, 245–310. [CrossRef]
35. Fontaine, D.D.; Paton, N.E.; Williams, J.C. The omega phase transformation in titanium alloys as an example of displace controlled reactions transformation. *Acta Matallurgica* **1971**, *19*, 1153–1162. [CrossRef]
36. Devaraj, A.; Nag, S.; Srinivasan, R.; Williams, R.E.A.; Banerjee, S.; Banerjee, R.; Fraser, H.L. Experimental evidence of concurrent compositional and structure instabilities leading to ω precipitation in titanium-molybdenum alloys. *Acta Mater.* **2012**, *60*, 596–609. [CrossRef]
37. Sanchez, J.M.; Fontaine, D.D. Anomalous diffusion in omega forming systems. *Acta Matallurgica* **1978**, *26*, 1083–1095. [CrossRef]
38. Furuhara, T.; Maki, T.; Makino, T. Microstructure control by thermomechanical processing in β-Ti-15-3 alloy. *J. Mater. Process. Technol.* **2001**, *117*, 318–323. [CrossRef]
39. Dong, R.; Li, J.; Kou, H.; Fan, J.; Zhao, Y.; Hou, H.; Wu, L. ω-Assisted refinement of α phase and its effect on the tensile properties of a near β titanium alloy. *J. Mater. Sci. Technol.* **2020**, *44*, 24–30. [CrossRef]
40. Alina, G.; Butler, P.; Cho, J.; Doucet, M.; Kienzle, P. SANS Analysis Software. Available online: www.sasview.org (accessed on 12 December 2020).
41. Feigin, L.A.; Svergun, D.I. *Structure Analysis by Small-Angle X-ray and Neutron Scattering*; Plenum Press: New York, NY, USA, 1987.
42. Shi, R.; Ma, N.; Wang, Y. Predicting equilibrium shape of precipitates as function of coherency state. *Acta Mater.* **2012**, *60*, 4172–4184. [CrossRef]
43. Ohmori, Y.; Ogo, T.; Nakai, K.; Kobayashi, S. Effect of ω-phase precipitation on β→α, α″ transformations in a metastable β titanium alloy. *Mater. Sci. Eng. A* **2001**, *312*, 182–188. [CrossRef]

metals

MDPI

Article

Phase Transformations upon Ageing in Ti15Mo Alloy Subjected to Two Different Deformation Methods

Kristína Bartha [1,*], Josef Stráský [1], Anna Veverková [1], Jozef Veselý [1], Jakub Čížek [2], Jaroslav Málek [3], Veronika Polyakova [4], Irina Semenova [4] and Miloš Janeček [1]

[1] Department of Physics of Materials, Charles University, 121 16 Prague, Czech Republic; josef.strasky@gmail.com (J.S.); veverkova@karlov.mff.cuni.cz (A.V.); vesely@gjh.sk (J.V.); janecek@met.mff.cuni.cz (M.J.)
[2] Department of Low Temperature Physics, Charles University, 180 00 Prague, Czech Republic; jcizek@nbox.troja.mff.cuni.cz
[3] Faculty of Mechanical Engineering, Czech Technical University in Prague, 121 35 Prague, Czech Republic; jardamalek@seznam.cz
[4] Institute of Physics of Advanced Materials, Ufa State Aviation Technical University, Ufa 450008, Russia; vnurik@gmail.com (V.P.); semenova-ip@mail.ru (I.S.)
* Correspondence: kristina.bartha@met.mff.cuni.cz; Tel.: +420-95155-1361

Abstract: Ti15Mo alloy was subjected to two techniques of intensive plastic deformation, namely high pressure torsion and rotary swaging at room temperature. The imposed strain resulted in the formation of an ultrafine-grained structure in both deformed conditions. Detailed inspection of the microstructure revealed the presence of grains with a size of around 100 nm in both conditions. The microstructure after rotary swaging also contained elongated grains with a length up to 1 μm. Isothermal ageing at 400 °C and 500 °C up to 16 h was applied to both conditions to investigate the kinetics of precipitation of the α phase and the recovery of lattice defects. Positron annihilation spectroscopy indicated that the recovery of lattice defects in the β matrix had already occurred at 400 °C and, in terms of positron trapping, was partly compensated by the precipitation of incoherent α particles. At 500 °C the recovery was fully offset by the formation of incoherent α/β interfaces. Contrary to common coarse-grained material, in which the α phase precipitates in the form of lamellae, precipitation of small and equiaxed α particles occurred in the deformed condition. A refined two-phase equiaxed microstructure with α particles and β grain sizes below 1 μm is achievable by simple rotary swaging followed by ageing.

Keywords: metastable β-Ti alloy; severe plastic deformation; rotary swaging; α phase precipitation; defect structure

Citation: Bartha, K.; Stráský, J.; Veverková, A.; Veselý, J.; Čížek, J.; Málek, J.; Polyakova, V.; Semenova, I.; Janeček, M. Phase Transformations upon Ageing in Ti15Mo Alloy Subjected to Two Different Deformation Methods. *Metals* **2021**, *11*, 1230. https://doi.org/10.3390/met11081230

Academic Editor: Maciej Motyka

Received: 30 June 2021
Accepted: 30 July 2021
Published: 2 August 2021

1. Introduction

Thanks to their excellent mechanical properties, titanium alloys have extensive use in the aerospace as well as in the biomedical industry [1,2]. Ti-6Al-4V alloy still belongs to the most used alloy in the biomedical field. However, in recent years, biological safety has become a crucial issue in the development of biomedical metallic materials [3]. In the human body, small amounts of harmful elements, such Al or V can trigger an allergic reaction [4]. Therefore, metastable β-Ti alloys, which have comparable mechanical properties to Ti-6Al-4V alloy but contain only nontoxic elements, are considered to be appropriate candidates for surgical and orthopaedic implant manufacturing [5]. Current research focuses on the development of biocompatible metastable β-Ti alloys and the improvement of their mechanical properties.

Methods of severe plastic deformation (SPD) can be applied to enhance the strength of these alloys [6,7]. Materials subjected to SPD methods possess a high density of lattice defects and often exhibit an ultrafine-grained structure (UFG) [8].

Another approach to improve the mechanical properties of metastable β-Ti alloys is to apply a thermal treatment. Precipitation of the α phase results in improving the strength and ductility of these alloys [9–12]. However, volume fraction, size, morphology, and distribution of the α precipitates in the β matrix need to be appropriately engineered. The α phase typically nucleates heterogeneously at β grain boundaries, phase boundaries, dislocations and inclusions in the matrix [13,14].

The combination of intensive plastic deformation (such as SPD) and subsequent heat treatment leads to the precipitation of small, equiaxed and finely dispersed α precipitates, as reported in our previous studies [15–17] and in other studies [18,19]. Moreover, some authors previously reported improved strength of these deformed and aged metastable β-Ti alloys [19,20].

In this study a metastable β-Ti alloy, Ti15Mo (in wt%) was investigated. The non-deformed (solution treated) condition contains a β matrix and metastable ω phase as it was reported in our previous study [15–17]. The material after solution treatment was deformed by high pressure torsion (HPT), which is a type of SPD method, and alternatively using cold-rotary swaging (RS). RS is a common and relatively simple deformation method, which produces material with a high density of dislocations (similarly to SPD). Unlike HPT, RS produces significantly larger samples, which can consequently be applied in product manufacturing.

This study aims to investigate the precipitation of α particles in the two deformed conditions, i.e., HPT and RS, and in particular, the effect of different types of lattice defects (dislocations, grain boundaries) on the nucleation of α particles. The potential use of RS instead of HPT as a deformation method is also discussed.

2. Materials and Methods

The examined Ti15Mo alloy was supplied by Carpenter Co. (Richmond, VA, USA). A rod with a 10 mm diameter was solution treated (ST) in an inert Ar atmosphere at a temperature of 810 °C for 4 h followed by water quenching. The ST alloy was deformed in two different ways: high pressure torsion and rotary swaging.

Cylindrical specimens cut from the ST rod were used for HPT processing. The deformation was performed at room temperature at Ufa State Aviation Technical University, Russian Federation. First, the cylindrical samples were pressed by 6 GPa to achieve disk-shaped samples with a diameter of 20 mm and thickness of approximately 1 mm. Subsequently, the specimens were deformed in torsion by rotating one of the anvils. The detailed description of the HPT method can be found in a different study [21]. For this investigation, specimens after $N = 1$ HPT rotation were prepared. Due to the nature of the deformation, the equivalent deformation is inhomogeneous and, according to the von Mises equation [22], can reach 36 (3600%) at the periphery of the sample.

The ST rod was also used to perform cold-rotary swaging at the company ÚJP Praha, Czech Republic. The final diameter of the rod after RS was 5 mm which corresponds to a 75% reduction of the area. Due to the nature of RS, the resulting structure is also inhomogeneous [23]. All measurements were undertaken on the peripheral part of the samples, where the deformation reaches its highest values.

Both materials after HPT and RS (hereafter referred as HPT and RS samples, respectively) were aged at temperatures of 400 °C and 500 °C for 1, 4 and 16 h. Ageing was performed by immersing the samples into a preheated salt-bath (i.e., with a very high heating rate and without air access) and terminated by water quenching. All samples were mechanically ground and polished by standard methods followed by three-step vibratory polishing.

A scanning electron microscope (SEM) Zeiss Auriga Compact CrossBeam) (Carl Zeiss AG, Jena, Germany) was used for SEM analysis. Standard back-scattered electron (BSE) observations were performed at an accelerating voltage of 10 kV.

A transmission electron microscope (TEM) JEOL 2200 FS (Jeol Ltd., Tokyo, Japan) equipped with selected area electron diffraction (SAED) and automated crystallographic orientation mapping (ACOM-TEM) was operated at an accelerating voltage of 200 kV and was used for detailed microstructure observation.

The evolution of lattice defects with ageing was determined by positron annihilation spectroscopy (PAS). Positron lifetime measurements were carried out using a high-resolution digital spectrometer, described in detail in a previous study [24]. The digital lifetime spectrometer is equipped with two Hamamatsu H3378 photomultipliers (Hamamatsu Photonics K.K., Hamamatsu City, Japan) coupled with BaF_2 scintillators. Detector pulses are sampled by two ultrafast Acqiris DC211 8-bit digitizers (Acqiris SA, Plan-les-Ouates, Switzerland) at a sampling frequency of 4 GHz. The digitized pulses are analyzed off-line by software using a new algorithm for integral constant fraction timing [25]. The time resolution of the spectrometer was 145 ps. At least 10^7 annihilation events were accumulated in each lifetime spectrum. A decomposition of the lifetime spectra into exponential components was performed by a maximum likelihood code described in [26].

3. Results

The severely deformed microstructure of the Ti15Mo alloy after HPT is shown in the bright field (BF) TEM micrograph (Figure 1a). Figure 1b shows a dark field (DF) image from a β phase reflection. The individual grains of the β phase are not fully recognizable due to severe fragmentation, but their size is clearly below 200 nm. The corresponding SAED pattern displayed in Figure 1c confirms that HPT material contains both β and ω phases. The distortion of the individual spots in the SAED pattern suggests a severely deformed and fragmented β matrix with a strong texture in the direction $[111]_\beta$.

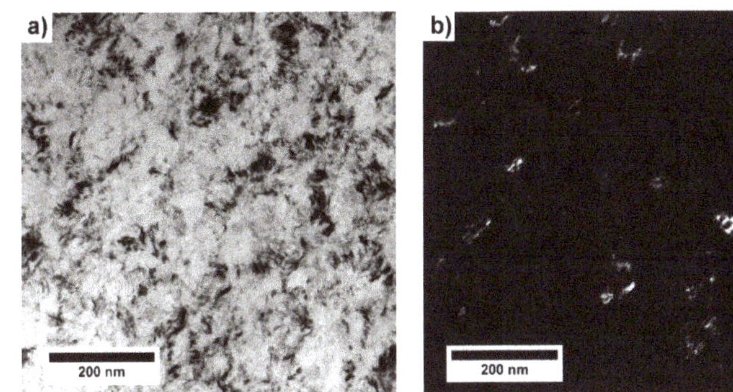
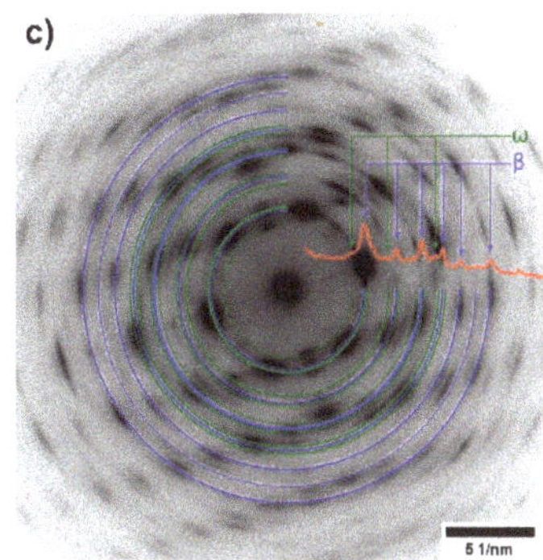

Figure 1. (**a**) TEM bright field image, (**b**) TEM dark field image and (**c**) the corresponding SAED pattern of the material after $N = 1$ turn of HPT deformation on the peripheral part of the sample. The dark field image was obtained from an arbitrarily selected part of the diffraction rings shown in Figure 1c.

The BF and DF images of the RS specimens in Figure 2a,b, respectively, indicate a highly deformed microstructure. The microstructure contains fragmented and partly elongated grains with sizes below 500 nm (note the lower magnification of the images in Figure 2 compared to Figure 1). Figure 2c, showing the corresponding SAED pattern, confirms that the RS material contains β and ω phases. However, the volume fraction of the ω phase seems to be higher in the HPT sample compared to the RS one. It can be caused by the deformation induced ω phase after SPD processing, as reported in study [27].

The microstructure of both the HPT and the RS specimens was also investigated using the ACOM-TEM technique. The inverse pole figure (IPF) map in Figure 3a also confirms the nanocrystalline character of the microstructure after HPT straining, with the majority of grains below 100 nm. Due to the character of HPT straining, most of the grains exhibit the typical ⟨111⟩ texture. The IPF map in Figure 3b shows the severely distorted microstructure of the RS specimen, containing elongated grains with a length of up to 1 μm and more equiaxed and smaller grains with sizes below 100 nm. The black areas of the image cannot be successfully indexed, which suggests a very high dislocation density. Note also that the majority of the grains, namely the larger elongated ones, exhibit preferred crystallographic orientation ⟨110⟩, parallel with the axial direction of the rod.

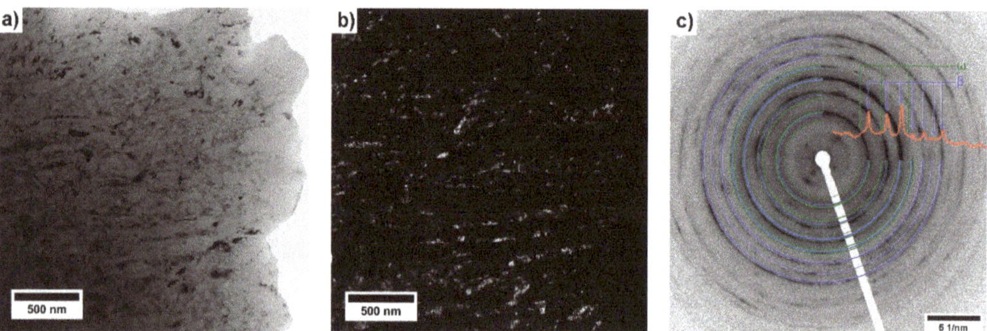

Figure 2. (**a**) TEM bright field image, (**b**) TEM dark field image and (**c**) the corresponding SAED pattern of the RS material. The dark field image was obtained from an arbitrarily selected part of the diffraction rings shown in Figure 2c.

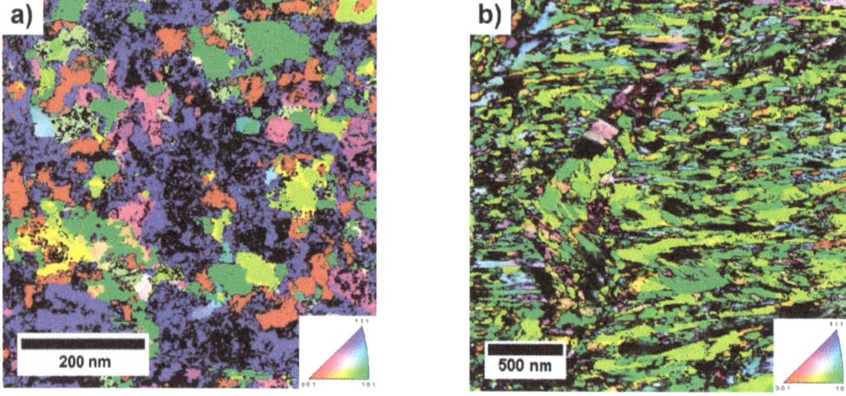

Figure 3. IPF map obtained using ACOM-TEM technique and the corresponding orientation triangle for BCC structure (**a**) HPT, (**b**) RS condition.

Figures 4 and 5 show the microstructure of the Ti15Mo alloy after isothermal ageing of the alloy previously deformed by HPT and RS, respectively. The contrast in the images is given by the chemical composition of the material—the so-called Z-contrast. As the α phase contains a lower amount of Mo than the β phase, it appears darker. In both materials, the particles of the α phase are small and equiaxed. These polygonal phase particles are observed in all conditions in both Figures 4 and 5. Their size and volume fraction increase with an increasing temperature and time of ageing; however, the precipitation is inhomogeneous even after ageing at 500 °C/16 h. The possible cause of the inhomogeneity is thoroughly discussed in the Discussion section.

Differences in the microstructures of HPT and RS samples can be observed already after ageing at 400 °C/1 h—more α phase particles precipitated in the HPT specimens compared to the RS ones.

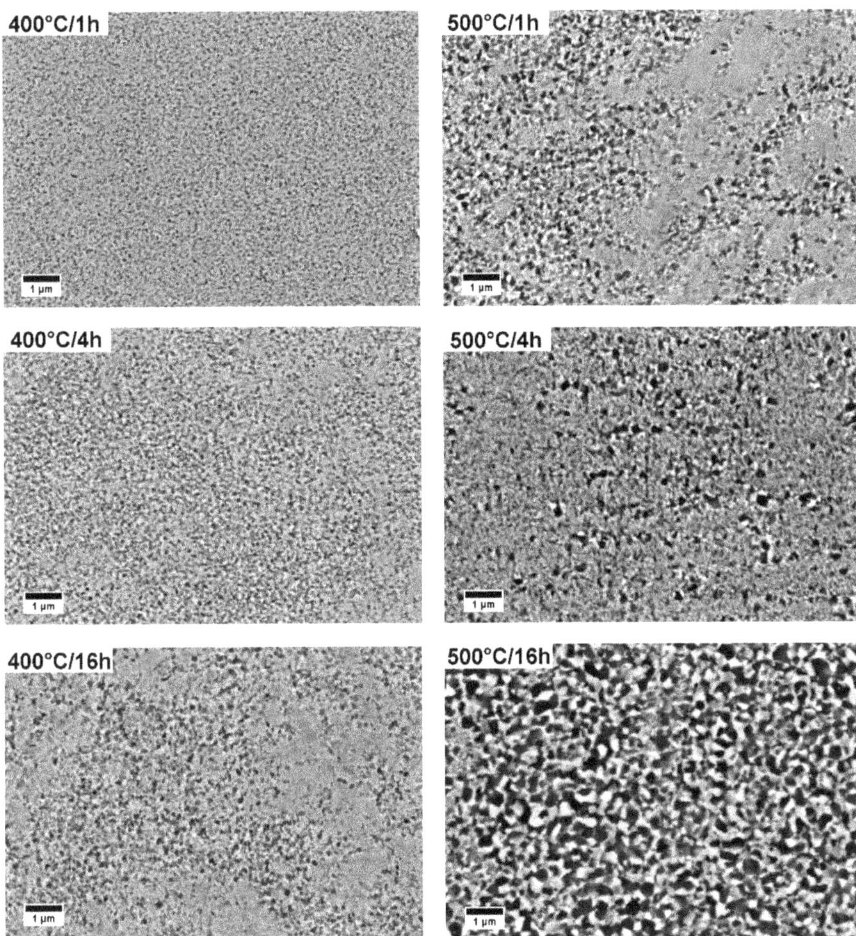

Figure 4. SEM–BSE images of the HPT samples after ageing at 400 °C and 500 °C for 1–16 h (ageing condition is indicated in the top left corner of each micrograph).

The evolution of defects in the HPT and RS samples after ageing were studied using PAS. The lifetime spectra of annihilated positrons are well described by two exponential components: the component with a shorter lifetime τ_1 represents a contribution of free positrons, while the component with a longer lifetime $\tau_2 \approx 170$ ps originates from positrons trapped at lattice defects, typically dislocations. The development of positron lifetimes after different ageing temperatures and times is shown in Figure 6a. The lifetime τ_2 is approximately constant for all aged conditions, suggesting that the average size of open-volume defects remains unchanged during ageing. The lifetime τ_1 for the RS sample does not change significantly with the ageing time for both ageing temperatures of 400 °C and 500 °C. Nonetheless, the lifetime corresponding to positrons annihilated in the free state in the HPT specimens slightly increases with the ageing time for an ageing temperature of 400 °C, suggesting a slight decrease in the concentration of lattice defects resulting in a higher fraction of positrons annihilated in the free state.

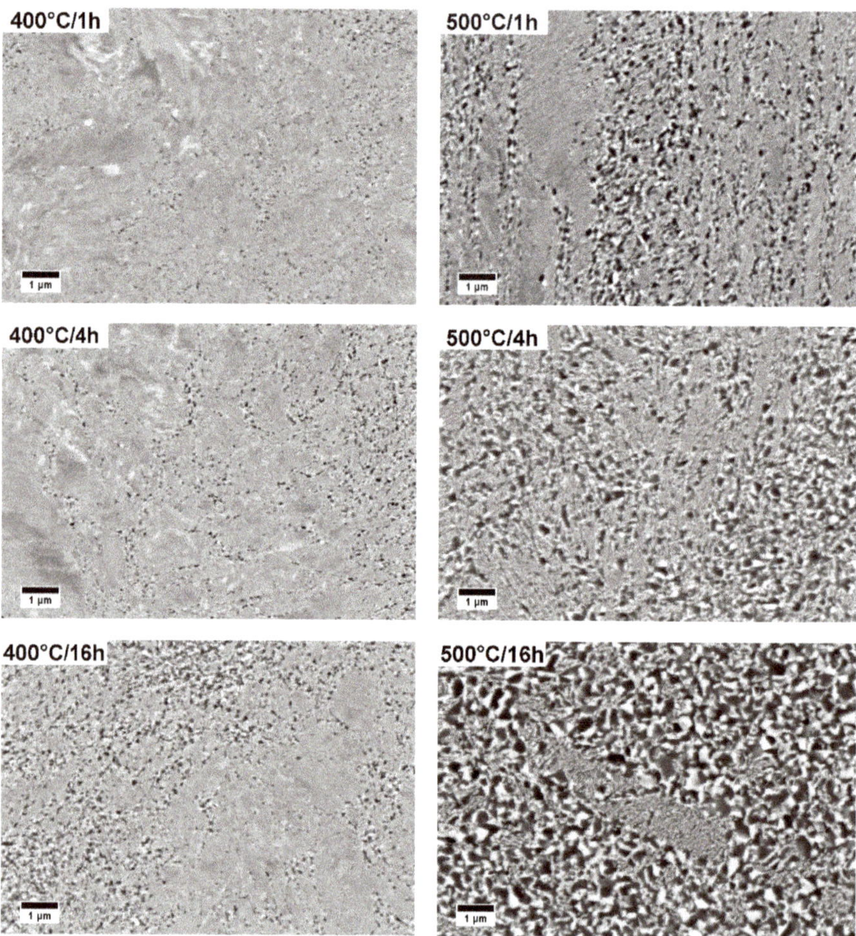

Figure 5. SEM-BSE images of the RS samples after ageing at 400 °C and 500 °C for 1–16 h (ageing condition is indicated in the top left corner of each micrograph).

Figure 6b shows the dependence of the intensity I_2 of positrons trapped at lattice defects on the ageing time for both the HPT and the RS samples for ageing temperatures of 400 °C and 500 °C. The component corresponding to positrons trapped at defects has a dominant intensity of around 98% for the HPT sample. This intensity gradually decreases with increasing ageing time at 400 °C due to the recovery of defects. However, the intensity I_2 for the ageing temperature of 500 °C remains constant within the experimental error, despite the fact that the recovery of defects introduced by deformation should also occur. This can be caused by an increasing fraction of incoherent α/β interfaces, which contain misfit defects also acting as traps for positrons [28]. During ageing of the material, precipitation of the α phase introduces new α/β interfaces and increases the concentration of these vacancy-like misfit defects. Since the lifetime of positrons trapped at misfit defects is comparable to the lifetime of positrons trapped at dislocations, distinguishing these components in the lifetime spectra is not possible and the component τ_2 contains contributions both from positrons trapped at dislocations and misfit defects at α/β interfaces.

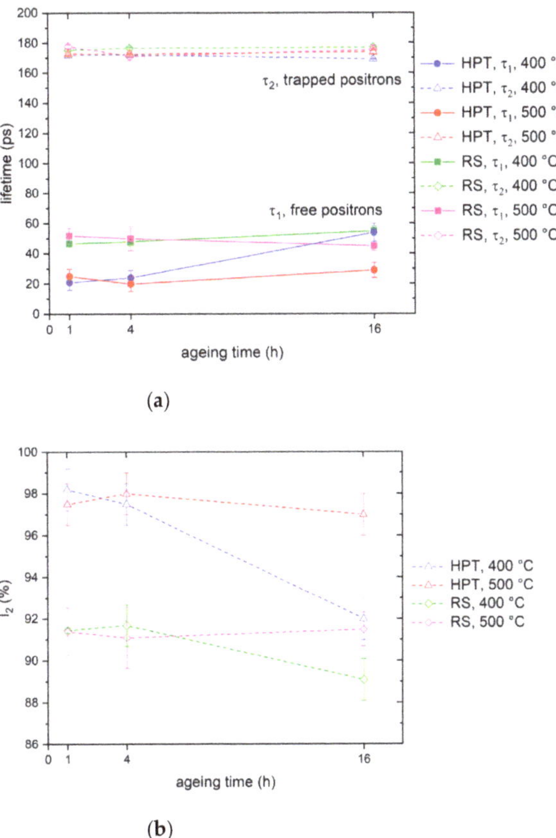

Figure 6. (a) Evolution of lifetime of positrons trapped at defects and of positrons annihilated in the free state with ageing time in the HPT and RS samples for ageing temperatures of 400 °C and 500 °C. (b) Dependence of the intensity I_2 of positrons trapped at defects in the HPT and RS samples on ageing time for ageing temperatures of 400 °C and 500 °C.

During aging of the RS specimens, the same trend as in the HPT ones was observed: the intensity I_2 decreases with increasing ageing time at 400 °C, which can be ascribed to the recovery of lattice defects. On the other hand, the I_2 for the RS samples aged at 500 °C does not change significantly with increasing ageing time—the recovery of defects and creation of new positron traps in the form of α/β interfaces act against each other. However, it should be noted that the intensity I_2 of the RS specimen is lower (~91%) when compared to the HPT specimen (~97%). This suggests a higher density of lattice defects in the HPT material.

4. Discussion

A Ti15Mo alloy was deformed by two different processes: high pressure torsion and cold-rotary swaging. The material after ST contains coarse β grains with the size of ~50 μm as it was described in our previous studies [15,16].

During HPT, the imposed deformation is 3600% in the peripheral part of the sample with a 20 mm diameter, as calculated using the von Mises equation [22]. The imposed deformation by RS is more than one order of magnitude lower, calculated by the finite element model reported in [23]—the authors claim that equivalent strain cannot reach 200% even on the rim of the sample processed by RS.

HPT straining results in the formation of a UFG structure of the Ti15Mo alloy [15,17,29–32]. In the rudimentary stage of the HPT deformation, a twinning process takes place, which is followed by significant grain refinement [31,33,34]. In contrast, deformation using RS creates shear bands with a high density of lattice defects and refined grains with an elongated character, as also reported in studies [35–38]. The shear bands are caused by the dislocation slip during RS, while the elongated character of the grains is the result of the combination of a dislocation slip and a grain boundary slip [37].

The RS specimen aged at 400 °C/1 h also contains small α particles. However, in comparison with the HPT sample, their volume fraction is lower. This is caused by the lower imposed strain in the RS specimen, which results in a lower density of precipitation-enhancing lattice defects.

Nucleation and growth of the α phase continues during longer ageing at both ageing temperatures in both prepared conditions. The size of α phase particles does not exceed 1 μm. They remain equiaxed in all aged conditions. A high density of α nuclei is caused by a high density of preferential nucleation sites. However, the growth of α nuclei is limited by two processes. First, each α particle causes a barrier for the growth of another α particle due to mutual impingement. Second, each α particle creates a barrier for the diffusion of β-stabilizing Mo in the β matrix. Therefore, the β matrix stabilized by Mo around each α particle limits the diffusional growth of surrounding particles. Consequently, α phase particles remain small and equiaxed in zones with a high density of lattice defects.

Microstructure observations clearly indicate that the precipitation of the α phase is inhomogeneous in both materials. This phenomenon can be caused by local chemical inhomogeneities in the material or by inhomogeneities in the density of lattice defects.

Ti15Mo alloy in a solution treated condition contains local chemical inhomogeneities as was proved by energy dispersive spectroscopy [15]. These chemical inhomogeneities (darker and lighter bands observed using backscattered electrons in SEM) seem to be extended during HPT deformation in the direction of the deformation [15]. Therefore, the nucleation of the α phase particles may be promoted in the areas depleted in Mo. Some of the authors also observed lighter and darker bands in the HPT-deformed alloys [19,20,39,40] which they ascribed to shear bands.

The formation of shear bands (areas with higher dislocation density) is typical in deformed alloys and can also cause an inhomogeneous precipitation of the α phase, as nucleation of the α phase is enhanced in areas with higher dislocation density [41]. This phenomenon has been described already in our previous studies [15–17] or in studies by other authors [18,20]. The growth of the α nuclei is diffusion-controlled and therefore is accelerated in the presence of a high density of dislocations as the pipe diffusion along dislocations is several orders of magnitude higher than the bulk diffusion [42].

Thus, the exact explanation of the origin of the lighter and darker bands in the deformed materials as well as the cause of the inhomogeneous precipitation of the α particles need further investigation.

Positron annihilation spectroscopy of severely deformed materials after ageing is usually utilized for monitoring the recovery of lattice defects during ageing [43,44]. The recovery of defects is manifested by an increasing intensity of the component related to the annihilation of free positrons (i.e., positrons not trapped in any lattice defect). A similar, though weak, trend is observed after ageing at 400 °C. On the other hand, ageing at 500 °C did not result in an increase of the free-positron component. This could be caused by the increasing fraction of incoherent α/β interfaces, which contain vacancy-like misfit defects which act as traps for positrons [28]. Since the lifetime of positrons trapped at misfit defects is comparable with that for dislocations, the distinction of these components in the lifetime spectra is not possible and the density of misfit defects and/or dislocations cannot be unambiguously determined. In conclusion, PAS measurement suggests that the recovery of lattice defects in the β matrix already occurs at 400 °C and in terms of positron trapping is partly compensated by the precipitation of incoherent α phase particles. At 500 °C, the precipitation is even more pronounced, and the recovery of positron trapping sites in the β matrix is fully offset by the formation of incoherent α/β interfaces.

5. Conclusions

Ti15Mo, a metastable β-Ti alloy was deformed by two techniques, namely, high pressure torsion (HPT) and rotary swaging (RS). The following conclusions can be drawn from this experimental study:

- Isothermal ageing resulted in the precipitation of the α phase in both deformed conditions. The nucleation and growth of the α phase is enhanced by lattice defects. These particles remain small and equiaxed even after ageing at 500 °C/16 h.
- Positron annihilation spectroscopy revealed the kinetics of the recovery of lattice defects during ageing. The recovery of lattice defects in the β matrix already occurs at 400°C, while the precipitation of the α phase forms new misfit defects which act as traps for positrons.
- Both dislocations and grain boundaries have a great influence on the nucleation and growth of the α phase.
- The precipitation of the α phase follows a similar trend in both deformed materials.

Author Contributions: Conceptualization, K.B. and J.S.; methodology, K.B., A.V., V.P.; investigation: K.B., J.V., J.Č., resources, K.B., J.M. and M.J.; data curation, K.B. and A.V.; writing—original draft preparation, K.B. and J.S.; writing—review and editing, K.B., J.S. and J.Č.; supervision, K.B., J.S. and I.S., project administration, J.S. and M.J.; funding acquisition, J.S. and M.J. All authors have read and agreed to the published version of the manuscript.

Funding: This research was funded by the Czech Science Foundation under the project No. 20-12624S. Partial financial support by the Ministry of Education, Youth and Sports, project no. LTAUSA 18045 and by ERDF under the project No. CZ.02.1.01/0.0/0.0/15 003/0000485 is also gratefully acknowledged. Computational resources were supplied by the project "e-Infrastruktura CZ" (e-INFRA LM2018140) provided within the program Projects of Large Research, Development and Innovations Infrastructures.

Institutional Review Board Statement: Not applicable.

Informed Consent Statement: Not applicable.

Data Availability Statement: The data presented in this study are available on request from the corresponding author.

Conflicts of Interest: The authors declare no conflict of interest. The funders had no role in the design of the study; in the collection, analyses, or interpretation of data; in the writing of the manuscript, or in the decision to publish the results.

References

1. Lütjering, G.; Williams, J.C. *Titanium; Engineering Materials, Processes*; Springer: Berlin/Heidelberg, Germany, 2007; ISBN 978-3-540-71397-5.
2. Niinomi, M. Mechanical Biocompatibilies of Titanium Alloys for Biomedical Applications. *J. Mech. Behav. Biomed. Mater.* **2008**, *1*, 30–42. [CrossRef]
3. Li, H.F.; Zheng, Y.F. Recent Advances in Bulk Metallic Glasses for Biomedical Applications. *Acta Biomater.* **2016**, *36*, 1–20. [CrossRef] [PubMed]
4. Chen, Q.; Thouas, G.A. Metallic Implant Biomaterials. *Mater. Sci. Eng. R Rep.* **2015**, *87*, 1–57. [CrossRef]
5. Eisenbarth, E.; Velten, D.; Müller, M.; Thull, R.; Breme, J. Biocompatibility of β-Stabilizing Elements of Titanium Alloys. *Biomaterials* **2004**, *25*, 5705–5713. [CrossRef] [PubMed]
6. Zehetbauer, M.J.; Steiner, G.; Schafler, E.; Koznikov, A.V.; Korznikova, E.A. Deformation Induced Vacancies with Severe Plastic Deformation: Measurements and Modelling. *Mater. Sci. Forum.* **2006**, *503*, 57–64. [CrossRef]
7. Estrin, Y.; Vinogradov, A. Extreme Grain Refinement by Severe Plastic Deformation: A Wealth of Challenging Science. *Acta Mater.* **2013**, *61*, 782–817. [CrossRef]
8. Valiev, R.Z.; Islamgaliev, R.K.; Alexandrov, I.V. Bulk Nanostructured Materials from Severe Plastic Deformation. *Prog. Mater. Sci.* **2000**, *45*, 103–189. [CrossRef]

9. Nag, S.; Banerjee, R.; Fraser, H.L. Intra-Granular Alpha Precipitation in Ti–Nb–Zr–Ta Biomedical Alloys. *J. Mater. Sci.* **2009**, *44*, 808–815. [CrossRef]
10. da Costa, F.H.; Salvador, C.A.F.; de Mello, M.G.; Caram, R. Alpha Phase Precipitation in Ti-30Nb-1Fe Alloys—Phase Transformations in Continuous Heating and Aging Heat Treatments. *Mater. Sci. Eng. A* **2016**, *677*, 222–229. [CrossRef]
11. Mantri, S.A.; Choudhuri, D.; Alam, T.; Viswanathan, G.B.; Sosa, J.M.; Fraser, H.L.; Banerjee, R. Tuning the Scale of α Precipitates in β-Titanium Alloys for Achieving High Strength. *Scr. Mater.* **2018**, *154*, 139–144. [CrossRef]
12. Jiang, B.; Emura, S.; Tsuchiya, K. Improvement of Ductility in Ti-5Al-5Mo-5V-3Cr Alloy by Network-like Precipitation of Blocky α Phase. *Mater. Sci. Eng. A* **2018**, *722*, 129–135. [CrossRef]
13. Duerig, T.W.; Terlinde, G.T.; Williams, J.C. Phase Transformations and Tensile Properties of Ti-10V-2Fe-3AI. *MTA* **1980**, *11*, 1987–1998. [CrossRef]
14. Furuhara, T.; Maki, T. Variant Selection in Heterogeneous Nucleation on Defects in Diffusional Phase Transformation and Precipitation. *Mater. Sci. Eng. A* **2001**, *312*, 145–154. [CrossRef]
15. Bartha, K.; Stráský, J.; Veverková, A.; Barriobero-Vila, P.; Lukáč, F.; Doležal, P.; Sedlák, P.; Polyakova, V.; Semenova, I.; Janeček, M. Effect of the High-Pressure Torsion (HPT) and Subsequent Isothermal Annealing on the Phase Transformation in Biomedical Ti15Mo Alloy. *Metals* **2019**, *9*, 1194. [CrossRef]
16. Bartha, K.; Veverková, A.; Stráský, J.; Veselý, J.; Minárik, P.; Corrêa, C.A.; Polyakova, V.; Semenova, I.; Janeček, M. Effect of the Severe Plastic Deformation by ECAP on Microstructure and Phase Transformations in Ti-15Mo Alloy. *Mater. Today Commun.* **2020**, *22*, 100811. [CrossRef]
17. Bartha, K.; Stráský, J.; Barriobero-Vila, P.; Šmilauerová, J.; Doležal, P.; Veselý, J.; Semenova, I.; Polyakova, V.; Janeček, M. In-Situ Investigation of Phase Transformations in Ultra-Fine Grained Ti15Mo Alloy. *J. Alloys Compd.* **2021**, *867*, 159027. [CrossRef]
18. Jiang, B.; Emura, S.; Tsuchiya, K. Formation of Equiaxed α Phase in Ti-5Al-5Mo-5V-3Cr Alloy Deformed by High-Pressure Torsion. *J. Alloys Compd.* **2018**, *738*, 283–291. [CrossRef]
19. Xu, W.; Edwards, D.P.; Wu, X.; Stoica, M.; Calin, M.; Kühn, U.; Eckert, J.; Xia, K. Promoting Nano/Ultrafine-Duplex Structure via Accelerated α Precipitation in a β-Type Titanium Alloy Severely Deformed by High-Pressure Torsion. *Scr. Mater.* **2013**, *68*, 67–70. [CrossRef]
20. Jiang, B.; Tsuchiya, K.; Emura, S.; Min, X. Effect of High-Pressure Torsion Process on Precipitation Behavior of α Phase in β-Type Ti–15Mo Alloy. *Mater. Trans.* **2014**, *55*, 877–884. [CrossRef]
21. Zhilyaev, A.P.; Langdon, T.G. Using High-Pressure Torsion for Metal Processing: Fundamentals and Applications. *Prog. Mater. Sci.* **2008**, *53*, 893–979. [CrossRef]
22. Valiev, R.Z.; Ivanisenko, Y.V.; Rauch, E.F.; Baudelet, B. Structure and Deformaton Behaviour of Armco Iron Subjected to Severe Plastic Deformation. *Acta Mater.* **1996**, *44*, 4705–4712. [CrossRef]
23. Moumi, E.; Ishkina, S.; Kuhfuss, B.; Hochrainer, T.; Struss, A.; Hunkel, M. 2D-Simulation of Material Flow During Infeed Rotary Swaging Using Finite Element Method. *Procedia Eng.* **2014**, *81*. [CrossRef]
24. Bečvář, F.; Čížek, J.; Procházka, I.; Janotová, J. The Asset of Ultra-Fast Digitizers for Positron-Lifetime Spectroscopy. *Nucl. Instrum. Methods Phys. Res. Sect. A Accel. Spectrometers Detect. Assoc. Equip.* **2005**, *539*, 372–385. [CrossRef]
25. Bečvář, F. Methodology of Positron Lifetime Spectroscopy: Present Status and Perspectives. *Nucl. Instrum. Methods Phys. Res. Sect. B Beam Interact. Mater. At.* **2007**, *261*, 871–874. [CrossRef]
26. Procházka, I.; Novotný, I.; Bečvář, F. Application of Maximum-Likelihood Method to Decomposition of Positron-Lifetime Spectra to Finite Number of Components. *Mater. Sci. Forum* **1997**, *255–257*, 772–774. [CrossRef]
27. Kilmametov, A.R.; Ivanisenko, Y.; Mazilkin, A.A.; Straumal, B.B.; Gornakova, A.S.; Fabrichnaya, O.B.; Kriegel, M.J.; Rafaja, D.; Hahn, H. The A→ω and B→ω Phase Transformations in Ti–Fe Alloys under High-Pressure Torsion. *Acta Mater.* **2018**, *144*, 337–351. [CrossRef]
28. Čížek, J. Characterization of Lattice Defects in Metallic Materials by Positron Annihilation Spectroscopy: A Review. *J. Mater. Sci. Technol.* **2018**, *34*, 577–598. [CrossRef]
29. Bartha, K.; Stráský, J.; Harcuba, P.; Semenova, I.; Polyakova, V.; Janeček, M. Heterogeneous Precipitation of the α-Phase in Ti15Mo Alloy Subjected to High Pressure Torsion. *Acta Phys. Pol. A* **2018**, *134*, 790–793. [CrossRef]
30. Václavová, K.; Stráský, J.; Zhǎňal, P.; Veselý, J.; Polyakova, V.; Semenova, I.; Janeček, M. Ultra-Fine Grained Microstructure of Metastable Beta Ti-15Mo Alloy and Its Effects on the Phase Transformations. *IOP Conf. Ser. Mater. Sci. Eng.* **2017**, *194*, 012021. [CrossRef]
31. Václavová, K.; Stráský, J.; Polyakova, V.; Stráská, J.; Nejezchlebová, J.; Seiner, H.; Semenova, I.; Janeček, M. Microhardness and Microstructure Evolution of Ultra-Fine Grained Ti-15Mo and TIMETAL LCB Alloys Prepared by High Pressure Torsion. *Mater. Sci. Eng. A* **2017**, *682*, 220–228. [CrossRef]
32. Václavová, K.; Stráský, J.; Veselý, J.; Gatina, S.; Polyakova, V.; Semenova, I.; Janeček, M. Evolution of Microstructure and Microhardness in Ti-15Mo β-Ti Alloy Prepared by High Pressure Torsion. Available online: https://www.scientific.net/MSF.879.2555 (accessed on 30 May 2018).
33. Wang, X.L.; Li, L.; Mei, W.; Wang, W.L.; Sun, J. Dependence of Stress-Induced Omega Transition and Mechanical Twinning on Phase Stability in Metastable β Ti–V Alloys. *Mater. Charact.* **2015**, *107*, 149–155. [CrossRef]
34. Zhou, X.; Min, X.A.; Emura, S.; Tsuchiya, K. Accommodative {332}<113> Primary and Secondary Twinning in a Slightly Deformed β-Type Ti-Mo Titanium Alloy. *Mater. Sci. Eng. A* **2017**, *684*, 456–465. [CrossRef]

35. Preisler, D.; Stráský, J.; Harcuba, P.; Halmešová, K.; Janeček, M. Cold Swaging and Recrystallization Annealing of Ti-Nb-Ta-Zr-O Alloy—Microstructure, Texture and Microhardness Evolution. *MSF* **2018**, *941*, 1132–1136. [CrossRef]
36. Cheng, J.; Wang, H.; Li, J.; Gai, J.; Ru, J.; Du, Z.; Fan, J.; Niu, J.; Song, H.; Yu, Z. The Effect of Cold Swaging Deformation on the Microstructures and Mechanical Properties of a Novel Metastable β Type Ti–10Mo–6Zr–4Sn–3Nb Alloy for Biomedical Devices. *Front. Mater.* **2020**, *7*. [CrossRef]
37. Guo, W.Y.; Xing, H.; Sun, J.; Li, X.L.; Wu, J.S.; Chen, R. Evolution of Microstructure and Texture during Recrystallization of the Cold-Swaged Ti-Nb-Ta-Zr-O Alloy. *Metall. Mater. Trans. A* **2008**, *39*, 672–678. [CrossRef]
38. Guo, W.; Quadir, M.Z.; Ferry, M. The Mode of Deformation in a Cold-Swaged Multifunctional Ti-Nb-Ta-Zr-O Alloy. *Met. Mat Trans. A* **2013**, *44*, 2307–2318. [CrossRef]
39. Xu, W.; Wu, X.; Stoica, M.; Calin, M.; Kühn, U.; Eckert, J.; Xia, K. On the Formation of an Ultrafine-Duplex Structure Facilitated by Severe Shear Deformation in a Ti–20Mo β-Type Titanium Alloy. *Acta Mater.* **2012**, *60*, 5067–5078. [CrossRef]
40. Zafari, A.; Xia, K. Formation of Equiaxed α during Ageing in a Severely Deformed Metastable β Ti Alloy. *Scr. Mater.* **2016**, *124*, 151–154. [CrossRef]
41. Porter, D.A.; Easterling, K.E.; Sherif, M.Y.A. *Phase Transformations in Metals and Alloys*, 3rd ed.; CRC Press: London, UK, 2009.
42. Legros, M.; Dehm, G. Obsevation of Giant Diffusivitiy along Dislocation Core. *Science* **2008**, *319*, 1646–1649. [CrossRef]
43. Bartha, K.; Zháňal, P.; Stráský, J.; Čížek, J.; Dopita, M.; Lukáč, F.; Harcuba, P.; Hájek, M.; Polyakova, V.; Semenova, I.; et al. Lattice Defects in Severely Deformed Biomedical Ti-6Al-7Nb Alloy and Thermal Stability of Its Ultra-Fine Grained Microstructure. *J. Alloys Compd.* **2019**, *788*, 881–890. [CrossRef]
44. Janeček, M.; Čížek, J.; Stráský, J.; Václavová, K.; Hruška, P.; Polyakova, V.; Gatina, S.; Semenova, I. Microstructure Evolution in Solution Treated Ti15Mo Alloy Processed by High Pressure Torsion. *Mater. Charact.* **2014**, *98*, 233–240. [CrossRef]

Article

Improving the Mechanical Properties of a β-type Ti-Nb-Zr-Fe-O Alloy

Vasile Danut Cojocaru [1], Anna Nocivin [2,*], Corneliu Trisca-Rusu [3], Alexandru Dan [1], Raluca Irimescu [1], Doina Raducanu [1] and Bogdan Mihai Galbinasu [4]

[1] Materials Science and Engineering Faculty, University Politehnica of Bucharest, 060042 Bucharest, Romania; dan.cojocaru@upb.ro (V.D.C.); alexandru_dan_ro@yahoo.com (A.D.); raluca.irimescu@stud.sim.upb.ro (R.I.); doina.raducanu@mdef.pub.ro (D.R.)

[2] Mechanical, Industrial and Maritime Faculty, Ovidius University of Constanța, 900527 Constanța, Romania

[3] National Institute for Research and Development in Micro-technologies, 077190 Bucharest, Romania; corneliu.trisca@nano-link.net

[4] Dental Medicine Faculty, University of Medicine and Pharmacy "Carol Davila" Bucharest, 020021 Bucharest, Romania; bogdan.galbinasu@yahoo.com

* Correspondence: anocivin@univ-ovidius.ro; Tel.: +40-241-660-431

Received: 8 October 2020; Accepted: 6 November 2020; Published: 9 November 2020

Abstract: The influence of complex thermo-mechanical processing (TMP) on the mechanical properties of a Ti-Nb-Zr-Fe-O bio-alloy was investigated in this study. The proposed TMP program involves a schema featuring a series of severe plastic deformation (SPD) and solution treatment (STs). The purpose of this study was to find the proper parameter combination for the applied TMP and thus enhance the mechanical strength and diminish the Young's modulus. The proposed chemical composition of the studied β-type Ti-alloy was conceived from already-appreciated Ti-Nb-Ta-Zr alloys with high β-stability by replacing the expensive Ta with more accessible Fe and O. These chemical additions are expected to better enhance β-stability and thus avoid the generation of ω, α′, and α″ during complex TMP, as well as allow for the processing of a single bcc β-phase with significant grain diminution, increased mechanical strength, and a low elasticity value/Young's modulus. The proposed TMP program considers two research directions of TMP experiments. For comparisons using structural and mechanical perspectives, the two categories of the experimental samples were analyzed using SEM microscopy and a series of tensile tests. The comparison also included some already published results for similar alloys. The analysis revealed the advantages and disadvantages for all compared categories, with the conclusions highlighting that the studied alloys are suitable for expanding the database of possible β-Ti bio-alloys that could be used depending on the specific requirements of different biomedical implant applications.

Keywords: β-Titanium alloys; thermo-mechanical processing; SEM; mechanical properties

---

1. Introduction

Good biomedical materials for orthopedic implants with long-term service need a combination of a low Young's modulus, close proximity to the human bone, and high strength to avoid the known "stress shielding effect" [1–5]. Among many investigated biocompatible materials, non-cytotoxic β-Ti alloys have been developed in recent decades [6–13], as these alloys have the most attractive combination of a low Young's modulus, high mechanical strength, and high ductility. This property combination can also ensure good processability (i.e., plastic deformation, machinability, welding, etc.) and good tribological characteristics, all of which are necessary for the long-term dimensional stability of an implant [14]. Compared to other possible Ti-bio-alloys, such as the frequently used (α) or (α + β) types, the special advantage of β-type Ti-alloys is their lower Young's modulus (45–65 GPa) [2,4,15–17]. Beta-Ti alloys are also reported to exhibit high fracture toughness [18] and a good response to heat treatment if the alloying is achieved using suitable β-stabilizing chemical elements [5,19–21], such as non-toxic Nb, Ta, Mo, and Zr [22–24]. Ti-Nb-Ta-Zr (TNTZ) alloys are highly regarded [25–30] because of their suitable combination of functional properties, both mechanical and biomedical. However, they exhibit some disadvantages, such as a much higher melting point of Ta compared to other related alloying elements ($T_{top}{}^{Ta}$ = 2996 °C, $T_{top}{}^{Nb}$ = 2468 °C, $T_{top}{}^{Zr}$ = 1855 °C, $T_{top}{}^{Ti}$ = 1660 °C), which decreases the casting alloy characteristics. Even though Nb also has a high melting point compared to Ti and Zr, it is still indispensable due to its strong β-stabilizing characteristics. In addition, Ta is an expensive chemical element [6]. Therefore, efforts have been made to replace Ta with more accessible Fe and O [31–36]. Fe is selected not only for economic considerations but also because it is a strong β stabilizer [37,38]. It was shown that the Young's modulus (E) can be reduced to 91 GPa by adding Fe [38], which is slightly lower than the E value for commonly used α/β Ti-64 (~110 GPa [39]) but higher than the E of cortical bone (~30 GPa) [1]. Similar results were reported in [40–44]. Oxygen, despite not being a β stabilizing element [20,21], may hamper the formation of α″ [1,2]. Recent research found that by adding oxygen to β-Ti alloys, both strength and ductility can be improved simultaneously [43,44]. Moreover, Fe and interstitial oxygen are beneficial for improving mechanical properties through solid-solution strengthening [18,31–33]. Therefore, both Fe and O are attractive elements for the development of high-strength and low-Young's-modulus β-Ti alloys.

Apart from the chemical composition being an important factor for enhancing the mechanical properties in β-Ti alloys, thermo-mechanical processing (TMP) is also a very important way to decrease the Young's modulus and increase the mechanical strength of β-Ti alloys [14,45]. The applied TMP can involve a combination of severe plastic deformations (SPD) and solution treatments (STs) to achieve grain refinement featuring ultra-fine or even nano-meter grain dimensions. Through this, the mechanical strength can increase significantly, and the Young's modulus can decrease. For example, Bertrand et al. [46] reported a very low Young's modulus for a Ti-25Ta-25Nb alloy (55 GPa, one of the lowest values for a β-Ti alloy) developed via TMP; the objective of the applied treatment (cold rolling + solution treatment) was to restore a fully recrystallized β phase microstructure from the cold rolled state using a decreased β grain size and a low Young's modulus.

Previous studies on TNTZ-Fe-O alloys [31–33,47,48] provided valuable information for developing novel alloys by tuning the composition and processing parameters to obtain the best mechanical and biomedical properties suitable for medical implants. These reports refer to alloys in a cold-rolling (CR) state or after a CR + ST combination, for which the Young's modulus can vary between 60 and 107 GPa, and the ultimate tensile strength-UTS can vary between 903 and 1370 MPa. The intent of the present work is also to elaborate a complex TMP program applied to a particular chemical composition from the TNTZ-Fe-O family-alloy (but without Ta and with Fe and O) to find a better combination of TMP parameters coupled with a suitable chemical composition and ultimately obtain better mechanical properties than those reported previously. Thus, a chemical composition of the Ti-Nb-Zr-Fe-O alloy was proposed for investigation, with sufficient β-stabilizing elements to obtain a stable β-Ti alloy. On the other hand, the proposed TMP program contains two distinct directions: One is formed from a CR series with a gradual increase in the applied deformation degree, coupled with an ST series featuring variable parameters; the second one combines a series of severe plastic deformation (SPD) and ST, both in the effort to achieve evident grain refinement. Thus, both previously reported modalities for increasing the strength of β-Ti alloys will be applied [49–51], including the addition of Fe and O and a complex TMP program with variable processing parameters. The experimental results will be compared with those reported in [31–33,47,48] by highlighting the advantages and disadvantages of the results obtained for the proposed alloy and corresponding TMP parameters. At the same time, the proposed investigations seek to describe the relationship between the microstructure and mechanical properties of this alloy under different heat treatment conditions. Alongside this main objective, the intention is also to enlarge the database of possible β-Ti bio-alloys, depending on the specific requirements of different implant applications.

2. Materials and Methods

2.1. Synthesis of the Studied Alloy

The nominal new chemical composition for the studied alloy, Table 1, is as follows:

Table 1. The chemical composition of the studied alloy (wt.%).

The Chemical Composition of the Studied Alloy (wt.%)	Ti	Nb	Zr	Fe	O
Ti-Nb-Zr-Fe-O	57.25	34.10	7.59	0.90	0.16

This alloy was obtained using a levitation induction melting furnace FIVE CELES-MP25 with a nominal power of 25 kW and a melting capacity of 30 cm^3 in a high vacuum of 10^{-4}–10^{-5} mbar. The alloy synthesis was conducted with an intense agitation of the melted alloy. The ingots were re-melted twice to achieve a high degree of chemical homogeneity.

2.2. Thermomechanical Processing Program of the New β-Ti Alloy

After the alloy synthesis and before the complex TMP program, the as-cast alloy was treated to obtain a quality-homogenised precursor (Figure 1). The applied treatment consisted of (a) Cold Rolling (CR) with a relative reduction of ε = 20% using a Mario di Maio LQR120AS rolling-mill (Mario di Maio Inc., Milano, Italy) with a 3 m/min rolling speed and no lubrication; before the CR process, the sample was cleaned using an ultrasonic bath at 60 °C in ethylic alcohol; (b) a Homogenization Treatment (HT) at 1223 K/950 °C (above the β-transus temperature) with a holding time of 20 min. and water quenching (w.q.) using a GERO SR 100 × 500 type oven (Carbolite-Gero Inc., Neuhausen, Germany) under a high vacuum. The final obtained homogenised alloy was named "the initial alloy" and was processed by a complex TMP program.

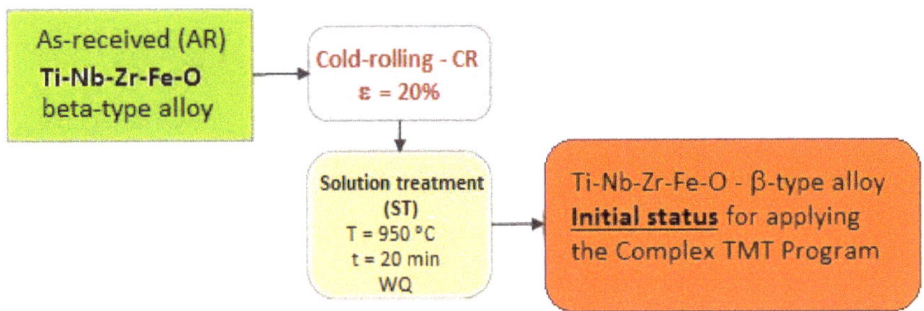

Figure 1. Schema for processing the as-cast sample to obtain a quality homogenised precursor named the initial alloy.

This program was structured and applied to two different research directions on a batch/group of samples obtained from the initial alloy (Figures 2 and 3). These two categories of samples were processed with different thermo-mechanical parameters for the final analysis, comparison, and recommendations for potential selection/application.

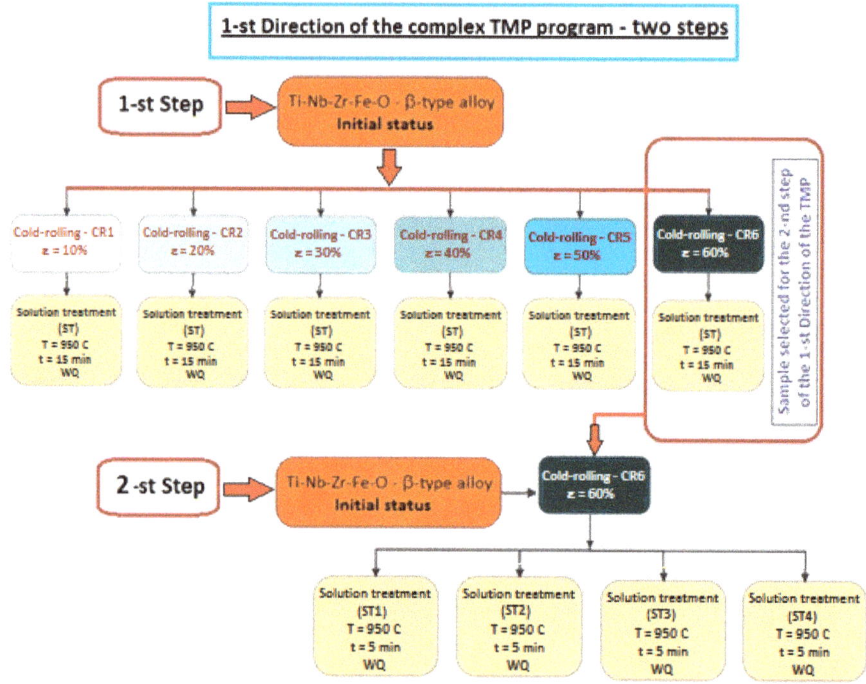

Figure 2. The 1st Direction of the complex TMP Program applied to the studied β-type Ti-Nb-Zr-Fe-O alloy.

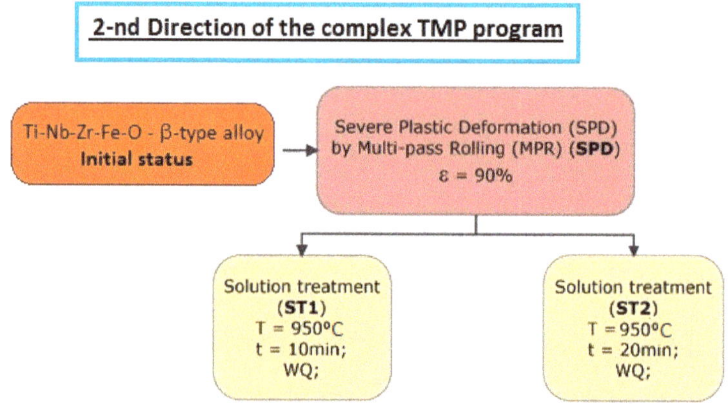

Figure 3. The 2nd Direction of the complex TMP Program applied to the studied β-type Ti-Nb-Zr-Fe-O alloy.

The first direction (Figure 2) included a two-step process: (1) The first step consisted of various CR cycles followed by identical STs for all CR samples—1223 K/15 min./w.q. CR was applied to six distinct samples with the initial state using six different total deformation degrees: $\varepsilon_{tot} = 10\%$—1 pass; $\varepsilon_{tot} = 20\%$—2 passes; $\varepsilon_{tot} = 30\%$—3 passes; $\varepsilon_{tot} = 40\%$—4 passes; $\varepsilon_{tot} = 50\%$—5 passes; $\varepsilon_{tot} = 60\%$—6 passes. (2) For the second step, from the six variants that were experimented on, the one with $((\varepsilon_{tot} = 60\%) + ST)$ was selected, as it presented the smallest grain size.

Four STs were applied to this selected sample after cold-rolling (ε_{tot} = 60%—6 passes) for different times of 5, 10, 15, and 20 min, while the other parameters remained unchanged.

The second direction (Figure 3) includes severe plastic deformation (SPD) using the multi-pass rolling (MPR) method, followed by two different STs. MPR processing was performed using 10 rolling passes with ε_{tot} = 90%. The two variants of the ST consisted of the same heating temperature (1223 K/950 °C) and cooling medium (water) but two distinct holding times (10 min. and 20 min.). All STs were performed using a GERO SR 100 × 500-type oven with a high vacuum—the same as that used for the initial treatment.

2.3. Micro-Structural and Mechanical Analysis of the Alloy Samples

A Metkon MICRACUT 200 type machine (Metkon Instruments Inc., Bursa, Turkey) with diamond cutting disks was used for cutting. The specimens were then fixed on a specific epoxy resin of a Buehler Sampl-Kwick type, abraded with 1200 grit SiC paper using a Metkon Digiprep ACCURA machine (Metkon Instruments Inc., Bursa, Turkey), and then mechanically polished using 6, 3, and 1 µm diamond paste and 0.03 µm colloidal silica on a Buehler VibroMet2 machine (Buehler Ltd., Lake Bluff, IL, USA).

The SEM images and analysis were realized using a scanning electron microscope—a TESCAN VEGA II—XMU (Tescan Orsay Holding, a.s., Brno, Czech Republic). The CR- and SPD-processed samples were examined mainly in the RD–ND cross-section (RD—rolling direction; ND—normal direction) to observe the grain deformation/texture degree evolution. To highlight the main microstructural characteristics, only the most representative images were selected.

The tensile tests, performed on the final states from both research directions, were achieved in the RD using a Gatan MicroTest-2000N-type machine (Gatan Inc., Pleasanton, CA, USA) with a strain rate of $1 \times 10^{-4}\ \mathrm{s}^{-1}$. Based on the obtained data, the following average values of the mechanical characteristics were determined: the ultimate tensile strength (σ_{UTS}); yield strength ($\sigma_{0.2}$); the elongation to fracture (ε_f); and the elastic modulus (E). The standard deviation (SD) was also calculated.

3. Results

3.1. Micro-Structural Analysis of the As-Cast and Homogenized Sample, Named the Initial State

The structures corresponding to the cast sample and the homogenized sample (by applying CR + HT) are indicated in Figure 4. Due to consistent β-stabilizing alloying elements (~42% in total, excluding oxygen), the alloy's structure was formed from only β-phase grains with an equiaxe shape. The measured average dimensions of the β grains were 121 µm for the as-cast sample and slightly larger for the homogenized sample (145 µm) due to grain growing during the applied heat treatment.

In metastable β-type Ti alloys, it is already known that the ω, α′, and α″ phases can be produced in the β matrix as a result of cold plastic deformation [52–55]. These phases are detrimental to the formability of the material, making it difficult to process. Therefore, the interest lies in designing β-Ti-type alloys containing a sufficient quantity of alloying elements with β-stabilizing characteristics. For that purpose, the main β-stabilizing element—Nb—should be no less than 35–38% to ensure good stability of the β phase in binary Ti-Nb alloys (see Hu et al. [56]). For this study, besides high Nb content (34.1 wt.%), the addition of Fe and Zr was considered, as both elements are reported to be effective for solid solution strengthening, while also playing a very important role from the perspective of β-phase stability by suppressing the generation of α′ and ω and by shifting the martensite starting temperature (Ms) to a lower one [47]. Fe also is a strong β stabilizer [37], and Zr, despite being considered by some reports to be a neutral element from a β-stabilization perspective [57], has been demonstrated by others to be beneficial for β-Ti-alloys because it suppresses the nucleation of isothermal ω [6] and also decreases the Ms temperature [58]. Therefore, it can be presumed that the absence of the ω, α′, and α″ phases may be a consequence of the combined stabilization of the β-phase under all the added alloying elements (~42%), a fact also reported in [4,59,60]. It can be also presumed that the design of the chemical composition of the studied alloy achieved the initially stated goal—obtaining a single β-phase for safe and easy mechanical processing. However, this assertion should be taken only as probably because, to prove the absence of the above secondary phases, a detailed XRD analysis should be performed, which represents a future objective for the following stages of this research work.

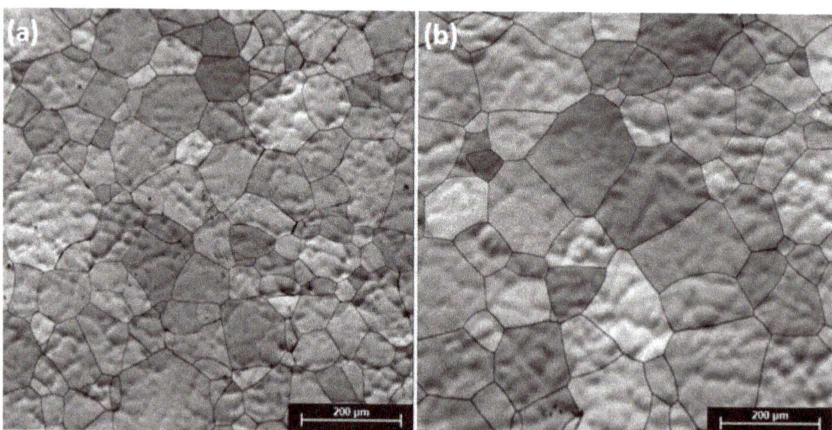

Figure 4. The SEM images of the Ti-Nb-Zr-Fe-O alloy microstructures in: (**a**) the as-cast state; (**b**) after the homogenisation treatment—named the initial state.

3.2. Micro-Structural and Mechanical Analysis of the Samples Processed by the First Direction of the TMP Program

Following the first direction schema of the TMP Program, a series of samples was processed by CR with different deformation degrees from $\varepsilon_{tot} = 10\%$ to $\varepsilon_{tot} = 60\%$ (Figure 2). Figure 5 shows the obtained microstructures. It can be observed that the β grains lengthened in the rolling direction step-by-step from sample to sample due to the increase in the applied total deformation degree. It can also be observed that the deformation products became increasingly visible and pronounced with an increase of the deformation degree (Figure 5a–c). These deformation products seemed to be kink bands, twin bands, or even shear bands [4], depending on the opening/value of the misorientation angle, which were smaller or larger between the β matrix and these deformation products. For example, if the misorientation angles between the kink bands and the β matrix for the Ti-22.4Nb-0.73Ta-2Zr-1.34O alloy during its compression straining, as reported by Yang et al. [59], are between 10–30°, the visible deformation products of the band-types in Figure 5a (CR at 10%) with similar misorientation angles can also be considered kink bands. Further, in Figure 5b,c, the misorientation angle increases between 50–55°. Since the misorientation angle between the {332} <113> β-mechanical twins and the β matrix is 50.5° in the <110> β direction [61], it can be presumed that, for the present case, the deformation products are evolving gradually from kink to twin bands (white arrows on the images). Concerning increases in the misorientation angle with an increase in plastic deformation, similar research results were reported for other β-Ti alloys [52,53]. Generally, it is already known that the twining phenomenon facilitates subsequent grain fragmentation/refinement during plastic deformation process development [62–65]. For this study, Figure 5b,c show the gradual grain refinement as a function of an increase in the deformation degree. In addition to this grain fragmentation, the structure acquires a clear textured appearance (Figure 5d,f); these last three images show the formation of shear bands due to severe deformation degrees, which were also reported in [59,64] for other similar β-type Ti-alloys.

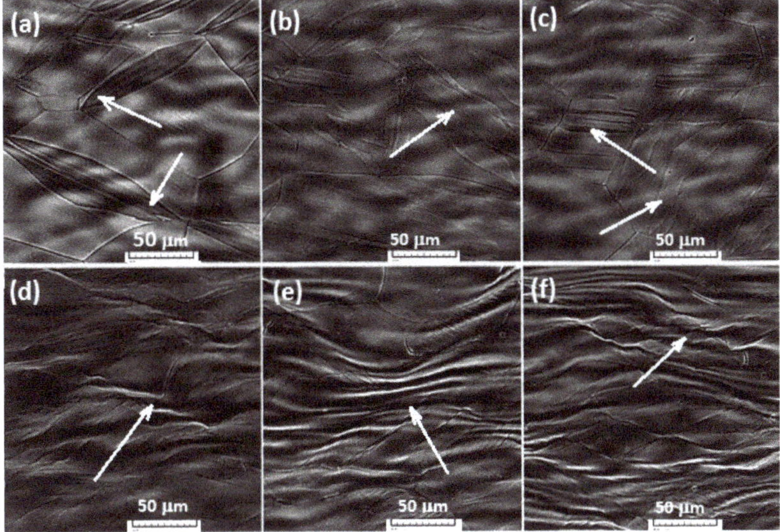

Figure 5. The SEM images of the Cold Rolling (CR) processed samples from the 1st step of the 1st direction of the TMP Program: (**a**) CR—ε_{tot} = 10%; (**b**) CR—ε_{tot} = 20%; (**c**) CR—ε_{tot} = 30%; (**d**) CR—ε_{tot} = 40%; (**e**) CR—ε_{tot} = 50%; (**f**) CR—ε_{tot} = 60%.

From Figure 5, it can also be observed that the shear bands are not homogeneously distributed and have a wavy shape. This wavy shape of the slip bands is reported to be typical for bcc metals [38], as in our case. Shear bands are considered to be formed from rotated and severely distorted crystals due to the localization/accumulation of a mass of shear stress on a slip plane [59].

In this study, the process starting with kinking, passing through {332} <113> β-mechanical twinning, and arriving to shear band formations provides the most probable deformation products that will gradually facilitate alloy deformation and texture during CR as the degree of deformation increases. However, this assertion will also have to be proven by a future XRD/TEM analysis, as noted previously.

After applying the same ST for all six CRed samples (Figure 2), the β-grains again became equiaxe (Figure 6), but with the grain size decreasing steadily depending on the ascending degree of the deformation that was previously applied. This decreasing phenomenon of the grain size is proven once again in Figure 7, which shows the measured average values of the obtained β-grains from 138 μm corresponding to the sample (CR1 + ST) and those at 75 μm corresponding to the sample (CR6 + ST). In this way, the β-grain dimensions became almost two times smaller than the initial dimensions (138/75 = 1.84) by applying a deformation degree that was six times greater (from 10% to 60%).

At this stage of the experiments, the sample (CR6 + ST) with the smallest β grains (75 μm) was selected to be further processed, following the 2nd step of the 1st direction of the TMP program. This second step (Figure 2) requires that for the sample selected (the one with the smallest β grains), after applying CR with the same ε_{tot} = 60%, the ST treatments will be varied/diversified by applying four different holding times of 5 min, 10 min, 15 min, and 20 min (same 1223 K/water quenching). The resulteding microstructures are shown in Figure 8, and the measured average values of the obtained β-grains are shown in Figure 9.

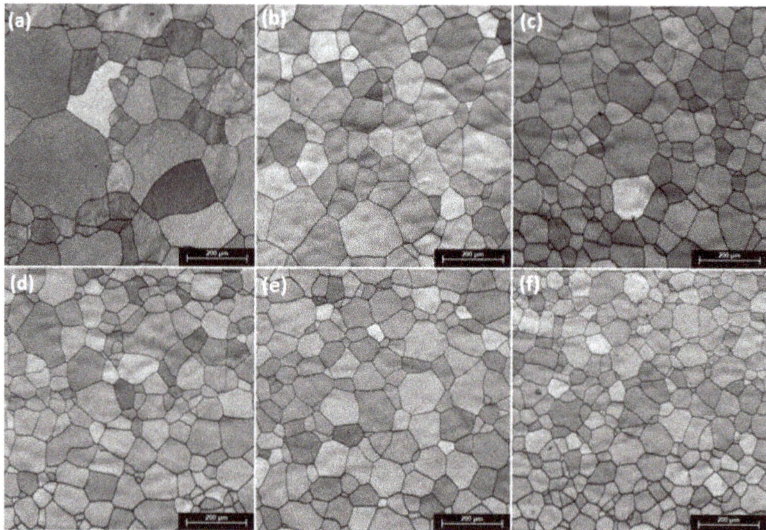

Figure 6. SEM images of the (CR + ST) processed samples corresponding to the 1st step of the 1st direction of the TMP Program, with various ε_{tot} values for CR and similar STs for all six samples: 1223 K/15 min./w.q.: (**a**) CR(ε_{tot} = 10%) + ST; (**b**) CR(ε_{tot} = 20%) + ST; (**c**) CR(ε_{tot} = 30%) + ST; (**d**) CR(ε_{tot} = 40%) + ST; (**e**) CR(ε_{tot} = 50%) + ST; (**f**) CR(ε_{tot} = 60%) + ST.

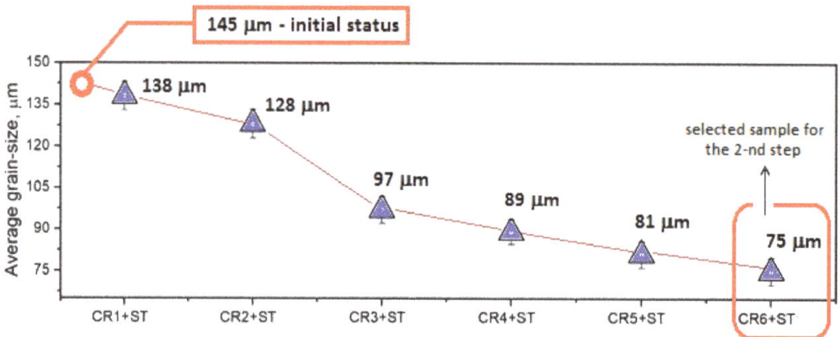

Figure 7. The evolution of the β grain average dimension as a function of the applied TMP variant corresponding to the 1st step of the 1st direction of the TMP Program.

The microstructures from Figure 8 also represent a series of β grains like those in Figure 6 but with visibly smaller grain dimensions. The grains are homogeneous and equiaxial without any traces of deformation products. Considering that the applied CR here was identical for all four samples (ε_{tot} = 60%), the variation of the β grain size was due to the use of a shorter or longer time for grain growing during the different holding times applied to the STs: The smallest grains correspond to the smallest holding time of 5 min., and vice versa. However, the resulting grains' dimensions (Figure 9) are much smaller than those in the anterior 1st step of the TMP program. This time, the smallest grains of 55 µm were obtained for the sample with CR (ε_{tot} = 60%) and ST1 (1223 K/5 min./w.q.).

Furthermore, tensile tests were applied to all four resulting samples. Figure 10 shows the strain–stress curves for the following samples: (a) the sample in the initial state; (b) the sample after applying the CR (ε_{tot} = 60%); and (c) the sample CR (ε_{tot} = 60%) + ST3 (1223 K/15 min./w.q.). For reasons of space, the inclusion of curves was waived for the other three CR+ST samples. Nevertheless, the mechanical characteristics were determined for all four samples. Table 2 indicates the average values for the ultimate tensile strength (σ_{UTS}); yield strength ($\sigma_{0.2}$); elongation to fracture (ε_f); and elastic modulus (E). Standard deviation is also included.

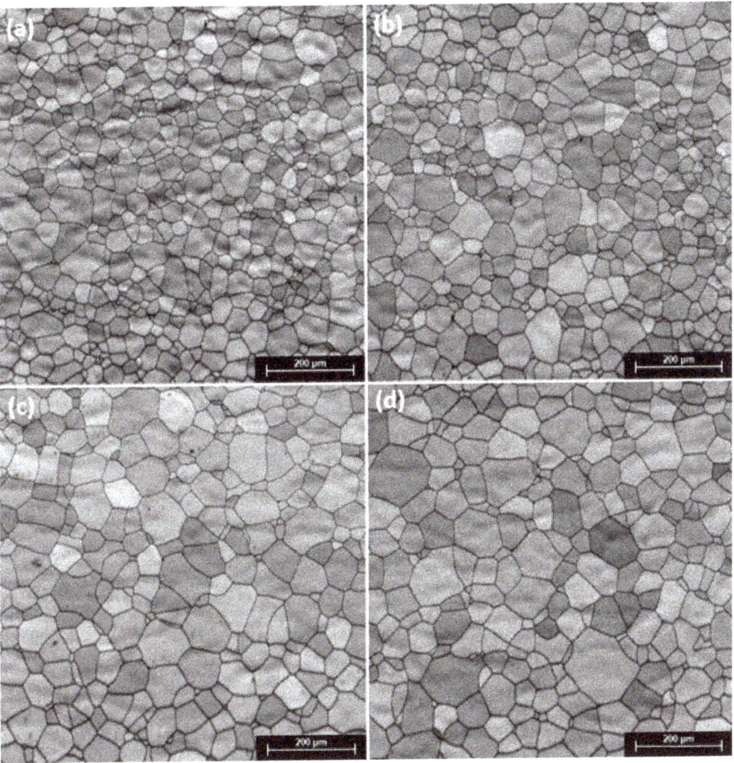

Figure 8. The SEM images of the (CR + solution treatment (ST) processed samples corresponding to the 2nd step of the 1st direction of the TMP Program with the same ε_{tot} = 60% for the applied CR but with various STs: (**a**) CR + ST1 (1223 K/5 min./w.q.); (**b**) CR + ST2 (1223 K/10 min./w.q.); (**c**) CR + ST3 (1223 K/15 min./w.q.); (**d**) CR + ST4 (1223 K/20 min./w.q.).

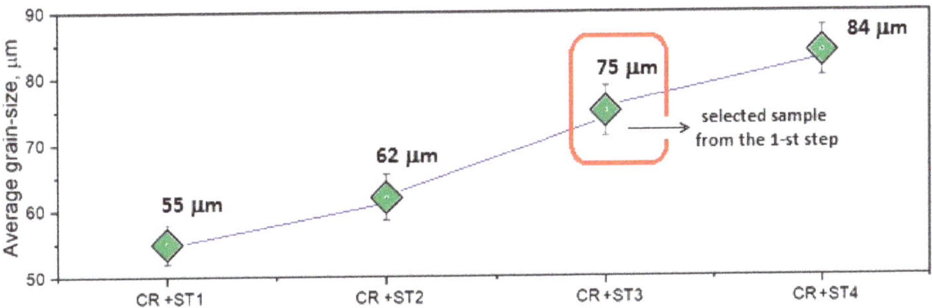

Figure 9. The evolution of the β grain average dimension as a function of the applied TMP variant, corresponding to the 2nd step of the 1st direction of the TMP Program.

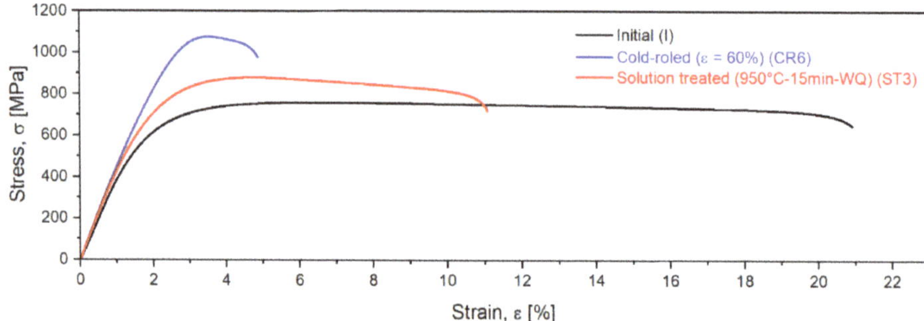

Figure 10. The stress–strain curves of the samples corresponding to the 2nd step of the TMT program (sample selected from the 1st step).

Table 2. Mechanical properties of the studied alloy at the end of the 1st direction of experiments from the TMP Program: Ultimate Tensile Strength (σ_{UTS}); Yield Strength ($\sigma_{0.2}$); Elongation to Fracture (ε_f); Elastic Modulus (E); SD—Standard Deviation.

Structural State	Mechanical Properties			
	σ_{UTS} (SD) (MPa)	$\sigma_{0.2}$ (SD) (MPa)	ε_f (SD) (%)	E (SD) (GPa)
Initial state (I)	759.7 (12.1)	546.1 (10.3)	20.9 (0.8)	49.1 (2.1)
CR—ε_{tot} = 60%	1076.1 (15.3)	808.1 (13.8)	4.9 (0.1)	48.8 (1.8)
CR + ST1 (1223 K-5 min-w.q.)	962.8 (14.5) (↑26.7%)	703.7 (12.6) (↑28.8%)	6.9 (0.2) (↓66.9%)	48.3 (1.6) (↓1.6%)
CR + ST2 (1223 K-10 min-w.q.)	901.7 (15.1) (↑18.7%)	631.4 (10.2) (↑15.6%)	8.2 (0.2) (↓60.7%)	47.6 (0.9) (↓3.0%)
CR + ST3 (1223 K-15 min-w.q.)	879.5 (13.7) (↑15.7%)	601.6 (10.4) (↑10.2%)	11.1 (0.4) (↓46.9%)	47.8 (1.1) (↓2.6%)
CR + ST4 (1223 K-20 min-w.q.)	809.0 (13.3) (↑6.5%)	574.4 (10.4) (↑5.2%)	15.9 (0.4) (↓23.9%)	49.3 (2.2) (↑0.4%)

A general characteristic for strain–stress curves, observed for bcc β-Ti alloys with soluble substitution elements and interstitial oxygen, is a flat zone lacking classic double yielding [60,66,67]. This characteristic is also visible in Figure 10. The flat zone indicates that the yield strength is very close to the ultimate tensile strength. Moreover, the stress values remain constant by increasing the strain. The explanation of this phenomenon reported in [30,42,60,68,69] relates to the heterogeneity of the deformation process for β-Ti alloys: dislocation sliding, β-phase twining, or "stress-induced martensitic transformation" β→α" (SIM) depending on the alloy's chemical composition, electronic parameters, deformation temperature, or even oxygen content. For the presently studied alloy, all these details regarding the deformation mechanisms are intended to be analysed in future research through TEM analyses specifically dedicated to this issue.

As shown in Table 2, it can be observed that the samples with CR + ST, compared to the initial state, feature increased values for UTS, between 6.5 and 26.7%, and YS provides values between 5.2 and 28.8%. The values of Young's modulus are almost the same (a very small decrease), but the elongation to fracture decreased significantly to between 23.9 and 66.9%. Considering that the grain dimension decreased from 145 μm (the initial state) to 55 μm (by about 62%), a preliminary conclusion here is that a combination of the applied parameters increased the UTS and YS for this first direction of the experiments without losing the acceptable level of a low Young's modulus (~48 GPa).

3.3. Micro-Structural and Mechanical Analysis of the Samples Processed by the 2nd Direction of the TMP Program

Following the 2nd Direction of the proposed experiments from the complex TMP Program, two samples of the initial state were processed by SPD-MPR with the same deformation degree of $\varepsilon_{tot} = 60\%$ (Figure 3). Figure 11 shows the obtained microstructures.

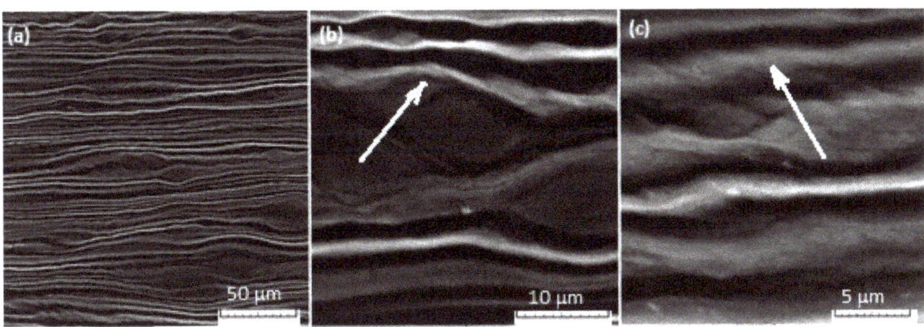

Figure 11. (**a–c**) SEM images of the sample processed by SPD-MPR ($\varepsilon_{tot} = 60\%$) corresponding to the 2nd direction of the TMP Program (**a–c**—same sample, with gradual increasing magnifications).

It can be observed that the β grains were strongly textured on the rolling direction. From the images with greater magnification, it can also be observed that the mechanism of deformation was similar to the mechanisms from the first category of the experimental samples, as discussed above. Here, due to the very high deformation degree being applied, the resulting deformation products are more like shear bands with very large misorientation angles, indicated by white arrows in Figure 11b. Moreover, the absence of any other phases can be observed—phases that, alongside β-grains, could have been formed during SPD (e.g., the ω, α′, and α″ phases). This means that the amount of the alloying elements with β-stabilizing effects was sufficient to secure the stability of the β phase and suppress the formation of secondary phases.

After SPD processing, both samples were subjected to ST using the same heating temperature and quenching medium (1223 K and water, respectively) as the first category of experiments but with two different holding times: 10 min and 20 min, respectively. The resulting microstructures are indicated in Figure 12, with the measured average β-grain dimensions indicated directly on the images. The first sample (SPD-60% + ST1(1223 K-10 min-w.q.)) had a similar β grain dimension (77 µm) to the sample (CR-60% + ST3(1223 K-15 min-w.q.)) from the first category of experiments (75 µm). The second sample (SPD-60% + ST2(1223 K-20 min-w.q.)) had a similar β-grain dimension (91 µm) to the sample (CR-40% + ST (1223 K/15 min/w.q.)) of 89 µm.

To assess the mechanical properties, tensile tests were applied to both the resulting SPD + ST samples. Similar to Figure 10, Figure 13 shows the strain–stress curves for initial, severely deformed, and solution-treated samples. Determination of the mechanical characteristics was also performed for both samples. Table 3 indicates the average values for the ultimate tensile strength (σ_{UTS}); yield strength ($\sigma_{0.2}$); elongation to fracture (ε_f); and elastic modulus (E).

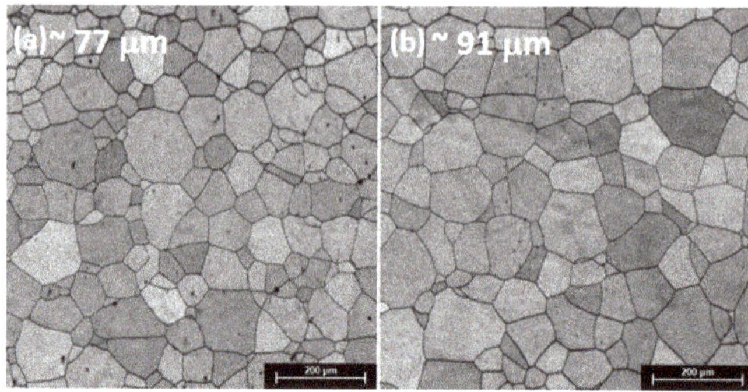

Figure 12. The SEM images of the two samples corresponding to the 2nd direction of the TMP Program: (**a**) SPD-MPR (ε_{tot} = 60%) + ST1 (1223 K-10 min-w.q.); (**b**) SPD-MPR (ε_{tot} = 60%) + ST2 (1223 K-20 min-w.q.).

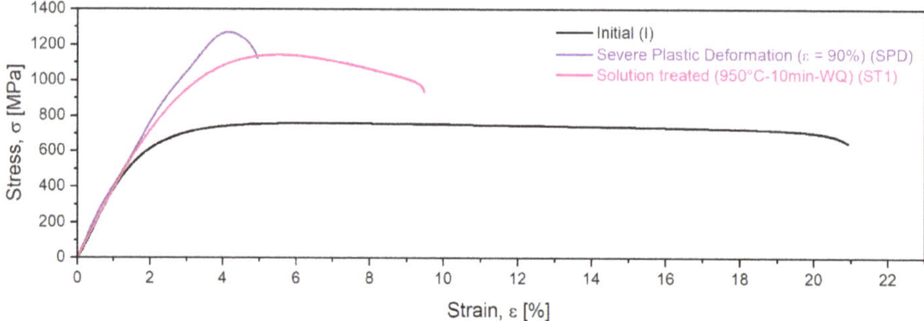

Figure 13. The stress–strain curve of the samples processed by the 2nd direction of the TMP program: SPD-MPR (ε_{tot} = 60%) + ST1 (1223 K-10 min-WQ).

Table 3. Mechanical properties of the studied alloy at the end of the 2nd direction of experiments: Ultimate Tensile Strength (σ_{UTS}); Yield Strength ($\sigma_{0.2}$); Elongation to Fracture (ε_f); Elastic Modulus (E); SD (Standard Deviation).

Structural State	Mechanical Properties			
	σ_{UTS} **(SD) (MPa)**	$\sigma_{0.2}$ **(SD) (MPa)**	ε_f **(SD) (%)**	**E (SD) (GPa)**
Initial state (I)	759.7 (13.3)	546.1 (10.6)	20.9 (0.6)	49.1 (1.6)
SPD by MPR—(ε_{tot} = 60%)	1270.7 (16,4)	1011.2 (11.2)	4.9 (0.1)	50.1 (1.4)
SPD + ST1 (1223 K-10 min-w.q.)	1143.8 (16.8) (↑50.6%)	772.6 (11.4) (↑41.5%)	9.5 (0.3) (↓54.5%)	48.7 (0.9) (↓0.8%)
SPD + ST2 (1223 K-20 min-w.q.)	1035.5 (15.9) (↑36.3%)	701.1 (10.1) (↑28.4%)	6.9 (0.2) (↓66.9%)	49.2 (1.1) (↑0.2%)

In Table 3, it can be observed that the two samples with SPD + ST provide much greater UTS and YS values (almost 50%) compared to results from the first category of experiments (CR + ST) and the initial state. This time, the UTS increased by 36.3% and 50.6%, respectively, while the YS increased by 28.4% and 41.5%, respectively. At the same time, the values of Young's modulus remained almost the same, and the elongation to fracture decreased significantly (by 54.5% and 66.9%, respectively). The obtained β-grain dimensions (77 µm and 91 µm, respectively) are situated in between the 1st step of the 1st experiments, at the bottom of the obtained values (138 → 128 → 97 → 89 → 81 → 75 µm—Figure 7), and those obtained from the 2nd steps of the 1st experiments, at the top of the obtained values (55 → 62 → 75 → 84 µm, Figure 9).

Comparing the mechanical properties obtained from the two research directions of the experimental program (Table 2 versus Table 3), the following observations can be underlined. The first direction of the experiments involved more processing steps than the second direction (two series of CR with two series of ST), but the result was a smaller β-grain dimension (55 µm versus 77 µm). The results for the Young's modulus were similar for both experimental directions (about 48–49 GPa), indicating a very good achievement. The results for YS and UTS were slightly higher (about 150–200 MPa) for the samples processed by SPD-MPR compared to those processed two times by CR. Nevertheless, the ε_f values were better/higher for the samples processed by CR. Thus, both processing schemas can provide good results, depending on the final specific application requirements.

3.4. Comparison Between the Experimental Alloy and Other Published Results for Similar Alloys

To better appreciate the obtained experimental results, this analysis includes not only a presentation and discussion of the obtained results but also a comparison with other published data for similar alloys. These data are shown in Table 4.

Table 4. Mechanical properties of the studied alloy and those of other similar alloys.

Studied Alloy: Ti-34.1Nb-7.59Zr-0.9Fe-0.16O	σ_{UTS} (MPa)	$\sigma_{0.2}$ (MPa)	ε_f (%)	E (GPa)
Samples from 1st direction of the TMP program:	-	-	-	-
CR + ST1-(1223 K-5 min-w.q.)	962.8	703.7	6.9	48.3
CR + ST2-(1223 K-10 min-w.q.)	901.7	631.4	8.2	47.6
CR + ST3-(1223 K-15 min-w.q.)	879.5	601.6	11.1	47.8
CR + ST4-(1223 K-20 min-w.q.)	809.0	574.4	15.9	49.3
Samples from 2nd direction of the TMP program:				
SPD + ST1-(1223 K-10 min-w.q.)	1143.8	772.6	9.5	48.7
SPD + ST2-(1223 K-20 min-w.q.)	1035.5	701.1	6.9	49.2
Other reported data:	-	-	-	-
Furuta et al. [47]: Ti-32Nb-2Ta-3Zr-0.5O (ST)	1370	-	12	55
Hussein et al. [31]: Ti-24.96Nb-17.15Ta-0.88Fe-0.25O (CR-50%)	851.1	-	11.1	60
M.A.-H. Gepreel [48]: Ti-20Zr-10Nb-3Ta-1Fe-1O (CR-90%) Ti-20Zr-10Nb-3Ta-1Fe-1O (ST)	- -	1198 784	- -	65 50
Strasky et al. [32]: Ti-35.3Nb-5.7Ta-7.3Zr-2Fe-0.4O (forged) Ti-35.3Nb-5.7Ta-7.3Zr-0.4O (forged) Ti-35.3Nb-5.7Ta-7.3Zr-0.7O (forged)	1130 903 1217	817 860 1017	28 16 21	107 81 80

By highlighting the positive achievements and possible negative points, the following observations can be underlined. The first observation, common to all anterior reported alloys, relates to the presence of expensive Tantalum, which is absent in the alloy that was studied here. The reported alloys were in an ST condition (like the present alloy) in only two studies—Furuta et al. [47] and M.A.-H. Gepreel [48]. Compared to these two cases, the obtained results are similar, i.e., they have very good limits. It can be observed that the modulus is even lower for the reported results (~48 GPa compared to 50–55 GPa). Further, compared with Hussein et al. [31] and Strasky et al. [32], who reported similar alloys (but in CR or forged states with Ta), the studied alloy in an ST state (both experimental research directions) provides superior characteristics, i.e., the ST state achieved almost the same positive values as those of the reported CR or forged alloys. Usually, β-type alloys in an ST state are expected to have lower values for YS and UTS than alloys in a severely deformed state; here, the obtained values were very close to similar alloys in a severely deformed state, proving that the addition of Fe and O is beneficial for improving mechanical properties [18,31–33]. In addition, for the studied alloy, the Young's modulus was much lower (~48 GPa) compared to the reported results (60, 80, or even 107 GPa).

As a conclusion, the mechanical parameters presented in Table 4 show that our experimental results are in good agreement with the values reported in the literature.

4. Conclusions

1. A stable β-Ti alloy–Ti-34.1Nb-7.59Zr-0.9Fe-0.16O—Was subjected to two TMP processing directions with variable parameters to obtain mechanical properties suitable for orthopedic implants with long-term service.
2. The amount of the β-stabilizing alloying elements was sufficient to obtain a single β-phase, without any subsequent generation of the secondary phases that are possible to be formed during complex TMP and can decrease the processability of the alloy.
3. The deformability of the studied alloy in the monophasic bcc β-status allowed the application of high deformation degrees for the CR process, as well as for SPD.
4. The ST parameters in both TMP variants led to the restoration of a fully recrystallized β phase microstructure from the cold rolled state, with a small β grain size and a reduced Young's modulus.
5. The structural deformation mechanisms during cold-rolling were predicted by SEM analysis, but these mechanisms must be proven by future detailed analyses.
6. The selected chemical composition and TMP parameters can obtain promising mechanical properties with high levels of YS (about 600–770 MPa) and UTS (about 900–1140 MPa), and low values for the Young's modulus (around 48 GPa, very close to the 30 GPa of the cortical bone), comparable to or even better than other data reported for similar alloys.
7. Both TMP methods applied to the studied alloy provided good final results, including a low β-grain size, low modulus, and high mechanical strength, all for a single β-phase, which can ensure adequate subsequent processability for the final implant shape.
8. In these obtained experimental results, the database for the mechanical properties of Ti-Nb-Zr-Fe-O family alloys was extended to a beneficial application as biomaterials for human implants with long-term service.

Author Contributions: Conceptualization, V.D.C.; methodology, A.N.; software, A.D.; validation, D.R. and A.N.; formal analysis, B.M.G.; investigation, C.T.-R.; writing—original draft preparation, A.N. and D.R.; writing—review and editing, A.N.; visualization, R.I.; supervision, V.D.C.; project administration, D.R. All authors have read and agreed to the published version of the manuscript.

Funding: This research was funded by the Romanian National Authority for Scientific Research CCCDI–UEFISCDI, grant no. 143/2020, within the Program PNCDI-III-EraNet-MANUNET-III.

References

1. Niinomi, M.; Yi, L.; Nakai, M.; Liu, H.; Hua, L. Biomedical titanium alloys with Young's moduli close to that of cortical bone. *Regener. Biomater.* **2016**, *3*, 173–185. [CrossRef] [PubMed]
2. Gepreel, M.A.-H.; Niinomi, M. Biocompatibility of Ti-alloys for long-term implantation. *J. Mech. Behav. Biomed. Mater.* **2013**, *20*, 407–415. [CrossRef] [PubMed]
3. Fu, Y.; Wang, J.; Xiao, W.; Zhao, X.; Ma, C. Microstructure evolution and mechanical properties of Ti–8Nb–2Fe-0.2O alloy with high elastic admissible strain for orthopedic implant applications. *Prog. Nat. Sci.-Mater. Int.* **2020**, *30*, 100–105. [CrossRef]
4. Ozan, S.; Lin, J.; Zhang, Y.; Li, Y.; Wen, C. Cold rolling deformation and annealing behavior of a β-type Ti–34Nb–25Zr titanium alloy for biomedical applications. *J. Mater. Res. Technol.* **2020**, *92*, 2308–2318. [CrossRef]

5. Raducanu, D.; Cojocaru, V.D.; Nocivin, A.; Cinca, I.; Serban, N.; Cojocaru, E.M. Surface Modifications of a Biomedical Gum-Metal Type Alloy by Nano Surface-Severe Plastic Deformation. *JOM* **2019**, *71*, 4114–4124. [CrossRef]

6. Kolli, R.P.; Devaraj, A. A Review of Metastable Beta Titanium Alloys. *Metals* **2018**, *8*, 506. [CrossRef]

7. Chen, W.; Li, C.; Feng, K.; Lin, Y.; Zhang, X.; Chen, C.; Zhou, K. Strengthening of a Near β-Ti Alloy through β Grain Refinement and Stress-Induced α Precipitation. *Materials* **2020**, *13*, 4255. [CrossRef] [PubMed]

8. Du, Z.; Ma, Y.; Liu, F.; Xu, N.; Chen, Y.; Wang, X.; Chen, Y.; Gong, T.; Xu, D. The Influences of Process Annealing Temperature on Microstructure and Mechanical Properties of near β High Strength Titanium Alloy Sheet. *Materials* **2019**, *12*, 1478. [CrossRef]

9. Wang, W.; Xu, X.; Ma, R.; Xu, G.; Liu, W.; Xing, F. The Influence of Heat Treatment Temperature on Microstructures and Mechanical Properties of Titanium Alloy Fabricated by Laser Melting Deposition. *Materials* **2020**, *13*, 4087. [CrossRef]

10. Lan, C.; Wu, Y.; Guo, L.; Chen, H.; Chen, F. Microstructure, texture evolution and mechanical properties of cold rolled Ti-32.5Nb-6.8Zr-2.7Sn biomedical β-Ti alloy. *J. Mater. Sci. Technol.* **2018**, *34*, 788–792. [CrossRef]

11. Dong, R.; Li, J.; Kou, H.; Fan, J.; Tang, B. Dependence of mechanical properties on the microstructure characteristics of a near β-Ti alloy Ti-7333. *J. Mater. Sci. Technol.* **2019**, *35*, 48–54. [CrossRef]

12. Lei, X.; Dong, L.; Zhang, Z.; Liu, Y.; Hao, Y.; Yang, R.; Zhang, L.-C. Microstructure, Texture Evolution and Mechanical Properties of VT3-1 Titanium Alloy Processed by Multi-Pass Drawing and Subsequent Isothermal Annealing. *Metals* **2017**, *7*, 131. [CrossRef]

13. Ma, Y.; Du, Z.; Cui, X.; Cheng, J.; Liu, G.; Gong, T.; Liu, H.; Wang, X.; Chen, Y. Effect of cold rolling process on microstructure and mechanical properties of high strength β titanium alloy thin sheets. *Prog. Nat. Sci. Mater. Int.* **2018**, *28*, 711–717. [CrossRef]

14. Mohammed, M.T.; Khan, Z.A.; Siddiquee, A.N. Beta Titanium Alloys: The Lowest Elastic Modulus for Biomedical Applications: A Review; World Academy of Science Engineering and Technology. *Int. J. Chem. Mol. Nucl. Mater. Metall. Eng.* **2014**, *8*, 822–827.

15. Gupta, A.; Khatirkar, R.K.; Kumar, A.; Parihar, M.S. Investigations on the effect of heating temperature and cooling rate on evolution of microstructure in an α + β titanium alloy. *Mater. Res. Soc.* **2018**, *33*, 946–957. [CrossRef]

16. Padmalatha, T.S.R.V.; Chakkingal, U. The effect of heat treatment and the volume fraction of the alpha phase on the workability of Ti-5Al-5Mo-5V-3Cr alloy. *J. Mater. Eng. Perform.* **2019**, *28*, 5352–5360. [CrossRef]

17. Xu, P.; Zhou, L.; Han, M.; Wei, Z.; Liang, Y. Flash-butt welded Ti6242 joints preserved base-material strength and ductility. *Mater. Sci. Eng. A* **2020**, *774*, 138915. [CrossRef]

18. Terlinde, G.; Fischer, G. Beta titanium alloys. In *Titanium and Titanium Alloys—Fundamentals and Applications*; Leyens, C., Peters, M., Eds.; Wiley-VCH: Weinheim, Germany, 2003; pp. 37–59.

19. Cordeiro, J.M.; Beline, T.; Ribeiro, A.L.R.; Rangel, E.C.; da Cruz, N.C.; Landers, R.; Faverani, L.P.; Vaz, L.G.; Fais, L.M.; Vicente, F.B.; et al. Development of binary and ternary titanium alloys for dental implants. *Dent. Mater.* **2017**, *33*, 1244–1257. [CrossRef]

20. Mohammed, M.T. Development of a new metastable beta titanium alloy for biomedical applications. *Karbala Int. J. Mod. Sci.* **2017**, *3*, 224–230. [CrossRef]

21. Niinomi, M.; Nakai, M.; Hieda, J. Development of new metallic alloys for biomedical applications. *Acta Biomater.* **2012**, *8*, 3888–3903. [CrossRef]

22. Mantri, S.A.; Banerjee, R. Microstructure and micro-texture evolution of additively manufactured β-Ti alloys. *Addit. Manuf.* **2018**, *23*, 86–98. [CrossRef]

23. Hafeez, N.; Liu, J.; Wang, L.; Wei, D.; Tang, Y.; Lu, W.; Zhang, L.C. Superelastic response of low-modulus porous beta-type Ti-35Nb-2Ta-3Zr alloy fabricated by laser powder bed fusion. *Addit. Manuf.* **2020**, *34*, 101264. [CrossRef]

24. Ozan, S.; Lin, J.; Li, Y.; Wen, C. New Ti-Ta-Zr-Nb alloys with ultrahigh strength for potential orthopedic implant applications. *J. Mech. Behav. Biomed. Mater.* **2017**, *75*, 119–127. [CrossRef]

25. Liu, H.; Niinomi, M.; Nakai, M.; Cho, K. β-Type titanium alloys for spinal fixation surgery with high Young's modulus variability and good mechanical properties. *Acta Biomater.* **2015**, *24*, 361–369. [CrossRef]

26. Zheng, Y.; Williams RE, A.; Nag, S.; Banerjee, R.; Fraser, H.L.; Banerjee, D. The effect of alloy composition on instabilities in the β phase of titanium alloys. *Scr. Mater.* **2016**, *116*, 49–52. [CrossRef]

27. Li, Y.; Yang, C.; Zhao, H.; Qu, S.; Li, X.; Li, Y. New Developments of Ti-Based Alloys for Biomedical Applications. *Materials* **2014**, *7*, 1709. [CrossRef]

28. Kalaie, M.A.; Zarei-Hanzaki, A.; Ghambari, M.; Dastur, P.; Málek, J.; Farghadany, E. The effects of second phases on super elastic behavior of TNTZ bio alloy. *Mater. Sci. Eng. A* **2017**, *703*, 513–520. [CrossRef]

29. Wu, C.; Zhan, M. Microstructural evolution, mechanical properties and fracture toughness of near β titanium alloy during different solution plus aging heat treatments. *J. Alloys Compd.* **2019**, *805*, 1144–1160. [CrossRef]

30. Nakai, M.; Niinomi, M.; Akahori, T.; Tsutsumi, H.; Ogawa, M. Effect of Oxygen Content on Microstructure and Mechanical Properties of Biomedical Ti-29Nb-13Ta-4.6Zr Alloy under Solutionized and Aged Conditions. *Mater. Trans.* **2009**, *50*, 2716–2720. [CrossRef]

31. Hussein, A.H.; Gepreel, M.A.-H.; Gouda, M.K.; Hefnawy, A.M.; Kandil, S.H. Biocompatibility of new Ti–Nb–Ta base alloys. *Mater. Sci. Eng. C* **2016**, *61*, 574–578. [CrossRef]

32. Strasky, J.; Harcuba, P.; Vaclavova, K.; Horvath, K.; Landa, M.; Srba, O.; Janecek, M. Increasing strength of a biomedical Ti-Nb-Ta-Zr alloy by alloying with Fe, Si and O. *J. Mech. Behav. Biomed. Mater.* **2017**, *71*, 329–336. [CrossRef]

33. Nocivin, A.; Cojocaru, V.D.; Raducanu, D.; Cinca, I.; Angelescu, M.L.; Dan, I.; Serban, N.; Cojocaru, M. Finding an Optimal Thermo-Mechanical Processing Scheme for a Gum-type Ti-Nb-Zr-Fe-O Alloy. *J. Mater. Eng. Perform.* **2017**, *26*, 4373–4380. [CrossRef]

34. Biesiekierski, A.; Lin, J.; Li, Y.; Ping, D.; Yamabe-Mitarai, Y.; Wen, C. Investigations into Ti-(Nb,Ta)-Fe Alloys for Biomedical Applications. *Acta Biomater.* **2016**, *32*, 336–347. [CrossRef]

35. Kopova, J.; Strasky, P.; Harcuba, M.; Landa, M.; Janecek, M.; Bacakova, L. Newly developed Ti-Nb-Zr-Ta-Si-Fe biomedical beta titanium alloys with increased strength and enhanced biocompatibility. *Mater. Sci. Eng. C* **2016**, *60*, 230–238. [CrossRef]

36. Zhang, D.C.; Mao, Y.F.; Li, Y.L.; Li, J.J.; Yuan, M.; Lin, J.G. Effect of ternary alloying elements on microstructure and superelasticity of Ti–Nb alloys. *Mater. Sci. Eng. A* **2013**, *559*, 706–710. [CrossRef]

37. Li, Q.; Miao, P.; Li, J.; He, M.; Nakai, M.; Niinomi, M.; Chiba, A.; Nakano, T.; Liu, X.; Zhou, K.; et al. Effect of Nb Content on Microstructures and Mechanical Properties of Ti-xNb-$_2$Fe Alloys. *J. Mater. Eng. Perform.* **2019**, *28*, 5501–5508. [CrossRef]

38. Ehtemam, H.S.; Attar, H.; Okulov, I.V.; Dargusch, M.S.; Kent, D. Microstructural evolution and mechanical properties of bulk and porous low-cost Ti–Mo–Fe alloys produced by powder metallurgy. *J. Alloys Compd.* **2020**, *853*, 156768. [CrossRef]

39. Park, C.H.; Park, J.W.; Yeom, J.T.; Chun, Y.S.; Lee, C.S. Enhanced mechanical compatibility of submicrocrystalline Ti-13Nb-13Zr alloy. *Mater. Sci. Eng. A* **2010**, *527*, 4914–4919. [CrossRef]

40. Hsu, H.C.; Hsu, S.K.; Wu, S.C.; Lee, C.J.; Ho, W.F. Structure and mechanical properties of as-cast Ti–$_5$Nb–xFe alloys. *Mater. Charact.* **2010**, *61*, 851–858. [CrossRef]

41. Cui, W.F.; Guo, A.H. Microstructure and properties of biomedical TiNbZrFe β-titanium alloy under aging conditions. *Mater. Sci. Eng. A* **2009**, *527*, 258–262. [CrossRef]

42. Gordin, D.M.; Ion, R.; Vasilescu, C.; Drob, S.I.; Cimpean, A.; Gloriant, T. Potentiality of the "Gum Metal" titanium-based alloy for biomedical applications. *Mater. Sci. Eng. C* **2014**, *44*, 362–370. [CrossRef] [PubMed]

43. Lei, Z.; Liu, X.; Wu, Y.; Wang, H.; Jiang, S.; Wang, S.; Hui, X.; Wu, Y.; Gault, B.; Kontis, P.; et al. Enhanced strength and ductility in a high-entropy alloy via ordered oxygen complexes. *Nature* **2018**, *563*, 546–550. [CrossRef]

44. Liu, H.; Niinomi, M.; Nakai, M.; Cong, X.; Cho, K.; Boehlert, C.J.; Khademi, V. Abnormal Deformation Behavior of Oxygen-Modified β-Type Ti-29Nb-13Ta-4.6Zr Alloys for Biomedical Applications. *Metall. Mater. Trans. A* **2016**, *48*, 139–149. [CrossRef]

45. Zafari, A.; Ding, Y.; Cui, J.; Xia, K. Achieving fine beta grain structure in a metastable beta titanium alloy through multiple forging-annealing cycles. *Metal. Mater. Trans. A* **2016**, *47*, 3633–3648. [CrossRef]

46. Bertrand, E.; Gloriant, T.; Gordin, D.M.; Vasilescu, E.; Drob, P.; Vasilescu, C.; Drob, S.I. Synthesis and characterisation of a new superelastic Ti–25Ta–25Nb biomedical alloy. *J. Mech. Behav. Biomed. Mater.* **2010**, *3*, 559–564. [CrossRef]

47. Furuta, T.; Kuramoto, S.; Hwang, J.; Nishino, K.; Saito, T.; Niinomi, M. Mechanical Properties and Phase Stability of Ti-Nb-Ta-Zr-O Alloys. *Mater. Trans.* **2007**, *48*, 1124–1130. [CrossRef]

48. Gepreel, M.A.-H. Improved elasticity of new Ti-Alloys for Biomedical Applications. *Mater. Today Proceed.* **2015**, *2*, S979–S982. [CrossRef]

49. Ozan, S.; Lin, J.; Li, Y.; Ipek, R.; Wen, C. Development of Ti–Nb–Zr alloys with high elastic admissible strain for temporary orthopedic devices. *Acta Biomater.* **2015**, *20*, 176–187. [CrossRef]

50. Biesiekierski, A.; Lin, J.X.; Munir, K.; Ozan, S.; Li, Y.C.; Wen, C.E. An investigation of the mechanical and microstructural evolution of a TiNbZr alloy with varied ageing time. *Sci. Rep.* **2018**, *8*, 5737. [CrossRef] [PubMed]

51. Shekhar, S.; Sarkar, R.; Kar, S.K.; Bhattacharjee, A. Effect of solution treatment and aging on microstructure and tensile properties of high strength β titanium alloy Ti-5Al-5V-5Mo-3Cr. *Mater. Des.* **2015**, *66*, 596–610. [CrossRef]

52. Ozan, S.; Lin, J.; Li, Y.; Zhang, Y.; Munir, K.; Jiang, H.; Wen, C. Deformation mechanism and mechanical properties of a thermo-mechanically processed β Ti–28Nb–35.4Zr alloy. *J. Mech. Behav. Biomed. Mater.* **2018**, *78*, 224–234. [CrossRef]

53. Ozan, S.; Li, Y.C.; Lin, J.X.; Zhang, Y.W.; Jiang, H.W.; Wen, C.E. Microstructural evolution and its influence on the mechanical properties of a thermos-mechanically processed beta Ti-32Zr-30Nb alloy. *Mater. Sci. Eng. A-Struct. Mater. Prop. Microstruct. Process* **2018**, *719*, 112–123. [CrossRef]

54. Tang, B.; Chu, Y.; Zhang, M.; Meng, C.; Fan, J.; Kou, H.; Li, J. The ω phase transformation during the low temperature aging and low rate heating process of metastable β titanium alloys. *Mater. Chem. Phys.* **2020**, *239*, 122–125. [CrossRef]

55. Ali, T.; Wang, L.; Cheng, X.; Liu, A.; Xu, X. Omega phase formation and deformation mechanism in heat treated Ti-5553 alloy under high strain rate compression. *Mater. Lett.* **2019**, *236*, 163–166. [CrossRef]

56. Hu, Q.-M.; Li, S.-J.; Hao, Y.-L.; Yang, R.; Johansson, B.; Vitos, L. Phase stability and elastic modulus of Ti alloys containing Nb, Zr, and/or Sn from first-principles calculations. *Appl. Phys. Lett.* **2008**, *93*, 121902. [CrossRef]

57. Correa, D.R.N.; Vicente, F.B.; Donato, T.A.G.; Arana-Chavez, V.E.; Buzalaf, M.A.R.; Grandini, C.R. The effect of the solute on the structure, selected mechanical properties, and biocompatibility of Ti–Zr system alloys for dental applications. *Mater. Sci. Eng. C* **2014**, *34*, 354–359. [CrossRef]

58. Banerjee, D.; Williams, J.C. Perspectives on Titanium Science and Technology. *Acta Mater.* **2013**, *61*, 844–879. [CrossRef]

59. Yang, Y.; Wu, S.Q.; Li, G.P.; Li, Y.L.; Lu, Y.F.; Yang, K.; Ge, P. Evolution of deformation mechanisms of Ti-22.4Nb-0.73Ta-2Zr-1.34O alloy during straining. *Acta Mater.* **2010**, *58*, 2778–2787. [CrossRef]

60. Besse, M.; Castany, P.; Gloriant, T. Mechanisms of deformation in gum metal TNTZ-O and TNTZ titanium alloys: A comparative study on the oxygen influence. *Acta Mater.* **2011**, *59*, 5982–5988. [CrossRef]

61. Furuta, T.; Kuramoto, S.; Hwang, J.; Nishino, K.; Saito, T. Elastic deformation behavior of multi-functional Ti-Nb-Ta-Zr-O alloys. *Mater. Trans.* **2005**, *46*, 3001–3007. [CrossRef]

62. Tan, M.H.C.; Baghi, A.D.; Ghomashchi, R.; Xiao, W.; Oskouei, R.H. Effect of niobium content on the microstructure and Young's modulus of Ti-xNb-7Zr alloys for medical implants. *J. Mech. Behav. Biomed. Mater.* **2019**, *99*, 78–85. [CrossRef]

63. Ozan, S.; Lin, J.; Weng, W.; Zhang, Y.; Li, Y.; Wen, C. Effect of thermomechanical treatment on the mechanical and microstructural evolution of a β-type Ti-40.7Zr-24.8Nb alloy. *Bioact. Mater.* **2019**, *4*, 303–311. [CrossRef]

64. Wang, L.; Lu, W.; Qin, J.; Zhang, F.; Zhang, D. Microstructure and mechanical properties of cold-rolled TiNbTaZr biomedical β-titanium alloy. *Mater. Sci. Eng. A* **2008**, *490*, 421–426. [CrossRef]

65. Okulov, I.V.; Wendrock, H.; Volegov, A.S.; Attar, H.; Kühn, U.; Skrotzki, W.; Eckert, J. High strength beta titanium alloys: New design approach. *Mater. Sci. Eng. A* **2015**, *628*, 297–302. [CrossRef]

66. Xie, K.Y.; Wang, Y.; Zhao, Y.; Chang, L.; Wang, G.; Chen, Z.; Cao, Y.; Liao, X.; Lavernia, E.J.; Valiev, R.B.; et al. Nanocrystalline β-Ti alloy with high hardness, low Young's modulus and excellent in vitro biocompatibility for biomedical applications. *Mater. Sci. Eng. C* **2013**, *33*, 3530–3536. [CrossRef]

67. Tahara, M.; Kim, H.Y.; Inamura, T.; Hosoda, H.; Miyazaki, S. Role of interstitial atoms in the microstructure and non-linear elastic deformation behavior of Ti–Nb alloy. *J. Alloys Compd.* **2013**, *577*, S404–S407. [CrossRef]

68. Wei, L.S.; Kim, H.Y.; Miyazaki, S. Effects of oxygen concentration and phase stability on nano-domain structure and thermal expansion behavior of Ti–Nb–Zr–Ta–O alloys. *Acta Mater.* **2015**, *100*, 313–322. [CrossRef]

69. Wei, Q.; Wang, L.; Fu, Y.; Qin, J.; Lu, W.; Zhang, D. Influence of oxygen content on microstructure and mechanical properties of Ti–Nb–Ta–Zr alloy. *Mater. Des.* **2011**, *32*, 2934–2939. [CrossRef]

Publisher's Note: MDPI stays neutral with regard to jurisdictional claims in published maps and institutional affiliations.

Article

Effect of Fe Content on the As-Cast Microstructures of Ti–6Al–4V–xFe Alloys

Ling Ding [1], Rui Hu [2], Yulei Gu [1], Danying Zhou [1], Fuwen Chen [1], Lian Zhou [1] and Hui Chang [1,*]

[1] Tech Institute for Advanced Materials & College of Materials Science and Engineering, Nanjing Tech University, Nanjing 210009, China; dingling2013@njtech.edu.cn (L.D.); 825721910@njtech.edu.cn (Y.G.); zhoudanying@njtech.edu.cn (D.Z.); fuwenchen@njtech.edu.cn (F.C.); zhoul@c-nin.com (L.Z.)

[2] State Key Laboratory of Solidification Processing, Northwestern Polytechnical University, Xi'an 710072, China; rhu@nwpu.edu.cn

* Correspondence: ch2006@njtech.edu.cn; Tel.: +86-13813916521

Received: 14 June 2020; Accepted: 18 July 2020; Published: 22 July 2020

Abstract: In this work, the evolution of the solidification microstructures of Ti–6Al–4V–xFe (x = 0.1, 0.3, 0.5, 0.7, 0.9) alloys fabricated by levitation melting was studied by combined simulative and experimental methods. The growth of grains as well as the composition distribution mechanisms during the solidification process of the alloy are discussed. The segregation of the Fe element at the grain boundaries promotes the formation of a local composition supercooling zone, thus inhibiting the mobility of the solid–liquid interface and making it easier for the grains to grow into dendrites. With the increase in Fe content, the grain size of the alloy decreased gradually, while the overall decreasing trend was mitigated. The segregation of Fe was more obvious than that of Al and V, and the increase in Fe content had less effect on the segregation of Al and V.

Keywords: titanium alloy; simulation; boundary; segregation

---

1. Introduction

Titanium alloys have become excellent structural materials in many fields in recent years, especially in the field of aerospace applications due to their high specific strength, corrosion resistance, and other advantages [1]. Fe, as a common β-eutectoid alloy element in titanium alloys, which is even stronger than Cr, has a great influence on the solid/liquid transformation point. The increase in Fe content may cause a β-spot. Generally, the Fe content in titanium alloy is less than 5.5 wt% [2].

According to the previous research, the mechanical properties of titanium alloys can be effectively improved by adding an appropriate amount of Fe [3,4]. Kudo et al. studied the influence of microstructure on the formability of a Ti–Fe alloy [5] and found that the formability of a Ti–Fe alloy increased with the decrease in the size of the prior β phase region. Bermingham et al. found that the addition of an appropriate amount of Fe can effectively refine the grains of titanium alloys [6]. It was considered that the segregation of Fe provided the undercooling needed to inhibit grain growth and activate adjacent nuclei. It is obvious that the addition of an appropriate amount of Fe in the titanium alloys can affect the morphology of the original beta grains during solidification, thus influencing the mechanical properties. Ehtemam designed and manufactured Ti–11Nb–xFe (x = 0.5, 3.5, 6, 9 wt%) alloys by cold crucible levitation melting to study the effect of Fe addition on its phase transformation, microstructure, and mechanical properties [7]. The results showed that the Ti–11Nb–0.5Fe alloy had a typical dual phase microstructure of α + β and the volume fraction of the β phase could be increased by increasing the Fe content. However, the formation and growth of the original beta grains during the solidification process of titanium alloys are difficult to observe experimentally, so it is not easy to verify the mechanism of Fe on the grain morphology.

Through the phase field simulation, the microstructure evolution during the solidification process as well as the influence of element content on the microstructure of the alloys can be examined. The phase field model is a powerful tool to describe the complex evolution of the interface between the matrix and new phases in the non-equilibrium state based on the unified control equations in the whole system [8,9], which is suitable for describing solid–liquid phase transformation [10]. However, the simulation of microstructure evolution with the phase field method relies on the data of the temperature field parameters and thermophysical parameters of the related elements. The electromagnetic-thermal coupled simulation conducted by Kermanpur et al. [11] and the multi physical field coupling simulation conducted by Li [12] provided the data needed for the temperature field of the microstructure simulation.

For the simulation of microstructure, Kundin et al. used the phase field method to simulate the solidification of the Ti–Fe alloy [13], and Gong et al. studied the microstructure evolution of a Ti–6Al–4V alloy by the phase field method [14–16]. As for the related thermophysical parameters, Nakajima used the tracer diffusion method and Mossbauer spectrum to study the diffusion of Fe in the β-titanium alloy [17]. It is considered that the diffusion mechanism of Fe in β-titanium alloy is an extremely rapid interstitial diffusion. Chen et al. used the DICTRA software (Thermo-Calc Software Solna, Sweden) to strictly evaluate the experimental diffusion data to determine the atomic mobility of the BCC phase in the Ti–Al–Fe system [18]. Through the comprehensive comparison between the calculated and the experimental diffusion coefficients, a better consistency is obtained. The developed mobility of atoms is verified by good prediction of the mutual diffusion behavior observed in the diffusion couple experiment in the existing literatures.

In this work, the effect of Fe content on the microstructure of Ti–6Al–4V–xFe (x = 0.1, 0.3, 0.5, 0.7, 0.9) alloys produced by levitation melting was studied by the phase field method and verified by experiments. Levitation melting is often used in the laboratory research of titanium alloys due to the small size and uniform composition of the ingot. The Ti–6Al–4V alloy is the most widely used titanium alloy ($\alpha + \beta$ type) with good comprehensive mechanical properties, which is composed of a vanadium rich BCC phase (body centered cubic, β) and aluminum rich HCP phase (hexagonal close packed, α) [19].

2. Model and Experiments

2.1. Phase Field Model

As the temperature change calculated according to the simulation is small, the following assumptions were made:

(1)   The diffusion coefficients of Al, V, and Fe in the solid phase and the liquid phase did not change in the simulation.
(2)   The temperature gradient and cooling rate in the whole process remained invariant.

Dendritic growth and grain growth models were established using MICRESS 6.3 (ACCESS e.V. Aachen, Germany) software. The dendrite growth model had a mesh size of 600 × 600, a cell resolution of 0.1 μm, and a minimum time step of 1×10^{-3} s. The initial condition was considered to be 1 for the initial grain. The grain growth model had a grid size of 1000 × 1000, a cell resolution of 5 μm, and a minimum time step of 1×10^{-2} s. Figure 1 shows a schematic of the modeled domain, which was in the middle of ingot. Set 10 initial grain levels to randomly generate grains according to grain radius and distribution density.

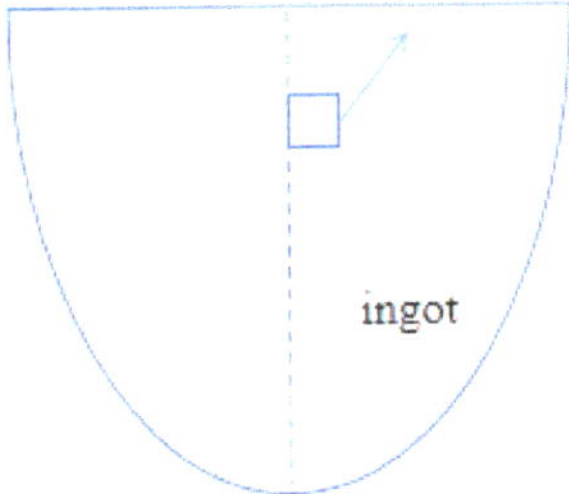

Figure 1. Schematic of the modeled domain.

Multiphase field theory is a computational method to describe the evolution of multiphase field parameters $\varphi_\alpha(\vec{x}, t)$ in time and space [20]. At the solid–liquid interface, 0 and 1 represent the liquid phase and solid phase, respectively, and φ_α changes continuously between 0 and 1 with an interface thickness η. Based on the principle of minimum free energy, the multiphase field equation of MICRESS was used [21]:

$$\dot{\varphi}_\alpha = \sum_\beta M_{\alpha\beta}(\vec{n})\left(\sigma^*_{\alpha\beta}(\vec{n})K_{\alpha\beta} + \frac{\pi}{\eta}\sqrt{\varphi_\alpha\varphi_\beta}\Delta G_{\alpha\beta}(\vec{c}, T)\right) \tag{1}$$

$$K_{\alpha\beta} = \varphi_\beta\nabla^2\varphi_\alpha - \varphi_\alpha\nabla^2\varphi_\beta + \frac{\pi^2}{\eta^2}(\varphi_\alpha - \varphi_\beta) \tag{2}$$

where $M_{\alpha\beta}$ is the mobility of the interface of the interface orientation, described by the normal vector $\vec{n}$. $\sigma^*_{\alpha\beta}$ is the effective anisotropic surface energy, and $K_{\alpha\beta}$ is about the local curvature of the interface. $\Delta G_{\alpha\beta}$ is the thermodynamic driving force, which is a function of the composition $\vec{c}$, and the diffusion equation can be described as:

$$\dot{\vec{c}} = \nabla \sum_{\alpha=1}^{N} \varphi_\alpha\vec{D}_\alpha\nabla\vec{c}_\alpha \tag{3}$$

where $\vec{D}_\alpha$ is the multicomponent diffusion coefficient matrix for phase α.

The boundary conditions are based on the symmetric boundary of the MICRESS software. The phase field value of the boundary element is defined to be the same as the second adjacent element in the analog domain, thereby revealing that a plane of symmetry is crossing through the center of the outermost element of the region. This condition is similar to an isolation condition that moves half a unit. The interface thickness is 5 cells.

The simulated interface energy can use common interface energy [22]. The phase diagram data required for the simulation are directly extracted from the Thermo-Calc 2015b (Thermo-Calc Software Solna, Sweden) TTTi3 database. The solid phase diffusion coefficients of the Al and V are calculated from the MOBTI1 database, and the liquid phase diffusion coefficients of Al and V are estimated. Since there are no diffusion data of Fe in the MOBTI1 database, a kinetic database containing Fe was prepared by Chen's study of β phase diffusion kinetics of a Ti–Al–Fe alloy [18], and the data obtained were imported into MICRESS to calculate the solid phase diffusion coefficient of Fe. The liquid phase diffusion coefficient of Fe was derived from the solid phase diffusion coefficient of Fe with reference to Kundin's study [13]. The partitial physical parameters are shown in Table 1.

Table 1. Partial physical parameters [13,18,22].

Physical Parameters	Ti–6Al–4V–xFe
Interface energy σ (J/cm^2)	2×10^{-5}
Al Liquid diffusion coefficient D_l (cm^2/s)	1.5×10^{-5}
Al Solid diffusion coefficient D_s (cm^2/s)	1.3×10^{-7}
V Liquid diffusion coefficient D_l (cm^2/s)	5×10^{-5}
V Solid diffusion coefficient D_s (cm^2/s)	6.9×10^{-7}
Fe Liquid diffusion coefficient D_l (cm^2/s)	1×10^{-4}
Fe Solid diffusion coefficient D_s (cm^2/s)	2×10^{-5}
Molar volume V (cm^3/mol)	11.2
Calculated temperature T (K)	1950
Anisotropic strength η	0.05

2.2. Experiments

The chemical composition of the Ti–6Al–4V–xFe samples (melted by Levitation melting to obtain a hemispherical ingot of about 800 g with a diameter of 90 mm, furnace cooling) is shown in Table 2.

Table 2. Mass fraction of each element.

Alloys	Al (wt%)	V (wt%)	Fe (wt%)	O (wt%)
Ti–6Al–4V	5.98	4.10	0.03	0.083
Ti–6Al–4V–0.1Fe	5.92	4.05	0.13	0.110
Ti–6Al–4V–0.3Fe	5.99	4.09	0.33	0.084
Ti–6Al–4V–0.5Fe	5.95	4.07	0.52	0.076
Ti–6Al–4V–0.7Fe	5.92	4.02	0.73	0.081
Ti–6Al–4V–0.9Fe	5.99	4.10	0.91	0.033

A 5 mm thick flat plate was cut by wire electrode cutting in the middle of the ingot. Six 15×15 mm squares were cut from the center of the ingots. The samples were electrolytic polished (using $HClO_4$:C_2H_5OH = 3:57 electrolyte) and quickly washed in alcohol and distilled water.

The metallographic photographs were obtained with an optical microscope (OM, Carl Zeiss, Jena, Germany). Line scan and surface scan images of the grain boundary of the Ti–6Al–4V–xFe alloys were obtained by electron probe micro analysis (EPMA, JEOL, Tokyo, Japan). As the primary β grain of the alloy is larger and the grain boundary is finer, the grain boundary is easily confused with the precipitated α lamellae structure, making it difficult to find the grain boundary in backscattered electron (BSE) mode. However, electropolishing (electropolishing is slightly corrosive) and secondary electron image (SEI) mode are used to find the original β grain boundary. Due to the precision limitation of EPMA, when the Fe content is low, it is difficult to measure it accurately. Therefore, Ti–6Al–4V–0.5Fe and Ti–6Al–4V–0.9Fe alloys were selected for surface scanning on the triangular crystal surface, and Ti–6Al-4V–0.5Fe, Ti–6Al–4V–0.7Fe, and Ti-6Al-4V-0.9Fe alloys were selected for line scanning through the grain boundary to obtain the corresponding element concentration distribution.

3. Results and Discussions

3.1. Effect of Fe Content on the Microstructure of Single Crystal

First, we studied the growth of the single grain. In the process of the alloy growing, the solute concentration in the liquid phase at the front of the solid–liquid interface decreased with the increase in distance from the interface, and the corresponding liquidus temperature T_L changed from low to high. When the curve of the liquidus temperature T_L was higher than the actual temperature T_Q line in the liquid phase, the composition supercooled zone will be formed in the liquid phase at the front of the solid–liquid interface.

With the solidification layer moving inward, the heat dissipation ability of the solid phase was gradually weakened. The internal temperature gradient tended to be gentle. The solute atoms in the liquid phase were enriched, so the component supercooling in front of the interface increased. As the distribution coefficient of Al and V elements is close to that of Ti and their content is relatively low, the alloy is similar to pure metal if there is no Fe element in the alloy. Therefore, the component supercooling was not obvious and the grain was nearly plane growth, as shown in Figure 2a (the color bar represents the field parameters field parameters φ, and 0 and 1 represent the liquid phase and solid phase respectively). When the Fe content increased to 0.9 wt%, the component supercooled region at the front of the interface was larger. The protruding part continued to grow into the supercooled liquid phase. At the same time, branches grew on its side, and the grain growth tended to be dendrite. With the increase in Fe content, the growth rate of the whole grain decreased. In a certain concentration range, Fe content has a great influence on the morphology of Ti–6Al–4V grains. As shown in Figure 3, in the early stage of solidification, the grain surface was relatively stable. The solid surface formed a bulge and gradually extended with time to the supercooled zone. Due to the small temperature gradient (5 K/cm) of the suspension melting, equiaxed grains were finally formed.

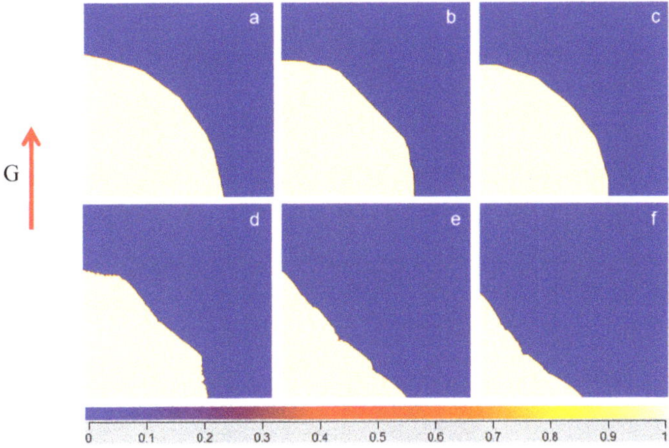

Figure 2. Effect of Fe content on the morphology of a single grain: (**a**) Ti–6Al–4V, (**b**) Ti–6Al–4V–0.1Fe, (**c**) Ti–6Al–4V–0.3Fe, (**d**) Ti–6Al–4V–0.5Fe, (**e**) Ti–6Al–4V–0.7Fe, (**f**) Ti–6Al–4V–0.9Fe.

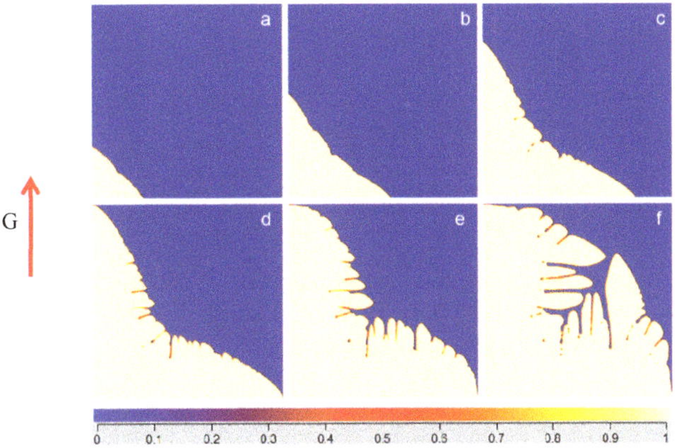

Figure 3. Grain growth with time of Ti–3Al–0.9Fe alloy: (**a**) 4 ms, (**b**) 8 ms, (**c**) 12 ms, (**d**) 16 ms, (**e**) 20 ms, (**f**) 30 ms.

The influence of the increase in Fe content on the component supercooling was discussed. The existence of supercooling zone depends on the temperature gradient at the solid–liquid interface determined by the external heat flux,

$$G = \frac{dT_L}{dx'}\bigg|_{x'=0} \tag{4}$$

where G is the temperature gradient at the solid–liquid interface determined by the external heat flux; T_L is the actual temperature of the liquid phase at the front of the interface; and x' is the direction of the temperature gradient. In equilibrium, there is $G = -mG_c$, where G_c is the concentration gradient. When $G \geq G_c$, the liquid phase in the front interface is in the state of component supercooling. According to the study of Kurz et al. [23], by assuming that there is no convection in the liquid phase and only diffusion, the critical condition of component supercooling can be rewritten as

$$\frac{G}{v} \geq \frac{mC_0(1-k_0)}{Dk_0} \tag{5}$$

where m is the slope of liquidus; D is the diffusion coefficient of liquid phase; v is the migration rate of interface; C_0 is the initial composition; and k_0 is the distribution coefficient. For the Ti–6Al–4V–xFe alloy in this paper, if Ti is the solvent and Al, V, and Fe are the solute, then the liquid surface is a function of the concentration of Al, V, and Fe in the liquid phase, $T_L = T_L(C_{Al}, C_V, C_{Fe})$. The temperature gradient of the liquid melting point at the solid–liquid interface is:

$$\frac{dT_l}{dx'}\bigg|_{x'=0} = m_{Al}\left(\frac{dC_{Al}}{dx'}\right)_{x'=0} + m_V\left(\frac{dC_V}{dx'}\right)_{x'=0} + m_{Fe}\left(\frac{dC_{Fe}}{dx'}\right)_{x'=0} \tag{6}$$

where m_{Al} is slope of the liquidus of C_{Al}, $m_{Al} = \frac{dT_L}{dC_{Al}}$, $m_V = \frac{dT_L}{dC_V}$, and $m_{Fe} = \frac{dT_L}{dC_{Fe}}$.

In the equilibrium state, the solute mass at the solid–liquid interface is conserved. Assuming that there is no interaction between Al, V, and Fe, there is

$$D_{Al}\left(\frac{dC_{Al}}{dx'}\right) = -v\left(\frac{C_{0Al}}{k_{Al}} - C_{0Al}\right) \tag{7}$$

$$D_V\left(\frac{dC_V}{dx'}\right) = -v\left(\frac{C_{0V}}{k_V} - C_{0V}\right) \tag{8}$$

$$D_{Fe}\left(\frac{dC_{Fe}}{dx'}\right) = -v\left(\frac{C_{0Fe}}{k_{Fe}} - C_{0Fe}\right) \tag{9}$$

where D_{Al}, D_V, and D_{Fe} are the liquid diffusion coefficients of the corresponding element; C_{0Al}, C_{0V}, and C_{0Fe} are the initial concentrations of the corresponding element; k_{Al}, k_V, and k_{Fe} are the partition coefficients of the corresponding elements. Substitute Equations (7)–(9) into Equation (6), and the actual temperature gradient G is greater than or equal to $\frac{dT_l}{dx'}\big|_{x'=0}$:

$$\frac{G}{v} \geq -\frac{m_{Al}C_{0Al}(1-k_{Al})}{D_{Al}k_{Al}} - \frac{m_V C_{0V}(1-k_V)}{D_V k_V} - \frac{m_{Fe}C_{0Fe}(1-k_{Fe})}{D_{Fe}k_{Fe}} \tag{10}$$

According to Equation (10), due to $k_{Fe} < 1$, the component supercooling is easier to achieve when C_{0Fe} increases. Therefore, the increase in Fe content will promote the formation of the component supercooling zone, which will affect the morphology of the grains.

3.2. Effect of Fe Content on the Microstructure of Multiple Grains

The growth of several grains with different Fe content was simulated by MICRESS. Figure 4 shows the effect of Fe content on grain size (the color bar represents the mass fraction of Al). With higher Fe content, the shape of grains is more complex and the grain size is more refined. Due to the low directional temperature gradient in the levitation melting, the overall appearance of equiaxed crystal appears. The crystal interface is always composed of crystal faces with smaller interface energy. The interface energy is smaller at the wide crystal face, while the energy of narrow crystal face at the edge is larger. Therefore, the crystal morphology tends to be spherical polyhedron in a stable state. Figure 5 shows as the time goes on, the liquid phase almost disappeared at 0.85 s, and an equiaxed crystal with larger grains was obtained. For the titanium alloy, the BCC phase of the cubic crystal system was first formed during solidification, and the optimal growth direction was the <001> crystal direction.

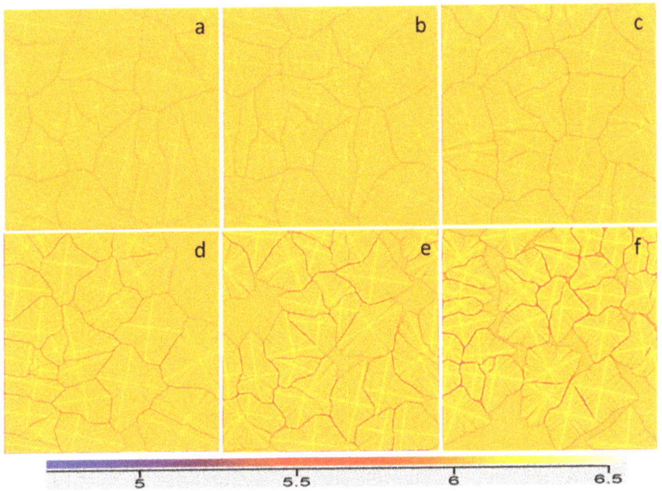

Figure 4. Effect of Fe content on grain size at 2 s: (**a**) Ti–6Al–4V–0Fe, (**b**) Ti–6Al–4V–0.1Fe, (**c**) Ti–6Al–4V–0.3Fe, (**d**) Ti–6Al–4V–0.5Fe, (**e**) Ti–6Al–4V–0.7Fe, (**f**) Ti–6Al–4V–0.9Fe.

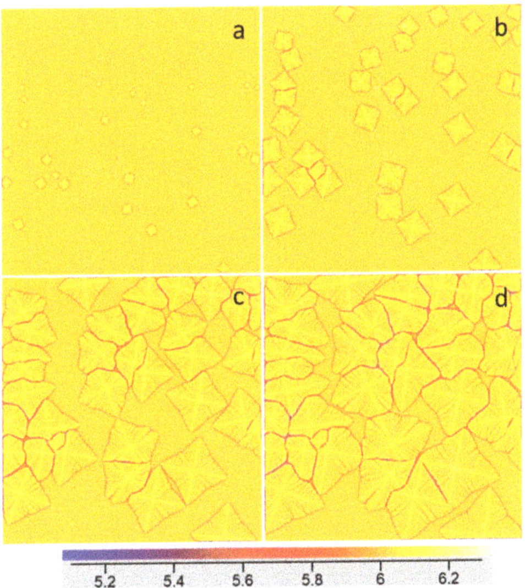

Figure 5. The grain growth with mass fraction of Fe is 0.9: (**a**) 0.3 s, (**b**) 0.8 s, (**c**) 1.5 s, (**d**) 2.0 s.

For the Ti–6Al–4V–xFe alloy, there was a large solute concentration gradient in the solid–liquid interface at the front edge of the polyhedron, and its diffusion rate was faster than that of the large plane crystal surface with a small solute concentration gradient at the front edge of the interface, resulting in the gradual change of the crystal from an octahedron to a star. This trend was more obvious at the region with a higher Fe content. Compared with Figure 4d, the segregation of Fe at the front of the solid–liquid interface in Figure 4f was stronger, and the resulting local supercooling slowed down the interface migration rate.

The microstructure of each direction was very different due to the different influence of the solute diffusion field and temperature diffusion field in four <001> directions. Figure 6 shows the effect of Fe content on the grain growth rate. When the Fe content exceeds 0.3 wt%, the growth rate of the alloy begins to decrease significantly. If the Fe content reaches 0.9 wt%, more time is needed for the liquid phase to disappear.

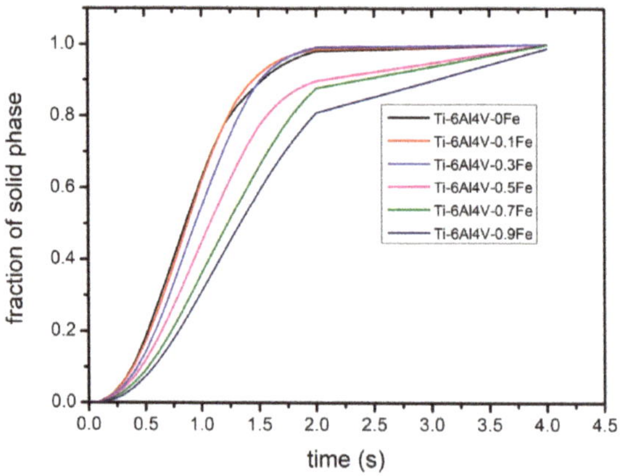

Figure 6. Effect of Fe content on grain growth rate.

In the growth process, the gap between the grains is large, and the growth speed of the grains is slow, which may provide more space for the growth of small grains and reduce the annexation of grains. Therefore, the increase of Fe in the experiment made the grains more refined.

Figure 7 shows the grain size of the alloy obtained by levitation melting. In a certain range, with the increase in Fe content, the grain size of the alloy gradually decreases. According to the number of grains and the cut-off area, the average grain radius is simply estimated, as shown in Figure 8. Compared with the simulated grain size, the experimental result was larger, which is due to the limited simulation time, while the experimental grains completed the grain growth. When there was no Fe in the alloy, as shown in Figure 7a, the grain size was the largest and the grain distribution was relatively uniform. The grain radius was about 2.29 mm, and the shape of the grain was close to circular. With the increase in Fe content, the grain size of the alloy decreased gradually, while the overall decreasing trend was mitigated. When the Fe content was 0.9 wt%, the average grain radius was the smallest (about 1.03 mm).

With the increase in Fe content, the distribution of grains was no longer uniform. Some small grains were distributed at the junction of larger grains, and the morphology of grains was close to a complex polygon. It can be considered that the addition of Fe changes the size and distribution of the grains and affects the shape of the grains, which verifies the simulation results.

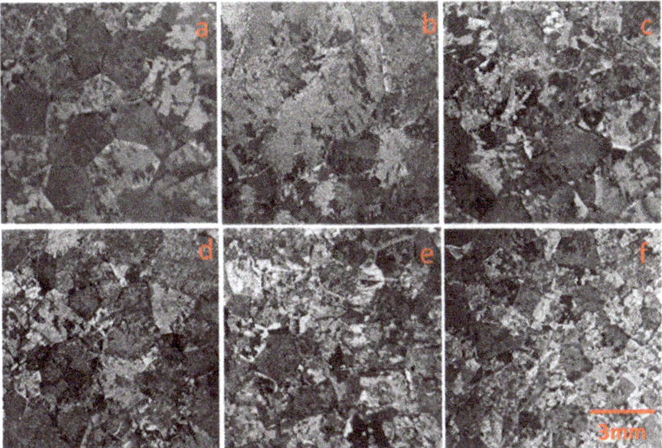

Figure 7. Effect of Fe content on grain size: (**a**) 0Fe, (**b**) 0.1Fe, (**c**) 0.3Fe, (**d**) 0.5Fe, (**e**) 0.7Fe, (**f**) 0.9Fe.

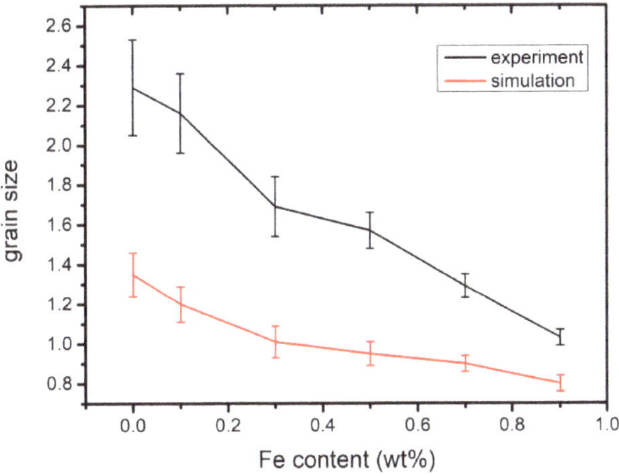

Figure 8. The change in the average grain size with Fe content.

3.3. Element Distribution in Ti–6Al–4V–xFe Alloy

As shown in Figure 9, the Fe composition distribution along the green lines was obtained and demonstrated in Figure 9e. As the Fe content increased from Figure 9a–d, the maximum solute concentration C_L^* of Fe in the liquid phase at the solid–liquid boundary continued to rise (here represented by mass fraction), which were 0.67, 1.12, 1.48, and 1.63 wt%, respectively, corresponding to the four peaks in Figure 9e. The segregation ratio S_R was 4.01, 4.15, 4.00, and 3.54, respectively, and the overall segregation trend was reduced. Within a certain range, the diffusion distance δ_n of Fe (Figure 9f) in the liquid phase had a linear relationship with the Fe content in the alloy, and the relationship can be fitted as:

$$\delta_n = 31.1C_0 + 29.5 \tag{11}$$

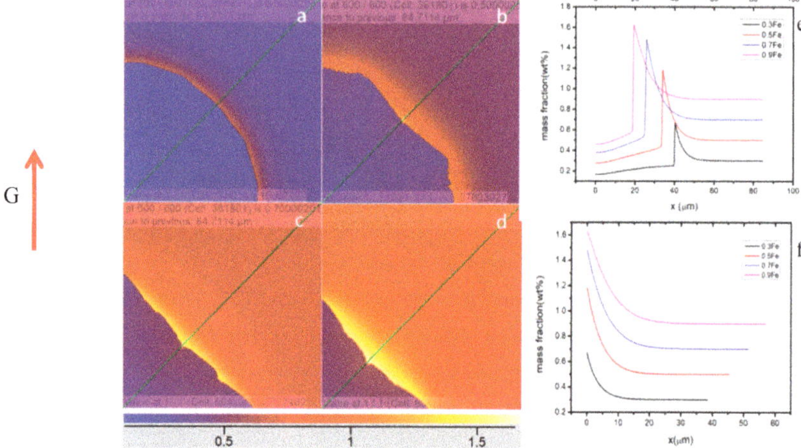

Figure 9. The change of Fe mass fraction in the direction perpendicular to the larger plane of grain under different Fe content: (**a**) 0.3Fe, (**b**) 0.5Fe, (**c**) 0.7Fe, (**d**) 0.9Fe, (**e**) liquid and solid phase, (**f**) liquid phase.

According to classical theory, for convective solute distribution, under directional solidification conditions, there is:

$$D_L \frac{d^2 C_L}{dx^2} + v \frac{dC_L}{dx'} = 0 \tag{12}$$

when $x = 0$, $C_L = C_L^* < C_0/k_0$, and when $x = \delta_n$, $C_L = C_0$.

Defining $\frac{dC_L}{dx} = z$, $\frac{dz}{dx} = \frac{d^2 C_L}{dx^2}$, then $\frac{dz}{z} = -\frac{v}{D_L}d$. After inserting the boundary conditions into the function, we can obtain:

$$C_L = \left(1 - \frac{1 - e^{-\frac{v}{D_L}x}}{1 - e^{-\frac{v}{D_L}\delta_n}}\right)(C_L^* - C_0) + C_0 \tag{13}$$

where k_0 is the partition coefficient; x is the diffusion distance; v is the interface moving rate; and D_L is the liquid diffusion coefficient.

Three assumptions were made: (1) there is only diffusion (no convection) in the liquid phase; (2) the diffusion distance δ_n at the thin solid–liquid interface are infinity; and (3) the components of the liquid phase outside the solute enrichment layer keep the original concentration C_0 unchanged during the solidification process. Under these assumptions, the maximum solute concentration is $C_L^* = C_0/K_0$ in the stable liquid phase, and the solute distribution equation of the stable state can be simplified as:

$$C_L = C_0 \left[1 + \left(\frac{1 - k_0}{k_0}\right)e^{-\frac{v}{D_L}x}\right] \tag{14}$$

The composition of the liquid phase outside the solute enrichment layer is no longer C_0, but gradually increases in the case of limited liquid volume due to convection in the outer diffusion layer during the actual solute redistribution process. As the actual $C_L^* < C_0/K_0$, the solute concentration calculated by Equation (13) is higher, as shown in Figure 10. In this work, since the solidification process of levitation melting was not directional solidification, the direction of the temperature gradient has little effect on the grain morphology. As shown in Figure 9, the grain growth speed was slow in the direction perpendicular to the larger plane of the grain, which had an angle of 45° with the relative temperature gradient direction. The solute distribution value should be between Equations (13) and (14). For the levitation melting of the Ti–6Al–4V–xFe alloy with a slow growth rate, the modified distribution equation of solute in the steady state can be proposed according to the results of phase field simulation:

$$C_L = 0.79C_0\left[1 + \left(\frac{1-k_0}{k_0}\right)e^{-\frac{v}{D_L}x}\right] + 0.11 \tag{15}$$

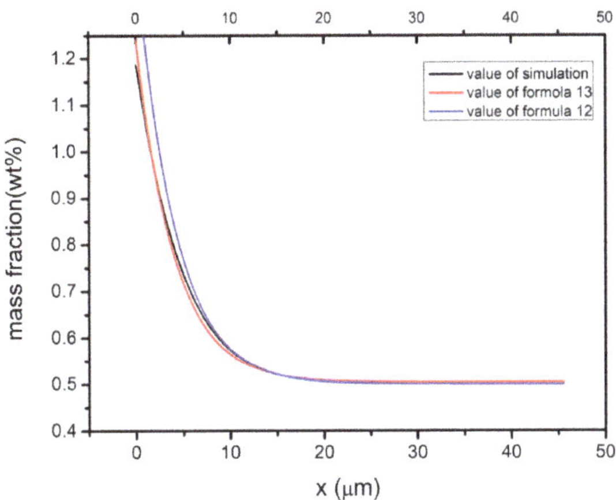

Figure 10. Solute distribution of Ti–6Al–4V–0.5Fe in the liquid phase.

The agreement of Equation (14) with the simulation results was close to 90%, while the agreement of the modified equation with the simulation results was close to 97%.

The composition at the triangular grain boundaries of Ti–6Al–4V–0.5Fe and Ti–6Al–4V–0.9Fe alloys were scanned by EPMA, and the results are shown in Figure 11. The segregation of Ti at the grain boundary was not obvious. The overall distribution presents a homogeneous contrast due to the matrix material of Ti. Compared with Figure 11a, Figure 11b shows that there was a certain segregation of the Al element in α lamellae. The most serious segregation was in the β grain boundary, while the lowest content was at the edges of the β grain boundary.

In the solidification process of the titanium alloy, the solid–liquid phase transformation first occurs, forming β original grains, and growing continuously with the decrease in temperature. The amount of liquid phase gradually decreases and concentrates at the boundary of β grains at the end of the solid–liquid phase transformation, as presented in Figure 5d. As the solidification proceeded, the liquid phase finally disappeared, forming the original β grain, as indicated in Figure 4a. With the slow decrease in temperature (i.e., non-quenching), the BCC phase in the high temperature state of the Ti alloy was gradually transformed into the HCP phase (i.e., β/α transformation, forming primary α phase). The α lamellar structure (about 0.5–2 μm) was formed in the original β grain; and the remaining β phase was distributed at the boundary of the α lamellar. The morphology of the β original grain was retained without any deformation in the end. As a result, the microstructure shown in Figure 11a was formed. During the cooling process, a relatively wide α lamellar structure (about 2–3 μm) was formed from the β grain boundaries. The remaining β phase was distributed at the edges.

In the same way, V and Fe, as β stable elements, concentration increased from the inner area to the edges of α lamellae. Due to the wide β grain boundary, the segregation at the edge of the β grain boundary was more obvious. Since Fe is a stronger β stable element than V, the segregation of Fe was more obvious. Comparing the β grain boundaries in Figure 11a,c, Figure 11c was finer (about 1–2 μm), which may be attributed to the grain refinement of Fe. The distribution trend of Figure 11b was the same as in Figure 11d.

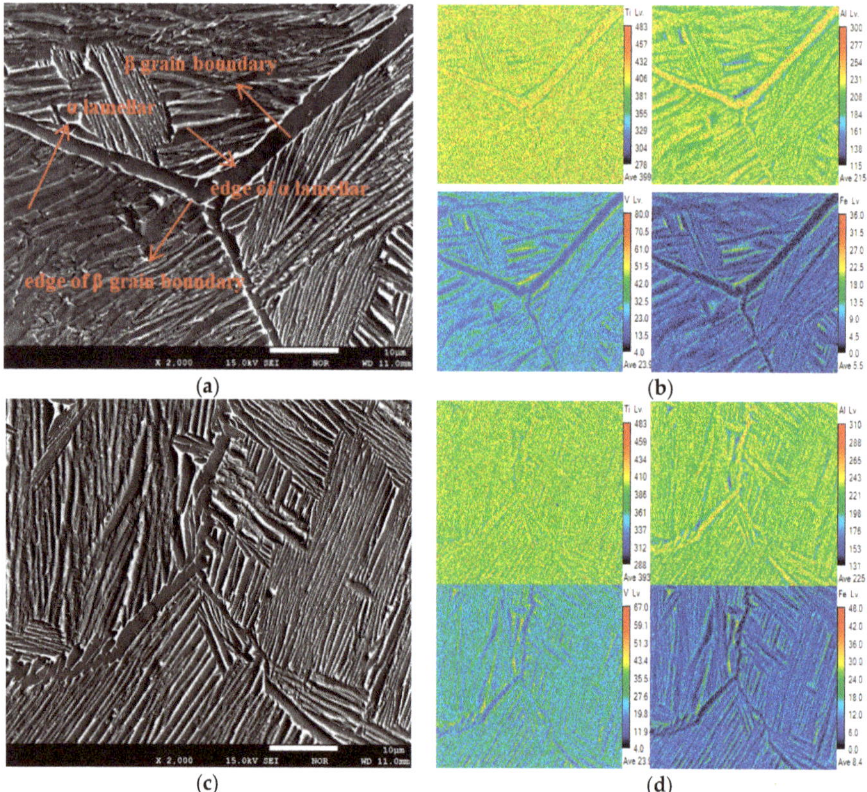

Figure 11. Results of the electron probe micro analysis surface scan. (**a**) Secondary electron image of Ti–6Al–4V–0.5Fe, (**b**) Concentration distribution of Ti–6Al–4V–0.5fe, (**c**) Secondary electron image of Ti–6Al–4V–0.9Fe, (**d**) Concentration distribution of Ti–6Al–4V–0.9Fe.

Figure 12 compares the simulated with the experimental values of the Fe composition distribution. As the simulation process does not complete the β/α transformation, the segregation of Fe is mainly concentrated in the residual liquid phase between β grains. Comparing the segregation degree of the simulated and experimental values, the segregation degree of Fe in the simulation was not more than three times that of the nominal composition, whereas the segregation degree of the experimental value was close to six times the nominal composition. This may be attributed to the decrease in β phase amount in β/α transformation and the further compression of the range of Fe segregation distribution.

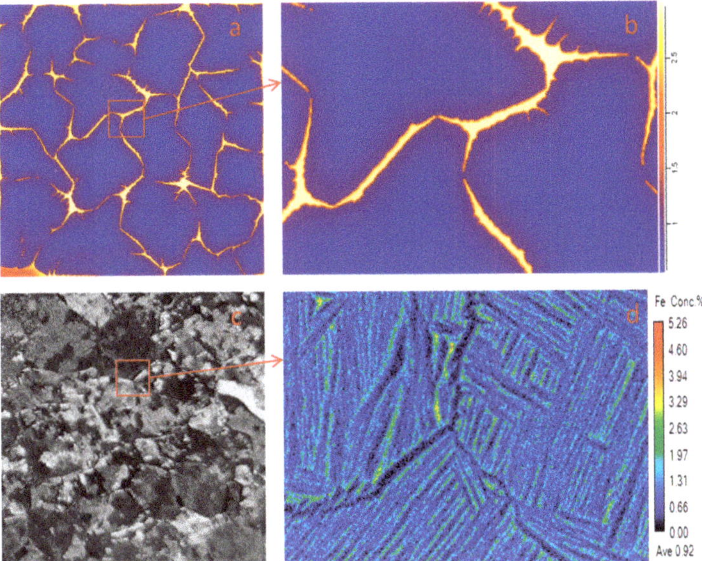

Figure 12. Comparison of simulated and experimental values of Fe concentration in Ti–6Al–4V–0.9Fe alloy. (**a**) Selection of simulation position, (**b**) Simulated value of Fe concentration, (**c**) Selection of experimental position, (**d**) Experimental value of Fe concentration.

Through line scans crossing the grain boundaries of Ti–6Al–4V–0.5Fe, Ti–6Al–4V–0.7Fe, and Ti–6Al–4V–0.9Fe alloys, it can be seen from Figure 13b that the fluctuation range of Al composition in the alloy ranged from 9.78 to 11.07 at%, V ranged from 2.19 to 7.98 at%, and Fe from 0.18 to 1.81 at%. Compared with the average composition, the fluctuation values of Al, V and Fe were 8%, 91%, and 202%, respectively. Obviously, the segregation of Fe was greater than V, while the segregation of V was greater than Al.

For three samples, the mean values of Fe composition were 0.60 at%, 0.76 at%, and 0.87 at% with the standard deviations of 0.31, 0.39, and 0.40, respectively. Considering that the grain boundary of the Ti–6Al–4V–0.9Fe sample was less and the overall element distribution was more uniform, the segregation of Fe was still slightly larger. It can be seen that in a certain range, with the increase in Fe content in the alloy, the segregation of Fe tended to increase; however, the influence of Fe content on the segregation of Al and V elements was negligible. With the increase in Fe content, the trends of Fe segregation in the simulation and experiment were the opposite. It is considered that the segregation of Fe mainly occurs in the stage of grain growth or solid-state phase transformation, which needs further study.

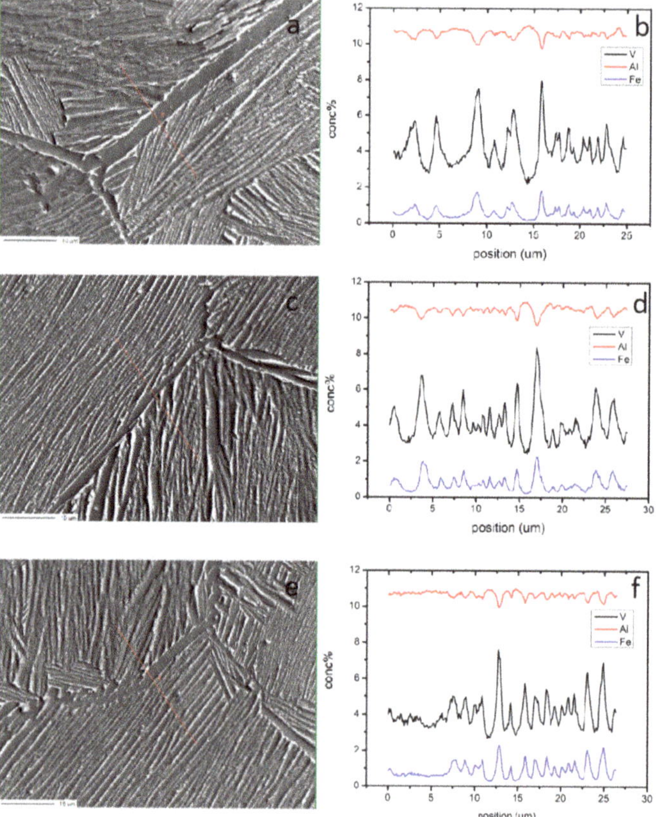

Figure 13. Results of the EPMA line scan. (**a**) Selected location of Ti–6Al–4V–0.5Fe, (**b**) Composition distribution of Ti–6Al–4V–0.5Fe, (**c**) Selected location of Ti–6Al–4V–0.7Fe, (**d**) Composition distribution of Ti–6Al–4V–0.7Fe, (**e**) Selected location of Ti–6Al–4V–0.9Fe, (**f**) Composition distribution of Ti–6Al–4V–0.9Fe.

4. Conclusions

In this work, the processes of levitation melting of five Ti–6Al–4V–xFe alloys were simulated, the effect of Fe content on the microstructure of single crystal and multi crystal was studied, and the distribution of elements in the Ti–6Al–4V–xFe alloy was discussed. Some simulation results were verified by experiments. The specific conclusions are as follows:

(1) The segregation of Fe element at the grain boundary of Ti–6Al–4V–xFe alloys can inhibit the interface mobility, thus promoting the formation of a local supercooling zone and making the grains easier to grow into dendrites.

(2) With the increase of Fe content, the grain size of the alloy decreased gradually. When there was no Fe in the alloy, the grain size was the largest (radius close to 2.29 mm), the grains were more uniform, and the shape of the grain was close to circular. The grain size decreased gradually with an increase in the Fe content and the overall decrease trend slowed down. When the Fe content was 0.9, the average grain radius was the smallest, which was about 1.03 mm.

(3) With the increase in Fe content, the distance of diffusion layer δ_n increased in the liquid phase. Within a certain range, there was a linear relationship between them.

(4)   The segregation of Fe was more obvious than that of Al and V. With the increase in Fe content, the segregation of Fe increased, but there was less of an effect on Al and V.

Author Contributions: Conceptualization, H.C.; data curation, Y.G.; formal analysis, L.D.; funding acquisition, L.Z.; investigation, D.Z.; methodology, F.C.; software, L.D.; supervision, R.H. All authors have read and agreed to the published version of the manuscript.

Funding: This work was supported by Primary Research and Development Plan of Jiangsu Province (BE2019119).

Conflicts of Interest: The authors declare no conflict of interest.

References

1.   Allen, P. Titanium alloy development. *Adv. Mater. Process.* **1996**, *150*, 35–37.

2.   Lütjering, G.; Williams, J.C. *Titanium*; Springer: Berlin/Heidelberg, Germany, 2007.

3.   Xin-ping, Z.; Si-rong, Y.; Zhen-ming, H.; Qiu-hua, H. Mechanical properties of new type ti-fe-mo-mn-nb-zr titanium alloy. *Chin. J. Nonferrous Met.* **2002**, *12*, 78–82.

4.   Hotta, S.; Yamada, K.; Murakami, T.; Narushima, T.; Iguchi, Y.; Ouchi, C. Beta. Grain refinement due to small amounts of yttrium addition in.Alpha.+.Beta. Type titanium alloy, sp-700. *ISIJ Int.* **2006**, *46*, 129–137. [CrossRef]

5.   Kudo, T.; Murakami, S.; Itsumi, Y. Influence of microstructure on formability in ti-fe alloy. *J. Mater. Process. Technol.* **2010**, *60*, 33–36.

6.   Bermingham, M.J.; Mcdonald, S.D.; Stjohn, D.H.; Dargusch, M.S. Segregation and grain refinement in cast titanium alloys. *J. Mater. Res.* **2009**, *24*, 1529–1535. [CrossRef]

7.   Ehtemam-Haghighi, S.; Liu, Y.; Cao, G.; Zhang, L.-C. Phase transition, microstructural evolution and mechanical properties of ti-nb-fe alloys induced by fe addition. *Mater. Des.* **2016**, *97*, 279–286. [CrossRef]

8.   Boettinger, W.J.; Warren, J.A.; Beckermann, C.; Karma, A. Phase-field simulation of solidification. *Ann. Rev. Mater. Res.* **2002**, *32*, 163–194. [CrossRef]

9.   Gyoon Kim, S.; Tae Kim, W.; Suzuki, T.; Ode, M. Phase-field modeling of eutectic solidification. *J. Cryst. Growth* **2004**, *261*, 135–158. [CrossRef]

10.   Suzuki, T.; Ode, M.; Kim, S.G.; Kim, W.T. Phase-field model of dendritic growth. *J. Cryst. Growth* **2002**, *237*, 125–131. [CrossRef]

11.   Kermanpur, A.; Jafari, M.; Vaghayenegar, M. Electromagnetic-thermal coupled simulation of levitation melting of metals. *J. Mater. Process. Technol.* **2011**, *211*, 222–229. [CrossRef]

12.   Li, H.; Wang, S.; He, H.; Huangfu, Y.; Zhu, J. Electromagnetic-thermal-deformed-fluid-coupled simulation for levitation melting of titanium. *IEEE Trans. Magn.* **2016**, *52*, 1–4. [CrossRef]

13.   Kundin, J.; Kumar, R.; Schlieter, A.; Choudhary, M.A.; Gemming, T.; Kühn, U.; Eckert, J.; Emmerich, H. Phase-field modeling of eutectic ti–fe alloy solidification. *Comput. Mater. Sci.* **2012**, *63*, 319–328. [CrossRef]

14.   Gong, X.; Chou, K. Phase-field modeling of microstructure evolution in electron beam additive manufacturing. *JOM J. Miner. Met. Mater. Soc.* **2015**, *67*, 1176–1182. [CrossRef]

15.   Sahoo, S.; Chou, K. Phase-field simulation of microstructure evolution of ti–6al–4v in electron beam additive manufacturing process. *Addit. Manuf.* **2016**, *9*, 14–24. [CrossRef]

16.   Wu, L.; Zhang, J. Phase field simulation of dendritic solidification of ti-6al-4v during additive manufacturing process. *JOM* **2018**, *70*, 2392–2399. [CrossRef]

17.   Nakajima, H.; Ohshida, S.; Nonaka, K.; Yoshida, Y.; Fujita, F.E. Diffusion of iron in β ti-fe alloys. *Scr. Mater.* **1996**, *34*, 949–953. [CrossRef]

18.   Chen, Y.; Li, J.; Tang, B.; Kou, H.; Segurado, J.; Cui, Y. Computational study of atomic mobility for bcc phase in ti–al–fe system. *Calphad* **2014**, *46*, 205–212. [CrossRef]

19.   Tiley, J.S.; Shiveley, A.R.; Pilchak, A.L.; Shade, P.A.; Groeber, M.A. 3d reconstruction of prior β grains in friction stir–processed ti–6al–4v. *J. Microsc.* **2014**, *255*, 71–77. [CrossRef]

20.   Böttger, B.; Eiken, J.; Apel, M. Phase-field simulation of microstructure formation in technical castings—A self-consistent homoenthalpic approach to the micro–macro problem. *J. Comput. Phys.* **2009**, *228*, 6784–6795. [CrossRef]

21. Eiken, J.; Bottger, B.; Steinbach, I. Multiphase-field approach for multicomponent alloys with extrapolation scheme for numerical application. *Phys. Rev. EStat. NonlinearSoft Matter Phys.* **2006**, *73*, 066122. [CrossRef] [PubMed]
22. Guo, J.; Li, X.; Su, Y.; Wu, S.; Li, B.; Fu, H. Phase-field simulation of structure evolution at high growth velocities during directional solidification of ti55al45 alloy. *Intermetallics* **2005**, *13*, 275–279. [CrossRef]
23. Kurz, W.; Fisher, D.J. Dendrite growth at the limit of stability: Tip radius and spacing. *Acta Metall.* **1981**, *29*, 11–20. [CrossRef]

Article

Corrosion and Tensile Behaviors of Ti-4Al-2V-1Mo-1Fe and Ti-6Al-4V Titanium Alloys

Yanxin Qiao [1], Daokui Xu [2,*], Shuo Wang [1,2,3], Yingjie Ma [2,*], Jian Chen [1], Yuxin Wang [1] and Huiling Zhou [1]

[1] School of Materials Science and Engineering, Jiangsu University of Science and Technology, Zhenjiang 212003, China; yxqiao@just.edu.cn (Y.Q.); 1910221@mail.neu.edu.cn (S.W.); jchen496@uwo.ca (J.C.); ywan943@163.com (Y.W.); zhouhl@just.edu.cn (H.Z.)

[2] Institute of Metal Research, Chinese Academy of Sciences, Shenyang 110016, China

[3] School of Materials Science and Engineering, Northeastern University, Shenyang 110004, China

* Correspondence: dkxu@imr.ac.cn (D.X.); yjma@imr.ac.cn (Y.M.); Tel.: +86-24-2392-8381 (D.X.)

Received: 12 October 2019; Accepted: 6 November 2019; Published: 11 November 2019

Abstract: X-ray diffraction (XRD), scanning electron microscope (SEM), immersion, electrochemical, and tensile tests were employed to analyze the phase constitution, microstructure, corrosion behaviors, and tensile properties of a Ti-6Al-4V alloy and a newly-developed low cost titanium alloy Ti-4Al-2V-1Mo-1Fe. The results showed that both the Ti-6Al-4V and Ti-4Al-2V-1Mo-1Fe alloys were composed of α and β phases. The volume fractions of β phase for these two alloys were 7.4% and 47.3%, respectively. The mass losses after 180-day immersion tests in 3.5 wt.% NaCl solution of these alloys were negligible. The corrosion resistance of the Ti-4Al-2V-1Mo-1Fe alloy was higher than that of the Ti-6Al-4V alloy. The tensile tests showed that the Ti-4Al-2V-1Mo-1Fe alloy presented a slightly higher strength but a lower ductility compared to the Ti-6Al-4V alloy.

Keywords: titanium alloy; corrosion; passive film; mechanical behavior

---

1. Introduction

Titanium and its alloys are widely used in aerospace [1,2], marine [3], chemical [4,5], and biomedical [6–8] fields due to their excellent mechanical properties, high corrosion resistance, and good biocompatibility. Their high corrosion resistance in aggressive environments is ensured by the formation of a compact and chemically-stable oxide film, mainly composed of titanium oxide, TiO_2, which spontaneously covers the metal surface to protect the metal substrate [9–11]. The corrosion behaviors of the commercial Ti-6Al-4V alloy vary with corrosive environments. Yue et al. [12] found that the Ti-6Al-4V alloy exhibited a poor corrosion behavior in a solution with a high Cl^- concentration or in acid environments with a local accumulation of Cl^- ions. Blanco-Pinzon et al. [13] reported that the Ti-6Al-4V alloy was susceptible to corrosion in H_2SO_4 solution but its corrosion resistance could be significantly improved by alloying Pd and Ni. Wang et al. [5] investigated the effects of alloying elements Pd, Mo, and Ni on the corrosion behavior of titanium alloy in H_2SO_4 solution containing fluoride ions, and found that these alloying elements had no influence on the interaction of the F^- ions with titanium matrix, and on the film composition. The addition of Ni accelerated the cathodic reactions while the addition of Mo retarded the anodic process [5]. Newman et al. [14,15] found that Mo located at defect sites preferentially dissolved, leading to the formation of stable Mo oxides to decrease the anodic dissolution rate [16].

Generally, element alloying is an effective method to improve the mechanical properties of titanium and its alloys [17,18]. Since Fe and Al elements are characteristic of low toxicity and cost [19–21], they are added into the titanium and titanium alloys, and the effect of Fe or Al alloying on the mechanical and corrosion properties of titanium alloys has been studied. It is reported that the Ti-4.5Al-3V-2Mo-2Fe alloy has superior mechanical properties to the Ti-6Al-4V alloy [17] due to its microstructural characteristics and element alloying (i.e., Mo and Fe). Lu et al. [22] investigated the mechanical properties and electrochemical corrosion behaviors of Ti-6Al, Ti-6Al-4V, and Ti-6Al-xFe alloys, and found that the Ti-6Al-4Fe alloy possessed the lowest Young's modulus and exhibits the highest strength:modulus ratios, and the Ti-6Al-xFe alloys exhibited a higher corrosion resistance in simulated human body fluid (SBF) than both the Ti-6Al and Ti-6Al-4V alloys. However, the corrosion resistance of the Ti-6Al-xFe alloys decreased with the increasing Fe content, suggesting that the content of Fe added into the titanium alloys needed to be controlled at a lower level to achieve a better corrosion performance. It is consistent with the results conducted by Pimenova et al. [21] and Hsu et al. [23,24]. When the concentration of Al was higher than 15 wt.%, Ti-xAl-yFe alloys underwent severe pitting corrosion due to the precipitates of β phase and uneven distribution of the alloying elements [21]. Thus, more work has to be performed to clarify the effect of Al and Fe alloying on the corrosion and mechanical performances of titanium alloys, especially when Mo is added to improve the resistance of localized corrosion.

In this research, the Ti-4Al-2V-1Mo-1Fe alloy developed by the Institute of Metal Research, Chinese Academy of Sciences (IMR), was investigated. This type of alloy has lower contents of Al and V but higher contents of Mn and Fe. The cost of this alloy is relatively low, providing a potential alternative to the Ti-6Al-4V alloy. The aim of this work is to investigate the corrosion behavior of this newly-developed alloy in a simulated marine environment (3.5 wt.% NaCl solution), and its tensile property. The difference of alloy property between Ti-6Al-4V and Ti-4Al-2V-1Mo-1Fe alloys was systematically studied.

2. Experimental Details

The materials used in the present study were the commercial Ti-6Al-4V alloy, and the Ti-4Al-2V-1Mo-1Fe alloy fabricated in the IMR, Shenyang, China. The chemical compositions (wt.%) of these two alloys are listed in Table 1. Prior to the study, the Ti-6Al-4V and Ti-4Al-2V-1Mo-1Fe alloys were heated to 750 °C for 3 h and then cooled to room temperature in air. Samples for the immersion test were cut into sheets with dimensions of 40 mm × 20 mm × 4 mm. Samples for the electrochemical test and microstructure observation were cut into square sheets (10 mm × 10 mm × 2 mm) and sealed in a mixture of epoxy and polyamide resins with an exposed surface of 1 cm^2. Then each specimen was gradually ground with SiC papers up to 1000 grit, polished with a diamond paste of 0.5 μm, then cleaned in ethanol, and finally dried with hot air.

Table 1. Chemical compositions of the tested alloys (wt.%).

Alloy	Al	V	Mo	Fe	Ti
Ti-6Al-4V	5.95	4.03	-	0.33	Bal.
Ti-4Al-2V-1Mo-1Fe	3.96	2.03	1.05	0.92	Bal

The immersion and electrochemical tests were performed in a 3.5 wt.% NaCl solution at 25 ± 1 °C (controlled by a thermostat water bath). The 3.5 wt.% NaCl solution was prepared using analytical-grade sodium chloride and distilled water. Immersion tests were carried out to investigate the long-term corrosion behaviors of the Ti-6Al-4V and Ti-4Al-2V-1Mo-1Fe alloys. Five samples were prepared for each test solution, with an immersion period of 180 days to ensure the reproducibility. Samples with dimensions of 10 mm × 10 mm × 2 mm were successively ground with abrasive papers to 1000 grit and then were immersed in aerated 3.5% NaCl solution (volume: 1.5 L) without stirring. The solutions were replaced every 10 days. The electrochemical behaviors of the tested alloys were measured using a CS350 (Wuhan Corrtest Instruments Corp., Ltd. Wuhan, China) electrochemical workstation and a three-electrode electrochemical cell, and the method was described in the literature [25]. Potentiodynamic polarization was performed at a scan rate of 0.1667 mV/s from

−500 mV$_{SCE}$ below the open circuit potential (OCP) and terminated at 2500 mV$_{SCE}$. After immersing in 3.5 wt.% NaCl solution at the OCP for 1 h, electrochemical impedance spectroscopy (EIS) was conducted with a sinusoidal potential perturbation of 10 mV and a frequency range from 10^5 to 10^{-2} Hz. All measurements were repeated at least three times in naturally-aerated 3.5 wt.% NaCl solution without stirring to ensure the reproducibility. Cview and Zview software were used to fit the electrochemical data. Tensile tests of these two alloys were carried out on an Instron-type testing machine with a stain rate of 3 mm·s^{-1} at 25 ± 1 °C [26]. To ensure the reliability of the measured data, at least three repeated measurements were carried out for each time. The standard deviation method was used to analyze the data and obtain the mechanical property parameters.

The crystal structures of the tested alloys were determined using a D/Max 2400 X-ray diffractometer (Rigaku Corporation, Tokyo, Japan) with Cu K$_\alpha$ radiation at 10 kV and 35 mA at a step size of 0.02° and a scan rate of 4°/min. The specimens used for the microstructure observation were etched in Kroll reagent (3 mL HF, 9 mL HNO$_3$ and 88 mL H$_2$O) for 10 s. The microstructure was observed by a Keyence VHX-700 (Keyence Co. Ltd., Osaka, Japan) (LM) and a scanning electron microscope (XL30-FEG ESEM, FEI, Hillsboro, OR., USA) equipped with EDS. Image-Pro Plus software was used to calculate the phase volume fraction of the tested alloys.

3. Results and Discussion

3.1. Microstructure Characterization

The XRD patterns of the Ti-6Al-4V and Ti-4Al-2V-1Mo-1Fe alloys are shown in Figure 1. The diffraction peaks in Figure 1 corresponded to the peaks of α and β phases, suggesting that both alloys have duplex structure. However, the proportions of these two phases in the Ti-6Al-4V and Ti-4Al-2V-1Mo-1Fe alloys were different. As seen in Figure 1, the peak intensity of α phase was higher, indicating a higher content of α phase in both alloys.

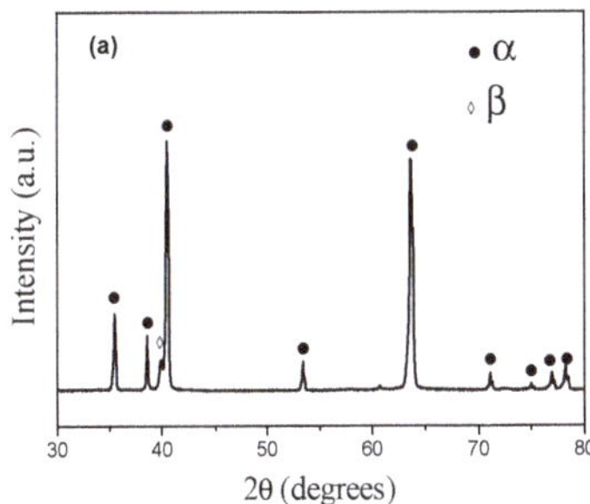
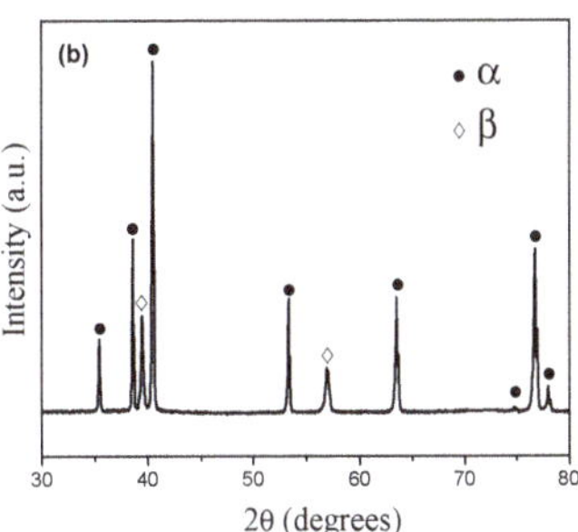

Figure 1. XRD patterns of the (**a**) Ti-6Al-4V and (**b**) Ti-4Al-2V-1Mo-1Fe alloys.

The LM images of the Ti-6Al-4V and Ti-4Al-2V-1Mo-1Fe alloys are shown in Figure 2. It was found that the two tested alloys had a bi-phase structure, consistent with the XRD results in Figure 1. The average grain size of the α phase in the Ti-6Al-4V alloy was ~25 μm, and some grains were larger than 50 μm, as shown in Figure 2a. In comparison, the grain size of the α phase in the Ti-4Al-2V-1Mo-1Fe alloy was smaller, with an average grain size of about ~8 μm, as shown in Figure 2b.

The SEM observations of the Ti-6Al-4V and Ti-4Al-2V-1Mo-1Fe alloys are shown in Figure 3. As seen in Figure 3, both alloys were composed of the dark α phase and the bright β phase. No other precipitates were observed either inside grains or at grain boundaries. As shown in Figure 3a, the β phase dispersed and scattered inside the equiaxed α phase. However, the distribution of the β phase in the Ti-4Al-2V-1Mo-1Fe alloy was more continuous. As shown in Figure 3b, clustered β phase distributed evenly inside the α phase in the Ti-4Al-2V-1Mo-1Fe alloy. The volume fraction of the β phase in the Ti-6Al-4V and Ti-4Al-2V-1Mo-1Fe alloys was about 7.4% and 47.3%, respectively. EDS analysis was conducted to investigate the composition of α and β phases of the tested alloys, and the results are shown in Table 2. It is reported that Fe, Mo, and V atoms have long been recognized as strong β-stabilizing elements [19,22,27], thus the atomic ratio of these elements in β phase is higher than that in α phase. This is mainly due to the higher element solid solubility and elemental diffusion rate in the β phase [27,28]. Based on the research [28], the volume fraction of the β phase in Ti-4Al-2V-1Mo-1Fe alloy may be higher than that of the Ti-6Al-4V alloy.

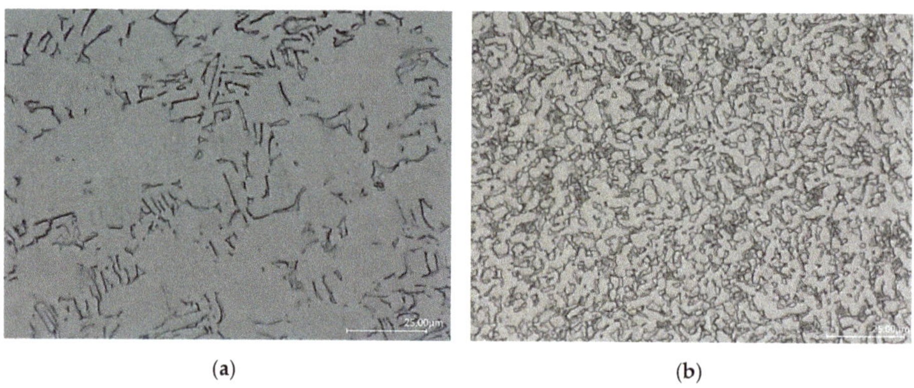

(a) (b)

Figure 2. LM of the (**a**) Ti-6Al-4V and (**b**) Ti-4Al-2V-1Mo-1Fe alloys.

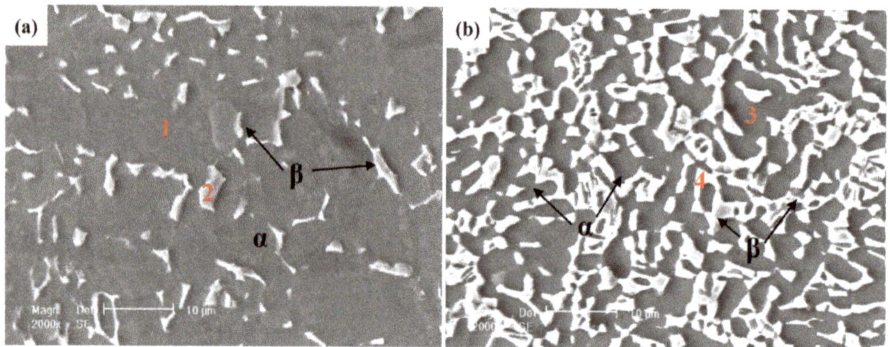

Figure 3. SEM microstructures of the (**a**) Ti-6Al-4V and (**b**) Ti-4Al-2V-1Mo-1Fe alloys.

Table 2. Chemical compositions (EDS) of α and β phases of the Ti-6Al-4V and Ti-4Al-2V-1Mo-1Fe alloys (wt.%).

Alloy		Ti	Al	V	Mo	Fe
Ti-6Al-4V	α (area 1)	88.6	6.9	4.5	-	-
	β (area 2)	80.7	2.4	15.9	-	1.0
Ti-4Al-2V-1Mo-1Fe	α (area 3)	93.5	5.3	1.2	-	-
	β (area 4)	85.4	2.6	4.7	4.4	2.9

3.2. Immersion Test

The microstructures of the Ti-6Al-4V and Ti-4Al-2V-1Mo-1Fe alloys after immersion in 3.5 wt.% NaCl solution for 180 days are shown in Figure 4. It can be seen that no traces of corrosion were observed, indicating the Ti-6Al-4V and Ti-4Al-2V-1Mo-1Fe alloys had excellent corrosion resistance in 3.5 wt.% NaCl solution.

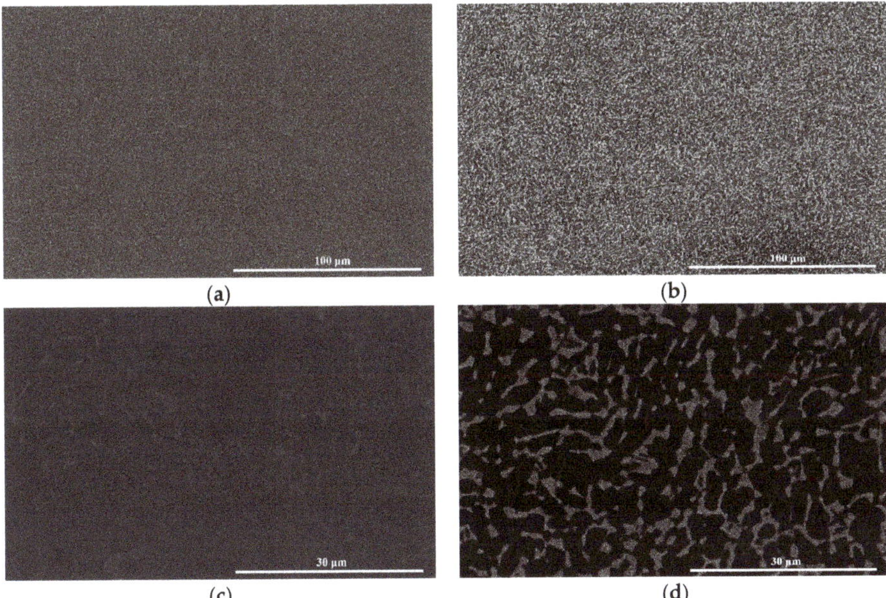

Figure 4. Microstructures of the (**a,c**) Ti-6Al-4V and (**b,d**) Ti-4Al-2V-1Mo-1Fe alloys after immersion in 3.5 wt.% NaCl solution for 180 days.

The mass loss rates of the Ti-6Al-4V and Ti-4Al-2V-1Mo-1Fe alloys after 180 days of immersion in 3.5% NaCl solution were calculated based on the mass loss, Δm, using Equation (1) shown as following:

$$\Delta m = \frac{m_0 - m_1}{S \times t} \tag{1}$$

where m_0 is the weight (mg) of the sample before the immersion test, m_1 is the weight (mg) of the sample after the immersion test, S is the surface area of the sample (cm^2), and t is the immersion time (180 days). The mass loss rate of the Ti-6Al-4V and Ti-4Al-2V-1Mo-1Fe alloys was 1.99×10^{-4} and 2.01×10^{-4} mg·cm^{-2}·day^{-1}, respectively. This indicates both behaved similarly, without any obvious occurrence of corrosion in the present tested solution.

3.3. Electrochemical Response

Once the Ti-6Al-4V and Ti-4Al-2V-1Mo-1Fe alloys were immersed in NaCl solution at OCP, the dissolution process of the naturally-formed oxide film (rutile TiO$_2$) in air began and the self-passivated film simultaneously formed [9]. The OCPs for the Ti-6Al-4V and Ti-4Al-2V-1Mo-1Fe alloys in 3.5 wt.% NaCl solution are shown in Figure 5. Since the specimens were exposed in ambient atmosphere for 1 h to allow the native growth of oxide film, the spontaneous OCP after the immersion indicated the stability of the naturally-formed oxide [5]. It is seen from Figure 5 that corrosion potential (E_{corr}) of the Ti-6Al-4V alloy shifted continuously to positive potential with the immersion time. As for the Ti-4Al-2V-1Mo-1Fe alloy, the E_{corr} gradually increased from −0.65 V$_{SCE}$ to more noble potential, and finally achieved steady-state potential of −0.40 V$_{SCE}$ when the immersion time was 1 h. The corrosion data in Figure 5 revealed that a shorter time was required for the steady-state potentials to be obtained in 3.5 wt.% NaCl solution for the Ti-4Al-2V-1Mo-1Fe alloy compared to the Ti-6Al-4V alloy.

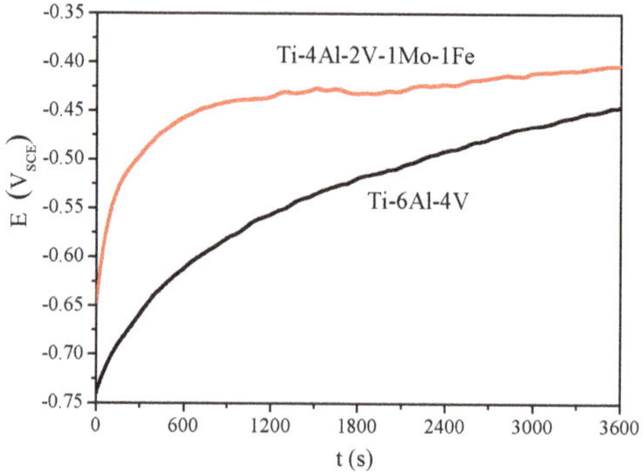

Figure 5. Evolution of open circuit potential (OCP) with immersion time for the Ti-6Al-4V and Ti-4Al-2V-1Mo-1Fe alloys in 3.5 wt.% NaCl solution.

To investigate the stability of the passive film formed on the Ti-6Al-4V and Ti-4Al-2V-1Mo-1Fe alloys, EIS measurements were carried out at OCP for both alloys in 3.5 wt.% NaCl solution. Figures 6 and 7 show the Nyquist and Bode plots for the Ti-6Al-4V and Ti-4Al-2V-1Mo-1Fe alloys for 1 h immersion in 3.5 wt.% NaCl solution. As seen from Figure 6, both the Ti-6Al-4V and Ti-4Al-2V-1Mo-1Fe alloys exhibited an unfinished single capacitive arc. It can be seen in Figure 7 that only one time constant was observed. The initial impedance (Z) recorded for the Ti-4Al-2V-1Mo-1Fe alloy was higher than that of the Ti-6Al-4V alloy, indicating that the Ti-4Al-2V-1Mo-1Fe alloy had a superior corrosion resistance [3]. The impedance data were analyzed using the equivalent circuit shown in Figure 8. The use of a constant phase element (CPE) was necessary [29–31] due to the distribution of relaxation times resulting from heterogeneities at the electrode surface. The CPE was used for the description of a frequency-independent phase shift between an applied AC potential and its current response [32], and has been extensively investigated [29,33]. The impedance of the CPE was given by:

$$Z_{CPE} = \frac{1}{Q}(j\omega)^{-n} \tag{2}$$

Therefore, the total impedance was [34]:

$$Z_{total} = R_s + \left(Q(j\omega)^n + \frac{1}{R_p}\right)^{-1} \tag{3}$$

where n was the depression angle (in degrees) that evaluated the semicircle deformation, R_s was the electrolyte resistance, R_p represented the charge transfer resistance, and Q corresponded to the pseudo-capacitance of the film, expressed using the CPE. The reason may be as follows: the CPE accounted for two contributions, one arising from double-layer capacitance (C_H) and one arising from semiconductor capacitance relating to the passive film (Csc). The capacitance of the double layer seems to be neglected according to the result reported by Hirschorn et al. [35], so the capacitance behavior of Ti-6Al-4V and Ti-4Al-2V-1Mo-1Fe alloys was dominated by the passive film [36]. Table 3 shows the values of the electric parameters obtained using the equivalent electric circuit to fit the EIS data. It is seen in Table 3 that the film resistance for the Ti-6Al-4V and Ti-4Al-2V-1Mo-1Fe alloys was 5.69×10^5 $\Omega \cdot cm^{-2}$ and 6.50×10^5 $\Omega \cdot cm^{-2}$, respectively. It is consistent with the generally accepted sense that a larger capacitive arc indicates higher corrosion resistance. It can be inferred that the stability of the passive film formed on the Ti-4Al-2V-1Mo-1Fe alloy in the present solution was slightly better than that of the Ti-6Al-4V alloy.

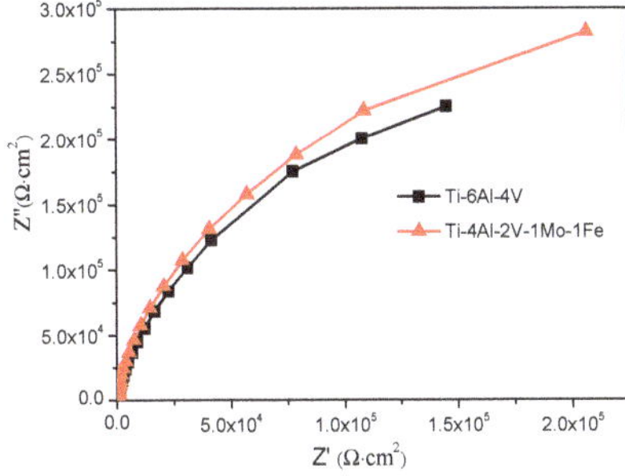

Figure 6. Nyquist plots for the Ti-6Al-4V and Ti-4Al-2V-1Mo-1Fe alloys for 1 h of immersion in 3.5 wt.% NaCl solution.

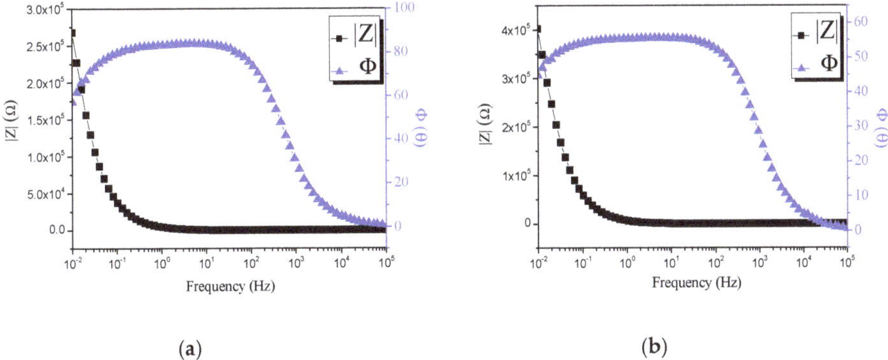

(**a**)   (**b**)

Figure 7. Bode plots for the (**a**) Ti-6Al-4V and (**b**) Ti-4Al-2V-1Mo-1Fe alloys for 1 h of immersion in 3.5 wt.% NaCl solution.

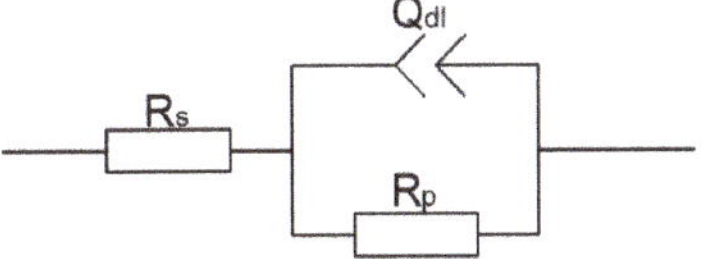

Figure 8. The equivalent circuit used for quantitative evaluation of electrochemical impedance spectroscopy (EIS).

Table 3. The EIS fitted data of Ti-6Al-4V and Ti-4Al-2V-1Mo-1Fe alloys.

Alloy	R_s ($\Omega \cdot cm^{-2}$)	R_p ($\Omega \cdot cm^{-2}$)	Q ($\Omega^{-1} \cdot S^n \cdot cm^{-2}$)	n
Ti-6Al-4V	10.89 ± 1.12	$5.69 \pm 0.13 \times 10^5$	$3.98 \pm 0.32 \times 10^{-5}$	0.93 ± 0.01
Ti-4Al-2V-1Mo-1Fe	8.71 ± 0.87	$6.50 \pm 0.35 \times 10^5$	$2.56 \pm 0.26 \times 10^{-5}$	0.95 ± 0.01

The potentiodynamic polarization curves for the Ti-6Al-4V and Ti-4Al-2V-1Mo-1Fe alloys in 3.5 wt.% NaCl solution are shown in Figure 9. It is seen from Figure 9 that the corrosion behaviors of the Ti-6Al-4V and Ti-4Al-2V-1Mo-1Fe alloys were similar. Both alloys were typically passive materials, displaying a wide passive region from 0.11 ± 0.03 V_{SCE} to 2.5 ± 0.05 V_{SCE}. After the potential was scanned to 2.5 V_{SCE} in 3.5 wt.% NaCl solution, no film breakdown was observed. This clearly indicates the passive film forming spontaneously on the Ti-6Al-4V and Ti-4Al-2V-1Mo-1Fe alloys surfaces was thermodynamically stable [37]. It is reported that Al or Fe could be oxidized and form a compact Al and Fe oxide layer on the top of passive film, inhibiting the dissolution of the oxide film [22]. The corrosion current density (i_{corr}) for the Ti-6Al-4V and Ti-4Al-2V-1Mo-1Fe alloys was $2.23 \pm 0.41 \times 10^{-7}$ A·cm^{-2} and $1.51 \pm 0.22 \times 10^{-7}$ A·cm^{-2}, respectively. Compared to the Ti-4Al-2V-1Mo-1Fe alloy, the lower E_{corr} and higher i_{corr} of the Ti-6Al-4V alloy suggested that the Ti-4Al-2V-1Mo-1Fe alloy had a higher corrosion resistance, consistent with the results of EIS test.

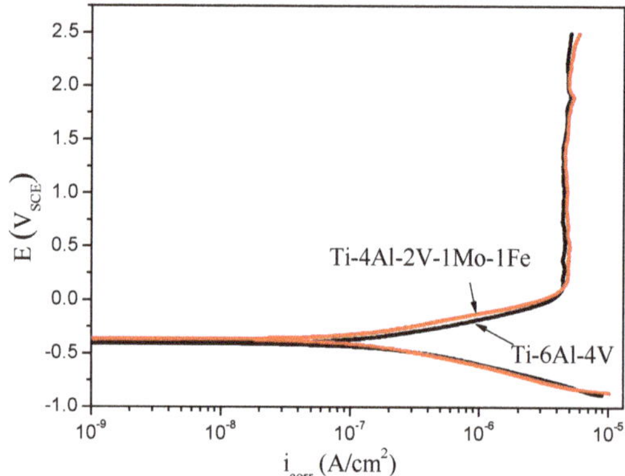

Figure 9. Potentiodynamic polarization curves of Ti-6Al-4V and Ti-4Al-2V-1Mo-1Fe alloys in 3.5 wt.% NaCl solution.

The corrosion resistance of titanium alloy relies on the stability of its passive film. The formation of passive layer requires the transfer of titanium and hydroxyl ions as follows [38]:

$$\text{Ti} \rightarrow \text{Ti}^{2+} + 2e^- \tag{4}$$

Since Ti^{2+} is unstable, it will react with H$_2$O and produce Ti^{3+} once it is formed;

$$2\text{Ti}^{2+} + 2\text{H}_2\text{O} \rightarrow 2\text{Ti}^{3+} + 2\text{OH}^- + \text{H}_2 \tag{5}$$

$$\text{Ti}^{2+} + 3\text{OH}^- \rightarrow \text{Ti(OH)}_3 \tag{6}$$

Transformation of Ti(OH)$_3$ might take place to hydrated TiO$_2$ layer in a dynamic equilibrium reaction as follows:

$$2\text{Ti(OH)}_3 \rightarrow \text{TiO}_2 \cdot \text{H}_2\text{O} + \text{H}_2 \tag{7}$$

In addition, the corrosion resistance of the Ti-4Al-2V-1Mo-1Fe alloy was slightly higher than the Ti-6Al-4V alloy, which was mainly due to the fact that the Ti-4Al-2V-1Mo-1Fe alloy contained Mo elements, making the passive film more stable [39].

3.4. Mechanical Properties

Figure 10 shows typical tensile curves of the Ti-6Al-4V and Ti-4Al-2V-1Mo-1Fe alloys. The specific mechanical property values including the tensile R_m, yield stress $R_{p0.2}$ strength, and elongation ε from the tensile curves, are summarized in Table 4. It shows that strength of the Ti-4Al-2V-1Mo-1Fe alloy was slightly lower but its elongation was higher compared to the Ti-6Al-4V alloy (yield strength of 838 MPa vs. 968 MPa, and the elongation of 15.8% vs. 13.8%). Since the volume fractions of the α and β phases were different in Ti-6Al-4V and Ti-4Al-2V-1Mo-1Fe alloys, these will have influences on their tensile properties. Moreover, the α phase had more slip systems and the β phases had limited slip systems according to their structure identified by XRD in Figure 1. Therefore, the stress concentration at α/β phase interfaces could be easily induced in the tensile tests due to the incompatible plastic deformation between two phases. Then, micro-cracks would preferentially initiate at the α/β phase interfaces. The fracture morphologies of the tensile test for the Ti-6Al-4V and Ti-4Al-2V-1Mo-1Fe alloys are displayed in Figure 11. Many dimples were observed in Figure 11, and the Ti-6Al-4V and Ti-4Al-2V-1Mo-1Fe alloys were typically transgranular with a dimple fracture [40].

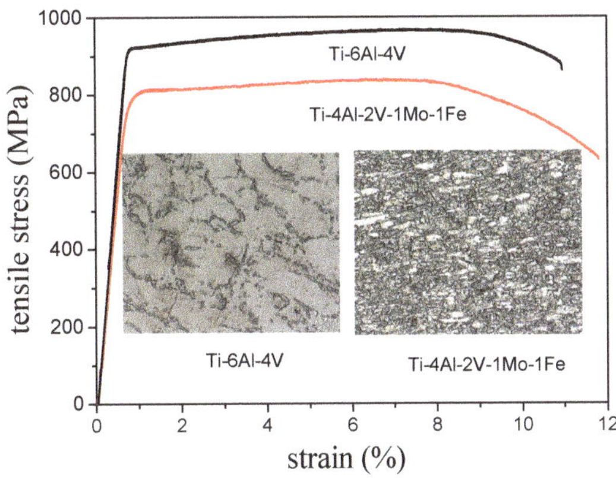

Figure 10. Tensile stress–strain curves of the Ti-6Al-4V and Ti-4Al-2V-1Mo-1Fe alloys.

Table 4. Mechanical parameters of the Ti-6Al-4V and Ti-4Al-2V-1Mo-1Fe alloys.

Alloy	R_m (MPa)	$R_{p0.2}$ (MPa)	ε (%)
Ti-6Al-4V	968 ± 22.1	921 ± 15.4	13.8 ± 1.1
Ti-4Al-2V-1Mo-1Fe	838 ± 16.3	796 ± 14.3	15.8 ± 1.3

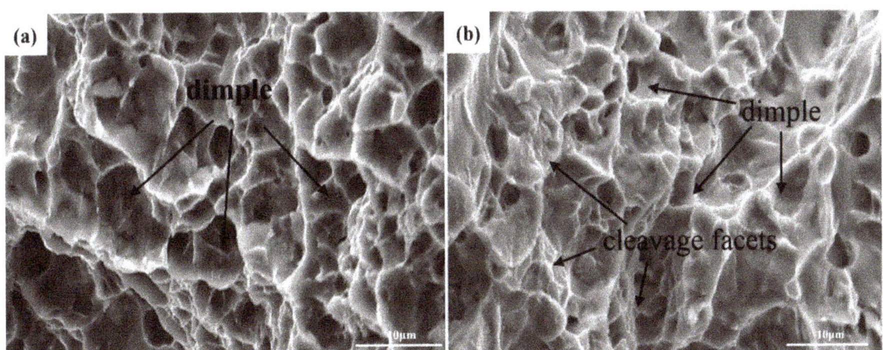

Figure 11. Fractography of the (**a**) Ti-6Al-4V and (**b**) Ti-4Al-2V-1Mo-1Fe alloys.

4. Conclusions

In this paper, the corrosion and tensile behaviors of the Ti-6Al-4V and Ti-4Al-2V-1Mo-1Fe alloys were investigated. The results were summarized as follows.

(1) Both Ti-6Al-4V and Ti-4Al-2V-1Mo-1Fe alloys were composed of the α and β phases. The volume fractions of the β phase in these two alloys were 7.4% and 47.3%.

(2) Both Ti-6Al-4V and Ti-4Al-2V-1Mo-1Fe alloys presented excellent corrosion resistance in 3.5 wt.% NaCl solution. No obvious corrosion was observed on the surface of the two alloys after immersion in 3.5 wt.% NaCl for 180 days. Compared to the Ti-6Al-4V alloy, the higher Mo content

in the Ti-4Al-2V-1Mo-1Fe alloy increased the stability of passivation film, and showed an increased corrosion resistance.

(3) Compared to the Ti-6Al-4V alloy, the Ti-4Al-2V-1Mo-1Fe alloy presented a slightly lower strength and higher ductility.

Author Contributions: Data curation, Y.W. and H.Z.; funding acquisition, Y.M.; methodology, D.X. and Y.M.; writing—original draft, Y.Q. and S.W.; writing—review and editing, J.C.

Funding: This research received no external funding.

Acknowledgments: The authors acknowledge the financial support of the National Natural Science Foundation of China (No. 51871225), National Key Research and Development Program of China [2018YFC0310400], and Natural Science Foundation of the Higher Education Institutions of Jiangsu Province, China [18KJB460007].

Conflicts of Interest: The authors declare no conflict of interest.

References

1. Boyer, R.R. An overview on the use of titanium in the aerospace industry. *Mater. Sci. Eng. A* **1996**, *213*, 103–144. [CrossRef]

2. Zhang, H.; Li, J.L.; Ma, P.Y.; Xiong, J.T.; Zhang, F.S. Study on microstructure and impact toughness of TC4 titanium alloy diffusion bonding joint. *Vacuum* **2018**, *152*, 272–277. [CrossRef]

3. Nady, H.; El-Rabiei, M.M.; Samy, M. Corrosion behavior and electrochemical properties of carbon steel, commercial pure titanium, copper and copper–aluminum–nickel alloy in 3.5% sodium chloride containing sulfide ions. *Egypt. J. Pet.* **2017**, *26*, 79–94. [CrossRef]

4. Cui, Z.Y.; Wang, L.W.; Zhong, M.Y.; Ge, F.; Gao, H.; Man, C.; Liu, C.; Wang, X. Electrochemical behavior and surface characteristics of pure titanium during corrosion in simulated desulfurized flue gas Condensates. *J. Electrochem. Soc.* **2018**, *165*, C542. [CrossRef]

5. Wang, Z.B.; Hu, H.X.; Zheng, Y.G.; Ke, W.; Qiao, Y.X. Comparison of the corrosion behavior of pure titanium and its alloys in fluoride-containing sulfuric acid. *Corros. Sci.* **2016**, *103*, 50–65. [CrossRef]

6. Contu, F.; Elsener, B.; Böhni, H. Serum effect on the electrochemical behaviour of titanium, Ti6Al4V and Ti6Al7Nb alloys in sulphuric acid and sodium hydroxide. *Corros. Sci.* **2004**, *46*, 2241–2254. [CrossRef]

7. Narayanan, R.; Seshadri, S.K. Point defect model and corrosion of anodic oxide coatings on Ti-6Al-4V. *Corros. Sci.* **2008**, *50*, 1521–1529. [CrossRef]

8. Tamilselvi, S.; Raman, V.; Rajendran, N. Evaluation of corrosion behavior of surface modified Ti-6Al-4V ELI alloy in hanks solution. *J. Appl. Electrochem.* **2010**, *40*, 285–293. [CrossRef]

9. Rai, S.; Dihingia, P.J. *Optoelectronics of Cu^{2+}-Doped TiO_2 Films Prepared by Sol.–Gel Method*; Springer India: New Delhi, India, 2015; pp. 581–589.

10. Hugot-Le Goff, A. Structure of very thin TiO_2 films studied by Raman spectroscopy with interference enhancement. *Thin Solid Films* **1986**, *142*, 193–197. [CrossRef]

11. Diamanti, M.V.; Souier, T.; Stefancich, M.; Chiesa, M.; Pedeferri, M.P. Probing anodic oxidation kinetics and nanoscale heterogeneity within TiO_2 films by conductive Atomic Force Microscopy and combined techniques. *Electrochim. Acta* **2014**, *129*, 203–210. [CrossRef]

12. Yue, T.M.; Yu, J.K.; Mei, Z.; Man, H.C. Excimer laser surface treatment of Ti–6Al–4V alloy for corrosion resistance enhancement. *Mater. Lett.* **2002**, *52*, 206–212. [CrossRef]

13. Blanco-Pinzon, C.; Liu, Z.; Voisey, K.; Bonilla, F.A.; Skeldon, P.; Thompson, G.E.; Piekoszewski, J.; Chmielewski, A.G. Excimer laser surface alloying of titanium with nickel and palladium for increased corrosion resistance. *Corros. Sci.* **2005**, *47*, 1251–1269. [CrossRef]

14. Newman, R.C. The dissolution and passivation kinetics of stainless alloys containing molybdenum- I. Coulometric studies of Fe-Cr and Fe-Cr-Mo alloys. *Corros. Sci.* **1985**, *25*, 331–339. [CrossRef]

15. Newman, R.C. The dissolution and passivation kinetics of stainless alloys containing molybdenum-II. Dissolution kinetics in artificial pits. *Corros. Sci.* **1985**, *25*, 341–350. [CrossRef]

16. Marcus, P. On some fundamental factors in the effect of alloying elements on passivation of alloys. *Corros. Sci.* **1994**, *36*, 2155–2458. [CrossRef]

17. Ouchi, C.; Fukai, H.; Hasegawa, K. Microstructural characteristics and unique properties obtained by solution treating or aging in β-rich α+β titanium alloy. *Mater. Sci. Eng. A* **1999**, *263*, 132–136. [CrossRef]

18. Erween Abd, R.; Safian, S. Investigation on tool life and surface integrity when drilling Ti-6Al-4V and Ti-5Al-4V-Mo/Fe. *JSME Int. J. Ser. C* **2006**, *49*, 340–345.

19. Lin, D.J.; Lin, J.H.C.; Ju, C.P. Structure and properties of Ti-7.5Mo-xFe alloys. *Biomaterials* **2002**, *23*, 1723–1730. [CrossRef]

20. Prodana, M.; Bojin, D.; Ioniță, D. Effect of hydroxyapatite on interface properties for alloy/biofluid. *UPB Sci. Bull. Ser. B Chem. Mater. Sci.* **2009**, *71*, 89–98.

21. Pimenova, N.V.; Starr, T.L. Electrochemical and corrosion behavior of Ti–xAl–yFe alloys prepared by direct metal deposition method. *Electrochim. Acta* **2006**, *51*, 2042–2049. [CrossRef]

22. Lu, J.W.; Zhao, Y.Q.; Niu, H.Z.; Zhang, Y.S.; Du, Y.Z.; Zhang, W.; Huo, W.T. Electrochemical corrosion behavior and elasticity properties of Ti-6Al-xFe alloys for biomedical applications. *Mater. Sci. Eng. C* **2016**, *62*, 36–44. [CrossRef] [PubMed]

23. Hsu, H.C.; Pan, C.H.; Wu, S.C.; Ho, W.F. Structure and grindability of cast Ti-5Cr-xFe alloys. *J. Alloy. Compd.* **2009**, *474*, 578–583. [CrossRef]

24. Hsu, H.C.; Hsu, S.K.; Wu, S.C.; Lee, C.J.; Ho, W.F. Structure and mechanical properties of as-cast Ti-5Nb-xFe alloys. *Mater. Charact.* **2010**, *61*, 851–858. [CrossRef]

25. Qiao, Y.X.; Cai, X.; Cui, J.; Li, H.B. Passivity and semiconducting behavior of a high nitrogen stainless steel in acidic NaCl solution. *Adv. Mater. Sci. Eng.* **2016**, *6065481*, 1–8. [CrossRef]

26. Qiao, Y.X.; Chen, J.; Zhou, H.L.; Wang, Y.X.; Song, Q.N.; Li, H.B.; Zheng, Z.B. Effect of solution treatment on cavitation erosion behavior of high-nitrogen austenitic stainless steel. *Wear* **2019**, *424–425*, 70–77. [CrossRef]

27. Huang, S.S.; Ma, Y.J.; Ping, Z.Y.; Zhang, S.L.; Yang, R. Influence of alloying elements partitioning behaviors on the microstructure and mechanical properties in $\alpha+\beta$ titanium alloy. *Acta Metall. Sin.* **2019**, *55*, 741–750.

28. Huang, S.S.; Zhang, J.H.; Ma, Y.J.; Zhang, S.L.; Youssef, S.S.; Qi, M.; Wang, H.; Qiu, J.K.; Xu, D.S.; Lei, J.F.; et al. Influence of thermal treatment on element partitioning in $\alpha+\beta$ titanium alloy. *J. Alloy. Compd.* **2019**, *791*, 575–585. [CrossRef]

29. Qiao, Y.X.; Tian, Z.H.; Cai, X.; Chen, J.; Wang, Y.X.; Song, Q.N.; Li, H.B. Cavitation erosion behaviors of a nickel-free high-nitrogen stainless steel. *Tribol. Lett.* **2019**, *67*, 1–9. [CrossRef]

30. Carnot, A.; Frateur, I.; Zanna, S.; Tribollet, B.; Dubois-Brugger, I.; Marcus, P. Corrosion mechanisms of steel concrete moulds in contact with a demoulding agent studied by EIS and XPS. *Corros. Sci.* **2003**, *45*, 2513–2524. [CrossRef]

31. Hitz, C.; Lasia, A. Experimental study and modeling of impedance of the her on porous Ni electrodes. *J. Electroanal. Chem.* **2001**, *500*, 213–222. [CrossRef]

32. Jeyaprabha, C.; Sathiyanarayanan, S.; Venkatachari, G. Influence of halide ions on the adsorption of diphenylamine on iron in 0.5 M H_2SO_4 solutions. *Electrochim. Acta* **2006**, *51*, 4080–4088. [CrossRef]

33. Qiao, Y.X.; Zheng, Y.G.; Ke, W.; Okafor, P.C. Electrochemical behaviour of high nitrogen stainless steel in acidic solutions. *Corros. Sci.* **2009**, *51*, 979–986. [CrossRef]

34. Grubač, Z.; Metikoš-Huković, M. EIS study of solid-state transformations in the passivation process of bismuth in sulfide solution. *J. Electroanal. Chem.* **2004**, *565*, 85–94. [CrossRef]

35. Hirschorn, B.; Orazem, M.E.; Tribollet, B.; Vivier, V.; Frateur, I.; Musiani, M. Determination of effective capacitance and film thickness from constant-phase-element parameters. *Electrochim. Acta* **2010**, *55*, 6218–6227. [CrossRef]

36. Wang, Z.B.; Hu, H.X.; Liu, C.B.; Zheng, Y.G. The effect of fluoride ions on the corrosion behavior of pure titanium in 0.05M sulfuric acid. *Electrochim. Acta* **2014**, *135*, 526–535. [CrossRef]

37. Cao, C.N. *Theory of Electrochemical*; Chemical Industry Press: Beijing, China, 2004; pp. 24–35.

38. Ibrahim, M.A.; Pongkao, D.; Yoshimura, M. The electrochemical behavior and characterization of the anodic oxide film formed on titanium in NaOH solutions. *J. Solid State Electrochem.* **2002**, *6*, 341–350. [CrossRef]

39. Wang, B.J.; Xu, D.K.; Wang, S.D.; Han, E.H. Recent progress in the research about fatigue crack initiation of Mg alloys under elastic stress amplitudes: A review. *Front. Mech. Eng.* **2019**, *14*, 113–127. [CrossRef]

40. Wang, B.J.; Xu, D.K.; Wang, S.D.; Sheng, L.Y.; Zeng, R.C.; Han, E.H. Influence of solution treatment on the corrosion fatigue behavior of an as forged Mg-Zn-Y.-Zr alloy. *Int. J. Fatigue* **2019**, *120*, 46–55. [CrossRef]

Review

Martensite Formation and Decomposition during Traditional and AM Processing of Two-Phase Titanium Alloys—An Overview

Maciej Motyka

Department of Materials Science, Rzeszow University of Technology, Al. Powstancow Warszawy 12, 35-959 Rzeszow, Poland; motyka@prz.edu.pl

Abstract: Titanium alloys have been considered as unique materials for many years. Even their microstructure and operational properties have been well known and described in details, the new technologies introduced—e.g., 3D printing—have restored the need for further research in this area. It is understood that martensitic transformation is usually applied in heat treatment of hardenable alloys (e.g., Fe alloys), but in the case of titanium alloys, it also occurs during the thermomechanical processing or advanced additive manufacturing. The paper summarizes previous knowledge on martensite formation and decomposition processes in two-phase titanium alloys. It emphasizes their important role in microstructure development during conventional and modern industrial processing.

Keywords: titanium alloys; martensitic formation; martensite decomposition; thermomechanical processing (TMP); additive manufacturing (AM)

Citation: Motyka, M. Martensite Formation and Decomposition during Traditional and AM Processing of Two-Phase Titanium Alloys—An Overview. *Metals* **2021**, *11*, 481. https://doi.org/10.3390/met11030481

Academic Editor: C. Issac Garcia

Received: 25 February 2021
Accepted: 12 March 2021
Published: 14 March 2021

Publisher's Note: MDPI stays neutral with regard to jurisdictional claims in published maps and institutional affiliations.

1. Introduction

Titanium alloys, due to their unique properties, are used by various industry branches for a wide spectrum of applications, from transportation (mainly in the aerospace industry) to medicine. The classification of titanium alloys is based on the content of α and β phases in their microstructure—the following main groups are distinguished: α, $\alpha + \beta$, and β alloys, with further subdivision into near-α and near-β alloys [1]. Microstructure, thereby mechanical properties, of titanium alloys are usually developed in combined plastic working and heat treatment processes—called thermomechanical processing (TMP) [2,3]. Generally, two extremely different microstructures are developed in two-phase $\alpha + \beta$ alloy, concerning the morphology of α phase: lamellar and globular (equiaxed) [4–6]. Quite unique mechanical properties can be achieved in the case of titanium alloys characterized by a mixture of the types of mentioned microstructures, i.e., bi-modal [5,6] or even tri-modal [6,7] microstructures, or a mixture of coarse and fine lamellae, i.e., bi-lamellar microstructure [6].

Martensitic microstructure in $\alpha + \beta$ alloys is considered to be rather temporary, and after tempering, it is transformed into fully fine lamellar microstructure [5,6]. Hence, heat-treated—i.e., hardened and tempered—martensitic titanium alloys (e.g., Ti-6Al-4V) exhibit high mechanical properties and are widely used for heavily loaded structural parts [8].

Semiatin's recent review [9] on TMP of $\alpha + \beta$ titanium alloys largely concerns the fragmentation and spheroidization processes of α lamellae and martensitic α' acicular phases. It was postulated that the acicular microstructure accelerated dynamic spheroidization, especially at low temperature. It was also mentioned that the martensitic phase plays an important role in superplastic deformation (SPD), during which it is fragmented into small equiaxed grains enabling enhanced plastic strain. Probably Inagaki [10] and Zherebtsov et al. [11] were the first to observe what was later confirmed and analyzed by other researchers [12,13]. Markovsky et al. [14] also demonstrated the benefit of initial martensitic microstructure in developing an ultrafine α grains in Ti-6Al-4V alloy.

131

Even if the TMP of titanium alloys seems to be well known and industrially verified, some modern technologies, like additive manufacturing (AM), intended to produce near-net shape structural parts, require a new approach which cannot be based on plastic deformation. Dutta and Froes [15,16], based on ASTM standards, classified AM technologies for metals into two categories: directed energy deposition (DED) and powder bed fusion (PBF). The DED technologies (direct metal deposition (DMD), laser engineered net shaping (LENS), shaped metal deposition or wire and arc additive manufacturing (WAAM)) offer larger build envelope and higher deposition rate compared with the PBF methods, but their ability to build hollow cooling passages and finer geometry is limited. The building of complex features and high precision parts is possible using the PBF technologies (selective laser sintering (SLS), direct metal laser sintering (DMLS), selective laser melting (SLM), and electron beam melting (EBM)) which, however, have other limitations related to the size of build envelope and horizontal layer building ability. In the case of titanium alloys, the AM technologies create the conditions for martensite formation in their microstructure. Yang et al. [17] analyzed such process in the SLM technique, where the metal powder layer is first heated and melted rapidly by the focused laser beam. They indicated that during SLM, the Ti-6Al-4V is subjected to a high temperature gradient (10^6 K/m) as well as rapid solidification and cooling rates (can reach 10^8 K/s). Under such rapid solidification and cooling rates, bcc β phase transforms completely into metastable hcp α' martensite. The resulted microstructure favors intergranular failure which deteriorates alloy ductility. He J. et al. [18] noticed that the martensitic α' phase in the microstructure of Ti-6Al-4V alloy resulted in the improvement of tensile strength and hardness and decrease in plasticity. Xu Y. et al. [19] considered that the fracture mechanisms of SLM-produced Ti-6Al-4V specimens were affected by the width of martensite needles. Martensitic transformation was also observed in titanium alloys produced by other AM methods, both DED and PBF ones—LENS [20], SLS [21], and EBM [22]. It is worth adding that the effect of martensite on mechanical properties of AM titanium alloys is not clear-cut. Zafari and Xia reported [23] a fully martensitic α' Ti-6Al-4V alloy, produced using SLM, exhibiting an impressive combination of high ductility and strength. In the same alloy, a two-phase $\alpha + \alpha'$ microstructure was developed by EBM process and a desirable increase in both strength and ductility was found [24]. This shows the research potential in this area.

Innovative technologies, like AM, offer new production possibilities for metallic materials which have been successfully processed for years by conventional methods. The novelty of the presented overview is an attempt to correlate commonly known physical properties of titanium and its alloys with the particular requirements of AM, especially in terms of martensite formation.

Further in this paper, principal data on martensite formation and decomposition are collected and presented (in Sections 2 and 3, respectively). The effect of martensite α' content in microstructure of two-phase $\alpha + \beta$ titanium alloys, processed by conventional (TMP) or modern (AM) methods, on the final mechanical properties is also analyzed. In Section 4, the role of martensitic phase in the development of microstructure of two-phase titanium alloys using TMP routes is discussed, while Section 5 is devoted to AM processes leading to unintentional martensite formation in titanium components.

2. Martensitic Transformation and Martensitic Phases

The possibility of martensite formation in titanium alloys results from allotropic transformation Tiα↔Tiβ, which is considered as a primary phase transformation in titanium. During the heating of titanium to the temperature above 882.5 °C, Tiα (hcp) transforms into Tiβ (bcc). In the case of cooling, the reverse transformation Tiβ→Tiα occurs. At low cooling rates, allotropic transformation Tiβ→Tiα proceeds by nucleation and growth of new phase crystals, whereas at ahigh cooling rate, the allotropic transformation has the features of martensitic transformation. In pure iodide titanium, after cooling at the rate of 100 K/s shear processes, related to diffusionless martensitic transformation, can be observed only on β phase grain boundaries. In titanium alloys, increase in alloying elements content (e.g.,

Cr, Mo, V, Nb, Sn, and Zr) reduce the martensite start temperature (T_{Ms}), but also the critical cooling rate, so fully martensitic microstructure in these alloys can be developed. In titanium alloys containing Mo, Mn, Nb, and V, martensitic transformation can be induced by plastic deformation caused by external load. The content of mentioned alloying elements must be high enough to reduce the T_{Ms} to a value close to room temperature. Moreover, the cooling rate should also be high enough to prevent precipitation of intermediate phases which preclude martensite formation through deformation [1,3,25].

Martensitic transformation in titanium alloys is related to atom movement and lattice deformation. Such a shear transformation process is believed to be realized by activation of the following shear systems: [111](11−2) and [111](−101) in β lattice or [2−1−13](−2112) and [2−1−13](−1011) in α lattice, also by twin formation in {111} or {112} plains [3,25]. According to Lütjering and Williams [4], the hexagonal martensite, designated as α′, exhibits two morphologies: massive (lath) and acicular ones. Massive martensite can be observed in pure and low-titanium alloy and titanium alloys which exhibit high T_{Ms}. Acicular martensite is characteristic for the alloys with higher solute content, reducing T_{Ms}.

Two martensitic phases are considered the most important in titanium alloys—α′ and α″. The α′ phase is a supersaturated solid solution of elements in Tiα allotropic form, and it has the same hcp crystal structure. It is usually obtained by fast cooling (water quenching) from the temperature range of β phase stability, but it can be formed by plastic deformation or during aging in zones of β phase depleted in elements stabilizing it (β-stabilizing elements). It is rather obvious that martensitic phases are compared to martensite in steels. In the case of titanium alloys, some features differentiate α′ phase from martensite in steel. The α′ phase is a supersaturated substitutional solid solution, whereas in steels, martensite is a supersaturated interstitial solid solution. Martensite content in microstructure results in lower strengthening effect compared to iron alloys and titanium alloys with fully martensitic microstructure have quite good plasticity, which is distinct from quenched steels [3].

As mentioned before, the martensitic α′ phase in titanium alloys is usually acicular (needle-like) (Figure 1). Besides alloy composition, the needle morphology and orientation in prior β grains (Figure 1a) are determined by heat treatment or plastic working at the temperature of β phase stability and even by minor changes of cooling conditions [1,3,4]. Transmission electron microscopy examination enables to reveal twins and high dislocation density in needles (Figure 1d).

Figure 1. *Cont.*

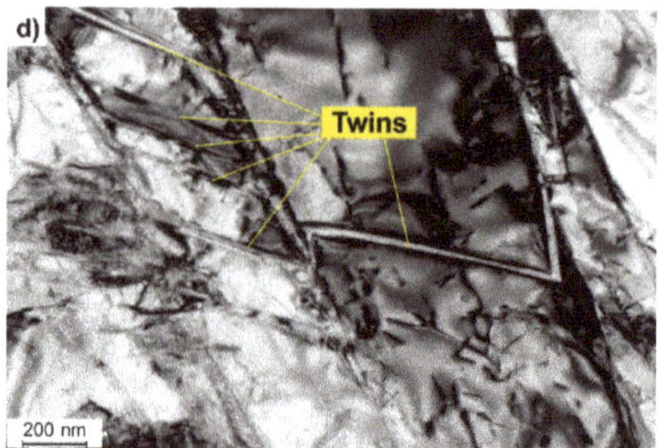

Figure 1. Martensitic microstructure in Ti-6Al-4V alloy after water quenching from the temperature of 1050 °C (β phase range): fully martensitic microstructure with visible prior β grains (LM) (**a**), common orientation of α′ needles in prior β grain (LM) (**b**), α′ needles growing perpendicular to the prior β grain boundary (LM/DIC) (**c**), twins and dislocation substructure in α′ needles (TEM) (**d**); LM—light microscopy, DIC—differential interference contrast, TEM—transmission electron microscopy.

In some martensitic titanium alloys containing transient elements (e.g., Mo, V, Nb, and Ta), another martensitic phase, α″, can be formed. It is believed that the following features of alloying elements—atom volume, concentration of electrons, and valence (higher than for titanium, i.e., >4)—determine α″ phase formation in the alloy. The α″ phase, same as the α′ phase, is a supersaturated solid solution of elements in Tiα phase, but it crystallizes in orthorhombic system. Morphology of both martensitic phases is similar (acicular), but they differ in needle size [3,25]. In Ti-Nb shape memory alloys, the structure of quenching-induced α″ martensite is often twinned—mainly by {111}-type twins [26]. Banerjee and Williams [27] analyzed the effect of β-stabilizing elements on metastable transformations in titanium alloys. They showed that T_{Ms} decreased with increasing β-stabilizing content and, moreover, the T_{Ms} for α″ phase is lower than that for the α′ phase (Figure 2).

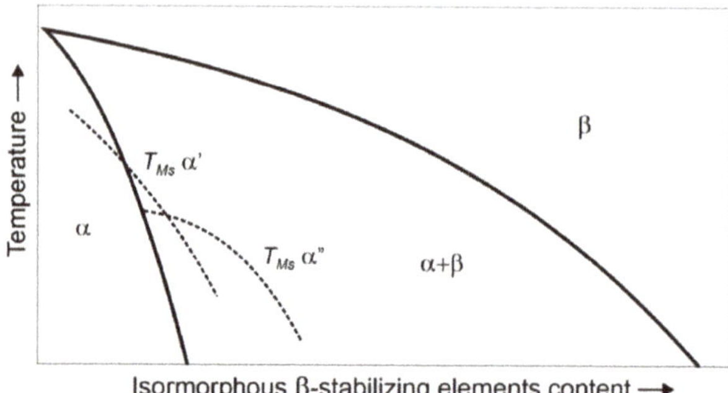

Figure 2. The effect of isomorphous β-stabilizing elements on martensite start temperature (TMs) of α′ and α″ phases (drawn based on [27]).

Moiseev et al. [28] further noted that in Ti-Mo and Ti-V titanium alloys, the α″ phase depleted in β-stabilizing elements had the same mechanical properties as α′ phase. They also found that the α″ phase was softer than that of α′. Even if the α′ phase in titanium alloys has no high hardness and strength, as martensite of steel has, formation of α′ phase can lead to their noticeable hardening [29]. In general, the hardness of such alloy increases as the rate of cooling [30] and the temperature of quenching [31] increases (both raise the martensite content). It is also worth adding that the grain size of prior β grains ($D_{prior β}$) determines room temperature mechanical properties of titanium alloys with fully martensitic microstructure. Chong et al. [32] indicated that for Ti-6Al-4V alloy, the ultimate tensile strength (UTS) and elongation for $D_{prior\beta}$ = 200 μm were 1047 MPa and 9.4%, while for $D_{prior\beta}$ = 8 μm, were 1298 MPa and 19.8%, respectively.

3. Martensite Decomposition

Both α' and α'' martensitic phases in two-phase structural titanium alloys are considered as intermediate phases, which during annealing (tempering), are transformed to lamellar $\alpha + \beta$ microstructure, giving them high mechanical properties [5]. Decomposition of martensitic phases is, therefore, a desirable process during heat treatment of titanium alloys applied for heavy load structural parts.

Martensite transformation in titanium alloys is most often associated with its decomposition into a mixture of thermodynamically stable α and β phases [4,5,33–36]. As reported by Bylica and Sieniawski [36], the α' and α'' martensite decomposition in titanium alloys takes place according to the following schemes:

$$\alpha' \rightarrow \alpha'_{depl} + \beta \rightarrow \alpha + \beta, \tag{1}$$

$$\alpha'' \rightarrow \alpha''_{enr} + \alpha_M \rightarrow \alpha + \beta_M \rightarrow \alpha + \beta, \tag{2}$$

wheres α'_{depl}—α' phase depleted of the β-stabilizing elements, α''_{enr}—α'' phase enriched in the β-stabilizing elements, α_M and β_M—metastable α and β phases, respectively. The mechanism of the mentioned processes depends mainly on the decomposition temperature and chemical composition of martensite. Decomposition of the α' phase (1) can be considered as a precipitation of β phase together with gradual depletion of α' phase until its chemical composition corresponds to an equilibrium α phase or as a precipitation of depleted α' phase and subsequent transition of enriched zones of α phase into β phase. Concerning decomposition of α'' phase (2), various mechanisms are also considered [36]:

- precipitation of metastable βM phase from martensitic α'' phase, and then α''_{depl} phase depleted in the β-stabilizing elements transforms into α' phase and, next, into α phase:

$$\alpha'' \rightarrow \alpha''_{depl} + \beta_M \rightarrow \alpha' + \beta_M \rightarrow \alpha + \beta; \tag{3}$$

- precipitation of α, causing gradual enrichment of α''_{enr} phase (in the β-stabilizing elements) and then its transition into metastable βM and, finally, β phase:

$$\alpha'' \rightarrow \alpha + \alpha''_{enr} \rightarrow \alpha + \beta_M \rightarrow \alpha + \beta; \tag{4}$$

- formation of zones with different concentration of alloying elements (α''_{depl} and α''_{enr}) which transform into metastable α''_{depl} and βM and, finally, stable α and β phases:

$$\alpha'' \rightarrow \alpha''_{depl} + \alpha''_{enr} \rightarrow \alpha''_{depl} + \beta_M \rightarrow \alpha + \beta. \tag{5}$$

It is believed that the efficiency of tempering titanium alloys having a specific chemical composition depends on the process temperature and time [37,38]. It should be noted that the effect of tempering—expected high strength—results from the formation of metastable phases, their decomposition, and precipitation of highly dispersed phases [37,39]. Therefore, the decomposition of martensitic phases (α', α'') can also be accompanied by the creation of intermetallic phases (e.g., Ti_3Al, TiFe).

Based on recent results concerning the continuous heating of hardened Ti–6Al–6V–2Sn alloy [40], it was found that the α'' martensite decomposition is diffusion-driven transformation of the supersaturated phase (at the temperature range of 400–600 °C). Moreover, precipitations of fine β particles (with diameter of ~5nm) were formed mainly along the boundaries of α laths, formed with the participation of nano-twinning process. Yu et al. [41] claimed that the $\alpha' \rightarrow \alpha + \beta$ decomposition in Ti-6Al-4V was an elemental diffusion transformation, and the shape and position of its products changed with time. They described a relationship between the shape and distribution of precipitates and the temperature of tempering. The precipitations were randomly distributed in the crystals and the grain boundaries or formed local aggregations at the grain boundaries.

Lee et al. [42] mentioned the spinodal decomposition of α'' martensite in Ti-Mo alloys, whose mechanism can be considered as a formation of zones varying in saturation of Mo ($\alpha''_{Mo\text{-}depl}$ and $\alpha''_{Mo\text{-}enr}$—depleted and enriched in Mo, respectively) according to formula (5). However, their work concerned the α'' martensite decomposition mechanism in Ti-Al-Fe-Si alloy. It was found that at low tempering temperature (<450 °C), Fe-rich α phase was transformed in the twins of α'' martensite, while at higher tempering temperature (>450 °C) the Fe-depleted α'' martensite (due to Fe atoms diffusion to the α''/α interface) transforms into α phase.

As was mentioned before, martensite decomposition during the tempering process is considered as a conventional way for developing specific mechanical properties through heat treatment. Other aspects, like the effect of factors other than tempering temperature on the martensite decomposition process, are not too often investigated. Some papers indicate the possibility of plastic deformation of titanium alloys in a hardened state [43–45]. Li et al. [43] analyzed the deformation mechanisms of martensite and found that deformation of α phase in Ti–3Al–4.5V–5Mo (VT16) alloy is accompanied by $\alpha'' \rightarrow \alpha'$ transformation [43]. In one study [44], the decomposition process of deformed martensite in the mentioned alloy was briefly analyzed during continuous heating up to 800 °C. Matsumoto et al. [45] investigated Ti-V-Al and Ti-V-Sn alloys, quenched and, next, cold groove rolled. They found that the microstructure of acicular martensite α' in as-quenched titanium alloy evolved into refined equiaxed cell structure with size less than 200 nm after cold groove rolling. Moreover, after the low temperature heat treatment at 300 °C of quenched and cold-deformed Ti-12V-2Al alloy, the α'' phase formed finally in the α' martensite matrix.

It should be also pointed that the hot deformation of martensite has also been analyzed as a mechanism for development of ultrafine-grained microstructure in two-phase titanium alloys [13,46] (discussed in more detail in Section 4).

Motyka et al. [47] analyzed the decomposition process of deformed martensitic phase in Ti-6Al-4V. The alloy was water-quenched from the temperature within stable β phase range (1050 °C) and, next, cold compressed up to about 20% strain. Deformed and undeformed specimens were tempered at 600, 750, and 900 °C for 1 and 2 h. It was found that the decomposition of martensitic phase at the temperature of 600 and 750 °C led to slight thickening of its laths—no meaningful differences between the morphology of tempered deformed and undeformed martensite were found. Significant microstructural changes were observed after tempering at the temperature of 900 °C (Figure 3a,b)—deformed martensite laths exhibit tendency towards fragmentation and spheroidization. Because the tempering temperature is in the lower range of $\alpha + \beta \rightarrow \beta$ phase transformation, the observed results could be caused by both martensite decomposition and phase transformation processes. Regardless of the tempering temperature, the product of martensite $\alpha'(\alpha'')$ decomposition was a mixture of α and β phases—both in the case of undeformed and deformed martensite. Motyka et al. [48] analyzed martensite decomposition in Ti-6Al-4V alloy during hot deformation process. They found that the microstructure is globular after compression at 900 °C (Figure 3c), similar to that obtained after tempering at the same temperature (Figure 3b). Martensite decomposition during hot deformation process seems to favor $\alpha + \beta \rightarrow \beta$ phase transformation—a higher volume fraction of β phase compared with "static" tempering at 900 °C (Figure 3b).

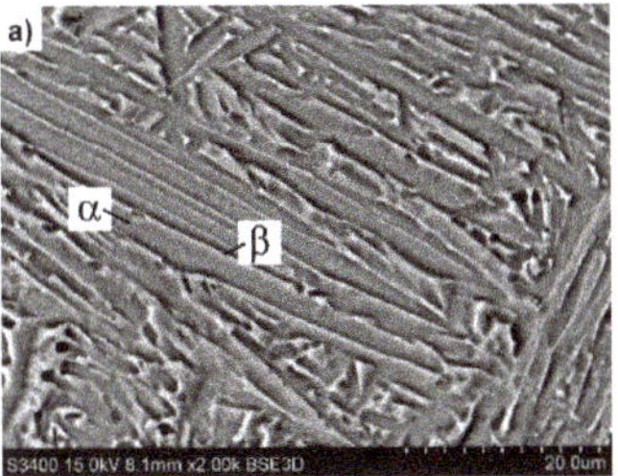
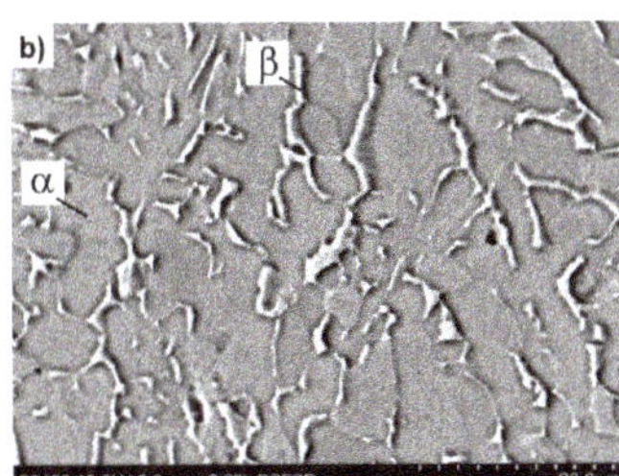
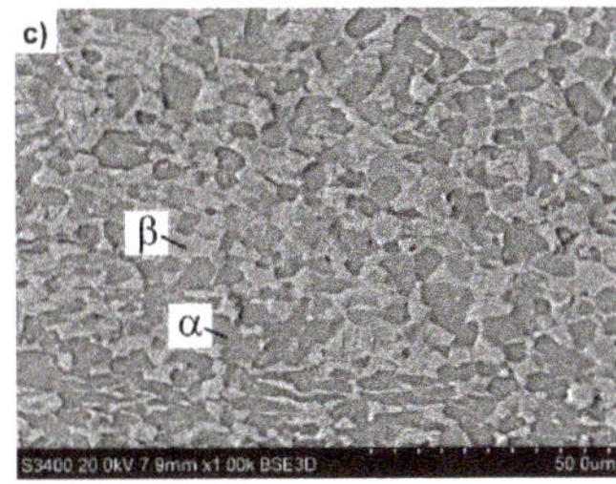

Figure 3. Morphology of α and β phases (SEM) in Ti-6Al-4V alloy having initial fully martensitic microstructure: after tempering at 900 °C without (**a**) and with preceding cold deformation (**b**) and after hot deformation at 900 °C (**c**).

4. The Role of α′(α″) Martensitic Phase in the Development of Microstructure of Two-Phase Titanium Alloys

As mentioned before, the development of the demanded microstructure of two-phase titanium alloys is usually realized in TMP, which is generally composed of three stages: I—initial heat treatment, II—hot deformation, and III—final heat treatment. Due to the limited possibilities of microstructure development by heat treatment methods only (resulting from unique features of Tiβ↔Tiα transition), hot working operation (at the temperature in the range of α + β→β transformation or in the range of stable β phase) is crucial in TMP. The initial heat treatment is applied for increasing plastic deformation effects (mainly grain refinement), whereas the final heat treatment operations are usually used for stabilization of microstructure (they restrict grain growth) [2–6,49,50].

Titanium alloys belong to the group of metallic materials for which superplastic forming (SPF) is applied, especially when forming parts can be simultaneously diffusion-bonded (SPF/DB). SPF is based on the fine-structured superplasticity, which demands fine-grained and equiaxed microstructure of deformed material [51,52]. Obviously, such material conditions may be met when severe plastic deformation (SPD) methods are used [53]. However, in the past, when SPD methods were not widespread, some attempts for development of fine-grained and equiaxed microstructure in TMP were made. Even though conditions for TMP developing a globular microstructure in two-phase α + β alloys were commonly known [4–6], the grain size of α phase was not satisfactorily small. Inagaki [10,54] investigated the Ti-6Al-4V alloy prepared by various routes of TMP. He found that the superplasticity of examined material can be enhanced by solution treatment 1050 °C/water and then severe hot rolling at temperatures below 750 °C. Inagaki observed that α′ martensite needles were broken up during hot rolling into stringers of tiny grains. This allowed obtaining a total elongation of 2100% in tensile test at 850 °C and strain rate of 10^{-2} s^{-1} (also in Ti-6Al-2Sn-4Zr-6Mo and Ti-6Al-2Sn-4Zr-2Mo alloys). Such behavior was further analyzed in detail using TEM methods. It was found that the initial microstructure of Ti-6Al-4V alloy, before the high-temperature tensile tests, consisted of fine, fragmented α + α′ martensite grains which, at earlier stages of the superplastic deformation, had decomposed into small, well-defined equiaxed α subgrains surrounded by thin β phase films.

Motyka et al. [12], based on above results, investigated superplasticity of Ti-6Al-4V alloy TM processed by water quenching (from the temperature of β stable region (1050 °C) and α + β range (950 °C)), hot forging at about 900 °C and recrystallization annealing (800 °C). Recrystallization annealing, the final heat treatment in TMP, caused spheroidization of α grains, though also their growth. Thus, the highest superplastic strain (a total elongation higher than 1600%) was found for the alloy water-quenched from the temperature of 1050 °C and hot forged—even the initial microstructure was not equiaxed. High elongation was achieved because highly deformed and elongated α grains transformed into globular ones (by fragmentation and spheroidization processes) during the heating to the deformation temperature and initial stage of superplastic deformation.

In summarizing, it can be assumed that spheroidization of elongated thin laths (formed from α' martensite needles) at the initial stage of superplastic deformation retards the growth of α grains during the process, while initial equiaxed grains grow more, which limits superplastic elongation to a greater extent.

As explained earlier, martensite decomposition during hot deformation process can lead to grain refinement in titanium alloys, enhancing their superplasticity. However, grain refinement methods are usually used for material strengthening. Warm or hot deformation of martensite was also analyzed in aspects of development of ultrafine-grained (UFG) microstructure in two-phase titanium alloys. Semiatin et al. [55,56] have postulated that the level of grain refinement in titanium alloys strongly depends on the initial microstructure and TMP route. They have shown that the final grain size can be significantly reduced through lowering of the deformation temperature and utilizing a finer initial microstructure, particularly by reducing the α platelet thickness. Zherebtsov et al. [11] observed that in Ti–6Al–4V billets processed by warm severe plastic deformation, the formation of a homogeneous UFG microstructure is promoted by initial microstructure which is either fully martensitic or globular. Park et al. [57,58] analyzed high-temperature deformation behavior and microstructural evolution process of ELI Ti-6Al-4V alloy having martensite microstructure. Hot compression tests were carried out at the temperature range of 700–950 °C. Fully dynamically globularized material exhibited high hardness due to the fine grain size and high dislocation density. Matsumoto et al. [59] also noticed that for Ti-6Al-4V alloy with an initial α' martensite microstructure, after compressing at 700 °C (strain rate of 10 s^{-1}), was characterized by UFG microstructure with an average grain size of 200 nm and a high fraction of high-angle grain boundaries. The behavior of the α' martensite microstructure was attributed to the considerable number of nucleation sites such as dislocations, interfaces of martensite variants and {101-1} twins, and the high-speed grain fragmentation along with subgrain formation in the α' starting microstructure during the initial stage of deformation. Chao et al. [13] obtained an even smaller grain size in Ti-6Al-4V alloy—150 nm—through TMP of a martensitic starting microstructure. They explained a mechanism of grain refinement which involves the development of substructure in the lath interiors at an early stage of deformation, which progressed into small high-angle segments with increasing strain. Observed grain refinement was caused by continuous dynamic recrystallization and decomposition of the supersaturated martensite. Similar effects were observed by Motyka et al. [48]. They found that martensite decomposition during compression at 600 °C enhanced grain refinement in Ti-6Al-4V alloy. The obtained UFG microstructure contained α grains having size even smaller than 200 nm. Gu et al. [46] investigated UFG $\alpha + \beta$ titanium alloy (Ti-6.5Al-3.3Mo-1.8Zr-0.26Si) produced by deformation at relatively low temperature or high strain rate. They noticed that during the compression of a martensite microstructure at 890 °C, both the α and β phases were dynamically recrystallized and the lamellae were broken up by means of grain boundary sliding and phase penetration along the subgrain boundaries. At lower temperature (650 °C), α phase was dynamically recrystallized, whereas β phase precipitated and grew in the α matrix. It was clearly suggested that the initial martensite microstructure should be preferred for the manufacturing of UFG titanium alloys by TMP routes.

Considering the role of martensitic phase in the development of microstructure of two-phase titanium alloys, thermohydrogen processing (THP) should be mentioned [60]. It is based on the modifying effect of hydrogen during heat treatment and forming processes, leading to refined microstructure and enhanced mechanical properties. It was known already in 1994 that THP combined with hot working enabled development of UFG microstructure in titanium alloys, although without participation of martensitic phase but needle-like metastable α_M phase [61]. According to Murzinova and co-authors [62], such processing enables the formation of nanocrystalline structure in two-phase titanium alloy. In order to control the effects of THP, it is important to know and understand the effect of hydrogen content on phase transformation in titanium alloys. Qazi et al. [63] investigated hot isostatically pressed Ti-6Al-4V-H alloy and observed, in its microstructure, the α' and α''

martensite laths for hydrogen content of 10 at.%. Zong et al. [64] found that the β transus, martensite start T_{Ms}, and finish T_{Mf} temperature decreased with increasing hydrogen content. In another paper, Qazi and co-workers [38] analyzed martensite decomposition in hydrogenated Ti-6Al-4V alloy. They found that tempering at a temperature below the β transus and above the Ms resulted in the transformation of the martensite to equilibrium α and β phases. In the case of tempering at a temperature below the Ms, martensite first partially transformed into metastable β phase, and then into α and β phases.

Martensite formation seems to play a crucial role in the microstructure development of titanium alloys during AM, so it will be described separately in Section 5.

5. Martensite Formation during Additive Manufacturing Processes

AM processes are considered as quite modern technologies opening up new possibilities for production of near-net shape structural parts. Even though they are used for well-known structural metallic materials, the obtained microstructure significantly differs from this, which is developed by conventional methods (i.e., cold or hot working). A characteristic feature of AM technologies is usually high cooling rate accompanying the building process, which favors martensite formation in the microstructure of titanium alloys.

It is generally accepted that depending on material and process, non-equilibrium phases evolve in the as-fabricated state, e.g., α' phase in titanium alloys. As heat conductivity is usually anisotropic during AM processes, with a significantly higher conductivity in the build direction through previously built layers, anisotropic microstructures with elongated grains are found that consequently lead to anisotropic properties. Obviously, the final operational properties can be achieved by using additional heat treatment. The thermal cycle in AM involves a reheating of already solidified layers, so in situ heat treatment is possible. If it is not sufficient, the final part microstructure may be adjusted within a wide range by subsequent ex situ heat treatment [65,66].

5.1. PBF Methods

Unique physical and thermal properties of titanium and its alloys, in combination with the high temperature gradients and complex thermal cycle usually involved in AM, make them one of the most interesting materials for investigation of the relationships between process conditions, microstructure, and mechanical properties. Among the titanium-based materials, mainly commercially pure titanium (CP Ti) and two-phase α + β Ti-6Al-4V alloy have been investigated by many research groups [65]. Qian et al. [67] analyzed the macro- and microstructural characteristics, defects, and tensile and fatigue properties of Ti-6Al-4V manufactured by SLM, laser metal deposition (both powder and wire), and selective EBM compared to conventionally produced bars (in mill-annealed condition). Authors also indicated the necessity for post-AM surface or/and heat treatments in most cases. They claimed that post-AM heat treatment, in general, was not needed for EBM-produced parts, but it was necessary for SLM-produced Ti-6Al-4V alloy for improving ductility and reducing anisotropy in mechanical properties. Regardless, the necessity for in situ or ex situ heat treatment of Ti-6Al-4V alloy after SLM, EBM, and laser metal deposition processes resulted from microstructure heterogeneity and α' martensite presence.

In the case of titanium alloys (especially Ti-6Al-4V alloy), SLM technology seems to be the most often described in the technical publications among AM methods [17–19,23,68–73]. According to their authors, the martensite formation and decomposition belong to the main factors determining mechanical properties of SLM-produced Ti-6Al-4V alloy. Yang and co-authors [17] observed that in the microstructure of SLM-produced samples, martensite laths seems to have a characteristic orientation related to heat history, material characteristics, and substructure features. Such martensitic structure was considered as a hierarchical one, where primary, secondary, tertiary, and quartic α' martensite laths were formed within columnar prior β grains. The primary α' martensite extended across the entire parent β phase. Fine secondary α' laths (with thickness of hundreds of nm) were parallel or perpendicular to the primary α' martensite. Tertiary and quartic α' martensite laths were even finer. It was found that the size of martensitic laths increases with hatch spacing, so it can be controlled by adjusting the SLM processing parameters (hatch spacing and scanning velocity). Hierarchical microstructure in SLM-produced Ti-6Al-4V alloy was also mentioned by Moridi et al. [68]. They distinguished a primary, secondary, and tertiary martensite as a result of cyclic heat treatment during the layer-wise SLM process which was partially tempered. Primary martensite in the prior β grains was more tempered (softer) compared with late (i.e., secondary and tertiary) martensite laths, which were finer because their growth was stopped by the boundaries of early martensite laths, and less tempered (harder). Even if we agree that the martensite decomposition is a clue for achieving high mechanical properties of produced parts, and it can take place during SLM process (called "autotempering" in [67]), the complete martensite decomposition requires in situ or ex situ heat treatment. He et al. [18] confirmed that at the beginning of the SLM process, the two-phase microstructure of Ti-6Al-4V transforms rapidly into a fully martensitic one. They noticed that the size of prior β columnar grains, filled by acicular martensite phase, was related to the width of laser scan track. Because the presence of martensitic α' phase in the microstructure resulted in the improvement of tensile strength and hardness and decrease in plasticity of the examined alloy, the authors proposed using a relieving heat treatment—annealing at 730 °C for 2 h in nitrogen gas conditions. It reduced α' phase content via its decomposition and enabled developing a fully $\alpha + \beta$ phase mixture in the alloy microstructure. Other researchers [19] also confirmed that the presence of acicular α' martensite and columnar prior β grains in the microstructure of SLM-processed Ti-6Al-4V alloy decreased its ductility. In this paper, it was mentioned that ductility could be improved by high-temperature preheating or heat treatment, but in the expense of the alloy strength and efficiency of the SLM process. The authors claimed that satisfactory mechanical properties of SLM-produced Ti-6Al-4V alloy could be achieved by some special temperature evolution conditions, controlled by processing parameters such as layer thickness, hatch spacing, energy density, and area ratios between the support structure and the component being manufactured. They demonstrated that a proper overall temperature and long cumulative residence time within the temperature window was favorable for in situ α' martensite decomposition. From the point of view of reducing the overall cost of AM produced parts, in situ heat treatment seems to be the appropriate operation enabling martensite decomposition in titanium alloy. In the work of [69], such treatment was applied during the SLM of Ti-6Al-4V alloy through the proper selection of processing parameters to minimize porosity with a tight hatch distance associated with long exposure periods at high temperature. Extensive martensite decomposition ($\alpha' \rightarrow \alpha + \beta$) along the building direction was found, resulting in the formation of a uniform, fine lamellar $\alpha + \beta$ microstructure. Because the temperature of the performed heat treatment was higher than T_{Ms}, a relatively thin martensitic layer (150–250 µm) was formed. The transition from α' plates to stable α lamellae proceeded along the building direction. The formation of the intermetallic α_2-Ti_3Al phase was also found along this direction using high-energy synchrotron XRD and TEM methods. Ali et al. [70] proposed high-temperature preheating during SLM, reducing thermal gradients and giving the possibility to control/tailor as-built mechanical properties. A high-temperature SLM powder bed capable of preheating to 800 °C was used during processing of Ti-6Al-4V feedstock. It was found that increasing the bed temperature to 570 °C significantly reduced residual stresses within components and enhanced yield strength and ductility. Preheating enabled the decomposition of martensite into an equilibrium $\alpha + \beta$ microstructure.

It should be mentioned that in the case of porous parts (e.g., for orthopedic purposes), thermal conditions during AM processing favor martensitic formation to a greater extent than in bulk elements. A route for α' decomposition in sub-β-transus ex situ heat treatment of SLM-fabricated parts made of Ti-6Al-4V alloy was proposed in [71]. The authors of this paper did not find any microstructural changes after annealing at 650 °C for 1 h, whereas after 2 h, fine precipitates of β phase along the α' needle boundaries were observed. The heat treatment performed at a higher temperature of 800 °C for either 1 or 2 h led to development of fine $\alpha + \beta$ microstructure, in which β phase was present as particles fewer in number and larger in size (compared with the parts annealed at 650 °C for 2 h). Ex situ heat treatment at 800 °C/2 h improved the ductility of SLM-processed Ti-6Al-4V alloy and slightly reduced its strength.

Ex situ, post-SLM, heat treatment of Ti-6Al-4V alloy was also analyzed by Caoa et al. [72] in terms of static coarsening mechanism of lamellar microstructure at the temperature range of 700–950 °C. It was found that high-temperature heat treatment facilitated martensite decomposition and promoted lamellae growth. Observed static coarsening behavior in SLM-produced Ti-6Al-4V alloy was interpreted by LSW (Lifshitz, Slyozov, and Wagner) theory. It was indicated that the coarsening mechanisms were bulk diffusion at 700–800 °C, and a combination of bulk diffusion and interface reaction at 900 and 950 °C.

Neikter et al. [73] focused attention on the binary nature of the microstructure developed in a SLM-produced Ti-6Al-4V alloy. The binary microstructure was found in the horizontal plane and was formed with various laser scan angles between adjacent layers. According to the authors, the fine microstructure zone separated the coarse microstructure zones. It was also found that after the post-SLM heat treatment, the binary microstructure was retained unless the heat treatment temperature reached the β transus, but when this temperature was exceeded, the binary microstructure disappeared.

Based on the published results mentioned above, martensite phase decomposition seems to be the main purpose of using ex or in situ heat treatment for SLM-produced two-phase $\alpha + \beta$ titanium alloys. However, some researchers argue that the presence of martensitic phase in the microstructure does not preclude high mechanical properties of SLM-produced parts. It was mentioned earlier that Zafari and Xia [23] achieved high yield strength (1150 MPa) and excellent tensile elongation (14–15%) in SLM-produced Ti-6Al-4V alloy with fully α' martensitic microstructure. According to them, it was critical to produce pure intersecting ultrafine α' plates free of β phase by optimizing SLM parameters in order to avoid stress concentration at the α'/β interface and easy crack propagation along this interface. It was also explained that plastic deformation would then proceed by dislocation slip, which cuts and reorients α' into randomly oriented nanograins until final ductile fracture with necking and dimple formation. The authors claimed that the usually observed poor ductility of SLM-produced parts was related to difficulty in preventing the formation of β phase and producing ultrafine α' laths, rather than to the brittleness of α' phase. Zafari et al., in their later work [74], confirmed that the energy density in SLM influenced the steady-state temperature in a deposited layer reached by balancing between heat input from the subsequent layers and heat loss in the previous ones. They determined a critical energy density necessary to reach for in situ α' decomposition in the produced parts.

Grain refinement, as a way for improving the mechanical properties of SLM-produced Ti-6Al-4V alloy, was also indicated by Xu et al. [75]. The lamellar microstructure composed of ultrafine (~200–300 nm) α laths and retained β phases was developed during SLM. The authors claimed that the heat treatment time accumulated from the thermal cycling effect during the process was sufficient for a near-complete transformation of α' martensite into ultrafine lamellar $\alpha + \beta$ mixture. As a consequence, the yield strength and total tensile elongation of the alloy were above 1100 MPa and 11.4%, respectively. It was noted that outstanding properties achieved were comparable with or better than in forged Ti–6Al–4V.

In the case of other PBF AM methods, SLS technology also promotes martensite formation in produced parts made of titanium alloys. In Kazantseva et al. [21], two martensitic phases, α' and α'', were found and distinguished in the microstructure of Ti-6Al-4V. The formation of the α' martensite was associated with high cooling rates, whereas the mechanisms of orthorhombic α'' phase formation with manufacturing conditions. It was explained that during PBF processes, including SLS, the temperature of the heating cycle can increase locally to $\alpha + \beta \rightarrow \beta$ range for a short time and repeatedly, which can lead to the activation of diffusional processes and result in the formation of areas enriched with vanadium. The authors concluded that the formation of martensitic phases in SLS-fabricated Ti-6Al-4V alloy strongly depend on the parameters of laser sintering and, thus, through the selection of SLS parameters, various microstructures and properties could be developed in the same material.

EBM is considered as an important metal 3D printing technology due to its high potential applications in aerospace and biomedical fields. Similarly to previously described PBF AM methods, the final mechanical properties of produced parts are also developed via martensite decomposition process [22,24,76]. Tan et al. [22] claimed that EBM processing of Ti-6Al-4V is related to α' martensitic transformation and α/β interface evolution. The authors noticed that α' martensite was formed regardless of printing geometries. It was found that by increasing in-fill hatched thickness, the elemental partitioning ratios were raised and the β volume fraction increased. Moreover, vanadium segregation and aluminum depletion at the interface front causing α/β interface widening were observed, which could weaken them and reduce the strength of fabricated parts as the printing thickness increased. The authors suggested that in this work, it was possible to modify the microstructure and optimize compressive mechanical properties for EBM-printed Ti-6Al-4V parts by varying printing temperature. Through the reduction of the average beam current of the EBM process, it was possible to achieve a lower printing temperature and then a mixed microstructure of $(\alpha' + \alpha/\beta)$ developed. It was also confirmed that martensitic formation and decomposition took place during EBM thermal cycling. Similarly to SLM, post-process heat treatment for improving the tensile properties of EBM-manufactured parts is also considered. Significant increase in strength and ductility in EBM-manufactured Ti-6Al-4V alloy was revealed after sub-transus annealing at 920 °C followed by water quenching, leading to the development of two-phase $\alpha + \alpha'$ microstructure [24]. According to the authors, the partial decomposition of the martensite during annealing induced a substantial hardening of the α' phase, which was attributed to fine-scale precipitation and solution strengthening. It was found that annealing at lower temperature—i.e., 500 °C—led to strengthening of α' phase by the dispersion of fine β precipitates, and plastic deformation was initiated in the α phase. At higher annealing temperature, the martensite fully decomposed, resulting in mechanical properties similar to conventionally TM-treated Ti-6Al-4V alloy. What is worth adding is that post-EBM heat treatment enabled development of a large variety of microstructures and to achieve a broad range of mechanical properties of produced parts.

High isostatic pressing (HIP) was also proposed as a post-EBM treatment [77]. According to the authors, HIP of the as-built parts, having α' martensitic microstructure, caused coarsening of martensite decomposition products. They claimed that the standard HIP at 920 °C, recommended by ASTM, might not be the optimal treatment. The HIP at 780 °C was instead recommended, which enabled to retain the fine microstructure of the as-built parts with their high strength, adequate elongation, good fatigue resistance, and narrow scatter of all the mentioned properties.

Some published data concern other AM techniques than those mentioned above, e.g., powder bed laser beam melting (PB-LBM) of the Ti-6Al-4V alloy in terms of its elastic behavior [78]. The elasticity of the α' martensitic phase was investigated by experimental and simulation methods. It was shown that the Young's modulus of the martensitic phase is lower than that of the pure Ti-α phase. Moreover, the effect of heat treatment, resulting in martensite decomposition into $\alpha + \beta$ mixture, on the elastic properties of PB-LBM-processed Ti-6Al-4V alloy was analyzed. It was found the elastic behavior of α' martensitic phase was more anisotropic than that of the α phase.

5.2. DED Methods

The nature of the DED processes significantly differs from PBF techniques. The DED methods allow simultaneously feeding different kinds of powders through multiple hoppers and, therefore, produced parts can be made of composite materials or compositionally graded materials. Liu and Shin [79] noticed that the thermal behavior of DED process resulted in an acicular α' martensitic formation in the microstructure of Ti-6Al-4V alloy, similarly to the SLM technique mentioned earlier. The presence of α' martensite significantly increased the ultimate tensile and yield strength but decreased the ductility of the as-built components. They claimed that the α' martensite laths in DED-processed alloy were also responsible for the lower crack propagation thresholds but higher fatigue limits as compared to EBM, wrought, forged, and heat-treated Ti-6Al-4V. Obviously, heat treatment for DED manufactured parts is considered. In [20], as-built and heat-treated specimens of LENS-processed Ti-6Al-4V alloy were compared. In both conditions, prior β grains aligned with the manufacturing direction were found, filled with acicular α' martensite in the case of as-built parts. During heat treatment, a two-phase $\alpha + \beta$ microstructure was developed, which increased material plasticity and reduced anisotropy of mechanical properties (with a decrease in yield stress and ultimate tensile stress). Generally, post-processing heat treatment leading to the creation of $\alpha + \beta$ microstructure is recommended to increase the mechanical properties of LENS-produced Ti-6Al-4V alloy [80]. For the components produced by direct laser deposition (DLD), like in the case of LENS, HIP was used [81] as a post-processing treatment, which effectively closed pores and fully transformed the martensite microstructure into lamellar $\alpha + \beta$ phase, resulting in ductility improvement and slight strength reduction.

The role of martensitic phase in achieving high-strength AM components made of titanium alloys seems to be not fully understood. Dependent on the AM method and post-processing heat treatment used, the presence of martensite fine laths or martensite decomposition determines the high mechanical properties of Ti-6Al-4V alloy, which seems to be the most often studied AM-processed titanium alloy. It is interesting that the martensite formation process is more sensitive to the AM method applied in the case of pure titanium. Attar et al. [82] noted that the microstructure of LENS-processed CP Ti was composed of plate-like (Widmanstätten) α phase, whereas the SLM samples showed only martensitic α' phase.

Martensitic phase can also be formed in titanium alloys during joining processes—like gas tungsten arc welding [83] or electron beam welding [84,85]—or surface engineering operations [86]. This kind of localized microstructure change is worth considering but needs to be studied separately.

6. Summary

Two-phase titanium alloys, due to their unique properties, are still attractive metallic materials processed by traditional and novel techniques. The $\alpha + \beta \leftrightarrow \beta$, $\beta \rightarrow \alpha'(\alpha'')$ and $\alpha'(\alpha'') \rightarrow \alpha + \beta$ phase transformations enable their microstructure and mechanical properties to be controlled quite freely. Martensite formation and decomposition have long been used in hardening and tempering operations, respectively, during thermomechanical processing of titanium alloys (Figure 4). In the case of additive manufacturing methods characterized by high temperature gradient and cooling rates, martensite formation is unintentional. Moreover, unintentional martensite formation already occurs during powder preparation [87,88]. Generally, the $\alpha'(\alpha'')$ martensite content in the microstructure of AM components is considered harmful and, therefore, its decomposition is induced during AM processing (in situ heat treatment) and/or afterwards (post-AM or ex situ heat treatment—Figure 4).

Martensitic microstructure in thermomechanically processed $\alpha + \beta$ alloys is considered to be rather temporary, and after tempering, is transformed into fine-grained $\alpha + \beta$ microstructure giving high mechanical properties. Martensitic microstructure, mainly in Ti-6Al-4V alloy, is also developed for further plastic deformation. Dependent on deformation conditions, significant grain refinement is achieved—improving the mechanical properties or increasing superplastic deformability.

AM methods seems to effectively extend the application range of structural titanium alloys, even though some production problems still need to be overcome. However, recent findings are promising, and the mechanical properties of the most popular Ti-6Al-4V alloy processed by SLM (yield strength > 1100 MPa, elongation > 10%) are comparable with those obtained using conventional methods.

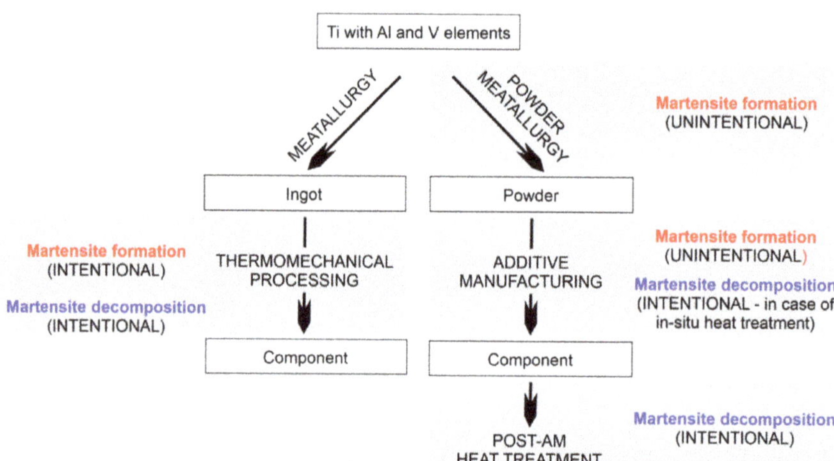

Figure 4. Martensite formation and decomposition at different stages of traditional and additive manufacturing processes.

The role of martensite content in AM components made of titanium alloys is not fully understood. Even in situ or ex situ heat treatment is used for strength increase by martensite decomposition, and high yield strength and excellent tensile elongation can be achieved in Ti-6Al-4V alloy with fully martensitic microstructure composed of ultrafine α' laths. This aspect is particularly important because AM is used for the manufacture of reliable structural elements—e.g., in the aerospace industry [89] or in medicine [90]. It should be added that ex situ (e.g., post-SLM) heat treatment causes phase transformation and beneficial microstructural changes, but the developed microstructure is still far from what is recommended for surgery implants according to ISO standards [91].

Funding: This research received no external funding.

Data Availability Statement: Not applicable.

Acknowledgments: The TEM image presented in the paper was made by Witold Chromiński (Warsaw University of Technology, Poland). I also want to thank Sergey Zherebtsov (Belgorod State University, Russia) for his valuable comments during manuscript preparation. **I devote this work to the memory of Krzysztof Kubiak,** who always shared his knowledge and supported me in the research on engineering materials, especially on titanium alloys.

Conflicts of Interest: The author declares no conflict of interest.

References

1. Leyens, P.; Peters, M. *Titanium and Titanium Alloys—Fundamentals and Applications*, 1st ed.; WILEY-VCH GmbH & Co. KGaA: Weinheim, Germany, 2003.
2. Motyka, M.; Ziaja, W.; Sieniawski, J. Introductory Chapter: Novel Aspects of Titanium Alloys' Applications. In *Titanium Alloys—Novel Aspects of Their Manufacturing and Processing*, 1st ed.; Motyka, M., Ziaja, W., Sieniawski, J., Eds.; IntechOpen: London, UK, 2019; pp. 3–6. [CrossRef]
3. Motyka, M.; Kubiak, K.; Sieniawski, J.; Ziaja, W. Phase transformations and characterization of $\alpha+\beta$ titanium alloys. In *Comprehensive Materials Processing*, 1st ed.; Hashmi, S., Ed.; Elsevier: Amsterdam, The Netherlands, 2019; Volume 2, pp. 7–36. [CrossRef]
4. Lütjering, G.; Williams, J.C. *Titanium*, 1st ed.; Springer: Berlin, Germany, 2003.
5. Peters, M.; Lütjering, G.; Ziegler, G. Control of microstructures of ($\alpha+\beta$) titanium alloys. *Z. Metallkd.* **1983**, *74*, 274–282.
6. Lütjering, G. Influence of processing on microstructure and mechanical properties of ($\alpha+\beta$) titanium alloys. *Mater. Sci. Eng. A* **1998**, *243*, 32–45. [CrossRef]
7. Wu, H.; Sun, Z.; Cao, J.; Yin, Z. Formation and evolution of tri-modal microstructure during dual heat treatment for TA15 Ti-alloy. *J. Alloys Compd.* **2019**, *786*, 894–905. [CrossRef]

8. Sieniawski, J.; Ziaja, W.; Kubiak, K.; Motyka, M. Microstructure and mechanical properties of high strength two-phase titanium alloys. In *Titanium Alloys—Advances in Properties Control*, 1st ed.; Sieniawski, J., Ziaja, W., Eds.; InTech: Rijeka, Croatia, 2013; Volume 2, pp. 69–80. [CrossRef]

9. Semiatin, S.L. An overview of the thermomechanical processing of α/β titanium alloys: Current status and future research opportunities. *Metall. Mater. Trans. A* **2020**, *51*, 2593–2625. [CrossRef]

10. Inagaki, H. Enhanced superplasticity in high strength Ti alloys. *Z. Metallkd.* **1995**, *86*, 643–650.

11. Zherebtsov, S.V.; Salishchev, G.A.; Galeyev, R.M.; Valiakhmetov, O.R.; Mironov, S.Y.; Semiatin, S.L. Production of submicrocrystalline structure in large-scale Ti–6Al–4V billet by warm severe deformation processing. *Scripta Mater.* **2004**, *51*, 1147–1151. [CrossRef]

12. Motyka, M.; Sieniawski, J.; Ziaja, W. Microstructural aspects of superplasticity in Ti–6Al–4V alloy. *Mater. Sci. Eng. A* **2014**, *599*, 57–63. [CrossRef]

13. Chao, Q.; Hodgson, P.D.; Beladi, H. Ultrafine grain formation in a Ti-6Al-4V alloy by thermomechanical processing of a martensitic microstructure. *Metall. Mater. Trans. A* **2014**, *45*, 2659–2671. [CrossRef]

14. Markovsky, P.E. Preparation and properties of ultrafine (submicron) structure titanium alloys. *Mater. Sci. Eng. A* **1995**, *203*, L1–L4. [CrossRef]

15. Dutta, B.; Froes, F.H. The additive manufacturing (AM) of titanium alloys. In *Titanium Powder Metallurgy—Science, Technology and Applications*, 1st ed.; Qian, M., Froes, F.H., Eds.; Butterworth-Heinemann: Waltham, MA, USA, 2015; pp. 447–468. [CrossRef]

16. Dutta, B.; Froes, F.H. The additive manufacturing (AM) of titanium alloys. *Met. Powder Rep.* **2017**, *72*, 96–106. [CrossRef]

17. Yang, J.; Yu, H.; Yin, J.; Gao, M.; Wang, Z.; Zeng, X. Formation and control of martensite in Ti-6Al-4V alloy produced by selective laser melting. *Mater. Des.* **2016**, *108*, 308–318. [CrossRef]

18. He, J.; Li, D.; Jiang, W.; Ke, L.; Qin, G.; Ye, Y.; Qin, Q.; Qiu, D. The martensitic transformation and mechanical properties of Ti6Al4V prepared via selective laser melting. *Materials* **2019**, *12*, 321. [CrossRef]

19. Xu, Y.; Zhang, D.; Guo, Y.; Hu, S.; Wu, X.; Jiang, Y. Microstructural tailoring of As-Selective Laser Melted Ti6Al4V alloy for high mechanical properties. *J. Alloys Compd.* **2020**, *816*, 152536. [CrossRef]

20. Szafrańska, A.; Antolak-Dudka, A.; Baranowski, P.; Bogusz, P.; Zasada, D.; Małachowski, J.; Czujko, T. Identification of mechanical properties for titanium alloy Ti-6Al-4V produced using LENS technology. *Materials* **2019**, *12*, 886. [CrossRef]

21. Kazantseva, N.; Krakhmalev, P.; Thuvander, M.; Yadroitsev, I.; Vinogradova, N.; Ezhov, I. Martensitic transformations in Ti-6Al-4V (ELI) alloy manufactured by 3D Printing. *Mater. Charact.* **2018**, *146*, 101–112. [CrossRef]

22. Tan, X.; Kok, Y.; Toh, W.Q.; Tan, Y.J.; Descoins, M.; Mangelinck, D.; Tor, S.B.; Leong, K.F.; Chua, C.K. Revealing martensitic transformation and α/β interface evolution in electron beam melting three-dimensional-printed Ti-6Al-4V. *Sci. Rep.* **2016**, *6*, 26039. [CrossRef] [PubMed]

23. Zafari, A.; Xia, K. High ductility in a fully martensitic microstructure: A paradox in a Ti alloy produced by selective laser melting. *Mater. Res. Lett.* **2018**, *6*, 627–633. [CrossRef]

24. De Formanoir, C.; Martin, G.; Prima, F.; Allain, S.Y.P.; Dessolier, T.; Sun, F.; Vivès, S.; Hary, B.; Bréchet, Y.; Godeta, S. Micromechanical behavior and thermal stability of a dual-phase $\alpha+\alpha'$ titanium alloy produced by additive manufacturing. *Acta Mater.* **2019**, *162*, 149–162. [CrossRef]

25. Zwicker, U. *Titanium and Titanium Alloys*, 1st ed.; Springer: Berlin, Germany, 1974. (In German)

26. Chai, Y.W.; Kim, H.Y.; Hosoda, H.; Miyazaki, S. Interfacial defects in Ti–Nb shape memory alloys. *Acta Mater.* **2008**, *56*, 3088–3097. [CrossRef]

27. Banerjee, D.; Williams, J.C. Perspectives on titanium science and technology. *Acta Mater.* **2013**, *61*, 844–879. [CrossRef]

28. Moiseev, V.N.; Polyak, É.V.; Sokolova, A.Y. Martensite strengthening of titanium alloys. *Met. Sci. Heat Treat.* **1975**, *17*, 687–691. [CrossRef]

29. Moiseyev, V.N. *Titanium Alloys—Russian Aircraft and Aerospace Applications*, 1st ed.; CRC Press Taylor & Francis Group: Boca Raton, FL, USA, 2006.

30. Venkatesh, B.D.; Chen, D.L.; Bhole, S.D. Effect of heat treatment on mechanical properties of Ti–6Al–4V ELI alloy. *Mater. Sci. Eng. A* **2009**, *506*, 117–124. [CrossRef]

31. Jovanović, M.T.; Tadić, S.; Zec, S.; Mišković, Z.; Bobić, I. The effect of annealing temperatures and cooling rates on microstructure and mechanical properties of investment cast Ti–6Al–4V alloy. *Mater. Des.* **2006**, *27*, 192–199. [CrossRef]

32. Chong, Y.; Bhattacharjee, T.; Yi, J.; Shibata, A.; Tsuji, N. Mechanical properties of fully martensite microstructure in Ti-6Al-4V alloy transformed from refined beta grains obtained by rapid heat treatment (RHT). *Scripta Mater.* **2017**, *138*, 66–70. [CrossRef]

33. Mantani, Y.; Tajima, M. Phase transformation of quenched α'' martensite by aging in Ti-Nb alloys. *Mater. Sci. Eng. A* **2006**, *438–440*, 315–319. [CrossRef]

34. Sato, K.; Matsumoto, H.; Kodaira, K.; Konno, T.J.; Chiba, A. Phase transformation and age-hardening of hexagonal α' martensite in Ti–12 mass%V–2 mass%Al alloys studied by transmission electron microscopy. *J. Alloys Compd.* **2010**, *506*, 607–614. [CrossRef]

35. Zeng, L.; Bieler, T.R. Effects of working, heat treatment, and aging on microstructural evolution and crystallographic texture of α, α', α'' and α phases in Ti-6Al-4V wire. *Mater. Sci. Eng. A* **2005**, *392*, 403–414. [CrossRef]

36. Bylica, A.; Sieniawski, J. *Tytan i Jego Stopy*, 1st ed.; PWN: Warsaw, Poland, 1985. (In Polish)

37. Gil Mur, F.X.; Rodríguez, D.; Planell, J.A. Influence of tempering temperature and time on the α''-Ti-6Al-4V martensite. *J. Alloys Compd.* **1996**, *234*, 287–289. [CrossRef]

38. Qazi, J.I.; Senkov, O.N.; Rahim, J.; Froes, F.H. Kinetics of martensite decomposition in Ti-6Al-4V-xH alloys. *Mater. Sci. Eng. A* **2003**, *359*, 137–149. [CrossRef]
39. Sieniawski, J.; Filip, R.; Kubiak, K. Influence of ageing time and temperature on mechanical properties of two-phase titanium alloy Ti-6Al-2Mo-2Cr at high temperature. *Inz. Materialowa* **1998**, *19*, 536–539.
40. Barriobero-Vila, P.; Oliveira, V.B.; Schwarz, S.; Buslaps, T.; Requena, G. Tracking the α'' martensite decomposition during continuous heating of a Ti-6Al-6V-2Sn alloy. *Acta Mater.* **2017**, *135*, 132–143. [CrossRef]
41. Yu, H.; Li, W.; Li, S.; Zou, H.; Zhai, T.; Liu, L. Study on transformation mechanism and kinetics of α' martensite in TC4 alloy isothermal aging process. *Crystals* **2020**, *10*, 229. [CrossRef]
42. Lee, S.W.; Park, C.H.; Hong, J.K.; Yeom, J.-T. Aging temperature dependence of α''-martensite decomposition mechanism in Ti-Al-Fe-Si alloy. *Metall. Mater. Trans. A* **2018**, *49*, 5913–5918. [CrossRef]
43. Li, X.; Sha, M.; Chu, J. Compressive deformability of hardened titanium alloy VT16. *Met. Sci. Heat Treat.* **2009**, *51*, 594–598. [CrossRef]
44. Illarionov, A.G.; Demakov, S.L.; Stepanov, S.I.; Illarionova, S.M. Structural and phase transformations in a quenched two-phase titanium alloy upon cold deformation and subsequent annealing. *Phys. Met. Metallogr.* **2015**, *116*, 267–273. [CrossRef]
45. Matsumoto, H.; Kodaira, K.; Sato, K.; Konno, T.J.; Chiba, A. Microstructure and mechanical properties of α' martensite type Ti alloys deformed under α' the processing. *Mater. Trans.* **2009**, *50*, 2744–2750. [CrossRef]
46. Gu, J.L.; Sun, X.J.; Bai, B.Z.; Chen, N.P. Microstructural evolution during fabrication of ultrafine grained alpha+beta titanium alloy. *Mater. Sci. Technol.* **2001**, *17*, 1516–1524. [CrossRef]
47. Motyka, M.; Baran-Sadleja, A.; Sieniawski, J.; Wierzbińska, M.; Gancarczyk, K. Decomposition of deformed $\alpha'(\alpha'')$ martensitic phase in Ti-6Al-4V alloy. *Mater. Sci. Technol.* **2019**, *35*, 260–272. [CrossRef]
48. Motyka, M.; Ziaja, W.; Baran-Sadleja, A.; Ślemp, K. The effect of plastic deformation on martensite decomposition process in Ti-6Al-4V alloy. In Proceedings of the 14th World Conference on Titanium (Ti 2019), Nantes, France, 10–14 June 2019; Volume 21, p. 12034. [CrossRef]
49. Kubiak, K.; Sieniawski, J. Development of the microstructure and fatigue strength of two phase titanium alloys in the processes of forging and heat treatment. *J. Mater. Process. Technol.* **1998**, *78*, 117–121. [CrossRef]
50. Motyka, M.; Sieniawski, J. The influence of initial plastic deformation on microstructure and hot plasticity of $\alpha+\beta$ titanium alloys. *Arch. Mat. Sci. Eng.* **2010**, *41*, 95–103.
51. Nieh, T.G.; Wadsworth, J.; Sherby, O.D. *Superplasticity in Metals and Ceramics*, 1st ed.; Cambridge University Press: Cambridge, UK, 1997. [CrossRef]
52. Jackson, M. Superplastic forming and diffusion bonding of titanium alloys. In *Superplastic Forming of Advanced Metallic Materials—Methods and Applications*, 1st ed.; Giuliano, G., Ed.; Woodhead Publishing: Cambridge, UK, 2011. [CrossRef]
53. Valiev, R.; Islamgaliev, R.; Semenova, I.; Yunusova, N. New trends in superplasticity in SPD-processed nanostructured materials. *Int. J. Mater. Res.* **2007**, *98*, 314–319. [CrossRef]
54. Inagaki, H. Mechanism of enhanced superplasticity in thermomechanically processed Ti-6Al-4V. *Z. Metallkd.* **1996**, *87*, 179–186.
55. Shell, E.B.; Semiatin, S.L. Effect of initial microstructure on plastic flow and dynamic globularization during hot working of Ti-6Al-4V. *Metall. Mater. Trans. A* **1999**, *30*, 3219–3229. [CrossRef]
56. Semiatin, S.L.; Bieler, T.R. The effect of alpha platelet thickness on plastic flow during hot working of TI–6Al–4V with a transformed microstructure. *Acta Mater.* **2001**, *49*, 3565–3573. [CrossRef]
57. Park, C.H.; Ko, Y.G.; Park, J.W.; Lee, C.S. High-temperature deformation behavior of ELI Grade Ti-6Al-4V alloy with martensite microstructure. *Mater. Sci. Eng. A* **2008**, *496*, 150–158. [CrossRef]
58. Park, C.H.; Park, K.T.; Shin, D.H.; Lee, C.S. Microstructural mechanisms during dynamic globularization of Ti-6Al-4V alloy. *Mater. Trans.* **2008**, *49*, 2196–2200. [CrossRef]
59. Matsumoto, H.; Bin, L.; Lee, S.-H.; Li, Y.; Ono, Y.; Chiba, A. Frequent occurrence of discontinuous dynamic recrystallization in Ti-6Al-4V alloy with α' martensite starting microstructure. *Metall. Mater. Trans. A* **2013**, *44*, 3245–3260. [CrossRef]
60. Senkov, O.N.; Jonas, J.J.; Froes, F.H. Recent advances in the thermohydrogen processing of titanium alloys. *JOM* **1996**, *48*, 42–47. [CrossRef]
61. Yoshimura, H.; Kimura, K.; Hayashi, M. Ultra-fine equiaxed grain refinement of titanium alloys by hydrogenation, hot working, heat treatment and dehydrogenation. *Nippon Steel Tech. Rep.* **1994**, *62*, 80–84.
62. Murzinova, M.A.; Salishev, G.A.; Afonichev, D.D. Formation of nanocrystalline structure in two-phase titanium alloy by combination of thermohydrogen processing with hot working. *Int. J. Hydrogen Energy* **2002**, *27*, 775–782. [CrossRef]
63. Qazi, J.L.; Rahim, J.; Senkov, O.N.; Froes, F.H. Phase transformations in the Ti-6Al-4V-H system. *JOM* **2002**, *54*, 68–71. [CrossRef]
64. Zong, Y.-Y.; Huang, S.-H.; Guo, B.; Shan, D.-B. In situ study of phase transformation in Ti-6Al-4V-xH alloys. *Trans. Nonferrous Met. Soc. China* **2015**, *25*, 2901–2911. [CrossRef]
65. Herzog, D.; Seyda, V.; Wycisk, E.; Emmelmann, C. Additive manufacturing of metals. *Acta Mater.* **2016**, *117*, 371–392. [CrossRef]
66. Qiao, B.; Dong, C.; Tong, S.; Kong, D.; Ni, X.; Zhang, H.; Wang, L.; Li, X. Anisotropy in α' martensite and compression behavior of Ti6Al4V prepared by selective laser melting. *Mater. Res. Express* **2019**, *6*, 126548. [CrossRef]
67. Qian, M.; Xu, W.; Brandt, M.; Tang, H.P. Additive manufacturing and postprocessing of Ti-6Al-4V for superior mechanical properties. *MRS Bull.* **2016**, *41*, 775–784. [CrossRef]

68. Moridi, A.; Demir, A.G.; Caprio, L.; Hart, A.J.; Previtali, B.; Colosimo, B.M. Deformation and failure mechanisms of Ti–6Al–4V as built by selective laser melting. *Mater. Sci. Eng. A* **2019**, *768*, 1384568. [CrossRef]

69. Barriobero-Vila, P.; Gussone, J.; Haubrich, J.; Sandlöbes, S.; Da Silva, J.C.; Cloetens, P.; Schell, N.; Requena, G. Inducing stable α + β microstructures during selective laser melting of Ti-6Al-4V using intensified intrinsic heat treatments. *Materials* **2017**, *10*, 268. [CrossRef]

70. Ali, H.; Ma, L.; Ghadbeigi, H.; Mumtaz, K. In-situ residual stress reduction, martensitic decomposition and mechanical properties enhancement through high temperature powder bed pre-heating of Selective Laser Melted Ti6Al4V. *Mater. Sci. Eng. A* **2017**, *695*, 211–220. [CrossRef]

71. Sallica-Leva, E.; Caram, R.; Jardini, A.L.; Fogagnolo, J.B. Ductility improvement due to martensite α' decomposition in porous Ti-6Al-4V parts produced by selective laser melting for orthopedic implants. *J. Mech. Behav. Biomed. Mater.* **2016**, *54*, 149–158. [CrossRef] [PubMed]

72. Caoa, S.; Hu, Q.; Huang, A.; Chen, Z.; Sun, M.; Zhang, J.; Fu, C.; Jia, Q.; Lim, C.V.S.; Boyer, R.R.; et al. Static coarsening behaviour of lamellar microstructure in selective laser melted Ti-6Al-4V. *J. Mater. Sci. Technol.* **2019**, *35*, 1578–1586. [CrossRef]

73. Neikter, M.; Huang, A.; Wu, X. Microstructural characterization of binary microstructure pattern in selective laser-melted Ti-6Al-4V. *Int. J. Adv. Manuf. Technol.* **2019**, *104*, 1381–1391. [CrossRef]

74. Zafari, A.; Barati, M.R.; Xia, K. Controlling martensitic decomposition during selective laser melting to achieve best ductility in high strength Ti-6Al-4V. *Mater. Sci. Eng. A* **2019**, *744*, 445–455. [CrossRef]

75. Xu, W.; Brandt, M.; Sun, S.; Elambasseril, J.; Liu, Q.; Latham, Q.; Xia, K.; Qian, M. Additive manufacturing of strong and ductile Ti-6Al-4V by selective laser melting via in situ martensite decomposition. *Acta Mater.* **2015**, *85*, 74–84. [CrossRef]

76. Yamanaka, K.; Mori, M.; Chiba, A. Martensitic transformation and its effects on microstructural evolution and mechanical properties in electron beam melting of commercially pure titanium and Ti-6Al-4V alloy. In Proceedings of the 13th World Conference on Titanium, San Diego, CA, USA, 16–20 August 2015; Venkatesh, V., Pilchak, A.L., Allison, J.E., Ankem, S., Boyer, R., Christodoulou, J., Fraser, H.L., Imam, M.A., Kosaka, Y., Rack, H.J., et al., Eds.; John Wiley & Sons, Inc.: Hoboken, NJ, USA, 2016. Chapter 218. [CrossRef]

77. Ganor, Y.I.; Tiferet, E.; Vogel, S.C.; Brown, D.W.; Chonin, M.; Pesach, A.; Hajaj, A.; Garkun, A.; Samuha, S.; Shneck, R.Z.; et al. Tailoring microstructure and mechanical properties of additively-manufactured Ti6Al4V using post processing. *Materials* **2021**, *14*, 658. [CrossRef]

78. Dumontet, N.; Connétable, D.; Malard, B.; Viguier, B. Elastic properties of the α' martensitic phase in the Ti-6Al-4V alloy obtained by additive manufacturing. *Scripta Mater.* **2019**, *167*, 115–119. [CrossRef]

79. Liu, S.; Shin, Y.C. Additive manufacturing of Ti6Al4V alloy: A review. *Mater. Des.* **2019**, *164*, 107552. [CrossRef]

80. Antolak-Dudka, A.; Płatek, P.; Durejko, T.; Baranowski, P.; Małachowski, J.; Sarzyński, M.; Czujko, T. Static and dynamic loading behavior of Ti6Al4V honeycomb structures manufactured by laser engineered net shaping (LENS^{TM}) technology. *Materials* **2019**, *12*, 1225. [CrossRef]

81. Qiu, C.; Ravi, G.A.; Dance, C.; Ranson, A.; Dilworth, S.; Attallah, M.M. Fabrication of large Ti–6Al–4V structures by direct laser deposition. *J. Alloys Compd.* **2015**, *629*, 351–361. [CrossRef]

82. Attar, H.; Ehtemam-Haghighi, S.; Kent, D.; Wu, X.; Dargusch, M.S. Comparative study of commercially pure titanium produced by laser engineered net shaping, selective laser melting and casting processes. *Mater. Sci. Eng. A* **2017**, *705*, 385–393. [CrossRef]

83. Murgau, C.C. Microstructure Model for Ti-6Al-4V Used in Simulation of Additive Manufacturing. Ph.D. Thesis, Luleå University of Technology, Luleå, Sweden, May 2016.

84. Adamus, J.; Lacki, P.; Motyka, M. EBW titanium sheets as material for drawn parts. *Arch. Civ. Mech. Eng.* **2015**, *15*, 42–47. [CrossRef]

85. Fan, Y.; Tian, W.; Guo, Y.; Sun, Z.; Xu, J. Relationships among the microstructure, mechanical properties, and fatigue behavior in thin Ti6Al4V. *Adv. Mater. Sci. Eng.* **2016**, 7278267. [CrossRef]

86. Molak, R.M.; Araki, H.; Watanabe, M.; Katanoda, H.; Ohno, N.; Kuroda, S. Warm spray forming of Ti-6Al-4V. *J. Therm. Spray Technol.* **2014**, *23*, 197–212. [CrossRef]

87. Chen, G.; Zhao, S.Y.; Tan, P.; Wang, J.; Xiang, C.S.; Tang, H.P. A comparative study of Ti-6Al-4V powders for additive manufacturing by gas atomization, plasma rotating electrode process and plasma atomization. *Powder Technol.* **2018**, *333*, 38–46. [CrossRef]

88. Birt, A.M.; Champagne, V.K.J.; Sisson, R.D.; Apelian, D. Microstructural analysis of Ti–6Al–4V powder for cold gas dynamic spray applications. *Adv. Powder Technol.* **2015**, *26*, 1335–1347. [CrossRef]

89. Ulhmann, E.; Kersting, R.; Klein, T.B.; Cruz, M.F.; Borille, A.V. Additive manufacturing of titanium alloy for aircraft components. *Proc. CIRP* **2015**, *35*, 55–60. [CrossRef]

90. Murr, L.E.; Quinones, S.A.; Gaytan, S.M.; Lopez, M.I.; Rodela, A.; Martinez, E.Y.; Hernandez, D.H.; Martinez, E.; Medina, F.; Wicker, R.B. Microstructure and mechanical behavior of Ti–6Al–4V produced by rapid-layer manufacturing, for biomedical applications. *J. Mech. Behav. Biomed. Mater.* **2009**, *2*, 20–32. [CrossRef] [PubMed]

91. Yadroitsev, I.; Krakhmalev, P.; Yadroitsava, I. Selective laser melting of Ti6Al4V alloy for biomedical applications: Temperature monitoring and microstructural evolution. *J. Alloys Compd.* **2014**, *583*, 404–409. [CrossRef]

metals

Article

Evolution of the Microstructure and Mechanical Properties of a Ti35Nb2Sn Alloy Post-Processed by Hot Isostatic Pressing for Biomedical Applications

Joan Lario *, Ángel Vicente and Vicente Amigó

DIMM ETSII, Universitat Politècnica de València, Camino de Vera s/n, 5E Building, 46022 Valencia, Spain; avicente@mcm.upv.es (Á.V.); vamigo@mcm.upv.es (V.A.)
* Correspondence: joalafe@upv.es

Abstract: The HIP post-processing step is required for developing next generation of advanced powder metallurgy titanium alloys for orthopedic and dental applications. The influence of the hot isostatic pressing (HIP) post-processing step on structural and phase changes, porosity healing, and mechanical strength in a powder metallurgy Ti35Nb2Sn alloy was studied. Powders were pressed at room temperature at 750 MPa, and then sintered at 1350 °C in a vacuum for 3 h. The standard HIP process at 1200 °C and 150 MPa for 3 h was performed to study its effect on a Ti35Nb2Sn powder metallurgy alloy. The influence of the HIP process and cold rate on the density, microstructure, quantity of interstitial elements, mechanical strength, and Young's modulus was investigated. HIP post-processing for 2 h at 1200 °C and 150 MPa led to greater porosity reduction and a marked retention of the β phase at room temperature. The slow cooling rate during the HIP process affected phase stability, with a large amount of α''-phase precipitate, which decreased the titanium alloy's yield strength.

Keywords: hot isostatic pressing; β-Type titanium alloy; biomaterial; phase transformation; powder metallurgy

Citation: Lario, J.; Vicente, Á.; Amigó, V. Evolution of the Microstructure and Mechanical Properties of a Ti35Nb2Sn Alloy Post-Processed by Hot Isostatic Pressing for Biomedical Applications. *Metals* **2021**, *11*, 1027. https://doi.org/10.3390/met11071027

Academic Editor: Maciej Motyka

Received: 8 June 2021
Accepted: 23 June 2021
Published: 25 June 2021

Publisher's Note: MDPI stays neutral with regard to jurisdictional claims in published maps and institutional affiliations.

1. Introduction

The low elastic modulus of titanium alloys has replaced other metallic alloys, such as Co-Cr alloys and stainless steels (316L or 307), for load-bearing orthopaedic and dental implants [1,2]. The bone reabsorption issue is related to the stress shielding effect, which occurs due to a stiffness mismatch between implant and bone. Beta titanium alloys present the lowest Young's modulus by both enhancing the stress transmission between the bone and the implant and inhibiting bone resorption [2–4].

The development of low-modulus high-strength beta titanium alloys is essential for the next generation of material prostheses to prolong implant lifetime beyond 15 years, which has been the standard for the past 50 years for Ti Cp and Ti6Al4V ELI alloys [5]. Mechanical properties are directly connected to their metallurgical processing routes, chemical composition, and stabilised phases. The refractory elements added to stabilise the titanium beta phase (Nb, Mo, Ta, Zr) have a high melting point and a large difference in specific gravity, which seriously complicate the melting process, which requires 6–10 remelting steps followed by a homogenising annealing treatment to increase chemical and phase homogeneity [6,7]. The vacuum melting technologies (VIM, VAR, EBM) employed to obtain titanium alloys are complex and expensive, use considerable energy, and often involve supplier delivery lead times of around 35 weeks [8–10].

Powder metallurgy (PM) technology is an alternative processing route to obtain beta titanium alloys, where the alloying elements used to stabilise the β phase can be incorporated into the solid state during sintering. Conventional PM titanium alloys present a remaining porosity after sintering (2–8 vol%) that can significantly reduce long-time

149

performance. Therefore, the next advanced titanium alloys for the biomedical sector will rely on HIP post-processing to remove residual porosity, which will allow powder metallurgy to be processed by plastic deformation techniques [9–17]. HIP has been used for several decades in the aerospace and automotive industries to improve fatigue, ductility, and fracture toughness [18]. Therefore, it is very important to predict materials' long-term behaviour to design and manufacture reliable implants. To avoid mechanical stress, corrosion, and fatigue problems, titanium alloys must not present porosity, and must have excellent chemical homogeneity. Rapid cooling HIP furnaces enable the HIP cycle to be combined with conventional heat treatment where high cooling rates are required to stabilise the beta phase at room temperature. During a conventional HIP process, titanium powders need to be encapsulated in containers, degassed, and sealed before the hot isostatic press cycle starts. However, in the present study, the beta powder titanium samples were pre-sintered under such conditions with residual porosity below 6% to allow the alloys to be hot isostatic-pressed directly without using containers.

The present paper proposes a processing route to obtain Ti35Nb2Sn based on conventional powder metallurgy (compact and vacuum sintering) and HIP post-processing. It aims to evaluate the effect of the HIP route on beta titanium powder metallurgy alloys (Ti35Nb2Sn) at different cooling rates.

2. Materials and Methods

2.1. Processing Conditions

A new beta titanium alloy with a nominal Ti35Nb2Sn composition was fabricated by using a conventional powder metallurgical route (press and vacuum sintering). Hydride–dehydride titanium powder (99.7 %wt purity), niobium powder (99.6 %wt), and tin powder (99.6 %wt), with a maximum particle size particle of 45 μm, were chosen as the raw materials. Powders were weighed and prepared in an argon chamber GP Campus (Jacomax, Dagneux, France) to avoid oxidation. The mixture was prepared in a shaker mixer Turbula T2F (Willy A Bachofen AG, Muttenz, Switzerland) for 2 h to increase homogeneity. Subsequently, rectangular samples (30 mm × 10 mm × 10 mm) were produced in a double-effect floating die press at 700 MPa pressure. Green compacts were sintered in an HVT 15/75/450 tube vacuum sintering furnace (Carbolite Gero Ltd., Parson, UK) in a high vacuum at $<10^{-4}$ mbars. Samples were heated to 800 °C at 15 °C/min, held at that temperature for 30 min, heated to 1350 °C at 10 °C/min, and finally held for 180 min and cooled at 10 °C/min.

Without using containers, some sintered samples were hot isostatic-pressed under two different conditions. HIP was performed by a Quintus model QIH21 (Quintus Technologies AB, Västerås, Sweden) under the following operation conditions: temperature of 1200 °C, argon gas pressure of 150 MPa, holding time of 2 h and two different cooling rates: 500 °C/min and 100 °C/min (Figure 1). Three scenarios (sintered, HIP1, HIP2) were designed to study the influence of the HIP process on the Ti35Nb2Sn alloys. The conditions for each experiment are found in Table 1.

Table 1. Parameters selected according to Ti35Nb2Sn processing step.

Processing Route	Pressure (MPa)	Maximum Temperature (°C)	Time (minutes)	Cooling Rate (°C/min)
Vacuum Sintering	N/A	1350	180	15
HIP 1 (Fast cooling)	150	1200	120	500
HIP 2 (Slow cooling)	150	1200	120	100

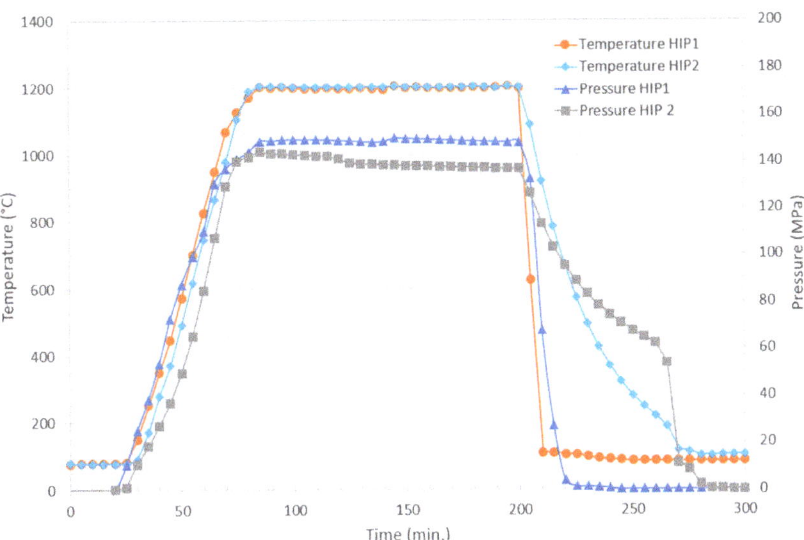

Figure 1. Temperature and pressure evolution during Hot Isostatic Pressing performed on Quintus QIH21 equipment.

2.2. Characterization

The relative density for each scenario was measured based on Archimedes' principle in accordance with ASTM standard C373-88. Tensile specimens were tested by a Shimadzu AG-X plus mechanical tester (Shimadzu, Kyoto, Japan) at a crosshead speed of 0.5 mm/min on three points bending test per standard ISO 3325. The ultrasonic technique was selected to measure Young's modulus, for which the Sonelastic CA-DP equipment was employed (ATCP Ingeniería Física, Ribeirao Preto, Brazil). At least five test specimens with size of 30 mm × 10 mm × 10 mm were measured from each processing condition.

The optical microscopy study (LV100 Nikon, Tokyo, Japan) was performed on the polished samples to observe internal residual porosity. To observe phases present in Ti35Nb2Sn according to the processing route, the mechanically polished samples were etched in Kroll's reagent to reveal their microstructure and to be viewed by optical microscopy. Grain size and orientation, morphology, and phase distribution surface were characterised by a Scanning Electron Microscope (SEM, ZEISS, Oberkochen, Germany) and an Electron Backscatter Diffraction (EBSD) detector. The EBSD operated at 20 kV and 5 nA, and step size was set at 0.05 µm on a sample tilted 70° from the horizontal for orientation mapping. The elemental analyses of the different processing routes were carried out by Energy Dispersive Spectroscopy (EDS) from Oxford Instruments Ltd. (Abingdon-on-Thames, UK).

3. Results and Discussion

The alloy's microstructure is one of the main factors controlling its tensile strength, wear resistance, and corrosion resistance. Table 2 shows the variation in density, phases, and average grains of the Ti35Nb2Sn alloy under different manufacturing conditions. The samples presented high densification, around 75% of the theoretical density after pressing, and approximately 98% after the sintering cycle, which was higher than 99% after HIP post-processing. As the percentage of the retained β phase was above 20% at room temperature, the three processing conditions for Ti35Nb2Sn allowed classification in the β titanium alloys category.

Table 2. Archimedes' and metallographic characterization of the Ti35Nb2Sn alloy.

Processing Route	Green Density (%)	Relative Density (%)	β Phase (%)	α + β Phase (%)	Grain Size (μm)
Vacuum Sintering		94.9 ± 0.1	38.4	56.5	19.1 ± 11.3
HIP 1 (Fast cooling)	80.2 ± 0.5	100.1 ± 0.2	95.5	3.7	57.2 ± 32.3
HIP 2 (Slow cooling)		100.2 ± 0.1	78.9	20.8	55.2 ± 33.5

The microstructures of the tested alloys are presented in Figure 2. The α-phase areas decreased with a rising HIP cooling rate. A heterogeneous phase composition was observed on the sintered and HIP 2 samples, composed predominantly by a β equiaxial grain, with small areas with the α + β phase (Figure 2A,C). As both temperature and pressure increased due to the HIP cycle, plastic deformation on the sintered samples took place. The HIP temperature above the beta transus temperature was required to reduce yield stress, and to enhance creep under pressure, to improve diffusion efficiency and remove residual porosity in a reasonable time (Table 2). The obtained data show that HIP post-processing gave relievable and repeatable results. The HIP process also significantly influenced the enlarged grain size. Initially for the sintered sample, the average grain size was around 19 μm and then increased to 55 μm after HIP post-processing increased (Table 2).

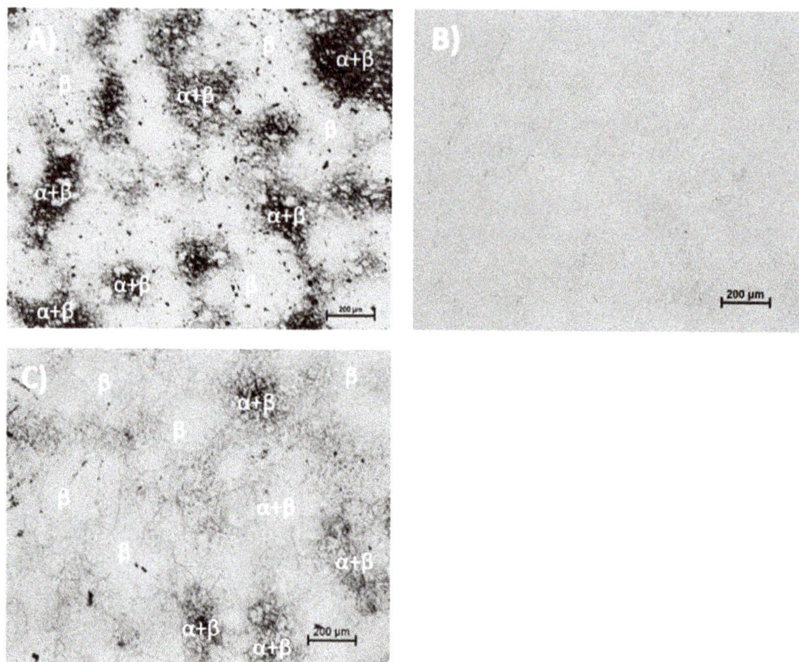

Figure 2. The microstructure of Ti35Nb2Sn according to its manufacturing condition. (**A**) Conventional vacuum sintering at 1350 °C for 180 min. (**B**) HIPed under 1200 °C, 150 MPa, and cooling rate 500 °C/min. (HIP 1) (**C**) HIPed under 1200 °C, 150 MPa, and cooling rate 100 °C/min (HIP 2).

Nowadays, thermo-mechanical deformation metallurgical processes (forge, cold rolling, hot drawing) are employed to modify grain size, ductility, and materials' tenacity [19,20]. This type of process cannot be performed on powder metallurgical material if residual porosity is higher than 2% because of the work hardening that materials present, which tends to fracture in the confirmation process. The present study corroborates that the inclusion of the HIP cycle after sintering on the beta titanium manufacturing route drastically reduced residual porosity, which allowed powder metallurgical titanium alloys to be submitted to the plastic deformation metallurgical process to obtain advanced titanium alloys. The mechanical–corrosion properties of titanium alloys depend on not only chemical and phase homogeneity, but also on morphology and grain size. This study shows that metastable β-type Ti35Nb2Sn alloys can be obtained when a 500 °C/min cooling rate is employed during HIP post-processing (Figure 2B).

Figure 3 shows the Ti35Nb2Sn microstructures after performing HIP processing under different conditions from various scales by SEM and EBSD. The EBSD images demonstrated that the previous α + β areas detected by optical microscopy were actually α″ precipitates at the β matrix (Figure 3B). The phases present at the microscopic level became more homogeneous, with no pronounced differences between the red (β phase) and yellow (α″ phase) areas after sintering and HIP processing. The EBSD analysis corroborated that the second HIP scenario, in which the titanium alloy was cooled more slowly from the β field to room temperature at 100 °C/min instead of at 500 °C/min, presented more intragranular α″ precipitates around 32%, compared with 5% quantified on the rapid cooling samples (HIP 1). During rapid cooling from the β-phase stability temperature range, diffusion can be restricted, and stable α phase formation prevented. Since the current Ti35Nb2Sn presents an elevated amount of beta alloying elements, α-martensite's hexagonal structure becomes distorted, and the crystalline structure loses its hexagonal symmetry to acquire an orthorhombic structure (α″-martensite).

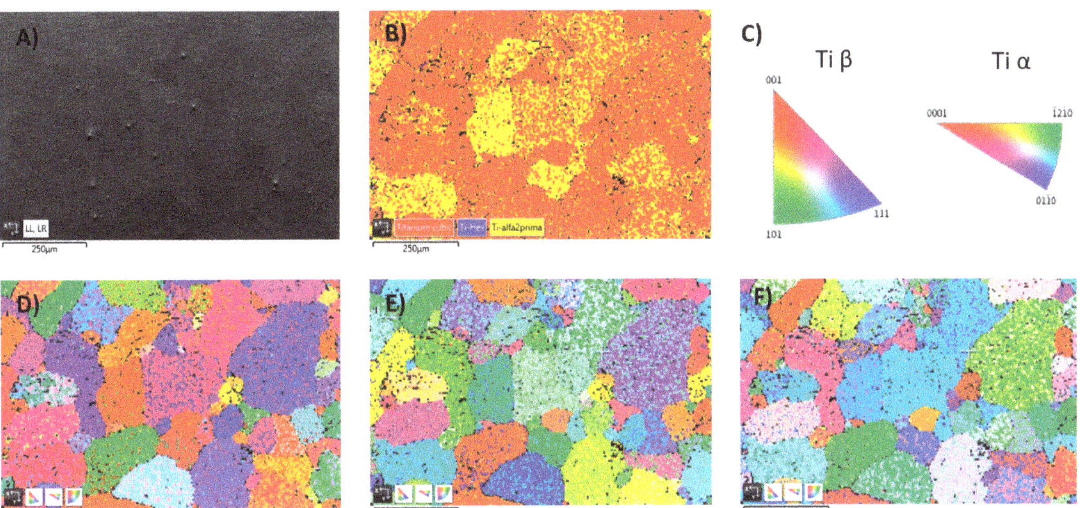

Figure 3. The microstructure of Ti35Nb2Sn after HIPping under 1200 °C, 150 MPa, and cooling rate 100 °C/min (HIP 2). (**A**) SEM image. (**B**) images showing the α″ precipitates. (**C**) crystallographic orientation for β and α phase. (**D–F**) Inverse pole figures (IPF) for X, Y, and Z reference direction maps highlighting grains.

The SEM images and corresponding EDS mappings presented more significant information about alloying element distribution and diffusion on the titanium matrix. The microstructural analysis performed in EDS revealed that some parts were not completely homogeneous (Figure 4C). These brighter areas presented low diffusion between niobium and the titanium matrix. The HIP cycle allowed the brighter areas to reduce due to the thermo-mechanical treatment, even though the titanium alloy's chemical homogeneity still had room for improvement.

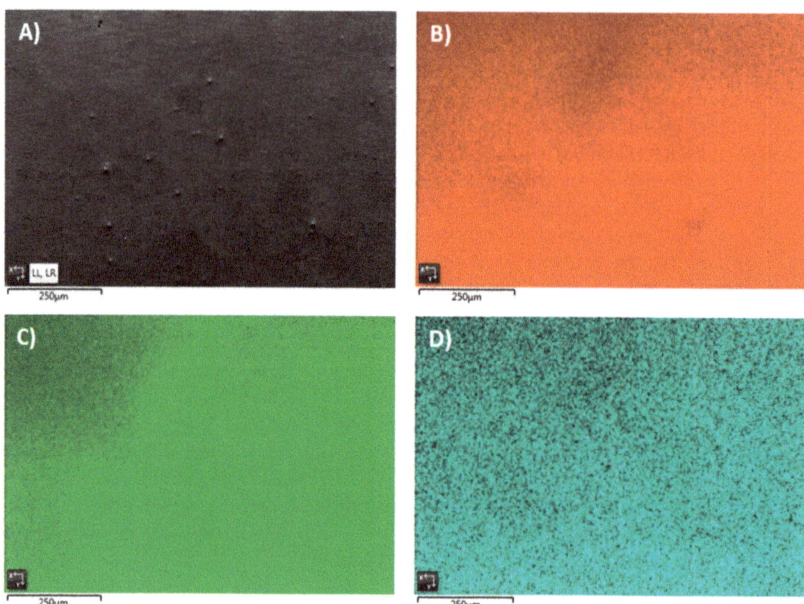

Figure 4. SEM and EDS mapping of Ti, Nb, and Sn respectively, for Ti35Nb2Sn HIPed at 1200 °C, 150 MPa and cooled at 100 °C/min (HIP 2). (**A**) SEM image. (**B**) Titanium EDS mapping. (**C**) Niobium EDS mapping. (**D**) Tin EDS mapping.

The α+β regions with fewer refractory elements displayed high chemical heterogeneity. The mechanical characteristics obtained on the Ti35Nb2Sn alloy are summarized in Table 3. The heterogeneity of the titanium phases (β, α, and α″), together with residual porosity, were the main causes for the differences found in the titanium alloy's mechanical properties. Residual porosity reduction increased the percentage of ductility, and also improved Ti35Nb2Sn's mechanical response. The microstructure had a stronger influence on ultimate tensile strength rather than porosity, as reflected by UTS, which lowered from 900 MPa to 700 MPa when the α″ content grew due to the 100 °C/min cooling rate employed in HIP 2, even though porosity decreased in relation to the sintered sample (see Table 3).

Table 3. Mechanical results of the Ti35Nb2Sn alloy.

Processing Route	UTS (MPa)	ε (%)	E (GPa)	G (GPa)	Poisson's Ratio
Vacuum Sintering	993 ± 82	3.44 ± 0.36	70.49 ± 1.09	25.62 ± 0.39	0.38 ± 0.04
HIP 1 (Fast cooling)	997 ± 63	3.31 ± 0.23	71.23 ± 2.04	24.62 ± 0.85	0.45 ± 0.06
HIP 2 (Slow cooling)	702 ± 80	2.65 ± 0.12	72.29 ± 1.04	23.77 ± 1.73	0.53 ± 0.13

Young´s modulus is one of the most crucial mechanical properties to control the load transfer between the implant and bone [1–3]. Given its low elastic modulus, excellent mechanical properties, and greater corrosion resistance, TixNbySn is a promising candidate for medical use [4]. The large amount of beta stabilisers (niobium) in the alloy's composition contributed to lower Young´s modulus, with values of 70 GPa, which improved the current 100 or 110 GPa that today's titanium alloys employed in prosthesis present (Ti CP and Ti6Al4V ELI), and may minimise bone atrophy due to the stress shielding effect, which increases implant durability.

L. Tan et al. (2018) studied the precipitation phenomenon, the prior particle boundary (PPB), that occurs during HIPing and justify the lower deformation values on titanium samples. Plastic deformation metallurgic process, such as hot forging, cold drawing, and ECAP, modifies the grain dimension and morphology, minimizing the impact from PPB precipitation on PM parts, and increasing the strength of the titanium alloys [18]. Li et al. (2011) found that Ti2448 alloys fabricated by powder metallurgy exhibited very limited ductility (~2%) during tensile tests. Their mechanical properties can be substantially improved simply by solution treatment at a suitable temperature, followed by water quenching. After heat treatment, Ti2448 elongation increased to 15% and its Young's modulus lowered to 58.3 MPa [20–22]. Severe plastic deformation fabrication methods draw a great of attention to the β-type biomedical alloys field as they add grain refinement, which significantly increases both strength and fatigue life expectancy [23–25].

Developing an appropriate microstructure with optimum mechanical properties and good corrosion resistance is a challenging problem in the low modulus β-type titanium alloys field, which should be addressed in the biomedical industrial sector. This would imply the inclusion of an advanced thermo-mechanical process, such as hot drawing and working or ECAP, heat, and electrochemical treatments [26–29].

4. Conclusions

This study confirms that field-assisted consolidation processes, such as HIP, can be employed to reduce residual porosity and to increase the chemical and phase homogeneity of beta powder metallurgy titanium alloys required for further developing advanced titanium alloys for the biomedical field. When applying HIP at 1200 °C and 150 MPa for 3 h to pre-sintered Ti35Nb2Sn samples, a high degree of densification occurs, which allows residual porosity to close and the density theoretical value to be obtained. As the cooling rate during the HIP process lowers from 500 °C/min (HIP 1) to 100 °C/min (HIP 2), the phase transformation from the β phase (bcc) to the α phase (hcp) is restricted and the supersaturated α″ phase appears, which diminishes Ti35Nb2Sn yield strength and elongation.

Author Contributions: J.L. and V.A. carried out the experiment. J.L. wrote the manuscript with support from V.A. The samples were fabricated by J.L. and Á.V.; Á.V. and V.A. helped supervise the project. All authors provided critical feedback and helped shape the research, analysis, and manuscript. All authors have read and agreed to the published version of the manuscript.

Funding: This research was funded by Spanish Ministry of Economy and Competitiveness, Research Project 553 Project RTI2018-097810-B-I00, and the European Commission via FEDER funds to purchase equipment for research purposes and the Microscopy Service at the Valencia Polytechnic University.

Institutional Review Board Statement: Not applicable.

Informed Consent Statement: Not applicable.

Data Availability Statement: Data available in a publicly accessible repository. The data presented in this study are openly available in FigShare repository at 10.6084/m9.figshare.14444570.

Acknowledgments: Thanks to Johannes Gårdstam and Mat Sjöstedt from Quintus Technologies AB, Sweden for the technical assistance and the realization of the HIP process.

Conflicts of Interest: The authors declare no conflict of interest. The funders had no role in the design of the study; in the collection, analyses, or interpretation of data; in the writing of the manuscript, or in the decision to publish the results.

References

1. Kuroda, D.; Niinomi, M.; Morinaga, M.; Kato, Y.; Yashiro, T. Design and mechanical properties of new β type titanium alloys for implant materials. *Mater. Sci. Eng. A* **1998**, *243*, 244–249. [CrossRef]
2. Long, M.; Rack, H.J. Titanium alloys in total joint replacement—A materials science perspective. *Biomaterials* **1998**, *19*, 1621–1639. [CrossRef]
3. Niinomi, M. Mechanical biocompatibilities of titanium alloys for biomedical applications. *J. Mech. Behav. Biomed. Mater.* **2008**, *1*, 30–42. [CrossRef]
4. Miura, K.; Yamada, N.; Hanada, S.; Jung, T.K.; Itoi, E. The bone tissue compatibility of a new Ti-Nb-Sn alloy with a low Young's modulus. *Acta Biomater.* **2011**, *7*, 2320–2326. [CrossRef]

5. Bjursten, L.M.; Rasmusson, L.; Oh, S.; Smith, G.C.; Brammer, K.S.; Jin, S. Titanium dioxide nanotubes enhance bone bonding in vivo. *J. Biomed. Mater. Res. Part A* **2010**, *92*, 1218–1224. [CrossRef]
6. de Mello, M.G.; Dainese, B.P.; Caram, R.; Cremasco, A. Influence of heating rate and aging temperature on omega and alpha phase precipitation in Ti–35Nb alloy. *Mater. Charact.* **2018**, *145*, 268–276. [CrossRef]
7. Málek, J.; Hnilica, F.; Veselý, J.; Smola, B. Heat treatment and mechanical properties of powder metallurgy processed Ti-35.5Nb-5.7Ta beta-titanium alloy. *Mater. Charact.* **2013**, *84*, 225–231. [CrossRef]
8. Cui, C.; Hu, B.M.; Zhao, L.; Liu, S. Titanium alloy production technology, market prospects and industry development. *Mater. Des.* **2011**, *32*, 1684–1691. [CrossRef]
9. Leyens, C.; Peters, M. *Titanium and Titanium Alloys. Fundamentals and Applications*; WILEY-VCH Verlag GmbH & Co. KGaA: Weinheim, Germany, 2003; ISBN 3-527-30534-3.
10. Li, Y.; Yang, C.; Zhao, H.; Qu, S.; Li, X.; Li, Y. New Developments of Ti-Based Alloys for Biomedical Applications. *Materials* **2014**, *7*, 1709–1800. [CrossRef]
11. Duan, W.; Yin, Y.; Zhou, J.; Wang, M.; Nan, H.; Zhang, P. Dynamic research on Ti6Al4V powder HIP densification process based on intermittent experiments. *J. Alloy. Compd.* **2019**, *771*, 489–497. [CrossRef]
12. Cao, L.; Wu, X.; Zhu, S.; Mei, J.; Wu, X.; Bettles, C. The effect of HIPping pressure on phase transformations in Ti-5Al-5Mo-5V-3Cr. *Mater. Sci. Eng. A* **2014**, *598*, 207–216. [CrossRef]
13. Dekhtyar, A.I.; Bondarchuk, V.I.; Nevdacha, V.V.; Kotko, A.V. The effect of microstructure on porosity healing mechanism of powder near-β titanium alloys under hot isostatic pressing in α + β-region: Ti-10V-2Fe-3Al. *Mater. Charact.* **2020**, *165*, 110393. [CrossRef]
14. Atkinson, H.V.; Davies, S. Fundamental aspects of hot isostatic pressing: An overview. *Metall. Mater. Sci.* **2000**, *31*, 2981–3000. [CrossRef]
15. Bolzoni, L.; Ruiz-Navas, E.M.; Zhang, D.; Gordo, E. Modification of sintered titanium alloys by hot isostatic pressing. *Key Eng. Mater.* **2012**, *520*, 63–69. [CrossRef]
16. Molaei, R.; Fatemi, A.; Phan, N. Significance of hot isostatic pressing (HIP) on multiaxial deformation and fatigue behaviors of additive manufactured Ti-6Al-4V including build orientation and surface roughness effects. *Int. J. Fatigue* **2018**, *117*, 352–370. [CrossRef]
17. Lopez, M.; Pickett, C.; Arrieta, E.; Murr, L.E.; Wicker, R.B.; Ahlfors, M.; Godfrey, D.; Medina, F. Effects of postprocess hot isostatic pressing treatments on the mechanical performance of EBM fabricated TI-6Al-2Sn-4Zr-2Mo. *Materials* **2020**, *13*, 2604. [CrossRef] [PubMed]
18. Samarov, V.; Seliverstov, D.; Froes, F.H. Fabrication of near-net-shape cost-effective titanium components by use of prealloyed powders and hot isostatic pressing. In *Titanium Powder Metallurgy: Science, Technology and Applications*; Elsevier Inc.: Amsterdam, The Netherlands, 2015. [CrossRef]
19. Yudin, S.N.; Kasimtsev, A.V.; Tabachkova, N.Y.; Sviridova, T.A.; Markova, G.V.; Volod'Ko, S.S.; Alimov, I.A.; Alpatov, A.V.; Titov, D.D. Features of β-Phase Decay in Ti–22Nb–6Zr Alloy. *Inorg. Mater. Appl. Res.* **2019**, *10*, 1115–1122. [CrossRef]
20. Zherebtsov, S.V.; Dyakonov, G.S.; Salem, A.A.; Malysheva, S.P.; Salishchev, G.A.; Semiatin, S.L. Evolution of grain and subgrain structure during cold rolling of commercial-purity titanium. *Mater. Sci. Eng. A* **2011**, *528*, 3474–3479. [CrossRef]
21. Tan, L.; He, G.; Liu, F.; Li, Y.; Jiang, L. Effects of temperature and pressure of hot isostatic pressing on the grain structure of powder metallurgy superalloy. *Materials* **2018**, *11*, 328. [CrossRef] [PubMed]
22. Li, X.; Zhou, Y.; Ebel, T.; Liu, L.; Shen, X.; Yu, P. The influence of heat treatment processing on microstructure and mechanical properties of Ti–24Nb–4Zr–8Sn alloy by powder metallurgy. *Materialia* **2020**, *13*, 100803. [CrossRef]
23. Li, Z.; Zheng, B.; Wang, Y.; Topping, T.; Zhou, Y.; Valiev, R.Z.; Shan, A.; Lavernia, E.J. Ultrafine-grained Ti-Nb-Ta-Zr alloy produced by ECAP at room temperature. *J. Mater. Sci.* **2014**, *49*, 6656–6666. [CrossRef]
24. Valiev, R.Z.; Semenova, I.P.; Latysh, V.V.; Rack, H.; Lowe, T.C.; Petruzelka, J.; Dluhos, L.; Hrusak, D.; Sochova, J. Nanostructured titanium for biomedical applications. *Adv. Eng. Mater.* **2008**, *10*, 8–11. [CrossRef]
25. Langdon, T.G. Processing of ultrafine-grained materials using severe plastic deformation: Potential for achieving exceptional properties. *Rev. Metal. (Madr.)* **2008**, *44*, 556–564. [CrossRef]
26. Amigó-Borrás, V.; Lario-Femenía, J.; Amigó-Mata, A.; Vicente-Escuder, Á. Titanium, Titanium Alloys and Composites. *Ref. Modul. Mater. Sci. Mater. Eng.* **2020**, *1*. [CrossRef]
27. Ahlfors, M.; Hjärne, J.; Shipley, J. Cost effective Hot Isostatic Pressing. A cost calculation study for AM parts. *Quintus Technol.* **2018**, *12*, 1–6.
28. Torralba, J.M.; Campos, M. Toward high performance in Powder Metallurgy. *Rev. Metal.* **2014**, *50*. [CrossRef]
29. Lario-Femenía, J.; Amigó-mata, A.; Vicente-escuder, Á.; Segovia-lópez, F. Desarrollo de las aleaciones de titanio y tratamientos superficiales para incrementar la vida útil de los implantes. *Rev. Metal.* **2016**, *52*, 1–13. [CrossRef]

 *metals*

Article

Effect of Strain Rate on Microstructure Evolution and Mechanical Behavior of Titanium-Based Materials

Pavlo E. Markovsky [1,*], Jacek Janiszewski [2], Vadim I. Bondarchuk [1], Oleksandr O. Stasyuk [1], Dmytro G. Savvakin [1], Mykola A. Skoryk [1], Kamil Cieplak [2], Piotr Dziewit [2] and Sergey V. Prikhodko [3]

[1] G.V. Kurdyumov Institute for Metal Physics of N.A.S. of Ukraine, 36 Academician Vernadsky Boulevard, UA-03142 Kyiv, Ukraine; vbondar77@gmail.com (V.I.B.); stasiuk@imp.kiev.ua (O.O.S.); savva@imp.kiev.ua (D.G.S.); mykolaskor@gmail.com (M.A.S.)

[2] Jarosław Dąbrowski, Military University of Technology, 2 gen. Sylwester Kaliski str., 00-908 Warsaw, Poland; jacek.janiszewski@wat.edu.pl (J.J.); kamil.cieplak@wat.edu.pl (K.C.); piotr.dziewit@wat.edu.pl (P.D.)

[3] Department of Materials Science and Engineering, University of California Los Angeles, Los Angeles, CA 90095, USA; sergey@seas.ucla.edu

[*] Correspondence: pmark@imp.kiev.ua; Tel.: +38-044-4243374

Received: 6 October 2020; Accepted: 19 October 2020; Published: 22 October 2020

Abstract: The goal of the present work is a systematic study on an influence of a strain rate on the mechanical response and microstructure evolution of the selected titanium-based materials, i.e., commercial pure titanium, Ti-6Al-4V alloy with lamellar and globular microstructures produced via a conventional cast and wrought technology, as well as Ti-6Al-4V fabricated using blended elemental powder metallurgy (BEPM). The quasi-static and high-strain-rate compression tests using the split Hopkinson pressure bar (SHPB) technique were performed and microstructures of the specimens were characterized before and after compression testing. The strain rate effect was analyzed from the viewpoint of its influence on the stress–strain response, including the strain energy, and a microstructure of the samples after compressive loading. It was found out that the Ti-6Al-4V with a globular microstructure is characterized by high strength and high plasticity (ensuring the highest strain energy) in comparison to alloy with a lamellar microstructure, whereas Ti6-Al-4V obtained with BEPM reveals the highest plastic flow stress with good plasticity at the same time. The microstructure observations reveal that a principal difference in high-strain-rate behavior of the tested materials could be explained by the nature of the boundaries between the structural components through which plastic deformation is transmitted: α/α boundaries prevail in the globular microstructure, while α/β boundaries prevail in the lamellar microstructure. The Ti-6Al-4V alloy obtained with BEPM due to a finer microstructure has a significantly better balance of strength and plasticity as compared with conventional Ti-6Al-4V alloy with a similar type of the lamellar microstructure.

Keywords: titanium alloy; high strain rate testing; split hopkinson pressure bar technique; microstructure influence; phase transformation; deformation mechanism; strain energy

---

1. Introduction

Titanium alloys are an important structural material used in modern aerospace, automotive, shipbuilding, and military fields, because of the high level of specific strength, fracture toughness, fatigue strength, corrosion resistance, non-magnetization, and other specific physical, mechanical, and service properties [1–4]. Since these alloys are relatively expensive, their advantages over other structural materials become more apparent when they are processed using different methods to as higher as possible values of specific strength. A separate important direction in the application of titanium alloys, which is increasingly being used, is the manufacture of armor elements [4–6], as well as individual elements of military equipment experiencing impact or explosive loading [4,7]. Traditionally the designers of new products are based on the mechanical properties of materials, which have been determined under certain standard, mainly quasi-static test conditions. However, in most cases of machines and devices, it does not correspond to the actual operating conditions of real parts, for example, emergency conditions of heavy-loaded structural elements of airplanes subjected to strong turbulence, or the landing gear during extremely hard landings. Therefore, to prevent the destruction of such structures, designers usually follow so-called strength reserve approach, which results in a significant increase in the mass of individual parts and, as a consequence, in an undesirable increase in the total weight of the products, their price and operating costs. Hence, a significant reduction in costs is expected, inter alia, through structure optimization with regard to strength and weight.

The mechanical behavior of materials strongly depends on the loading conditions. The influence of a quasi-static strain rate ranging up to $10\,s^{-1}$ on the mechanical behavior of different titanium alloys has been extensively studied [8–13]. A lot of attention has also been paid to the similar studies on a high-strain-rate (dynamic) testing of Ti-based materials [14–20]. However, some important issues such as an influence of the chemical and phase composition, microstructure, as well as the variation of the strain rate within a wide range, and especially, the combined effect of these have not been studied sufficiently so far. Thus, the goal of present work is a systematic study of a strain rate influence on the microstructure evolution and the mechanical behavior of the commonly used two-phase $\alpha + \beta$ titanium alloy Ti-6Al-4V (wt.%) under a loading condition of compression split Hopkinson pressure bar (SHPB). The results are also compared with the data obtained during the quasi-static compression. Moreover, the behavior of commercial pure titanium, considered as single-phase (h.c.p. lattice) material, is studied in similar conditions compared to Ti-6-4 alloy.

2. Materials and Methods

2.1. Materials and Materials Processing

The Ti-6Al-4V-alloy was studied in two different states, the first state was the alloy obtained via a conventional cast and wrought technology [1]. The second state alloy was produced using a cost-efficient blended elemental powder metallurgy (BEPM) approach [18,19] and was designated in this study as Ti64BEPM. The cast and wrought alloy was purchased in a shape of 10-mm diameter rods from Perryman Company (Houston, PA, USA). Two different metallurgical states of the Ti-based alloy were obtained using conventional heat treatments. Annealing of the alloy at 850 °C for 2 h resulted in obtaining a globular microstructure (GL), whereas annealing at 1100 °C for 0.5 h resulted in obtaining a lamellar structure (LM). The alloys with two different microstructural conditions were marked as Ti64GL and Ti64LM, respectively. Ti64BEPM alloy was obtained by blending the titanium hydride TiH_2 powder (particles size <100 μm) and 60Al-40V (wt.%) master alloy in a powder form (particles size <63 μm) and then cold pressing of the blend in a die at 640 MPa followed by sintering at 1250 °C for 4 h under the vacuum of 10^{-3} Pa. The used sintering conditions provide removal of hydrogen from the material to an admissible level (0.002–0.003%) and transformation of the powder compacts into bulk homogeneous alloy. More details on materials fabrication using the adopted BEPM protocol can be found in [21,22]. The sintered Ti64BEPM samples were bars with dimensions of $9 \times 9 \times 60$ mm. In order to assist the results evaluation and understand an influence of phase and chemical composition on mechanical behavior of the main object, the Ti64 alloy, a comparative analysis was performed on the commercial pure titanium (c.p.Ti) Grade 1. The alloy was purchased in the shape of rods with a diameter of 10 mm from Titan Ltd. (Kyiv, Ukraine).

2.2. Quasi-Static and High-Strain-Rate Tests

The quasi-static tensile properties were determined using INSTRON 3376 strength machine following ASTM E8 standard with specimens having gage diameter 4 mm and gage length 25 mm, whereas the quasi-static compressive tests were carried out with the use of MTS C45 strength machine. For both types of compression tests, quasi-static and dynamic, cylindrical specimens with a diameter and height of 5 mm were used. The Young, shear moduli, and Poisson's ratio of materials were measured with resonance-frequency-damping analysis (RFDA) apparatus (IMCE, Belgium) using impulse excitation technique in accordance with ASTM E1876-15 Standard.

The high-strain-rate compression tests were performed with the use of a split Hopkinson pressure bar (SHPB), also called a Kolsky bar technique [23–25]. The basic parameters of the SHPB system are shown in Figure 1. The length of the input and output bars was 1200 mm, the length of the striker bar was 250 mm, the diameter of all bars was 12 mm. The bars were made of maraging steel (heat-treated MS350 grade: yield strength—2300 MPa; elastic wave speed—4960 m/s). The striker bar was driven by a compressed air system with the barrel length of 1200 mm and the inner diameter of 12.1 mm. The impact striker bar velocities applied during the experiments were in the range from 10 to 25 m/s, which ensures strain rates in the range of $1100 - 3320$ s^{-1} for dimensions of the used specimens.

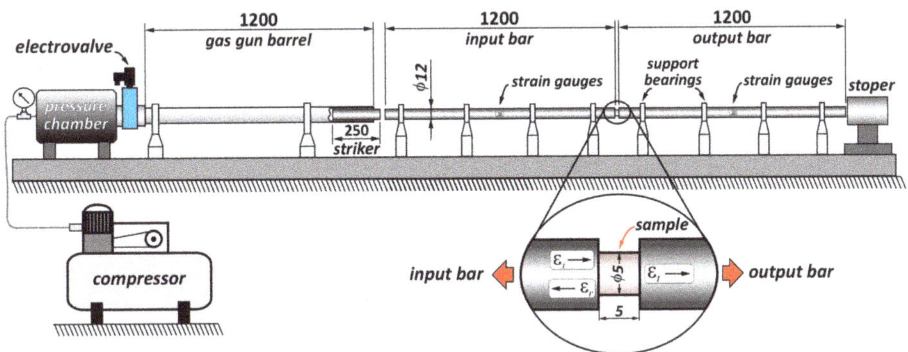

Figure 1. The schematics of the split Hopkinson pressure bar (SHPB) system used in this study.

The plastic flow stress, strain, and a strain rate of the sample were determined according to the classical Kolsky theory [24,25] based on the one-wave analysis method, which assumes stress equilibrium for a specimen under given testing conditions. Wave signals, incident (ε_i), transmitted (ε_t), and reflected (ε_r), were measured by pairs of strain gages (gauge length—1.5 mm; resistance—350 Ohm) glued at the half-length of the input and output bars (Figure 1). The signals from the strain gauges were conditioned with data-acquisition system, composed of Wheatstone bridge, amplifier, and a digital oscilloscope, allowing a high cut-off frequency of 1 MHz.

Strain and a strain rate in the specimens were calculated based on the profile of reflected wave (ε_r) using Equations (1) and (3), respectively, whereas the stress was determined based on transmitted wave (ε_t) (Equation (2))

$$\varepsilon(t) = -\frac{2c_b}{L} \int_0^t \varepsilon_r(t)dt \tag{1}$$

$$\sigma(t) = \frac{A_b E}{A_s} \varepsilon_t(t) \tag{2}$$

$$\dot{\varepsilon}(t) = -\frac{2c_b}{L} \varepsilon_r(t) \tag{3}$$

where: L, A_s are the length and the cross-section area of the specimen; E, A_b, c_b are the Young's modulus, the cross-sectional area, and the elastic wave velocity of the pressure bar, respectively.

To minimize dispersion of the wave by damping the Pochhammer-Chree high frequency oscillations and to facilitate the stress equilibrium, a pulse shaping technique was used [26]. The technique consists in placing a small disc made of soft material on the impact end of the input bar. The disc is often called a pulse shaper or a wave shaper. Plastic deformation of the pulse shaper physically filters out the high frequency components in the incident pulse and modifies its profile. The pulse shaper size needs to be chosen for the given striker impact velocity and mechanical response of tested material. The tests were performed to find the proper size of the pulse shaper in order to obtain stress equilibrium during high-strain-rate deformation of titanium alloy specimens. It was found that for a given SHPB test condition, the copper pulse shaper with a diameter of 3 mm and thicknesses in the range from 0.1 to 0.4 mm (depending on impact striker velocity) guarantees damping of the high frequency oscillations and achieving the dynamic stress equilibrium in the specimen (Figure 2). The typical raw signals recorded during the SHPB experiment are shown in Figure 2a, whereas Figure 2b presents the stresses on the front and back face of the Ti64LM specimen cracked under applied compressive loading. As it can be observed in Figure 2a, the presence of pulse shaper limits significantly a stress fluctuation on incident wave profile. The reflected and transmitted signals are also almost smooth without large oscillations. In turn, the stress-state equilibrium condition presented in Figure 2b is satisfied, i.e., front and back stresses are close throughout the loading history, except a peak at the top of the initial rise and a gap occurring at the end of the loading time. This gap is the result of a crack in the Ti64LM specimen.

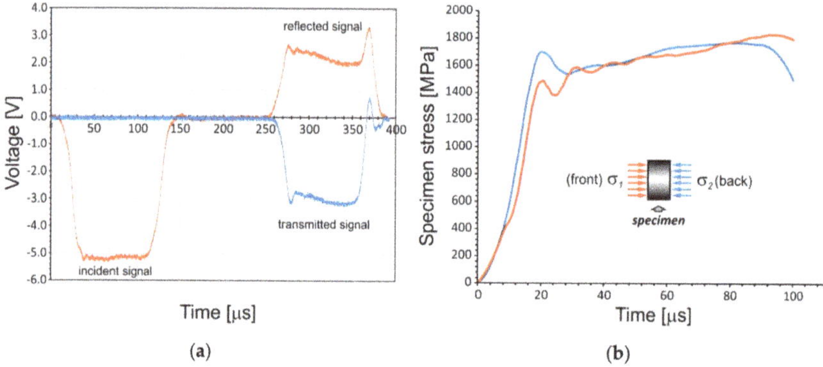

Figure 2. (**a**) Typical raw wave signals from SHPB experiment for Ti64LM; (**b**) stresses on the front and back end of Ti64LM specimen (dynamic stress equilibrium).

Moreover, the grease composed of mineral oil, lithium soap and molybdenum disulfide (MoS_2) was applied to the interfaces between the specimen and the bars to minimize the interfacial friction.

2.3. Microstructural Characterization

The structure of all materials was studied with light optical microscopy (LOM) using XL70 (Olympus, Shinjuku, Tokyo, Japan) and scanning electron microscopy (SEM) using Vega 3 and Mira 3 machines (both from Tescan, Czech Republic). SEM Mira 3 was also used to study the fracture surfaces and electron backscatter diffraction (EBSD) was used to evaluate local crystallographic orientation of different microstructural elements. Metallographic specimens were prepared according to standard grinding and polishing methods [1]. For SEM and EBSD, the final polishing was carried out using Saphir Vibro polisher (ATM, Germany). Some samples were additionally ion polished/etched using PECS model 682 (Gatan, CA, USA) or chemically etched using standard Kroll's solution [1]. The gas content in the sintered specimens was measured using a gas analyzer OH900 (Eltra, Germany).

3. Results and Discussion

3.1. Initial Microstructure Characterization

Typical microstructures of the studied materials are presented in Figure 3, and their chemical compositions are listed in Table 1. Pure titanium c.p.Ti is characterized by relatively coarse α-grain structure with an average size of about 600 μm (Figure 3a). The intragrain substructure is highly developed and is characterized both by the presence of twins and the dislocation network, including formation of cells. Such an extensive structure of defects may result from relatively fast cooling rates used after deformation or annealing (Figure 3a).

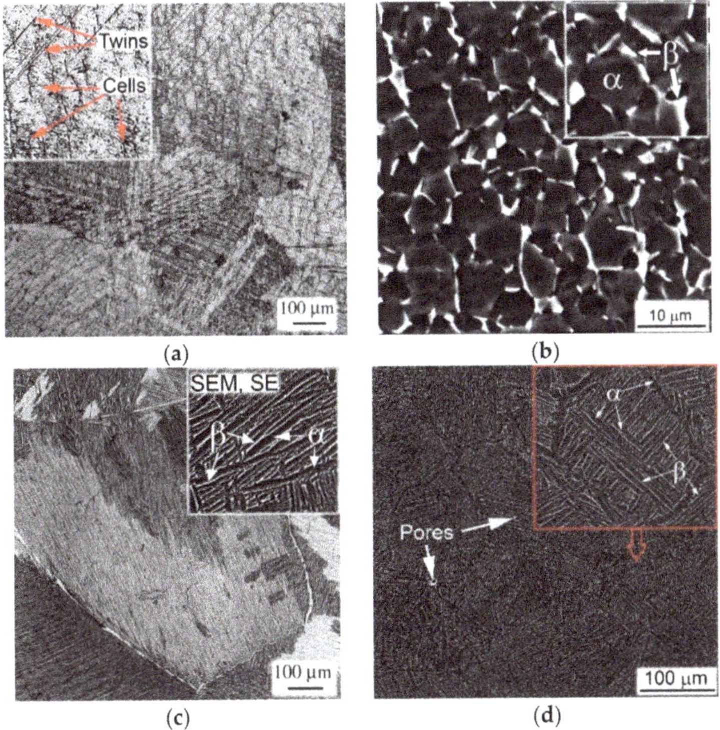

Figure 3. Microstructure of Ti-based materials in the initial state: (**a**) c.p.Ti; (**b**) Ti64GL; (**c**) Ti64LM; (**d**) Ti64BEPM; (**a,c**)—LOM; (**b,c** (in the corner),**d**)—SEM, SE (secondary electron image).

Table 1. Chemical composition of studied materials.

	Alloying Elements, wt.%					
	Al	**V**	**Fe**	**O**	**N**	**Ti**
c.p.Ti	<0.2	-	<0.08	0.01	0.007	Base
Ti64LM, GL	5.8	3.96	0.21	0.016	0.008	Base
Ti64 BEPM	5.94	4.06	0.16	0.2	0.03	Base

The Ti64GL specimens were of uniform and fine globular microstructure with an average size of α-globules of about 7 μm (Figure 3b), which explains a good balance of the quasi-static tensile strength and ductility (Table 2, #2). The microstructure of the Ti64LM consisted of rather coarse β-grains (average size 800 μm) with coarse (up to 500 μm in some grains) colonies of α-lamellae inside (Figure 3c), which caused a noticeable decrease in both tensile strength and ductility

(Table 2, #3). The Ti64BEPM alloy was also of coarse-grained lamellar microstructure (Figure 3d); however, both β-grains (average size of about 100 μm) and colonies of shorter α-plates were much finer (compare Figure 3c,d), because of a pinning role of residual pores in the grain boundary movement, which prevents from grain coarsening during sintering [18,19]. The Ti64BEPM demonstrates a higher strength compared to the Ti64LM (Table 2, # 4 vs. #3), because of a higher content of impurities (Table 1, #3), and shows lower ductility, which may be also related to the increased impurities content and presence of about 1.5–2 vol.% of residual pores.

Table 2. Mechanical properties of tested materials (tension rate 8×10^{-4} s^{-1}).

##	Tensile Yield Stress [MPa]	Ultimate Tensile Stress [MPa]	El. [1] [-]	RA [2] [-]	Young Module [GPa]	Shear Module [GPa]	Poisson's Ratio	Vickers Hardness [HV]
#1 c.p.-T	345	408	0.38	0.59	111.5	46	0.253	117
#2 Ti64GL	988	993	0.19	0.42	121.7	47	0.275	312
#3 Ti64LM	824	865	0.15	0.31	121.7	47	0.275	309
#4 Ti64BEPM	932	1033	0.08	0.21	123.0	N/A	N/A	339

[1] El.—elongation, [2] RA—reduction in area.

3.2. Mechanical Response

3.2.1. Stress–Strain Behavior

To characterize the base mechanical properties of the tested Ti-based materials, the quasi-static tensile and elastic characteristics are listed in Table 2. The Ti64GL presents the highest yield stress equal to 988 MPa, whereas the Ti64LM presents the lowest one—824 MPa, which is almost 2.4 times higher than the one of c.p.Ti (345 MPa). The Ti64GL material exhibits also the highest ductile properties (elongation 0.19), whereas the Ti64BEPM is characterized by the lowest elongation equal to 0.08. Ductility of all Ti-6-4 alloys tested is relatively low compared to the ductility of c.p.Ti (elongation 0.38).

Stress–strain behavior of the materials tested under uniaxial compression was, as predicted, slightly different. Generally, a compression yield point at a quasi-static strain rate for all materials was at the same or slightly lower level (349, 918, 938, and 879 MPa, for ## 1, 2, 3, and 4 in Table 2, respectively). An exception was the Ti64LM, which demonstrated a higher yield stress in compression than in tension. In turn, fracture of the specimens at the quasi-static compressive regime occurred at significantly higher strain values, i.e., Ti64LM and Ti64BEPM cracked at strain of 0.28 and 0.42, respectively, whereas c.p.Ti and Ti64GL did not fracture before reaching the strain of 0.5, at which the compression was stopped.

Similar dependencies in the specimen damage behavior under compression was observed in dynamic testing conducted at strain rates in the range from 1250 to 3320 s^{-1}. With the exception of c.p.Ti, all other materials cracked under dynamic loading; however, cracking occurred at lower strain values compared to the corresponding quasi-static test results, and it will be discussed in the further part of the paper. In Figure 4a, a few high-speed video frames are presented to illustrate typical successive stages of the specimen deformation process, i.e., the start (Figure 4a-1), the uniform deformation (Figure 4a-2), onset of specimen barreling (Figure 4a-3), intensive local heating, fracturing crack, and the spark flash at the final stage of the deformation (Figure 4a-4). In turn, Figure 4b shows a typical view of the cracked specimens after the quasi-static and dynamic compression.

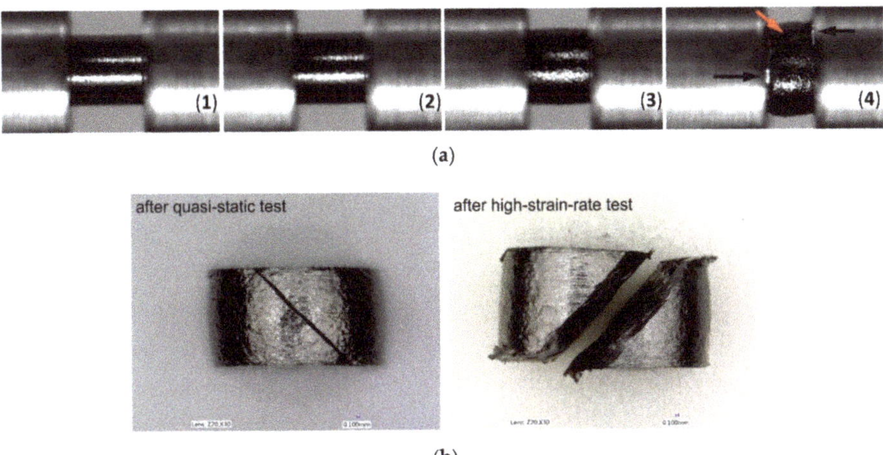

Figure 4. (**a**) High-speed video frames illustrating plastic deformation and fracture of the Ti64BEPM specimen during of the SHPB test at the strain rate of 2100 s^{-1}: the start—(**1**), uniform strain—30 µs (**2**), barreling onset—60 µs (**3**), specimen cracking—120 µs (**4**)—arrows indicate: fracture crack—red arrow, and a areas of intensive local heating on the contact surface with the bar—black arrow; (**b**) view of the cracked specimen of Ti64BEPM after quasi-static and high-strain-rate tests in compression.

The true stress–strain curves of the tested Ti-based materials compressed at quasi-static and high-strain-rates ranges are shown in Figure 5. A number of important observations can be based on these results. First of all, the initial peak stress and oscillations visible on the curve for the highest strain rate are not a real mechanical response of the material, but they result from technical limitations of the SHPB technique. However, the stress–strain curves oscillations were significantly reduced through applying a pulse shaper technique and, in the case of the SHPB experiments with lower strain rates, the obtained stress–strain curves are smooth and almost without oscillation. Second, the quasi-static stress–strain curves reveal differences in the strain hardening behavior of the tested materials (Figure 5e). The strain hardening coefficient n, calculated from a slope of a fitting line of the true stress–strain curve plotted on a logarithmic scale (assumed ranges of plastic strain—0.05–0.2 or 0.05–0.4), is the highest for the Ti64GL (0.148). The values of n for the Ti64LM and Ti64BEPM are relatively lower, and equal to 0.052 and 0.068, respectively. The c.p.Ti demonstrates the highest value of n coefficient, 0.334, as it was expected. Third, the work hardening behavior of the materials tested under high-strain-rate loading is similar to quasi-static loading; however, strain hardening effect is slightly reduced through the heat generation during the dynamic compression, which leads to the flow softening at high strains. This phenomenon seems to be the most pronounced for the Ti64GL and Ti64BEPM. It should be noted here that conversion of the deformation energy to the thermal energy is not uniform throughout the specimen volume, particularly at the end of the sample deformation stage. Careful observation of the high-speed camera films allowed detection of the intense heating and the strain localization near the contact surfaces of the specimen with the front surfaces of the bars (bright areas marked with black arrows in Figure 4a-4). This intense local heating is also manifested by the outflow of a part of the material on the sample side surface (see the image on the right side in Figure 4b).

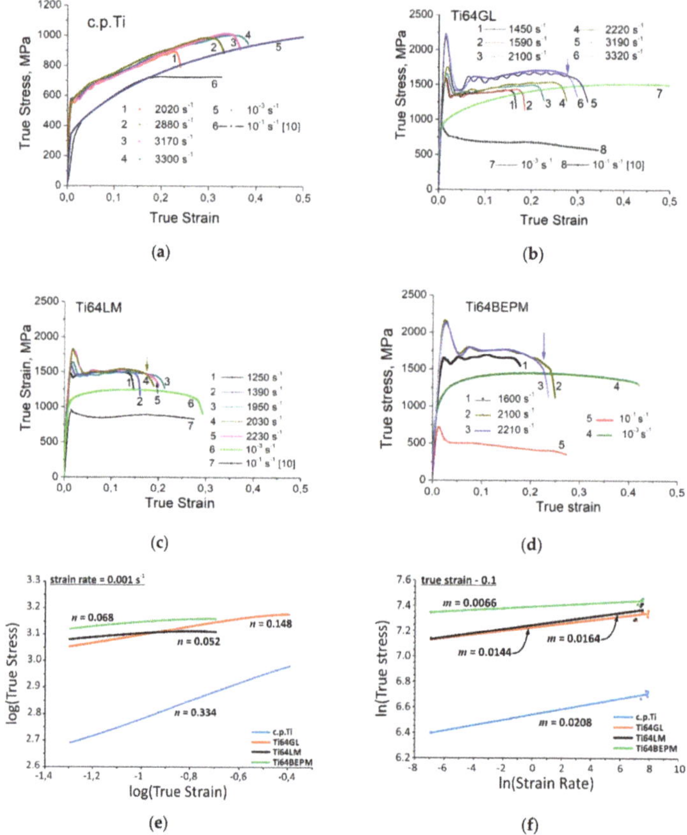

Figure 5. Quasi-static and dynamic stress–strain curves and the calculated mechanical data for: (**a**) c.p.Ti; (**b**) Ti64Gl; (**c**) Ti64LM; (**d**) Ti64BEPM (arrows in (**b**,**d**) indicate possible moment of fracture for relevant curves); (**e**) strain hardening coefficient—*n*; (**f**) strain rate sensitivity exponent—*m*.

As it was expected, the plastic flow stress levels at dynamic regime are significantly higher in comparison to the ones at quasi-static regime. The highest peak flow stress (a maximum stress in the plastic range of deformation) is presented by the Ti64BEPM (1770 MPa at strain of 0.17) and the Ti64GL (1700 MPa at strain of 0.23), whereas the Ti64LM reveals the lowest peak flow stress equal to 1540 MPa at 0.12. In the case of c.p.Ti, the peak flow stress, corresponding to unloading of incident wave, is at the level of 1000 MPa at strain of 0.35. In turn, analysis of values of the strain rate sensitivity exponent *m* (Figure 5f), calculated from the slopes of linear regressions (Equation (4)), shows the highest flow stress increase with an increasing strain rate for the Ti64LM (*m* = 0.0164). The strain rate sensitivity of Ti64GL is slightly lower (*m* = 0.0144), whereas the Ti64BEPM demonstrates the lowest one (*m* = 0.0066). The value of *m* factor for c.p.Ti is relatively high (0.0208) compared to the other alloys examined.

$$m = \mathrm{d}(\ln(\sigma_\mathrm{t}))/\mathrm{d}\left(\ln(\dot{\varepsilon})\right) \tag{4}$$

where σ_t is the true flow stress at strain of 0.1, and $\dot{\varepsilon}$ is average strain rates.

As it was noted earlier, cracking the Ti alloys tested under the dynamic loading occurred at lower strain values compared to the corresponding quasi-static test results. However, it was observed (see Table 3) that specimens deformed with the critical strain rates (at which cracking occurred) damage at lower strains (ε_{cr}) than specimens tested with slightly lower strain rates without cracking (strain designation—ε_{max}). For example, the Ti64GL specimen tested at a strain rate of 3190 s^{-1} achieved a strain value equal to 0.30 (curve #5 in Figure 5b), while an analogous specimen deformed at 3320 s^{-1} cracked at a slightly lower strain equal to 0.28 (curve #6 in Figure 5b). The same dependency was found for Ti64LM and Ti64BEPM.

Table 3. Comparison of ε_{cr} and ε_{max} for the tested Ti-alloys.

	Ti64GL	**Ti64LM**	**Ti64BEPM**
strain ε_{cr} at $\dot{\varepsilon}$ (s^{-1})	0.28 (3320)	0.17 (2030)	0.23 (2210)
strain ε_{max} at $\dot{\varepsilon}$ (s^{-1})	0.30 (3190)	0.19 (1950)	0.24 (2100)

Based on the data listed in Table 3, it can be concluded that the Ti64LM alloy exhibits the lowest value of ε_{cr} at the strain rates slightly above 2000 s^{-1} compared to Ti64BEPM and Ti64GL materials, which crack at strains of 0.23 and 0.28 and at strain rates of 2210 and 3320 s^{-1}, respectively. It should be emphasized that resistance to cracking of the Ti64GL alloy under the dynamic deformation is significantly higher compared to other Ti-alloys tested.

3.2.2. Material Strain Energy

In order to carry out more in-depth assessment of the mechanical behavior of the materials tested under dynamic loading, an additional parameter, i.e., strain energy (*SE*), was used. It is a convenient parameter that allows comparing the mechanical response of materials tested with various methods and strain rates [11–13]. The *SE* is defined as the internal work performed to deform a material specimen through an action of the externally applied forces. The *SE* was determined by integrating the area under the stress–strain curve (for integration limits from zero to ε_{upper}). In the case of the cracked specimens, a value of ε_{upper} corresponded to a value of strain at fracture, whereas for the non-cracked specimens, ε_{upper} was assumed to be equal to strain at the moment of the specimen unloading (sharp drop in stress–strain curve). The upper integration limit ε_{upper} for the non-cracked specimens under quasi-static loading was assumed to be 0.5.

As it can be seen in Figure 6a, the cast and the wrought alloy Ti64 in both the globular and the lamellar microstructural states have almost the same values of *SE* at the strain rates up to 2000 s^{-1} (curves 2 and 3 in Figure 6a). However, above this strain rate level, the Ti64LM cracked in contrast to the Ti64GL, which demonstrated the highest level of *SE* among all materials studied.

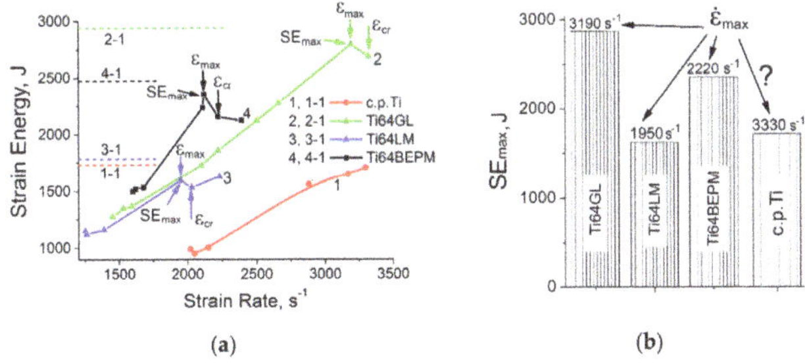

Figure 6. The strain energy for the studied Ti-based materials: (**a**) *SE*-strain rate dependence; (**b**) comparison of the maximum strain energy values (*SE*$_{max}$) corresponding to the maximal strain rate $\dot{\varepsilon}_{max}$ (indicated by arrows), for which specimens do not crack during the SHPB tests.

A change from cast and wrought to BEPM in the method for manufacturing the Ti64 alloy significantly affected the measured *SE* values. The Ti64BEPM material revealed a higher ability to store mechanical energy at strain rates up to 2300 s^{-1}. The *SE* values were noticeably higher compared to Ti64GL, and at the strain rate range of 2100–2300 s^{-1}, at which Ti64BEPM specimens broke (compare curves 4, 2 and 3; Figure 6a).

It is also interesting to compare the *SE* values calculated from the quasi-static and the high-strain-rate tests data. The maximal *SE* value determined at quasi-static tests (horizontal (1-1) line in Figure 6a) is approximately equal to the SE_{max} level obtained for the maximum strain rate ($\dot{\varepsilon}_{max}$) for c.p.Ti only (curve #1 in Figure 6a). In turn, *SE* values for Ti64 alloys in all microstructural states from the quasi-static tests (horizontal lines (2-1), (3-1), and (4-1) in Figure 6a) are slightly higher than the SE_{max} values obtained for the whole range of strain rates, at which specimen cracking occurred (marked by arrows on curves 2–4 in Figure 6a).

Since the main difference between the studied Ti64 structures was the β- grain size, the SE_{max} values (Figure 6b) were related to the grain size of the tested Ti-alloy. From the data presented in Table 4 it can be seen that a larger β-grain size in Ti64 alloy structure causes a decrease in SE_{max}.

Table 4. Dependency between SE_{max} parameter and β-grain size of the tested Ti64 alloys.

	Ti64GL	**Ti64LM**	**Ti64BEPM**
grain size [μm]	7	800	160
strain energy SE_{max} [J]	2795	1594	2354

It is also worth noting that c.p.Ti has—as predicted—the lowest *SE* value among all studied materials and at all strain rates. The *SE* value for c.p.Ti is almost twice lower compared to the cast and wrought alloys, despite the pure titanium shows very high plasticity, because of which specimen fracture does not occur at the applied strain and strain rates (curve 1 in Figure 6a).

In view of the fact that titanium alloys are often used instead of other structural materials in various critical applications, the results of the present study were additionally compared to similar data of other commonly used structural materials [27] (Figure 7). The lowest level of the *SE* parameter (Figure 7a) demonstrates the aluminum alloy B95 (curve 1), as it was expected. The high-strength steels ARMOX 600T and Docol 1500M are characterized by significantly higher *SE* values (curves 2, and 3), which exceed the corresponding values for Ti64GL and Ti64BEPM alloys at the same strain rates (curves 5, and 6). However, when these curves were converted considering the density of tested materials (Al alloy B95—2850 kg/m^3 [28], for all Ti-based materials—4500 kg/m^3 [1], and for steels of 7850 kg/m^3 [29]), results revealed another dependency between considered materials (Figure 7b).

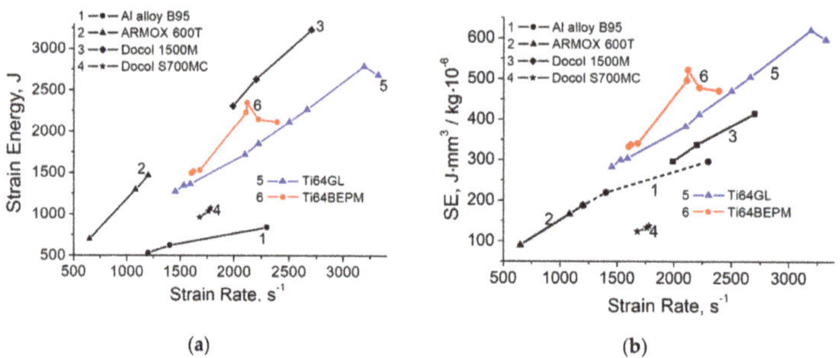

Figure 7. Comparison of the SE-strain rate dependencies for different materials: (**a**) absolute values of the SE; (**b**) relative (specific) values of the SE. The data used for the curves 1–4 were taken from [27].

It can be clearly seen that the aluminum alloy at the strain rates of about 1500 s^{-1} is not worse than ARMOX 600T steel (curves 1, and 2), whereas titanium-based materials are undeniably better than the high-strength steels (curves 5–7). The only, rather serious, drawback of the BEPM made titanium materials is that they could be cracked at relatively low strain rates. However, such a shortcoming could be overcome, at least partially, by incorporating these materials into multilayer structures combining them with high ductility Ti64 alloy layers in the optimized configuration, as it was shown in a few studies published earlier [30,31].

3.3. Deformed Microstructures Investigation

Analysis of the microstructure of materials formed during plastic deformation is a primary step to evaluate a difference in their mechanical behavior. The initial examination of the structural features of all the specimens after quasi-static tests and high-strain-rate SHPB deformation allows identification of four main zones distinguished by the stress state and, as a result, having specific differences in the microstructure (Figure 8). Zone I is quite narrow and is characterized by tangential shear stresses caused by the interaction on the contact surfaces of the specimen and SHPB bars. Zone II tracks across the entire specimen at an angle of approximately 45° to the vertical axis of the cylinder and it corresponds to the plane of the maximum shear stresses in accordance with Schmid law [32]. This zone is characterized not only by the maximum shear stresses, but also by the greatest strain localization, due to stress collapse, adiabatic shear band (ASB) initiation, temperature rise and crack formation [11,16,19]. Zone III is adjacent to zone II and it corresponds to the secondary strain localization, where ASBs and secondary (smaller) cracks were also observed. Zone IV is located away from the fields of intense stresses and strain localization; however, the stress state in this zone is more complicated, involving the compressive stress along the vertical axis of the cylinder sample and tensile (or shear) perpendicular to it.

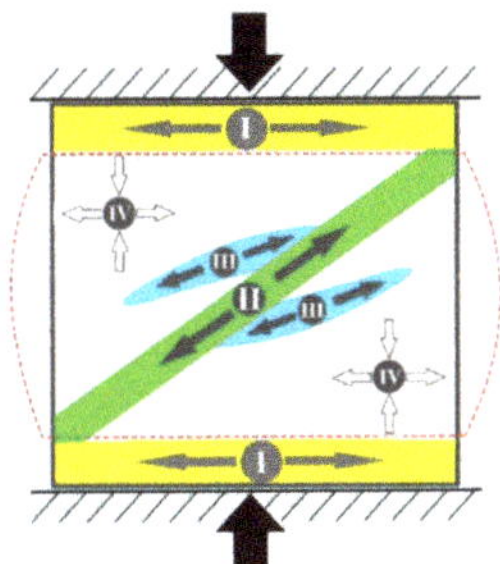

Figure 8. Schematic representation of specific zones distinguished in the longitudinal section of tested cylindrical specimen (Roman number denote a given zone; the arrows indicate the direction of compressive force).

3.3.1. Microstructure Analysis of c.p.Ti

The microstructure of c.p.Ti after SHPB tests is shown in Figure 9a,b. The numerous plastic deformation traces, in the form of slip bands, twins, and well-developed substructure inside α-phase plates, are observed in all zones near the specimen-to-bar contact surfaces (Figure 9a) as well as across the entire bulk of the specimen (Figure 9b). A comparison with the initial not-deformed state (Figure 3a) shows a significant increase in defects density, while a general character of the substructure remains approximately the same. Assuming that titanium with single-phase h.c.p. lattice could easily deform by twinning at sufficiently low temperatures [1,33,34], it was expected that defected areas should contain mainly twins. However, given the probability of a significant local increase in temperature during the deformation [18,35], a few other scenarios should not be excluded, namely, the formation of a well-developed dislocation substructure and the possibility of the phase transformations, including the martensite formation, as it was reported in [36,37], where c.p.Ti was subjected to fast heating and cooling. Eventually, because of the high plasticity of c.p.Ti, introduction of some deformation defects could not be completely excluded for this alloy as a result of sample preparation (after delicate ion polishing and/or etching), although their presence causes a "background" effect in all the samples made of this alloy. Nevertheless, the areas of localized deformation and ASB were not found in c.p.Ti specimens, even in those tested at the highest strain rates (Figure 9b).

The microstructure after the quasi-static tests seems to be uniform in all zones. It is modified by plastic deformation, showing a higher density of deformation defects and a smaller size of cells and twins (Figure 9c,d). This results from different strain rates used at quasi-static tests ($0.001\ s^{-1}$) and SHPB tests ($3200\ s^{-1}$); difference is more than 6 orders. Therefore, under the quasi-static tests, there are more possibilities and longer time for material relaxation compared to SHPB experiments.

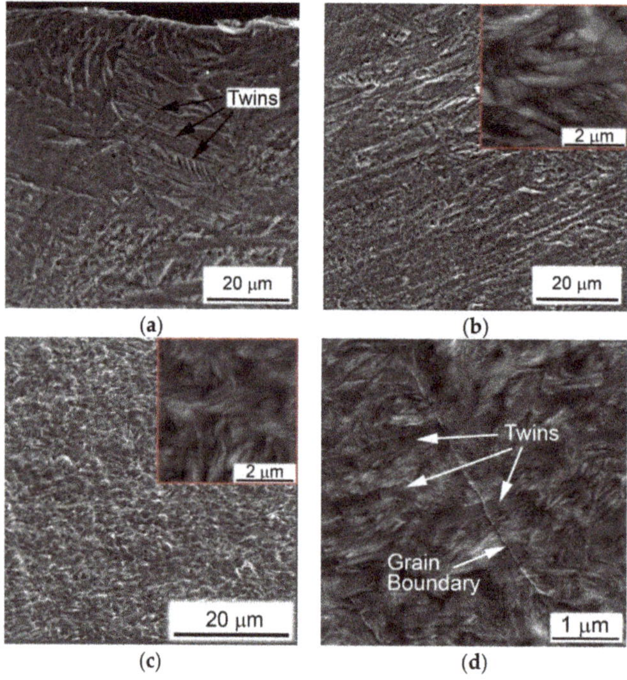

Figure 9. SEM images of c.p.Ti specimens sectioned along the longitudinal direction of the cylinders after: (**a**,**b**) SHPB test at the strain rate of $3200\ s^{-1}$—(**a**) near the zone I, (**b**) in the center of zone II; (**c**,**d**); quasi-static test with the strain rate $0.001\ s^{-1}$—zone 4. SEM, SE.

3.3.2. Microstructure Analysis of Ti64GL

Typical images of Ti64GL microstructure after SHPB tests, which did not cause fracture of specimens, can be seen in Figure 10. In general, the microstructure of this material did not change significantly in bulk after compression at $2100\ s^{-1}$, without failure, compared to the initial state. In some local micro-volumes, a couple of twins in the separate globules were observed closer to the specimen core (zones II and IV, Figure 10b). More significant traces of plastic deformation were observed in thin (5–7 µm) surface layers in zones I (Figure 10a), which could be a consequence of localized deformation, due to interaction between the surfaces of the specimens and the bars. An increase in a compression rate up to $2500\ s^{-1}$ (Figure 10c) or even to $2660\ s^{-1}$ (Figure 10d–f) also did not cause failure of the specimens, but led to a noticeable change in their microstructure. In this case, more than a half of all α-globules (about 65%) contain deformational twins, and they are observed in all zones throughout the bulk of the specimen. According to [1,33,34], during compression, the {11–22} twins are first formed inside the grains with the c-axis oriented parallel to the loading direction, and afterwards the formation of the {10–11} twins occurs in the grains oriented differently. Therefore, it appears that with an increase in compression strain rates, new twinning planes are activated inside the α-globules. Delicate etching of the deformed specimens reveals, in addition to twins, fine dislocation cells (their size is not more than a few tens of nanometers) observed in both α- and β- phases (Figure 10e,f).

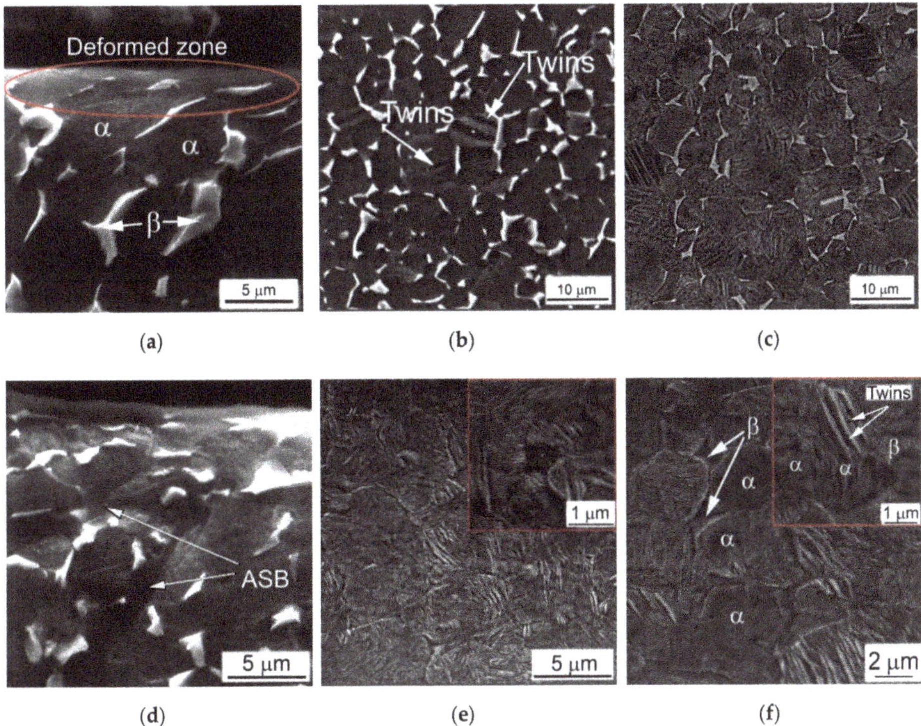

Figure 10. SEM images of Ti64GL specimens after the dynamic tests at the following strain rates: (**a**,**b**) 2100 s^{-1}; (**c**) 2500 s^{-1}; (**d**–**f**) 2660 s^{-1}—(**a**,**d**) show zone I; (**b**,**c**)—zone IV; (**e**,**f**)—zone III; (images (**a**,**b**,**d**–**f**) are taken in SE mode from the etched surface, whereas image (**c**) is in the BSE (back scattered electrons) mode from just polished surface.

A further increase in a strain rate up to 3320 s^{-1} caused cracking of the Ti64GL specimen, and appearance on the fracture surface (Figure 11a) small melted areas (Figure 11b) and dimples (Figure 11c), which are characteristic for ductile fracture. It should be emphasized that the main crack propagated in the unchanged direction (Figure 11a), which is clearly noticeable by the configuration of ductile grooves on the entire fracture surface (Figure 11c). There is also noticeable secondary or lateral crack propagation (indicated by B in Figure 11a) on both sides of the main crack fracture (indicated by A) and forms distinct relief.

The internal microstructure of the samples is distinct from the above-discussed cases of testing with low strain rates. There are multiple primary ASBs observed in zone II, on the edges along the entire main crack (Figure 11d,e). Finer secondary ASBs spreading out of the main crack at a certain direction were also found in zone III (Figure 11f). A much fewer number of twins are observed in zone III (Figure 11d,g) compared to the cases of deformation with low strain rates. Moreover, a lot of structural elements of completely different morphology were revealed. The extremely fine (not larger than 0.1 μm × 2.2 μm) needle-shaped elements are formed within individual globules, as it is shown in the top right corner of Figure 11d. Such crystals could be a result of crystallographically ordered transformation, as it was reported in [34]. Their shape and orientation suggest martensitic needles, considering that the individual needles are located at an angle of approximately 60° relative to each other. It is probable that in proximity of zones of the localized extreme heat, where the temperature can reach the single-phase region (about 1000 °C) [18,19], and primary α- phase was transformed into high-temperature β-phase. In such a case, the transformation takes place without redistribution of alloying elements between neighboring initial crystallites of α- and β- phases, because of extremely high heating and cooling rates. A similar mechanism of the structure formation takes place during laser heating [38,39] and, because of subsequent fast cooling, this metastable β-phase can transform into low-temperature α′-martensite.

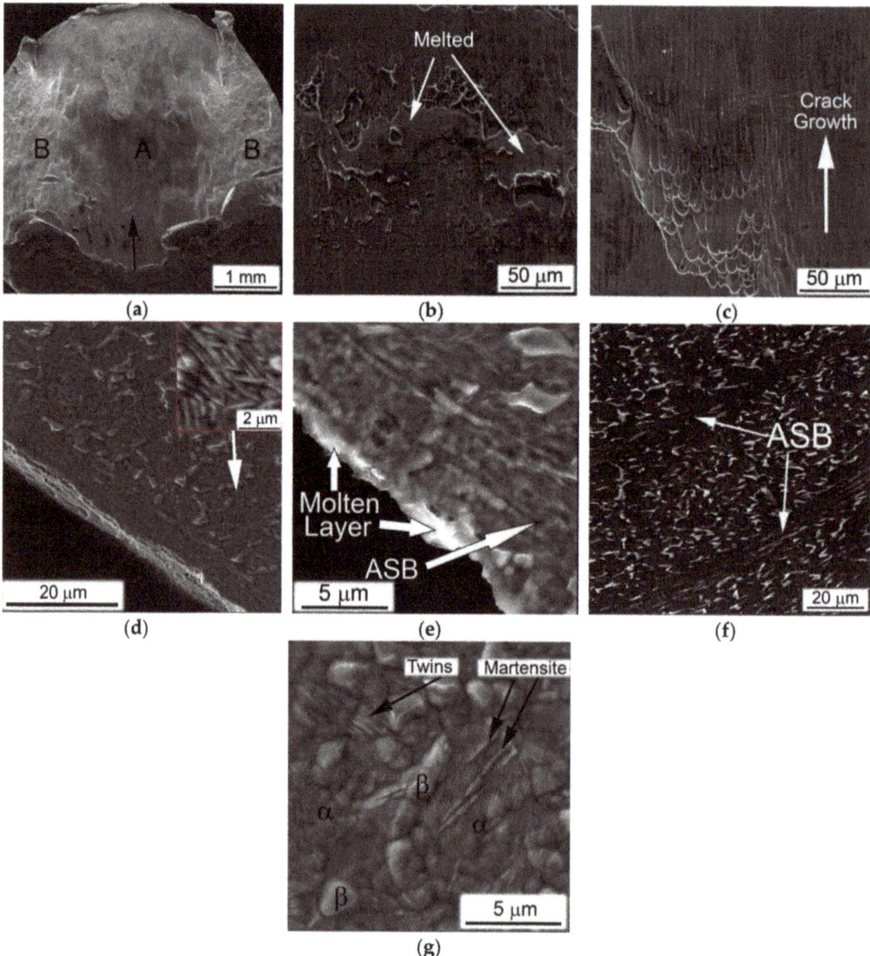

Figure 11. SEM images of the fracture surface (**a–c**), and the internal microstructure (**d–g**) of the Ti64GL specimen tested under the strain rate of 3330 s^{-1}; A—main crack spread, B—secondary cracking; (**d**) and (**e**)—zone II; (**f**)—zone III with secondary ASBs coming from the main crack; (**g**)—zone IV. The arrow in (**a**) indicates the direction of the crack propagation from its nucleation site. SEM, SE.

Bearing in mind a small size of tested specimens and their contact with massive input and output bars, the cooling process should be relatively rapid. It is probable that a pure (neat shear without any contribution of diffusion) martensitic transformation under these thermo-mechanical conditions does not take place, and a transformation of bainite type could easily take place, when diffusion of alloying elements could have happened during phase transformation, but most importantly Ti atoms move in an ordered way. A similar case was described for the Ti-6Al-4V alloy before experiments when the cooling rate from single β-phase temperatures was changed gradually; however, complete suppression of diffusion was established upon cooling at 400 °C per second [40]. When the cooling rates are lower, the shear transformation occurred with the involvement of diffusion and redistribution of alloying elements. This fact may explain a slightly unusual morphology of the observed martensite crystals (Figure 11d,g).

Microstructure of Ti64GL after the quasi-static tests presents distinct differences compared to the SHPB tests structure. The existed phase constituents, α-phase globules and β-phase interlayers, are flattened perpendicularly to the loading direction through almost the entire specimen volume (Figure 12). All the tested specimens did not crack, and thin bands of localized slip/shear were found in zone II of maximum localized strain (Figure 12a). The microstructure shows the same character of the flattened phase constituents in all other locations (zones III and IV); however, the images of high magnification reveal greater plastic deformation in α-phase compared to β interlayers (Figure 12b).

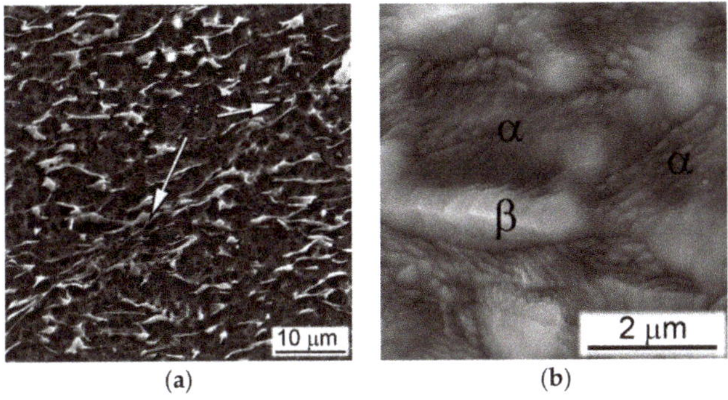

Figure 12. Microstructure of the non-cracked Ti64GL specimen after the quasi-static test at the strain rate of 0.001 s^{-1}: (**a**)—zone II, (**b**) zone IV. SEM, SE. Arrows in (**a**) indicate a slip localization band.

3.3.3. Microstructure Analysis of Ti64LM

The Ti64LM specimens deformed at strain rates for which fracture occurred show distinct plastic deformation in zone I on one side of the specimen (Figure 13a) as well as cracks on the other side (Figure 13b). It is probable that these cracks are initiated at β-grain boundaries reaching the specimen fracture surface and decorated [12] by α-phase layer. It should be noted that outside Zone I there is no significant evidence of plastic deformation in α-plates and their packets; however, small cracks were found within Zones II and IV (Figure 13c,d, respectively). These cracks are not associated with any structural elements such as β-grains boundaries, α-phase colonies, or α/β interlayers. These cracks easily crosscut the α-phase plates, partially deflecting on the α/β interlayers (Figure 13c), and finally move outside the individual plates (Figure 13d). The observed crack propagation insensitivity to structural elements appears to be remarkable since the microstructure of titanium alloys usually plays a pivotal role in the crack nucleation and growth under conditions of the quasi-static tension [1,11–13,41].

The fracture of the Ti64LM specimens under the SHPB test condition was observed at a strain rate of 2030 s^{-1}. A typical image of the crack surface and the microstructure in its vicinity are shown in Figure 14. The surface of the crack shows a typical ductile fracture disclosing multiple tear-off/shear dimples (Figure 14a). However, the zone adjacent to the crack surface on the section cut perpendicular to the crack shows rather unique features (Figure 14b). There are no evidences of plastic deformation at all, even close to the edge of the crack. Except slightly melted edge of the sample in the zone II, the structure in bulk of the sample is unchanged compared to initial condition (compare Figures 14b and 3c). This observation suggests that plastic deformation of the Ti64LM alloy becomes highly localized with an increase in the strain rate. The Ti64LM specimen deformation reaches its critical values slightly above 2000 s^{-1}, when the fracture occurs.

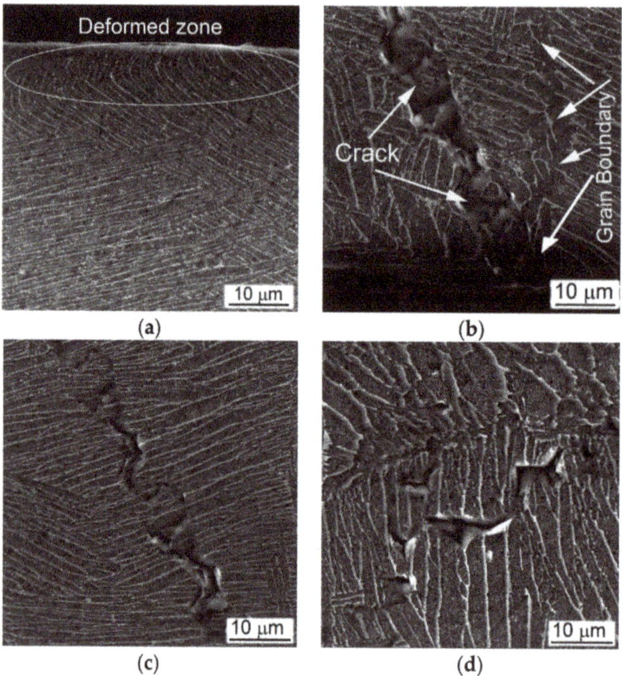

(a)

(b)

(c)

(d)

Figure 13. Microstructure of the Ti64LM specimen after SHPB test at the strain rate of 1390 s^{-1}: (**a,b**)—zone I, (**c**)—zone II, (**d**)—zone IV. SEM, SE.

The following conclusions can be drawn from comparison analysis of the Ti64GL and the Ti64LM structures subjected to the dynamic loading conditions. The principal difference between these two structural states is defined by configuration and extent of a/β interphase boundaries. The Ti64GL, similarly to c.p.Ti, has predominantly α/α boundaries that appear to facilitate a relatively free propagation of plastic deformation (flow) through the material even at high strain rates despite the fact that majority of neighboring globules present essentially different crystallographic orientation [1]. On the contrary, the Ti64LM boundaries between microstructural elements are almost on 100% presented by interphase α/β boundaries that remarkably prevent a free spread of deformation due to hindering of dislocation movement. It results in an increased strain localization and concentration of the stress, which may enable a cavity nucleation and, conclusively, fracture at an increased strain rate.

The SEM-EBSD analysis of Ti64LM specimen after the SHPB test with a strain rate of 2030 s^{-1} is presented in Figure 15. Kikuchi patterns, required to obtain EBSD data, were distinctive in many points near the contact (fracture) surface. However, Kikuchi lines were not identified by the software, which may result from significant distortions caused by high residual stresses remained in the material after the dynamic plastic deformation. Since the annealing, required to relieve these stresses, would inevitably lead to significant changes in the fine structure of the material resulting in polygonization or recrystallization, it was not carried out. The orientation map shows substantially non-uniform plastic deformation of the structure resulting from the test and clearly reveals few zones where deformation is localized. The β-phase cannot be resolved because of its size and morphology. It is represented by thin layers (from few tens to hundreds of nm) in between relatively thicker (a few m) α-lamellae and considering their likely tilt toward the analyzed surface the β-phase resolution becomes an unsolvable challenge for EBSD even the one operating with field-emission gun SEM, which was the case [42]. A significant number of the noise pixels, mostly aligned along the α-lamellae, result from unsuccessful orientation measurement of the β-phase. Because of the same reason, the individual α-lamellae also cannot be distinguished on the orientation map; however, it is clearly seen in the band contrast image, nonetheless the packers are easier to be traced on the orientation map plot. Some of the packets demonstrate significant misorientation of lamellae within the packet. For instance, a pink and purple packet, 40-m thick, on the left edge of the image shows approximately 10 deg. bent within a few m span. A big yellow and pink packet at the bottom of the image shows even bigger misorientation suggesting big dislocation density accumulated within some small zones. This image also demonstrates short secondary cracks originating outside of the main crack (Figure 15a), which indicate that crack nucleation is not related to such important microstructural elements as grain or interphase boundaries, and crack can nucleate inside a single α-lamella (see the left crack in Figure 15b).

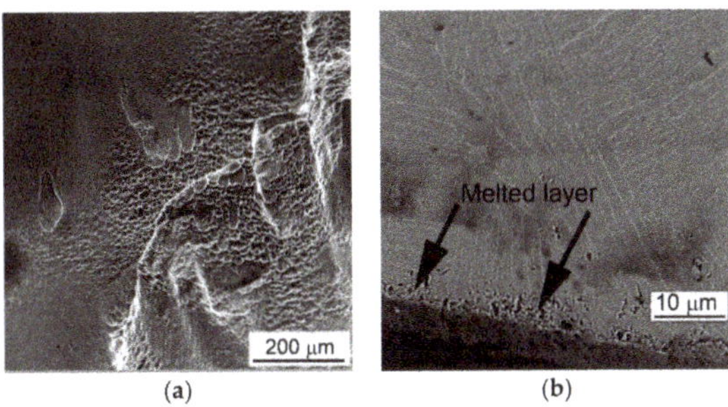

(a)   (b)

Figure 14. Microstructure of the cracked Ti64LM specimen after SHPB tested at the strain rate of 2030 s^{-1}: (**a**) fracture surface (SEM, SE); (**b**) microstructure in zone II (polished, not etched) (SEM, BSE).

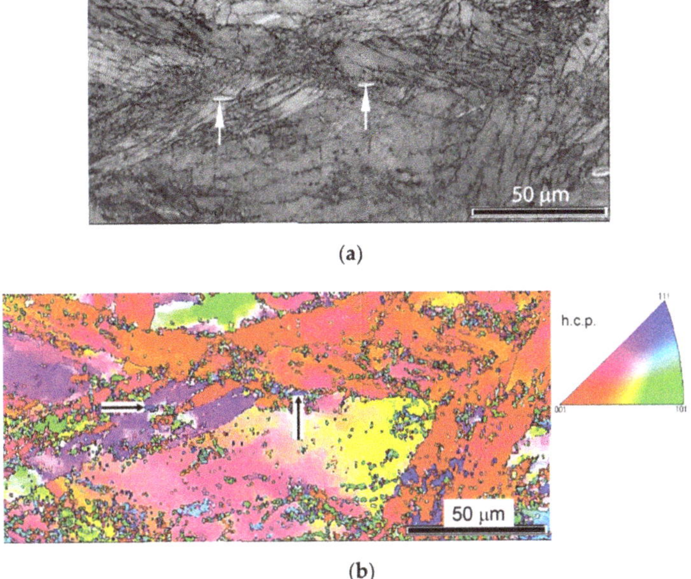

(a)

(b)

Figure 15. The SEM-EBSD images of Ti64LM specimen SHPB tested at 2030 s^{-1}: (**a**) band contrast image, (**b**) EBSD orientation map - arrows indicate small secondary cracks appeared inside α-lamellas.

The decisive impact of the compressive strain rate on plastic deformation localization is underlined by the results of the quasi-static tests of Ti64LM specimens (Figure 16). The plastic strain localization becomes more visible during compression at the strain rate of 0.001 s^{-1}. It resulted in the appearance of a considerably narrow zones crossing the sample at an angle of 45°, where maximum strain was localized, and finally causing the main crack nucleation which, in turn, initiated the secondary cracks (Figure 16a). These secondary cracks often cut and shift the individual grains. The α-layer, labeled with A in Figure 16b, wrapping the initial β-grain, was sheared by the crack in the direction perpendicular to this α-layer. The microstructure after the test (Figure 16c) is not much different compared to the initial structure (Figure 3c), except a few cracks seen in zones IV; however, a detailed examination reveals a complex substructure of multiple twins and dislocation slip traces inside the α-phase plates (Figure 16d).

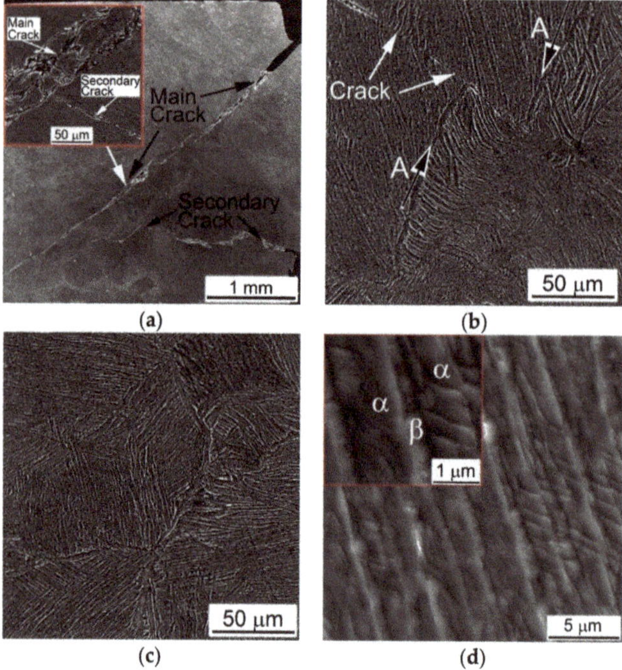

Figure 16. Microstructure of the Ti64LM specimen after the quasi-static test at the strain rate of 0.001 s^{-1}: (a) general view with zones I and II, (b)—zone III, (c,d)—zone IV. SEM, SE.

The processes of plastic deformation on micro- (inside separate α- and β-phases crystallites) and macro-level (at least a group of crystallites involved in cooperative response) should be distinguished which means that the dissipation of the total strain energy can be prioritized differently depending on the real structure. The alloy Ti64LM under slow quasi-static compression technically deforms in the same way as in the case of dynamic impact loading. Under such a condition, a smaller strain energy portion is apprehended at the micro-level in a form of plastic deformation localized inside α-lamellas, while its major fraction dissipates at the macro-level and is localized within a small zone. On the contrary, in the Ti64GL alloy, the strain energy is more apprehended at the micro-level, and no distinction between micro- and macro-level of energy dissipation is observed in c.p.Ti.

3.3.4. Microstructure Analysis of Ti64BEPM

Similarly to the cast and wrought Ti64LM, the Ti64BEPM alloy deformed at high strain rates demonstrates plastic deformation mainly in zone I, primarily on the samples that did not crack (Figure 17a). Moreover, the intense plastic deformation also took place locally, for example, in grain boundaries, and it was recognized as shear deformation (Figure 17b). Furthermore, the deformed (collapsed) residual pores were observed in various locations (Figure 17a). The diagonal zone II across the specimens was not found.

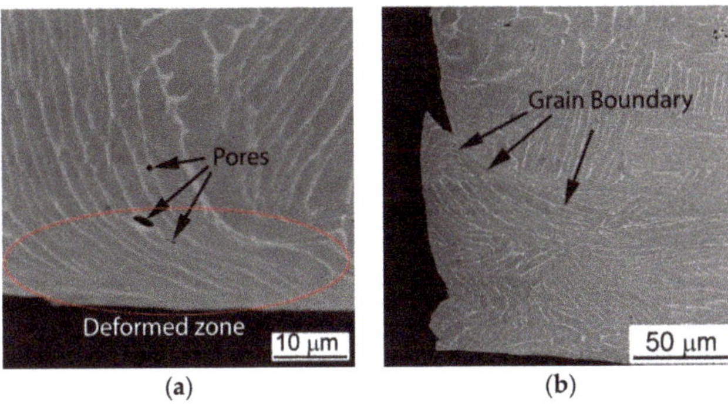

Figure 17. Microstructure of the non-cracked Ti64BEPM specimen tested at the strain rate of 2100 s^{-1}: (**a**) plastic deformation in zone I; (**b**) shear deformation in the vicinity zone I. SEM, BSE.

The microstructure of the Ti64BEPM alloy, after the test, was almost the same as the one of the Ti64LM discussed before. In zone I, similarly to the previous cases of the Ti64LM and Ti64BEPM (not fractured specimen—Figure 17), a plastically deformed layer of approximately the same depth (up to 20–30 m) was observed (Figure 18c). The main crack propagates in zone II through the entire specimen at an angle of 45° toward its vertical axis. Zone II itself is thin, not more than 5–6 m, and the ASB are formed on the sample fracture edges (Figure 18d). A number of pores collapsed and multiple α-lamellae and β-interlayers are curved due to the plastic flow of the material in vicinity of the crack (Figure 18d). Small secondary cracks initiated on the β-grain boundaries and α-colony boundaries are observed in zone III (Figure 18e). There is no evident plastic deformation in zone IV, except some slightly deformed pores partially flattened perpendicular to the direction of the applied load (Figure 18f). However, more detailed images of the slightly etched ion-beam samples reveal a fine needles microstructure inside the α-plates (Figure 18g,h). The formation of such an inner plate substructure is unique, and the observation of the uniform triangle arrangement of the needles with a uniform 60° angle between them (Figure 18h) is typical for martensite in h.c.p. lattice metals [37]. As it was mentioned before, a similar microstructure was also observed in the Ti64GL specimen fractured at the strain rate of 3330 s^{-1} (Figure 11d). Such a structure may result from rapid heating of the material during the impact, when the α-phase plates transformed into the high-temperature β-phase in such a fast way that the redistribution of alloying elements between the initial α- and β-phases did not occur. It is confirmed by the clear interphase boundary between the β-phase and the α-phase lamellar region, which includes the needles (Figure 18g). The area outlined by the circle in Figure 18h is of particular interest since it shows the intersection of the needles lying in the plane of the polished surface and perpendicular to it.

A general view of the fracture surface of the cracked Ti64BEPM specimen compressed at the strain rate of 2220 s^{-1} (Figure 18a) is similar to the fractures of the Ti-6-4 alloy with both globular (Figure 11a) and lamellar microstructure (Figure 14a). The fracture also demonstrates a rectilinear zone of the main crack growth (A in Figure 18a) and two side zones of the secondary crack propagation (B, ibid.). The images of higher magnification show the residual pores (Figure 18b), and their presence

in the structure makes Ti64BEPM alloy essentially different from the cast and wrought Ti64LM alloy
(Figure 14a).

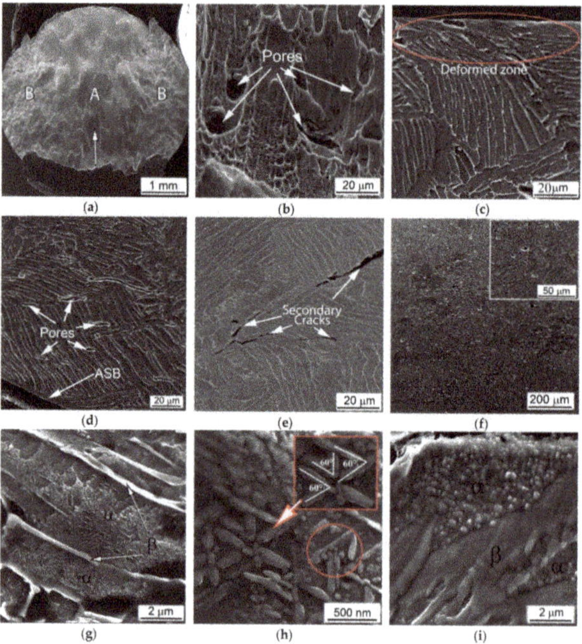

Figure 18. The SEM images of the fracture surface of the Ti64BEPM specimen tested at the strain rate of
$2220\ \text{s}^{-1}$: (**a**); general view; (**b**) residual pores at fracture surface; (**c–i**) internal microstructure; (**c**) zone
I, (**d**) zone II; (**e,f**) zone III; (**g–i**) zone IV; (the arrow in (**a**): indicates direction of the crack growth from
nucleation site; (A) and (B) label the fields of main and secondary cracks propagation, respectively).
(**a–e,g–i**) SE; (**f**) BSE.

It should be noted that the needles, due to the ion etching, do not demonstrate perfectly smooth
boundaries as they should for the martensite structure. The formation of a martensite structure during
plastic deformation at the strain rate of up to $2000\ \text{s}^{-1}$ was also reported on titanium Ti-8.5Cr-1.5Sn
alloy [35], although in that case the observed structure was deformation-induced high-alloyed
α''-martensite that formed simultaneously with the twins. The present study case is also different from
the case of a fast laser heating [35], because formation of martensitic needles appeared inside of the
α-plates with their apparently unchanged outlines (Figure 18g). The observed result can be explained
by the fact that, in addition to rapidly changing temperature, a complex stress state also acted on
particular zones within the specimen. Such a complex and combined effect is probably responsible for
the formation of the martensite in localized volumes and, as a result, at least three variants of Burgers
orientation relationships from 12 allowed are clearly seen. It should be emphasized that martensite-like
needle crystals were not frequently well revealed. They do not appear to be distinct and they are
similar to those shown in Figure 18i. However, it is implicit that the clear view of the martensite needles
depends on the "successful" coincidence of planes of their location with the prepared specimen's
section that is not highly probable.

The microstructures of the Ti64BEPM alloy after the quasi-static test were characterized by the main
crack appeared in zone II surrounded by the secondary cracks appeared on different microstructural
elements (Figure 19a). The other deformation features uniformly distributed throughout the specimen
were the collapsed residual pores and strongly bent α-lamellae (Figure 19b). Moreover, a cellular
dislocation substructure was observed inside the curved lamellae of the α- phases (Figure 19c), similar to

that observed in the case of the Ti64LM specimen after the identical quasi-static test (Figure 16b). This was expected since there is no significant heating taking place in material deformed at a low strain rate, and there is also enough time for more complete stress relaxation in the areas outside of the zone II.

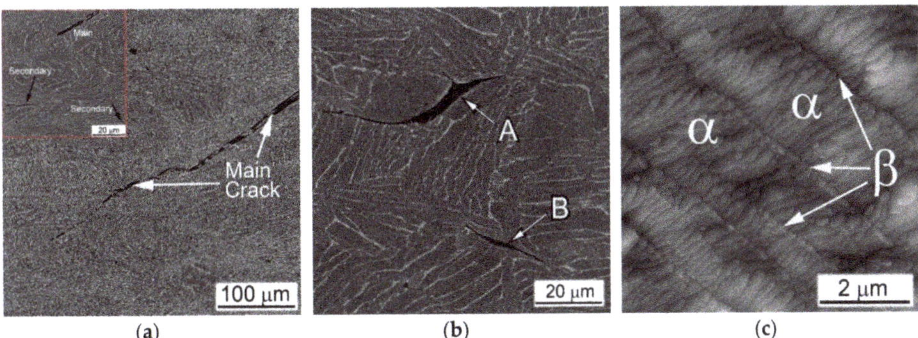

Figure 19. The SEM images of the fractured Ti64BEPM specimen after the quasi-static tests at the strain rate of 0.001 s^{-1}: (**a**) main crack—zone II; (**b**) collapsed residual pores—zone II; (**c**) cellular dislocation substructure inside the curved lamellae of the α-phases—zone IV. (**a**,**b**) BSE, (**c**) SE.

As it was shown in Section 3.1, there is a significant difference in the mechanical behavior of the Ti64LM and Ti64BEPM alloys. Based on the microstructural analysis presented above, different mechanical behavior of the tested Ti-based materials could be explained as follows. Both alloys present a similar type of microstructure, which is represented by relatively coarse β-grains with the colonies of lamellar α-phase inside. However, in detail, they are considerably different with a number of features, such as different sizes of both β- grains (more than 500 m in the Ti64LM vs. 100 m in the Ti64BEPM) and intragrain α-lamellae, the presence of residual pores in the Ti64BEPM, as well as the content of impurities (see Table 1). The high oxygen content as well as high nitrogen content in the Ti64BEPM led to an increase in strength, compared to the Ti64LM, and combined with residual pores of the Ti64BEPM significantly reduces its ductility under the tensile loading (Table 2). However, under quasi-static and dynamic compression the porous Ti64BEPM alloy shows noticeably better characteristics compared to the Ti64LM (Figure 5d,c). It could be explained by the fact that, under the tension loading, the pores work as the stress concentrators and/or crack initiation sites, particularly, when the pore shape is not globular, and the matrix alloy lacks the ductility. However, the pores impact on a deformation mechanism and the material damage process under compression is slightly different. The pores become flattened or even completely collapsed and, thence, play the role of additional "soft" phase without generating stress concentration zones, therefore general compressive plasticity of the porous material is determined by the plasticity of the matrix. Moreover, according to studies [43–45], a positive effect of porosity on the structure sustainability under compression can be amplified by their higher content; the most effective deformation energy absorption was reported at a porosity value of around 60% [46,47].

4. Conclusions

Based on presented experimental data and their analysis the following conclusions are drawn:

(a) Compressive mechanical behavior of titanium alloys is strongly dependent on the phase composition and microstructure of both the studied materials and the applied strain rate level.

(b) The mechanical behavior of a two-phase α + β Ti-6-4 alloy strongly depends on the type and coarseness of the microstructure. The fine-grained Ti-6-4 alloy with a globular (equiaxed) microstructure is more ductile and has the high reserve of plasticity, which allows it to deform without fracture at the strain rate below 3320 s^{-1}. The critical compression strain rate, at which the fracture occurred, falls to 2030 s^{-1}, when the microstructure changed from globular to coarse-grained lamellar. The observed significantly different mechanical behavior of two structures can be explained by the nature of the interface boundaries between the structural constituents involved in plastic deformation transmission, i.e., the α/α interphase boundaries are prevalent in the globular microstructure, while α/β boundaries are predominant in the lamellar microstructure.

(c) The Ti-6-4 alloy fabricated using BEPM demonstrates the reduced the size of β-grains and intragrain α- lamellae compared to the alloy with a coarse-grained lamellar microstructure produced using a conventional cast and wrought approach. The Ti64BEPM alloy demonstrates a considerably better balance of strength and plasticity under the quasi-static and dynamic compression tests, because of its finer microstructure despite of the presence of about 2% (vol.) of residual pores and higher content of impurities (oxygen and nitrogen). The residual pores do not play any negative role under compression loading in contrary to tension, since they do not work as stress concentrators.

(d) Strain energy was used as a parameter to compare mechanical behavior of the studied materials. It was established that the two-phase Ti-6-4 alloy with a globular microstructure demonstrates the highest value of SE_{max}, which implies the largest reserve of deformability of this alloy under the compression impact at the strain rates. The Ti64 alloy produced using BEPM demonstrates a lower value of the SE_{max} parameter. The Ti64 alloy with coarse lamellar microstructure reveals the lowest values SE_{max}.

(e) It was found that the strain rates increase up to 2200 s^{-1} cause a change in the strain localization mechanism in Ti64BEPM alloy from the macro-level (plastic flow in the sample volume, formation of adiabatic shear bands and cracks) to the micro-level (deformation within individual α-phase lamellae).

(f) The structures of all the studied materials demonstrate more uniform plastic deformation and the absence of its micro-level strain localization after quasi-static compression compared to the dynamic loaded structures.

(g) The Ti-6-4 alloys with a globular microstructure, fabricated using ingot metallurgy, and the Ti64BEPM alloy demonstrate higher relative (specific) SE values than B95 aluminum alloy, ARMOX 600T armor steel, or AHSS steel Docol 1500M.

Author Contributions: Conceptualization, validation and project administration, P.E.M., J.J.; methodology, P.E.M., J.J., D.G.S., M.A.S.; formal analysis, V.I.B., S.V.P.; experiments and investigation, P.E.M., J.J., O.O.S., V.I.B., K.C., P.D., M.A.S.; writing—original draft preparation, P.E.M.; writing—review and editing, J.J., D.G.S., V.I.B., S.V.P.; visualization, O.O.S.; All authors have read and agreed to the published version of the manuscript.

Funding: This research received no special funding. Present studies were performed according the Agreement of Cooperation between G.V. Kurdyumov Institute for Metal Physics of N.A.S. of Ukraine and Jarosłław Dąbrowski Military University of Technology, Poland. Some separate works were financed by N.A.S. of Ukraine within the frames of the research project #III-09-18. S.V.P. acknowledges funding from the NATO Agency Science for Peace and Security (#G5030).

Conflicts of Interest: The authors declare no conflict of interest.

References

1. Luetjering, G.; Williams, J.C. *Titanium*, 2nd ed.; Springer: Berlin/Heidelberg, Germany, 2007.

2. Williams, J.C.; Boyer, R.R. Opportunities and Issues in the Application of Titanium Alloys for Aerospace Components. *Metals* **2020**, *10*, 705. [CrossRef]

3. Niinomi, M. Recent metallic materials for biomedical applications. *Met. Mater. Trans. A* **2002**, *33*, 477–486. [CrossRef]

4. Fanning, J. Military Application for b Titanium Alloys. *J. Mater. Eng. Perform.* **2005**, *14*, 686–690. [CrossRef]

5. Montgomery, J.S.; Wells, M.G.H.; Roopchand, B.; Ogilvy, J.W. Low-cost titanium armors for combat vehicles. *JOM* **1997**, *49*, 45–47. [CrossRef]

6. Fanning, J. Ballistic Evaluation of Titanium Alloys Against Handgun Ammunition. In *Ti-2007, Science and Technology, Proceedings of the 11th World Conference on Titanium, Kyoto, Japan, 3–7 June 2007*; The Japan Institute of Metals Publish.: Tokyo, Japan, 2007; Volume 1, pp. 487–490.

7. Gooch, W. Potential Applications of Titanium Alloys in Armor Systems. In *Titanium-2011*; International Titanium Association: San Diego, CA, USA, 2011; Available online: https://www.researchgate.net/publication/292328353_Potential_Applications_of_Titanium_Alloys_in_Armor_Systems_-2011 (accessed on 20 October 2020).

8. Stefansson, N.; Weiss, I.; Hutt, A.J. *Titanium'95: Science and Technology*; Blenkinsop, P.A., Evans, W.J., Flower, H.M., Eds.; The University Press: Cambridge, UK, 1996; Volume 2, pp. 980–987.

9. Bhattacharjee, A.; Ghosal, P.; Gogia, A.; Bhargava, S.; Kamat, S. Room temperature plastic flow behaviour of Ti–6.8Mo–4.5Fe–1.5Al and Ti–10V–4.5Fe–1.5Al: Effect of grain size and strain rate. *Mater. Sci. Eng. A* **2007**, *452*, 219–227. [CrossRef]

10. Markovsky, P.; Matviychuk, Y.; Bondarchuk, V. Influence of grain size and crystallographic texture on mechanical behavior of TIMETAL-LCB in metastable β-condition. *Mater. Sci. Eng. A* **2013**, *559*, 782–789. [CrossRef]

11. Markovsky, P.; Bondarchuk, V.; Herasymchuk, O. Influence of grain size, aging conditions and tension rate on the mechanical behavior of titanium low-cost metastable beta-alloy in thermally hardened condition. *Mater. Sci. Eng. A* **2015**, *645*, 150–162. [CrossRef]

12. Markovsky, P.; Bondarchuk, V.I. Influence of Strain Rate, Microstructure and Chemical and Phase Composition on Mechanical Behavior of Different Titanium Alloys. *J. Mater. Eng. Perform.* **2017**, *26*, 3431–3449. [CrossRef]

13. Markovsky, P.E. Mechanical Behavior of Titanium Alloys under Different Conditions of Loading. *Mater. Sci. Forum* **2018**, *941*, 839–844. [CrossRef]

14. Peirs, J.; Verleysen, P.; Degrieck, J.; Coghe, F. The use of hat-shaped specimens to study the high strain rate shear behaviour of Ti–6Al–4V. *Int. J. Impact Eng.* **2010**, *37*, 703–714. [CrossRef]

15. Zheng, C.; Wang, F.; Cheng, X.; Liu, J.; Fu, K.; Liu, T.; Zhu, Z.; Yang, K.; Peng, M.; Jin, D. Failure mechanisms in ballistic performance of Ti–6Al–4V targets having equiaxed and lamellar microstructures. *Int. J. Impact Eng.* **2015**, *85*, 161–169. [CrossRef]

16. Morrow, B.; Lebensohn, R.; Trujillo, C.; Martinez, D.T.; Addessio, F.; Bronkhorst, C.A.; Lookman, T.; Cerreta, E. Characterization and modeling of mechanical behavior of single crystal titanium deformed by split-Hopkinson pressure bar. *Int. J. Plast.* **2016**, *82*, 225–240. [CrossRef]

17. Yin, W.; Xu, F.; Ertorer, O.; Pan, Z.; Zhang, X.; Kecskes, L.; Lavernia, E.J.; Wei, Q. Mechanical behavior of microstructure engineered multi-length-scale titanium over a wide range of strain rates. *Acta Mater.* **2013**, *61*, 3781–3798. [CrossRef]

18. Zhou, T.; Wu, J.; Che, J.; Wang, Y.; Wang, X. Dynamic shear characteristics of titanium alloy Ti-6Al-4V at large strain rates by the split Hopkinson pressure bar test. *Int. J. Impact Eng.* **2017**, *109*, 167–177. [CrossRef]

19. Guo, Y.; Ruan, Q.; Zhu, S.; Wei, Q.; Lu, J.; Hu, B.; Wu, X.; Li, Y. Dynamic failure of titanium: Temperature rise and adiabatic shear band formation. *J. Mech. Phys. Solids* **2020**, *135*, 103811. [CrossRef]

20. Sreenivasan, P.R.; Ray, S.K. Mechanical Testing at High Strain Rates. In *Encyclopedia of Materials: Science and Technology*, 2nd ed.; Elsevier: New York, NY, USA, 2001; pp. 5269–5271.

21. Ivasishin, O.M.; Anokhin, V.M.; Demidik, A.N.; Savvakin, D.G. Cost-Effective Blended Elemental Powder Metallurgy of Titanium Alloys for Transportation Application. *Key Eng. Mater.* **2000**, *188*, 55–62. [CrossRef]

22. Ivasishin, O.; Moxson, V. Low-cost titanium hydride powder metallurgy. *Titan. Powder Metall.* **2015**, *8*, 117–148. [CrossRef]

23. Chen, W.; Song, B. *Split Hopkinson (Kolsky) Bar: Design, Testing and Applications*; Springer: Berlin/Heidelberg, Germany, 2011.

24. Kolsky, H. Propagation of Stress Waves in Linear Viscoelastic Solids. *J. Acoust. Soc. Am.* **1965**, *37*, 1206. [CrossRef]

25. Kolsky, H. Stress waves in solids. *J. Sound Vib.* **1964**, *1*, 88–110. [CrossRef]

26. Panowicz, R.; Janiszewski, J.; Kochanowski, K. The non-axisymmetric pulse shaper position influence on SHPB experiment data. *J. Theor. App. Mech.* **2018**, *56*, 873–886. [CrossRef]

27. Janiszewski, J. *Unpublished Experimental Data Report*; Jarosław Dąbrowski Military University of Technology: Warsaw, Poland, 2020.

28. Database of Steel and Alloy (Marochnik). Available online: http://www.splav-kharkov.com/en/e_mat_start. php?name_id=1448 (accessed on 21 September 2020).

29. Nilsson, M. *Constitutive Model for Armox 500T and Armox 600T at Low and Medium Strain Rates*; Technical Report FOI-R-1068-SE; Swedish Defence Research Agency: Stockholm, Sweden, 2003; Available online: https://www.foi.se/rest-api/report/FOI-R--1068--SE (accessed on 21 September 2020).

30. Markovsky, P.E.; Savvakin, D.G.; Ivasishin, O.M.; Bondarchuk, V.I.; Prikhodko, S.V. Mechanical Behavior of Titanium-Based Layered Structures Fabricated Using Blended Elemental Powder Metallurgy. *J. Mater. Eng. Perform.* **2019**, *28*, 5772–5792. [CrossRef]

31. Prikhodko, S.V.; Ivasishin, O.M.; Markovsky, P.E.; Savvakin, D.G.; Stasiuk, O.O. Chapter 13: Titanium Armor with Gradient Structure: Advanced Technology for Fabrication. In *Advanced Technologies for Security Applications*; Springer: Dodrecht, The Netherlands, 2020; pp. 127–140.

32. Schmid, E.; Boas, W. *Plasticity of Crystals, Special Reference to Metals*; Springer US: New York, NY, USA, 1968.

33. Partridge, P.G. The crystallography and deformation modes of hexagonal close-packed metals. *Metall. Rev.* **1967**, *12*, 169–194. [CrossRef]

34. Ma, C.; Wang, H.; Hama, T.; Guo, X.; Mao, X.; Wang, J.; Wu, P. Twinning and detwinning behaviors of commercially pure titanium sheets. *Int. J. Plast.* **2019**, *121*, 261–279. [CrossRef]

35. Yang, H.; Wang, D.; Zhu, X.; Fan, Q. Dynamic compression-induced twins and martensite and their combined effects on the adiabatic shear behavior in a Ti-8.5Cr-1.5Sn alloy. *Mater. Sci. Eng. A* **2019**, *759*, 203–209. [CrossRef]

36. Markovsky, P.; Semiatin, S. Microstructure and mechanical properties of commercial-purity titanium after rapid (induction) heat treatment. *J. Mater. Process. Technol.* **2010**, *210*, 518–528. [CrossRef]

37. Banerjee, S.; Mukhopadhyay, P. Phase Transformations: Examples from Titanium and Zirconium Alloys. *Pergamon Mater. Series* **2007**, *12*, 1–813.

38. Markovsky, P.E. Two-stage transformation in ($\alpha + \beta$) titanium alloys on non-equilibrium heating. *Scr. Met. Mat.* **1991**, *25*, 2705–2710. [CrossRef]

39. Semiatin, S.L.; Obstalecki, M.; Payton, E.J.; Pilchak, A.L.; Shade, P.A.; Levkulich, N.C.; Shank, J.M.; Pagan, D.C.; Zhang, F.; Tiley, J.S. Dissolution of the Alpha Phase in Ti-6Al-4V During Isothermal and Continuous Heat Treatment. *Met. Mater. Trans. A* **2019**, *50*, 2356–2370. [CrossRef]

40. Gridnev, V.N.; Ivasishin, O.M.; Markovsky, P.E.; Svechnikov, V.L. The role of the cooling rate in the formation of the structure of titanium alloys thermally hardened with incomplete homogenization of the β phase. *Metallofiz (Phys. Met.)* **1985**, *7*, 37–44. (In Russian)

41. Lütjering, G. Influence of processing on microstructure and mechanical properties of ($\alpha + \beta$) titanium alloys. *Mater. Sci. Eng. A* **1998**, *243*, 32–45. [CrossRef]

42. Isabell, T.C.; Dravid, V.P. Resolution and sensitivity of electron backscattered diffraction in a cold field emission gun SEM. *Ultramicroscopy* **1997**, *67*, 59–68. [CrossRef]

43. Kumar, P.; Chandran, K.S.R.; Cao, F.; Koopman, M.; Fang, Z.Z. The Nature of Tensile Ductility as Controlled by Extreme-Sized Pores in Powder Metallurgy Ti-6Al-4V Alloy. *Met. Mater. Trans. A* **2016**, *47*, 2150–2161. [CrossRef]

44. Biswas, N.; Ding, J. Numerical study of the deformation and fracture behavior of porous Ti6Al4V alloy under static and dynamic loading. *Int. J. Impact Eng.* **2015**, *82*, 89–102. [CrossRef]

45. Banhart, J. Manufacture, characterization and application of cellular metals and metal foams. *Prog. Mater. Sci.* **2001**, *46*, 559–632. [CrossRef]

46. Suzuki, A.; Kosugi, N.; Takata, N.; Kobashi, M. Microstructure and compressive properties of porous hybrid materials consisting of ductile Al/Ti and brittle Al3Ti phases fabricated by reaction sintering with space holder. *Mater. Sci. Eng. A* **2020**, *776*, 139000. [CrossRef]

47. Garcia-Avila, M.; Portanova, M.; Rabiei, A. Ballistic performance of composite metal foams. *Compos. Struct.* **2015**, *125*, 202–211. [CrossRef]

Publisher's Note: MDPI stays neutral with regard to jurisdictional claims in published maps and institutional affiliations.

Communication

Grain Refinement of Ti-15Mo-3Al-2.7Nb-0.2Si Alloy with the Rotation of TiB Whiskers by Powder Metallurgy and Canned Hot Extrusion

Jiabin Hou [1,2], Lin Gao [1], Guorong Cui [1,*], Wenzhen Chen [1], Wencong Zhang [1] and Wenguang Tian [3]

[1] School of Materials Science and Engineering, Harbin Institute of Technology, Weihai 264209, China; houjiabinwh@163.com (J.H.); gaolinhit@hotmail.com (L.G.); nclwens@hit.edu.cn (W.C.); zwinc@hitwh.edu.cn (W.Z.)

[2] Naval Architecture and Marine Engineering College, Shandong Jiaotong University, Weihai 264209, China

[3] Oriental Bluesky Titanium Technology Co., LTD, Yantai 264003, China; tianwenguang@obtc.cn

* Correspondence: cuiguorong2010@126.com; Tel.: +86-631-5687209

Received: 8 December 2019; Accepted: 12 January 2020; Published: 15 January 2020

Abstract: In situ synthesized TiB whiskers (TiBw) reinforced Ti-15Mo-3Al-2.7Nb-0.2Si alloys were successfully manufactured by pre-sintering and canned hot extrusion via adding TiB_2 powders. During pre-sintering, most TiB_2 were reacted with Ti atoms to produce TiB. During extrusion, the continuous dynamic recrystallization (CDRX) of β grains was promoted with the rotation of TiBw, and CDRXed grains were strongly inhibited by TiBw with hindering dislocation motion. Eventually, the grain sizes of composites decreased obviously. Furthermore, the stress transmitted from the matrix to TiBw for strengthening in a tensile test, besides grain refinement. Meanwhile, the fractured TiBw and microcracks around them contributed to fracturing.

Keywords: TiB whiskers; dynamic recrystallization; grain refinement; strengthening

1. Introduction

Metastable β titanium alloys such as Ti-15Mo-3Al-2.7Nb-0.2Si are a promising candidate applied in aerospace and automotive industries, which have the advantages of high specific strength, excellent hot and cold workability, deep hardenability and oxidation resistance [1–3]. However, during hot working, metastable β titanium alloys coarsen rapidly at elevated temperatures, which could weaken their thermal stability and mechanical properties [4]. Therefore, it is necessary to reduce the grain sizes of metastable β titanium alloys and restrict their growth. In order to overcome the drawback, lots of studies about beta-titanium alloys reinforced by intermetallic particles (or eutectic structures) were carried out [5–12]. The Ni and Co elements were partially segregated to the interdendritic region to form TiNi and TiCo intermetallic phases, resulting in fine interdendritic precipitates or eutectic structures to obtain high strength [5–7,13]. Meanwhile, TiBw was considered to be one of the best reinforcements for Ti matrix with high elastic modulus, clean bonding interface and similar thermal expansion coefficient with matrix [14]. Huang et al. [15] reported that the accumulation of strain around TiBw provided a nucleation site for dynamic recrystallization (DRX) in hot deformation process, which results in some small and fine α of Ti60. Moreover, Feng et al. [16] reported that the growth of the recrystallization primary β was strongly restricted by the TiBw during the extrusion process, which reduced the grain sizes of Ti64. Okulov et al. [17] reported that multicomponent Ti alloys was refined by adding boron during casting. The TiB needle-shape particles distributed along primary β-Ti dendrites, and reduced the secondary dendrite arm spacing of the β-Ti phase, resulting in grain refinement. At present, there are fewer scholars researching metastable β titanium alloys reinforced by

TiBw (TiBw/metastable β titanium) in the hot deformation process, particularly the grain refinement mechanism of TiBw/metastable β titanium.

The grain-refinement mechanism of TiBw/Ti-15Mo-3Al-2.7Nb-0.2Si is worthy of exploring. Above all, composites were fabricated by low energy milling, pre-sintering and canned hot extrusion via adding 2.6 vol % TiB_2.

2. Materials and Methods

The spherical Ti-15Mo-3Al-2.7Nb-0.2Si (alloy) powders (β transus ~ 827 °C) and prismatic TiB_2 powders were chosen as raw materials to produce the as-extruded bars of TiBw reinforced alloys (composites) and alloys. Alloy powders were approximately 120 μm, and TiB_2 powders were approximately 3 μm. Alloy powders (97.4 vol %) and TiB_2 powders (2.6 vol %) were mixed at a speed of 100 rpm for a period of 6 h by low energy milling (LEM) in a planetary ball mill under the protective atmosphere of argon. The weight ratio between balls and powder mixture was 5:1. The mixed powders and alloy powders were weld-sealed into a 45# steel cup (with dimensions of 52 mm in outside diameter, 40 mm in inside diameter, 50 mm in height) [16]. Then billets were pre-sintered in high-temperature box furnace at 1100 °C for 1 h and then cooled by air to room temperature. The pre-sintered billets were heated at 800, 900 and 1000 °C for 30 min and then subsequently extruded by hydraulic machine, with the extrusion ratio of 10.6:1. The as-extruded bars with a diameter of 16 mm and a length of 500 mm were obtained.

The Phases of composites were identified by X-ray diffraction (XRD) (Rigaku Corporation, Tokyo, Japan). The microstructure was observed by scanning electron microscopy (SEM) (Zeiss-MERLIN, Zürich, Switzerland) and equipped with electron back scattered diffraction (EBSD) system. The tensile specimens with gauge lengths of 15 mm, widths of 4 mm and thickness of 2 mm were cut along the extrusion direction (ED) and then polished by metallographic sandpaper. All specimens were tested with a speed of 0.5 mm/min, and the ductility was measured by extensometer.

3. Results and Discussion

In order to identify the phases clearly, the SEM and XRD of the pre-sintered and as-extruded composites are shown in Figure 1. According to XRD in Figure 1a,b, the composites contained TiB, α-Ti and β-Ti. Noting that no TiB_2 was detected, indicating that most of TiB_2 were reacted with Ti to produce TiB. Ma et al. [18] reported that when the ratio of B atoms below 49% to 50 at %, Ti atom would react with TiB_2 to form TiB. The phases composition, morphology and distribution were studied by the SEM in Figure 1c–f. According to the SEM images, TiBw, β-Ti and α-Ti were identified. During pre-sintering, TiBw emerged in the β grains boundaries. After extrusion, the sizes of TiBw increased and distributed along the ED. Meanwhile, the average sizes of β grains increased with the increasing extrusion temperature.

The inverse pole figure (IPF) mapping of as-extruded alloys and composites are shown in Figure 2. The black lines represent high-angle grain boundaries (HAGBs with misorientations > 15°), and the white lines represent low-angle grain boundaries (LAGBs with misorientations between 2° and 15°) [16,19]. The red lines distributed around phases, including α phase and TiBw. The coarse elongated β grains distributed along ED. Meanwhile, a large volume fraction of fine equiaxed β grains distributed inhomogeneously between elongated β grains, indicating dynamic recrystallization (DRX) occurred during extrusion [20,21]. Some β gains were distributed around TiBw and elongated α grains, typically as indicated by the black ellipse. Moreover, the necklace structure was distributed along the grain boundaries of elongated β grains, consisting of fine β grains and bulged HAGBs, as shown in the white ellipse. During deformation, the bulging grain boundaries severed as the nucleus of recrystallized grains, and recrystallized grains grew with the migration of HAGBs, resulting in necklace structure (DRX process on the bulged grain boundaries), that is discontinuous dynamic recrystallization (DDRX) [22].

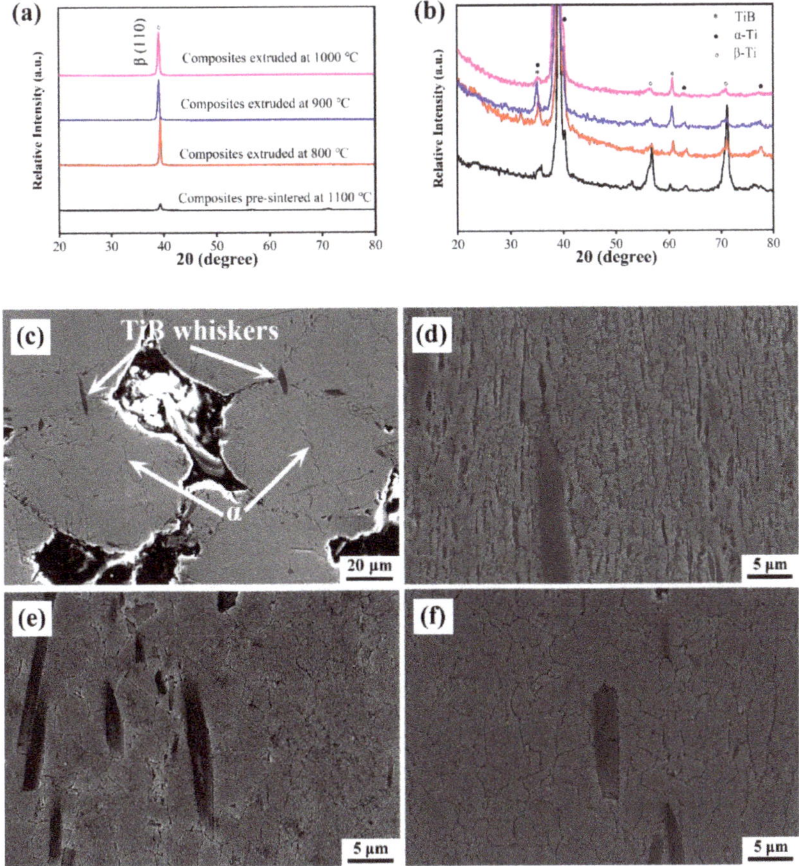

Figure 1. The XRD of pre-sintered and as-extruded composites (**a**), high magnification of XRD (**b**); SEM images of pre-sintered composites (**c**), composites extruded at 800 °C (**d**), composites extruded at 900 °C (**e**), composites extruded at 1000 °C (**f**).

In order to detect the orientation and texture of as-extruded alloys and composites, the discrete plot and texture along ED of alloys and composites extruded at 800 °C were shown in Figure 3a,b. The basal planes {0001} of α grains ($\langle 2\bar{1}\bar{1}0 \rangle$ (point A1) and $\langle 10\bar{1}0 \rangle$ (point A3)) almost paralleled to ED, which contributed to $\langle 2\bar{1}\bar{1}0 \rangle$ α texture in Figure 3a,b and $\langle 10\bar{1}0 \rangle$ α texture in Figure 3b. During deformation, the prismatic glide of primary α grains (α_p) is {$10\bar{1}0$} $\langle 11\bar{2}0 \rangle$ [16]. Therefore, the basal planes {0001} of α_p would turn to parallel to ED. The α grain ($\langle 10\bar{1}1 \rangle$ (point A2)) contributed to another center of α texture, which nucleated in the grain boundaries and grains according to the Burgers relationship {0001}//{110} and $\langle 111 \rangle$//$\langle 11\bar{2}0 \rangle$ during $\beta \rightarrow \alpha$ process [23]. Moreover, $\langle 101 \rangle$ β texture was maximum center in Figure 3a,b, which was commonly found in the fully recrystallized β grains of the deformed β phase due to the orientated nucleation mechanism [24,25].

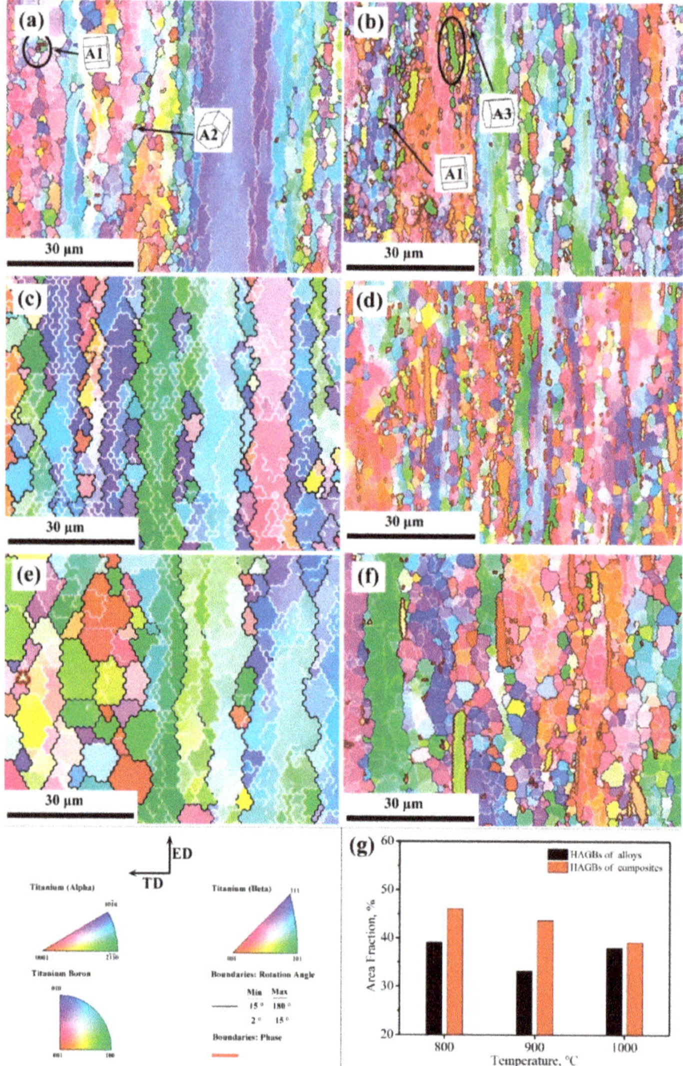

Figure 2. IPF mapping of alloys extruded at (**a**) 800 °C, (**c**) 900 °C, (**e**) 1000 °C and composites extruded at (**b**) 800 °C, (**d**) 900 °C, (**f**) 1000 °C; (**g**) histogram high-angle grain boundaries (HAGBs) counts of as-extruded alloys and composites.

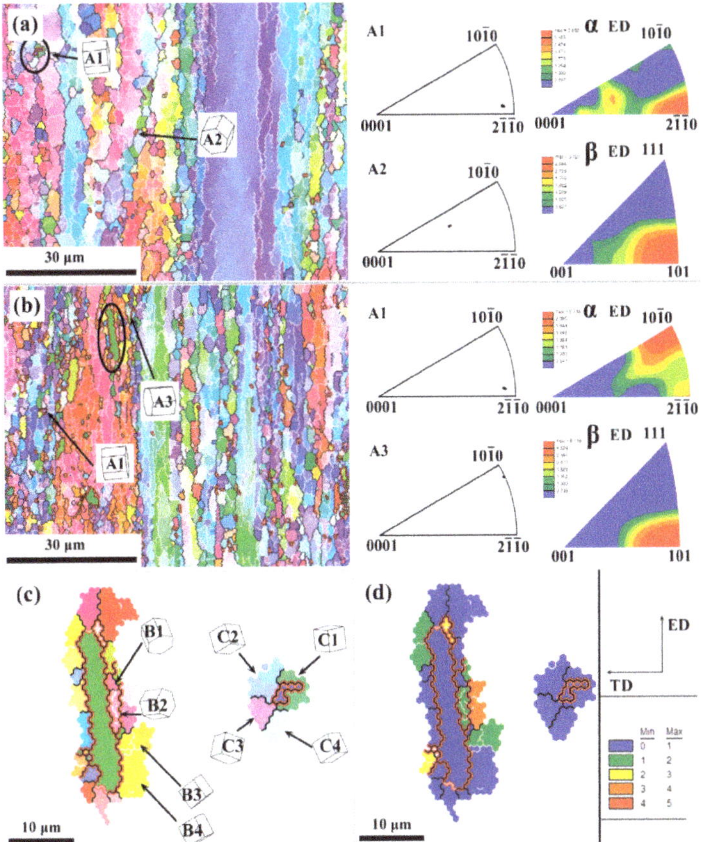

Figure 3. (**a**) IPF mapping, discrete plot and texture of alloy extruded at 800 °C; (**b**) IPF mapping, discrete plot and texture along extrusion direction (ED) of composite extruded at 800 °C; (**c**) highlighted mapping of (**a**,**b**); (**d**) grain orientation spread (GOS) mapping of (**c**).

The IPF and grain orientation spread (GOS) mapping of marked TiBw and α_p in Figure 3a,b are shown in Figure 3c,d. From Figure 3c, TiBw was surrounded by α grains and equiaxed β grains. Meanwhile, the GOS values of them are shown in Figure 4d. Basu et al. [26] revealed that the GOS values of the recrystallized grains were less than 1°, indicating β grains with blue color occurred DRX in Figure 3d. TiBw was a brittle phase with high elastic modulus, resulting in inharmonious deformation [27]. During extrusion, TiBw rotated to parallel to ED with the metal flow, and high extrusion stress concentrated around them by inharmonious deformation [24]. TiBw hindered dislocations motion, resulting in the piled up of dislocation around TiBw, which had been revealed by Park [28]. LAGBs emerged in the adjacent grains (point B2) and transformed into subgrains (point B3) with the accumulation of driving force. Subgrains rotated and formed equiaxed grains as further deformation, resulting in continuous dynamic recrystallization (CDRX) [25]. Meanwhile, the LAGBs evolved into HAGBs, which could be proved by the HAGBs volume fraction of composites increased by TiBw in Figure 2g. It should be pointed out that equiaxed β subgrains (point B4) occurred partial dynamic recrystallization without sufficient driving force [25]. A similar phenomenon was observed in Figure 2b,d,f, which may be attributed to a small rotation angle of TiBw.

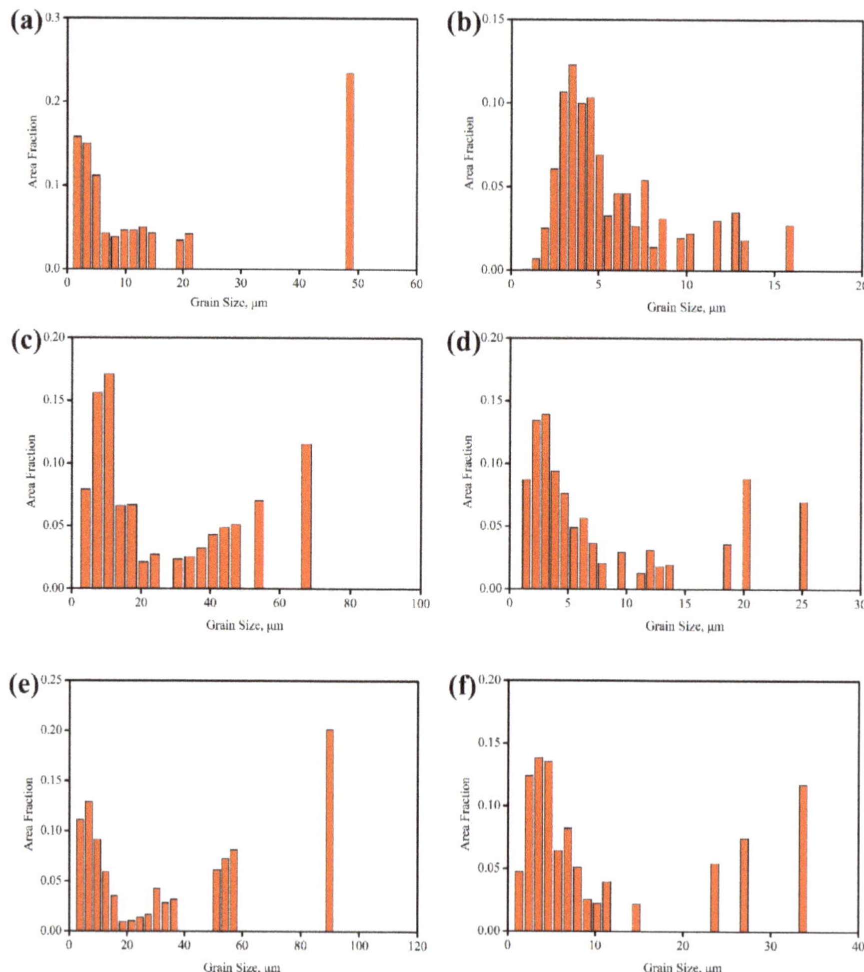

Figure 4. Histogram size counts of (**a**) alloys extruded at 800 °C, (**b**) composites extruded at 800 °C, (**c**) alloys extruded at 900 °C, (**d**) composites extruded at 900 °C, (**e**) alloys extruded at 1000 °C, (**f**) composites extruded at 1000 °C.

During extrusion at 800 °C, the basal planes {0001} of α_p would rotate to parallel to ED with the metal flow [24]. Meanwhile, the slip systems of α were limited with a hexagonal crystal structure (HCP), which led to inharmonious deformation. The relative high dislocation density would be accumulated in the grain boundaries between adjacent β grains and α_p [20], and the dislocation substructure formed [29]. The LAGBs in the adjacent β grains were contributed from the dislocation substructure slip and transformed into subgrains with the accumulation of deformation. Finally, the subgrains rotated and formed fine equiaxed β grains as further deformation, resulting in equiaxed β grains with random orientation (point C1, C2, C3, C4) [30]. Meanwhile, the LAGBs evolved into HAGBs, which also could be proved by that the HAGBs volume fraction of alloys reduced as α_p disappeared by extrusion temperature increased from 800 °C to 900 °C in Figure 2g.

The grain sizes distribution of alloys and composites extruded at different temperatures were shown by a histogram, as shown in Figure 4. The average sizes of β grains reduced obviously with the precipitation of TiBw, particularly the volume fraction of β grains less than 10 μm increased obviously. Moreover, the average sizes of β grains increased with the increasing of extrusion temperature, resulting from more energy was obtained to promote the migration of grain boundaries for DDRX and growth with higher extrusion temperature [25]. During deformation, the rotation of TiBw promoted CDRX of β grains in inharmonious deformation, and the growth of CDRXed β grains was strongly inhibited by them, resulting in the higher volume fraction of fine β grains. Noting that coarse elongated β grains were shown in Figure 2. Figure 4 shows the sizes of coarse grains increased with the increasing extrusion temperature, resulting from the migration of grain boundaries promoted by more energy. The DDRXed and CDRXed β grains and coarse elongated β grains were attributed to bimodal grain size distribution in Figure 4 [31].

The room temperature tensile curves of alloys and composites extruded at 800, 900 and 1000 °C are shown in Figure 5a,b. The tensile strength of alloys decreases from 1086 MPa to 925 MPa with the increasing of extrusion temperature. During a tensile test, α_p distributed along grain boundary trapped dislocation for strengthening, which had been revealed by Liu [32]. Meanwhile, small grains were beneficial for strengthening effect [19]. In a word, when the alloy was extruded at 800 °C, α_p and refined grains contributed mainly for strengthening, with a strength of 1086 MPa. Meanwhile, α_p triggered massive stress concentration and served a nucleation of crack [32], which led to fracture, with 3.5% elongation.

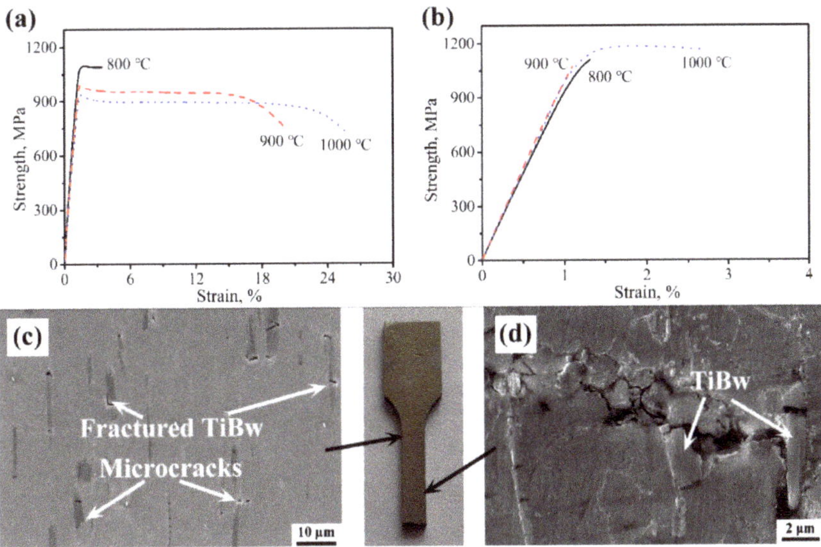

Figure 5. Room temperature tensile curves of (**a**) alloys extruded at 800, 900 and 1000 °C, (**b**) composites extruded at 800, 900 and 1000 °C; SEM of the longitudinal sections of the tensile test specimen of composites extruded at 1000 °C far away from the fracture surface (**c**), near the fracture surface (**d**).

During a tensile test for composites, the stress transfers from the matrix to TiBw due to inharmonious deformation, resulting in strengthening [27]. The findings for the composites extruded at 1000 °C will be discussed in the following in more detail and in Figure 5c,d [5]. Massive fractured TiBw and microcracks around TiBw were shown in the SEM image (Figure 5c) far away from the fracture surface., and the crack path included lots of cracks along TiBw and β grains boundaries in the SEM image (Figure 5d) near the fracture surface. These observations indicate that TiBw fractured with the accumulation of load transfer due to the inherent brittleness of TiBw [17], and then, microcracks emerged in the interface of TiBw and matrix. With further deformation, the cracks extended from microcracks to the β grains boundaries of Ti matrix, leading to the final fracture. When the composite was extruded at 1000 °C, TiBw and grain refinement contributed for strengthening, with a strength of 1200 MPa, raising 27.9%.

4. Conclusions

In this work, the grain refinement and strengthening of TiBw/Ti-15Mo-3Al-2.7Nb-0.2Si alloy fabricated by powder metallurgy and canned hot extrusion were studied. The following conclusions were drawn:

(1) The β grains of composites were refined with the rotation and inhibition of TiBw. In inharmonious deformation, the dislocation motion was inhibited by TiBw, and CDRX of β grains was promoted with the rotation of TiBw. Meanwhile, the growth of CDRXed β grains was strongly inhibited.

(2) During extrusion below β phase region, the basal planes {0001} of α_p rotated to parallel to ED, resulting in grain refinement. The high dislocation density was accumulated in α_p grain boundaries, and inharmonious deformation supplied driving force to promote CDRX of adjacent β grains. Meanwhile, α_p slipped along the prismatic glide of $\{10\bar{1}0\} \langle 11\bar{2}0\rangle$, resulting in $\langle 2\bar{1}\bar{1}0\rangle$ and $\langle 10\bar{1}0\rangle$ α texture.

(3) The strength of composites extruded at 1000 °C was improved. TiBw loaded the stress transmitted from matrix until fracture, and grain refinement contributed to strengthening. Meanwhile, the microcracks initiated from the fractured TiBw.

Author Contributions: Conceptualization, J.H. and G.C.; methodology, L.G.; software, W.C.; validation, G.C., W.Z. and W.T.; investigation, L.G.; resources, W.Z.; data curation, L.G.; writing—original draft preparation, J.H. and L.G.; writing—review and editing, G.C., W.Z. and W.T.; project administration, W.Z.; funding acquisition, G.C. All authors have read and agree to the published version of the manuscript.

Funding: This research was funded by the Sci-tech Major Project in Shandong Province, grant number 2018GGX102013.

Conflicts of Interest: The authors declare no conflict of interest.

References

1. Zheng, Y.; Williams, R.E.; Wang, D.; Shi, R.; Nag, S.; Kami, P.; Sosa, J.M.; Banerjee, R.; Wang, Y.; Fraser, H.L. Role of ω phase in the formation of extremely refined intragranular α precipitates in metastable β-titanium alloys. *Acta Mater.* **2016**, *103*, 850–858. [CrossRef]

2. Yao, T.; Du, K.; Wang, H.; Huang, Z.; Li, C.; Li, L.; Hao, Y.; Yang, R.; Ye, H. In situ scanning and transmission electron microscopy investigation on plastic deformation in a metastable β titanium alloy. *Acta Mater.* **2017**, *133*, 21–29. [CrossRef]

3. Xiao, J.; Nie, Z.; Tan, C.; Zhou, G.; Chen, R.; Li, M.; Yu, X.; Zhao, X.; Hui, S.; Ye, W. The dynamic response of the metastable β titanium alloy Ti-2Al-9.2 Mo-2Fe at ambient temperature. *Mater. Sci. Eng. A* **2019**, *751*, 191–200. [CrossRef]

4. Cherukuri, B.; Srinivasan, R.; Tamirisakandala, S.; Miracle, D.B. The influence of trace boron addition on grain growth kinetics of the beta phase in the beta titanium alloy Ti–15Mo–2.6Nb–3Al–0.2Si. *Scr. Mater.* **2009**, *60*, 496–499. [CrossRef]

5. Okulov, I.; Bönisch, M.; Okulov, A.; Volegov, A.; Attar, H.; Ehtemam-Haghighi, S.; Calin, M.; Wang, Z.; Hohenwarter, A.; Kaban, I. Phase formation, microstructure and deformation behavior of heavily alloyed TiNb-and TiV-based titanium alloys. *Mater. Sci. Eng. A* **2018**, *733*, 80–86. [CrossRef]

6. Okulov, I.; Kühn, U.; Marr, T.; Freudenberger, J.; Schultz, L.; Oertel, C.-G.; Skrotzki, W.; Eckert, J. Deformation and fracture behavior of composite structured Ti-Nb-Al-Co (-Ni) alloys. *Appl. Phys. Lett.* **2014**, *104*, 071905. [CrossRef]

7. Okulov, I.; Kühn, U.; Marr, T.; Freudenberger, J.; Soldatov, I.; Schultz, L.; Oertel, C.-G.; Skrotzki, W.; Eckert, J. Microstructure and mechanical properties of new composite structured Ti–V–Al–Cu–Ni alloys for spring applications. *Mater. Sci. Eng. A* **2014**, *603*, 76–83. [CrossRef]

8. Okulov, I.; Okulov, A.; Soldatov, I.; Luthringer, B.; Willumeit-Römer, R.; Wada, T.; Kato, H.; Weissmüller, J.; Markmann, J. Open porous dealloying-based biomaterials as a novel biomaterial platform. *Mater. Sci. Eng. C* **2018**, *88*, 95–103. [CrossRef]

9. Okulov, I.; Okulov, A.; Volegov, A.; Markmann, J. Tuning microstructure and mechanical properties of open porous TiNb and TiFe alloys by optimization of dealloying parameters. *Scr. Mater.* **2018**, *154*, 68–72. [CrossRef]

10. Okulov, I.; Soldatov, I.; Sarmanova, M.; Kaban, I.; Gemming, T.; Edström, K.; Eckert, J. Flash Joule heating for ductilization of metallic glasses. *Nat. Commun.* **2015**, *6*, 7932. [CrossRef]

11. Okulov, I.; Volegov, A.; Attar, H.; Bönisch, M.; Ehtemam-Haghighi, S.; Calin, M.; Eckert, J. Composition optimization of low modulus and high-strength TiNb-based alloys for biomedical applications. *J. Mech. Behav. Biomed. Mater.* **2017**, *65*, 866–871. [CrossRef] [PubMed]

12. Okulov, I.V.; Wendrock, H.; Volegov, A.S.; Attar, H.; Kühn, U.; Skrotzki, W.; Eckert, J. High strength beta titanium alloys: New design approach. *Mater. Sci. Eng. A* **2015**, *628*, 297–302. [CrossRef]

13. Okulov, I.; Bönisch, M.; Volegov, A.; Shahabi, H.S.; Wendrock, H.; Gemming, T.; Eckert, J. Micro-to-nano-scale deformation mechanism of a Ti-based dendritic-ultrafine eutectic alloy exhibiting large tensile ductility. *Mater. Sci. Eng. A* **2017**, *682*, 673–678. [CrossRef]

14. Zhang, W.; Wang, M.; Chen, W.; Feng, Y.; Yu, Y. Evolution of inhomogeneous reinforced structure in TiBw/Ti-6AL-4V composite prepared by pre-sintering and canned β extrusion. *Mater. Des.* **2015**, *88*, 471–477. [CrossRef]

15. Wang, B.; Huang, L.J.; Liu, B.X.; Geng, L.; Hu, H.T. Effects of deformation conditions on the microstructure and substructure evolution of TiBw/Ti60 composite with network structure. *Mater. Sci. Eng. A* **2015**, *627*, 316–325. [CrossRef]

16. Feng, Y.; Zhang, W.; Cui, G.; Wu, J.; Chen, W. Effects of the extrusion temperature on the microstructure and mechanical properties of TiBw/Ti6Al4V composites fabricated by pre-sintering and canned extrusion. *J. Alloy. Compd.* **2017**, *721*, 383–391. [CrossRef]

17. Okulov, I.; Sarmanova, M.; Volegov, A.; Okulov, A.; Kühn, U.; Skrotzki, W.; Eckert, J. Effect of boron on microstructure and mechanical properties of multicomponent titanium alloys. *Mater. Lett.* **2015**, *158*, 111–114. [CrossRef]

18. Ma, X.; Li, C.; Du, Z.; Zhang, W. Thermodynamic assessment of the Ti–B system. *J. Alloy. Compd.* **2004**, *370*, 149–158. [CrossRef]

19. Dyakonov, G.; Mironov, S.; Semenova, I.; Valiev, R.; Semiatin, S. EBSD analysis of grain-refinement mechanisms operating during equal-channel angular pressing of commercial-purity titanium. *Acta Mater.* **2019**, *173*, 174–183. [CrossRef]

20. Lin, P.; Sun, Y.; Zhang, S.; Zhang, C.; Wang, C.; Chi, C. Microstructure and texture heterogeneity of a hot-rolled near-α titanium alloy sheet. *Mater. Charact.* **2015**, *104*, 10–15. [CrossRef]

21. Su, J.; Sanjari, M.; Kabir, A.S.H.; Jung, I.-H.; Yue, S. Dynamic recrystallization mechanisms during high speed rolling of Mg–3Al–1Zn alloy sheets. *Scr. Mater.* **2016**, *113*, 198–201. [CrossRef]

22. Belyakov, A.; Sakai, T.; Miura, H.; Kaibyshev, R.; Tsuzaki, K. Continuous recrystallization in austenitic stainless steel after large strain deformation. *Acta Mater.* **2002**, *50*, 1547–1557. [CrossRef]

23. Wang, S.C.; Aindow, M.; Starink, M.J. Effect of self-accommodation on α/ α boundary populations in pure titanium. *Acta Mater.* **2003**, *51*, 2485–2503. [CrossRef]

24. Huang, G.; Han, Y.; Guo, X.; Qiu, D.; Wang, L.; Lu, W.; Zhang, D. Effects of extrusion ratio on microstructural evolution and mechanical behavior of in situ synthesized Ti-6Al-4V composites. *Mater. Sci. Eng. A* **2017**, *688*, 155–163. [CrossRef]

25. Doherty, R.D.; Hughes, D.A.; Humphreys, F.J.; Jonas, J.J.; Rollett, A.D. Current issues in recrystallization: A review. *Mater. Sci. Eng. A* **1997**, *238*, 219–274. [CrossRef]

26. Basu, I.; Al-Samman, T. Twin recrystallization mechanisms in magnesium-rare earth alloys. *Acta Mater.* **2015**, *96*, 111–132. [CrossRef]

27. Panda, K.B.; Chandran, K.S.R. First principles determination of elastic constants and chemical bonding of titanium boride (TiB) on the basis of density functional theory. *Acta Mater.* **2006**, *54*, 1641–1657. [CrossRef]

28. Park, J.G.; Keum, D.H.; Lee, Y.H. Strengthening mechanisms in carbon nanotube-reinforced aluminum composites. *Carbon* **2015**, *95*, 690–698. [CrossRef]

29. Murty, S.V.S.N.; Nayan, N.; Kumar, P.; Narayanan, P.R.; Sharma, S.C.; George, K.M. Microstructure–texture–mechanical properties relationship in multi-pass warm rolled Ti–6Al–4V Alloy. *Mater. Sci. Eng. A* **2014**, *589*, 174–181. [CrossRef]

30. Dong, R.; Li, J.; Kou, H.; Tang, B.; Hua, K.; Liu, S. Characteristics of a hot-rolled near β titanium alloy Ti-7333. *Mater. Charact.* **2017**, *129*, 135–142. [CrossRef]

31. He, J.; Jin, L.; Wang, F.; Dong, S.; Dong, J. Mechanical properties of Mg-8Gd-3Y-0.5 Zr alloy with bimodal grain size distributions. *J. Magnes. Alloy.* **2017**, *5*, 423–429. [CrossRef]

32. Liu, C.; Lu, Y.; Tian, X.; Liu, D. Influence of continuous grain boundary α on ductility of laser melting deposited titanium alloys. *Mater. Sci. Eng. A* **2016**, *661*, 145–151. [CrossRef]

 *metals*

Article

Reaction Layer Analysis of In Situ Reinforced Titanium Composites: Influence of the Starting Material Composition on the Mechanical Properties

Isabel Montealegre-Meléndez [1], Cristina Arévalo [1,*], Ana M. Beltrán [1], Michael Kitzmantel [2], Erich Neubauer [2] and Eva María Pérez Soriano [1]

[1] Escuela Politécnica Superior, Universidad de Sevilla, Calle Virgen de África, 7, 41011 Sevilla, Spain; imontealegre@us.es (I.M.-M.); abeltran3@us.es (A.M.B.); evamps@us.es (E.M.P.S.)
[2] RHP Technology GmbH, 2444 Seibersdorf, Austria; m.ki@rhp.at (M.K.); e.ne@rhp.at (E.N.)
* Correspondence: carevalo@us.es; Tel.: +34-954-482-278

Received: 30 January 2020; Accepted: 15 February 2020; Published: 18 February 2020

Abstract: This study aims at the analysis of the reaction layer between titanium matrices and reinforcements: B_4C particles and/or intermetallic Ti_xAl_y. Likewise, the importance of these reactions was observed; this was particularly noteworthy as regard coherence with the obtained results and the parameters tested. Accordingly, five starting material compositions were studied under identical processing parameters via inductive hot pressing at 1100 °C for 5 min in vacuum conditions. The results revealed how the intermetallics limited the formation of secondary phases (TiC and TiB) created from the B and C source. In this respect, the percentages of TiB and TiC slightly varied when the intermetallic was included in the matrix as prealloyed particles. On the contrary, if the intermetallics appeared in situ by the addition of Ti-Al powder in the starting blend, their content was lesser. The mechanical properties values and the tribology behaviour might deviate, depending on the percentage of the secondary phases formed and its distribution in the matrix.

Keywords: in situ composites; inductive hot pressing; tribology; titanium composites; reaction layer; intermetallics

---

1. Introduction

The interest of materials with high specific properties and good tribological behaviour raises the needs in terms of improving the properties of materials like titanium and its alloys. In the fields of aerospace, military industry and biomedicine, the use of these materials is very popular [1–4]; however, nowadays, there are limitations regarding their mechanical and poor tribological properties. Therefore, titanium matrix composites (TMCs) are valuated as materials that combine low density with mechanical properties [5–8]. Presently, there is a great variety of reinforcements employed in these composites. The most popular ones showed in diverse research works tend to increase the hardness, Young's modulus and the tribological behaviour without incrementing the density. As examples, it may include TiB_2, TiB, TiC, B, B_4C, carbon nanotubes, graphite and nanodiamonds [9–12]. Among these materials, B_4C ceramic particles have been presented in investigations as an optimal form to obtain B and C sources to origin in situ TiB and TiC [13–16]. In this regard, these secondary phases play a key role in the strengthening of the titanium matrices [17]. The main advantage of these in situ phases resides in the existence of a good interfacial bonding of the reinforcement-matrix. Moreover, the ceramic reinforcements formed during in -situ processing are finer, more thermodynamically stable and uniform in size distribution in the metal matrix [18,19]. Therefore, the study of the reaction layer between the matrix and the particles is important in order to achieve a better understanding of the cause that could promote or not these reactions and their products. While there is considerable literature on the grounds of strengthening and the reaction mechanisms [20,21], there are only a few studies where the reaction layer is analysed. It is also investigated how the presence of Ti_xAl_y as intermetallic in the titanium matrices could affect the final appearance of these secondary phases (TiB and TiC) [22–24]; however, the reaction layer in the presence of intermetallic, as well as its decomposition, has been little studied [25,26]. Hence, here lies the importance for understanding and determining the evolution of the reaction layer with B_4C particles and intermetallics Ti_xAl_y and how these layers and the final properties of the consolidated composites could be affected by the combination of the starting materials. Therefore, this research aims not only to study the reaction layer but also to investigate and characterise TMCs from five different blends. These blends were designed and processed considering interesting combinations of starting powders, in order to observe the regarding properties and the above-mentioned reaction layer phenomenon. Through the selection of the diverse intermetallic starting powders, variations in the behaviour of the TMCs may be expected.

In several investigations concerning the manufacture of titanium composites via powder metallurgy (PM) [27,28], inductive hot pressing (iHP) is considered as a suitable fabrication option due to its flexibility and short cycles. For that reason, and thanks to the experience of the authors on TMCs manufacturing, the fabrication route of the specimens was through iHP. In preliminary studies [29,30], an inflexion temperature, at which secondary phases were formed in situ, was observed. The analysis of the reaction layer has been carried out in specimens produced at this inflexion temperature (1100 °C) in vacuum conditions.

Hence, the five specimens were in detail characterised through scanning and scanning-transmission electron microscopies (SEM and (S)TEM) and by X-ray diffraction analysis (XRD). Furthermore, their tribological and physical properties were measured and evaluated. In this regard, the relation between the starting materials and the final properties was thoroughly investigated.

2. Materials and Methods

2.1. Materials

The starting blends were five different combinations of powders, using only one titanium matrix powder, as listed in Table 1.

Table 1. Summary of the composition of the tested specimens.

Blend	Materials	B_4C [Volume %]	Ti:Al [Volume %]
1	Ti + B_4C	30	-
2	Ti + B_4C + Ti-Al(1)	30	20
3	Ti + Ti-Al(1)	-	20
4	Ti + B_4C + Ti-Al(2)	30	20
5	Ti + Ti-Al(2)	-	20

The powders employed in this research were previously characterised through particle size distribution and morphological analysis, in order to verify their supplied information. The Mastersizer 2000 (Malvern Instruments, Malvern, UK) equipment was used to determinate the average particle size, and their morphology study was performed by SEM from FEI Teneo images (FEI, Eindhoven, Netherlands).

Titanium powder was produced by TLS GmbH (Bitterfeld, Germany); it showed a spherical morphology in agreement with the manufacturer's information, being its mean diameter of 109 μm. The B_4C particles were manufactured by ABCR GmbH & Co KG (Karlsruhe, Germany) with an average size diameter of 64 μm; the morphology of these ceramic particles was irregular and with slightly sharp edges.

Concerning the intermetallic powders, there were substantial differences between the two selected. The one named Ti-Al(1) was a prealloyed Ti_3Al and $TiAl_3$ intermetallic powder manufactured by TLS

GmbH (Bitterfeld, Germany). This Ti-Al(1) showed a spherical morphology, and its mean diameter was around 75 µm. The other intermetallic powder used was named Ti-Al(2). It was made from a blend of elementary Al (NMD GmbH, Heemsen, Germany; 9 µm) and Ti fine powder with a mean diameter of 29 µm in molar ratio 1:1 by TLS GmbH (Bitterfield, Germany) [22,30].

In addition to the starting materials' characterisations, XRD analysis was accomplished—Bruker D8 Advance A25 (Billerica, MA, USA) with Cu-K$_\alpha$ radiation. Figure 1 shows the patterns of the powders from which the five blends were prepared. Based on the resulted diffraction spectra, the composition of the powders agreed with the suppliers' information in the case of Ti powder, B$_4$C and Ti-Al(1) in Figure 1a–c, respectively. In the case of Ti-Al(2), the remarked peaks corresponded to the Ti and Al elements, matching the elementary blending of Ti and Al, whose molar ratio was 1:1 (Ti:Al), as it was above-mentioned.

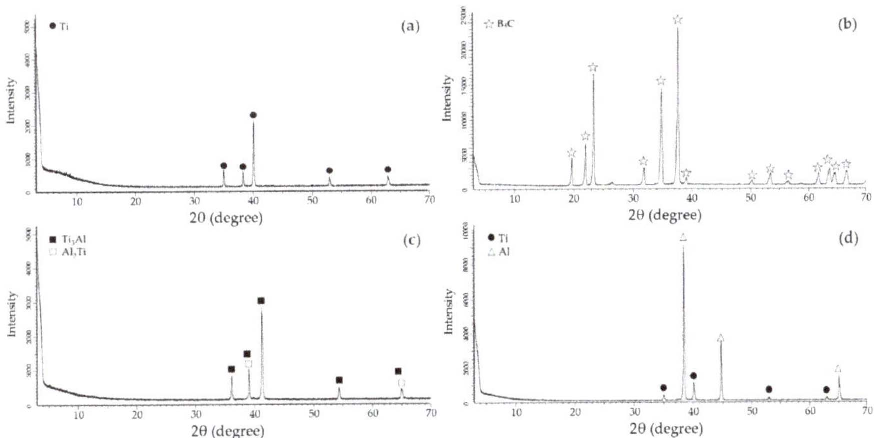

Figure 1. XRD (X-ray diffraction) patterns of the starting powders: (**a**) Ti powder, (**b**) B$_4$C particles, (**c**) Ti-Al(1) powder and (**d**) Ti-Al(2) powder.

2.2. Methods

After the blending preparation, similar as in previous authors' works [30], the consolidation of the specimens via iHP was performed, employing a self-made inductive hot-pressing machine; equipment from RHP-Technology GmbH (Seibersdorf, Austria).

This machine provided the time of the operational cycles to be reduced thanks to its advantageous high heating rate. The time set was 5 min in vacuum conditions at 1100 °C and 80 MPa. The same procedure to fill the die was carried out for the consolidation of each specimen [30]. The graphite die for all the iHP cycles had a diameter of 20 mm. A detailed description of the manufacturing process was reported in previous authors' work [22].

The consolidated composites were studied at length. After thorough metallographic preparation, their microstructures were examined by SEM at 15 kV. The (S)TEM characterisation was carried out using a FEI Talos F200S microscope (FEI, Eindhoven, Netherlands) operating at an accelerating voltage of 200 kV and equipped with a super-X energy-dispersive X-ray spectrometry (EDX) system, which includes two silicon drift detectors. The elemental mapping experiments were accomplished by combining high-angle annular dark-field imaging (HAADF) and EDX acquisition in (S)TEM mode. The mechanical grinding and ion milling procedures were performed for the (S)TEM studies following standard procedure for TEM lamella preparation. Moreover, an XRD analysis was conducted to identify the diverse crystalline phases in the TMCs. Ultrasonic method (Olympus 38 DL, Tokyo, Japan) was used to calculate Young's modulus by measuring longitudinal and transverse propagation velocities of acoustic waves [31]. The densification of the specimens was measured according to the Archimedes' method [32]. The relative density was computed as the ratio of the compacts' density to their theoretical values of the given material, determined by the rule of mixtures. A test model, Struers-Duramin A300 (Ballerup, Germany), was used to ascertain the Vickers hardness (HV2). The hardness measurements took place on the polished cross-section of the specimens. The reported values were the average of eight indentations.

After the characterisation of the specimens, the tribological behaviour was studied. Before running the tests, the specimens were prepared (grinded and polished), cleaned with acetone in an ultrasonic bath and, after that, dried. The wear behaviour of the TMCs was conducted in a ball-on-disc tribometer (Microtest MT/30/NI, Madrid, Spain) using alumina balls with a diameter of 6 mm. At room temperature, the normal load of 3 N on the ball with a sliding speed of 125 mm/s was employed to measure the wear properties. It was tested at a sliding distance of 500 m on the specimens' surface, with a circular path of 3 mm in radius. The results were analysed at similar conditions in order to compare and evaluate the influence of the starting materials. The morphology of the worn surfaces was characterised by optical microscopy with a Leica Zeiss DMV6 (Leica Microsystems, Heerbrugg, Switzerland).

3. Results and Discussion

3.1. Microstructural Study and XRD Analysis

The microstructural study revealed significant differences related to the starting powders employed. Figure 2 displayed a microstructure general overview of the specimens. The circular backscattered (CBS) SEM images in Figure 2a,b,d showed how the B_4C particles were homogeneously distributed in the titanium matrices.

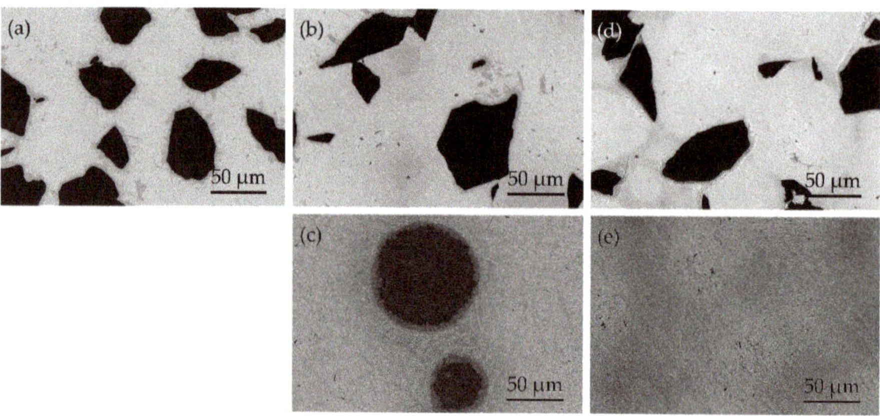

Figure 2. CBS-SEM (circular backscattered-scanning electron microscopies) images: (**a**) Ti + 30 vol. % B_4C, (**b**) Ti + 30 vol. % B_4C + 20 vol. % Ti-Al(1), (**c**) Ti + 20 vol. % Ti-Al(1), (**d**) Ti + 30 vol. % B_4C + 20 vol. % Ti-Al(2) and (**e**) Ti + 20 vol. % Ti-Al(2).

The different shades of grey close to these ceramic particles suggested the presence of in situ secondary phases. However, at this scale, that was not easily appreciated. Conversely, in the case of the specimen made from Ti only with Ti-Al(1) (Figure 2c), the dark grey areas corresponded to the intermetallic phases (Figure 2c). The intermetallics in the matrix were visible and recognised as spherical precipitates with degradation around. However, if there were Ti-Al powders employed in the starting mixing as Ti-Al(2), the location of the phases rich in Al was not appreciated (Figure 2e). Although, in the bibliography has been reported the formation of new phases due to the liquid metal dealloying reaction [33,34], in this specimen fabricated at this temperature, it seems that no intermetallics were in situ formed, having the aluminium in solid solution with the titanium matrix. These observations were verified by the XRD analysis.

The microstructural characterisation at higher magnification is presented in Figure 3, revealing several phases depending on the starting materials. Evaluating the area surrounding the B_4C particles, there were significant variations in the volume and thickness of the in situ secondary phases formed in the reaction layer. In the TMC without intermetallic addition, more in situ secondary phases were formed (Figure 3a). The incorporation of Ti_xAl_y in the starting powders of the TMCs not only modified the microstructure of the specimens but also affected the secondary phases formed, whereupon the final properties of the specimens could suffer variations. A detailed study of the TMCs' microstructure evolution contributed to a better understanding of in situ formed phases, as well as the distribution of the Ti_xAl_y in the titanium matrices. In Figure 3a,b,d, the reaction layer around the B_4C could be observed, being mainly composed of B and TiB. These TiB phases could also appear further from the B_4C particles when no intermetallic was used, as well as TiC (Figure 3a,b). It suggested that the employ of Ti-Al intermetallic in the initial blending involved a slight obstacle in the origin of in situ formed phases. The Al could act as a barrier, blocking the reactions between the Ti and the B and C sourced by the B_4C particles. This phenomenon was more visible in the sample with Al as an elementary powder (TiAl(2), Figure 3d). The morphology of in situ phases will be discussed later.

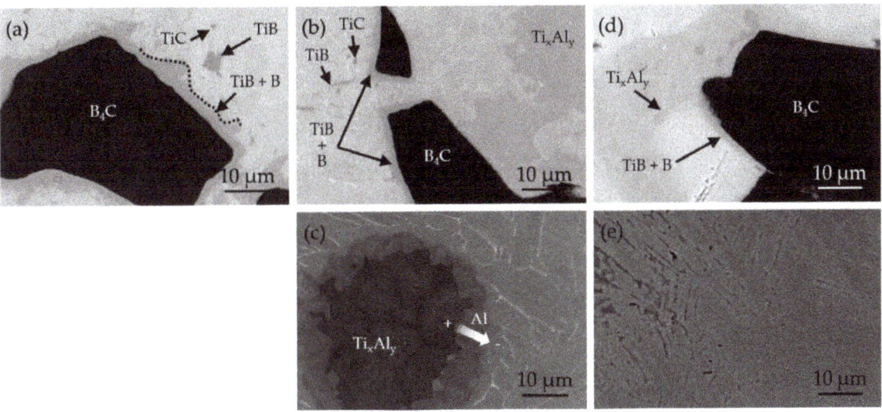

Figure 3. Phases identification via CBS-SEM images: (**a**) Ti + 30 vol. % B_4C, (**b**) Ti + 30 vol. % B_4C + 20 vol. % Ti-Al(1), (**c**) Ti + 20 vol. % Ti-Al(1), (**d**) Ti + 30 vol. % B_4C + 20 vol. % Ti-Al(2) and (**e**) Ti + 20 vol. % Ti-Al(2).

As previously mentioned, when no B_4C was added, how introducing Ti_xAl_y in the matrix caused differences. This can be clearly seen in Figure 3c,e. There was a reaction layer surrounding the intermetallic phases when prealloyed intermetallic was introduced as a starting powder (TiAl(1)). This reaction layer was due to the decomposition of the intermetallic being Al-incorporated into the matrix.

In order to understand TMCs with B_4C particles in detail, the reaction layer and secondary phases were analysed by TEM and (S)TEM (Figures 4–6). In the reaction layer, the in situ formed precipitates could be recognised. The more the presence of Al in the matrix, the thicker the reaction layer was observed. The reason, as previously discussed, could be the difficulted diffusion of B and C elements into the matrix. Next to the B_4C particles, the size of TiB and TiC in situ secondary phases created were smaller to grow as moving away from the ceramic particles. When Al was not present in the matrix, secondary phases were bigger in size (Figure 4a).

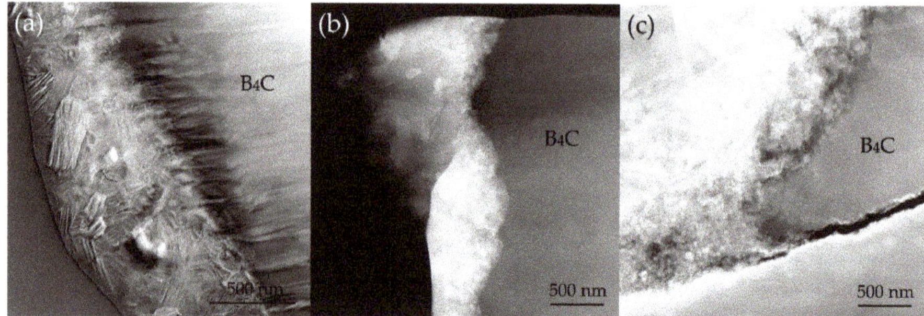

Figure 4. TEM (transmission electron microscopies) images: (**a**) Ti + 30 vol. % B$_4$C, (**b**) Ti + 30 vol. % B$_4$C + 20 vol. % Ti-Al(1) and (**c**) Ti + 30 vol. % B$_4$C + 20 vol. % Ti-Al(2).

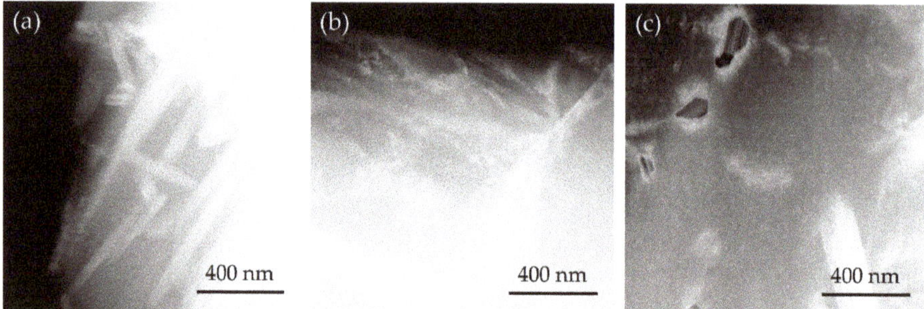

Figure 5. (S)TEM images 80,000x: (**a**) Ti + 30 vol. % B$_4$C, (**b**) Ti + 30 vol. % B$_4$C + 20 vol. % Ti-Al(1) and (**c**) Ti + 30 vol. % B$_4$C + 20 vol. % Ti-Al(2).

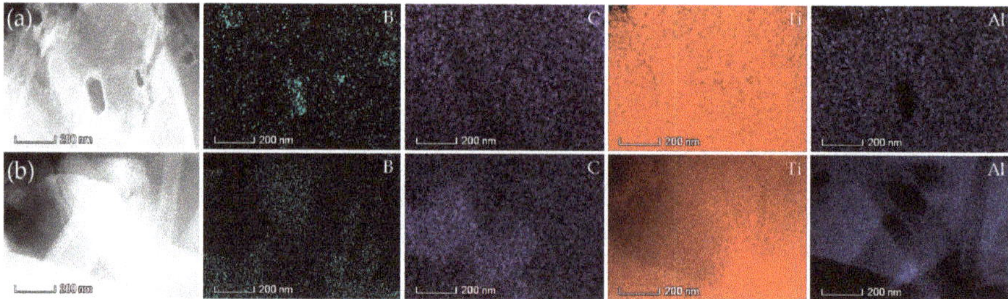

Figure 6. TEM images and compositional mappings: (**a**) Ti + 30 vol. % B$_4$C + 20 vol. % Ti-Al(1) and (**b**) Ti + 30 vol. % B$_4$C + 20 vol. % Ti-Al(2).

In Figures 4 and 5, the observed phases consisted in TiB and TiC were recognised due to their well-known morphology, as whiskers in the case of TiB and as isolated grey areas for TiC from its dendritic formation [30]. In TMCs reinforced only by B$_4$C, the morphologies of TiB and TiC phases could be more easily identified (Figure 4a), while for TMCs with B$_4$C and intermetallic, these precipitates were not clearly observed (Figure 4b,c). Figure 5 shows the formation of precipitates far from the B$_4$C particles. The lower the content of Al as a solid solution in the matrix, the higher the concentration of secondary phases in these areas was found. In this way, this was more significant with Ti-Al(2) (Figure 5c) than with Ti-Al(1) (Figure 5b), confirming SEM microscopy.

Concluding the microstructural characterisation, a compositional mapping of the main present elements in samples with B$_4$C and intermetallic was performed (Figure 6). The aim was to study the composition of the smallest precipitates to check if there were other formed secondary phases, apart from the expected TiB and TiC. The results confirmed that there was no reaction between the B and C particles with Al.

The XRD patterns of the specimens are illustrated in Figure 7. The one of TMCs without intermetallic (Figure 7a) was intended to serve as a reference on the comparison in terms of precipitates formation regarding the specimens with Ti-Al(1) (Figure 7b) and Ti-Al(2) (Figure 7d). The peaks of B_4C could clearly be identified in these three patterns. The variations in the peaks of the in situ formed TiB and TiC that were related to the intermetallic powder employed as the starting powder were meaningful in the results. It suggested that the reaction of the matrix and the ceramic particles was more decelerated if the starting powder was made from an elementary blending of Ti-Al. The peaks corresponding to the $TiAl_3$ phase were not visible in the XRD patterns. These results were in agreement with the microstructures observed by SEM and (S)TEM.

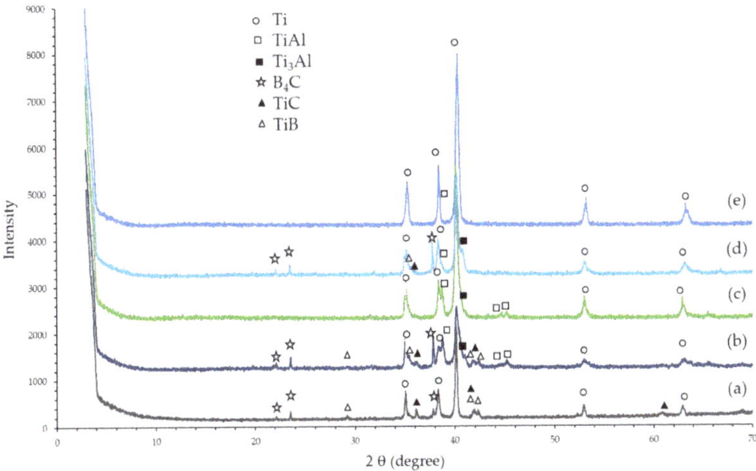

Figure 7. XRD patterns of the specimens: (**a**) Ti + 30 vol. % B_4C, (**b**) Ti + 30 vol. % B_4C + 20 vol. % Ti-Al(1), (**c**) Ti + 20 vol. % Ti-Al(1), (**d**) Ti + 30 vol. % B_4C + 20 vol. % Ti-Al(2) and (**e**) Ti + 20 vol. % Ti-Al(2).

When the starting blend was made without B_4C, while the pattern of the specimen produced from intermetallic powder Ti-Al(1) revealed the peaks of the Ti_3Al and TiAl phases (Figure 7c), the pattern of the specimens made from Ti-Al(2) powder showed mainly the displaced peaks of Ti and a small quantity of TiAl (Figure 7e). This could be due to the Al migrated into the titanium crystal lattice, avoiding the formation of Ti_xAl_y phases. The presence of B_4C and the secondary phases affected the Al diffusion; therefore, the peak of Ti_3Al was slightly detected in specimens made from B_4C and Ti-Al(2) (see Figure 7d).

3.2. Physical and Tribological Properties

The values of hardness in the matrix and Young's modulus are plotted in Figure 8. These results agreed with the previously commented microstructural study and XRD analysis. As it was expected, the B_4C addition led to the onset of TiB and TiC becoming in an increase of the hardness and Young's modulus. Whereas there was more evidence of the TiB and TiC precipitates in the composites made from Ti-Al(1), the hardness and Young's modulus of these specimens were similar to the composites made from elementary Ti-Al powder (Ti-Al(2)). The Al diffusion phenomenon into the titanium crystal lattice was helping to the matrix strengthening in specimens made from Ti-Al(2). Comparing samples without B_4C, the same trend was observed in the hardness and Young's modulus values (Figure 8). Therefore, this verified how, from different starting powders and diverse reactions phenomenon, the final properties could be, to some extent, similar; other authors have reported some positive or negative effects on the mechanical behaviour of the titanium alloys with other intermetallic additions [35,36].

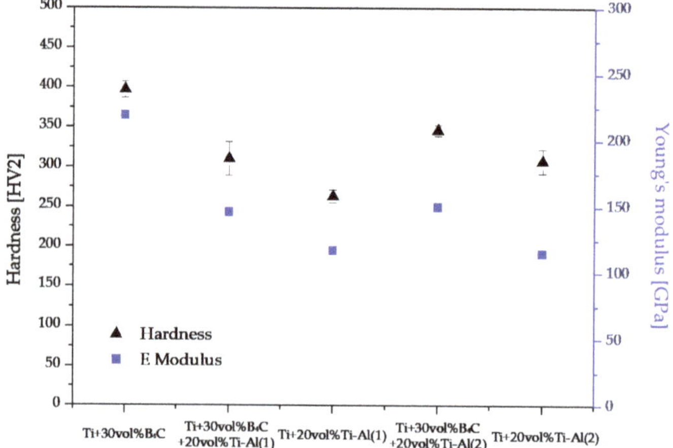

Figure 8. Hardness Vickers (HV2) and Young's modulus of the specimens.

As regards the tribological behaviour, the wear resistance and the coefficient of friction (COF) were measured and evaluated (Figures 9 and 10). The wear loss (mg) after sliding for 66.3 min (500 m, 400 rpm) is shown in Figure 9. This loss results differed significantly in their value due to the combination of B_4C and intermetallic content. The enhancement in fretting wear with only the addition of B_4C was relevant. It is important to note that the minor weight loss resulted in the TMC reinforced only with B_4C. The intermetallic incorporation in the starting powder involved that the weight loss increased by 50% and 53% in composites made from B_4C with Ti-Al(2) and Ti-Al(1), respectively. These phenomena were related to the amount of in situ formed TiB and TiC; if there were an obstacle to origin these secondary phases, the tribological properties could decline. Therefore, when the Al diffused into the matrix blocking these secondary reactions, there was less formation of TiB and TiC, which was reflected in the final tribological behaviour of the composites. These results were in agreement with the ones obtained by microscopy and XRD analysis.

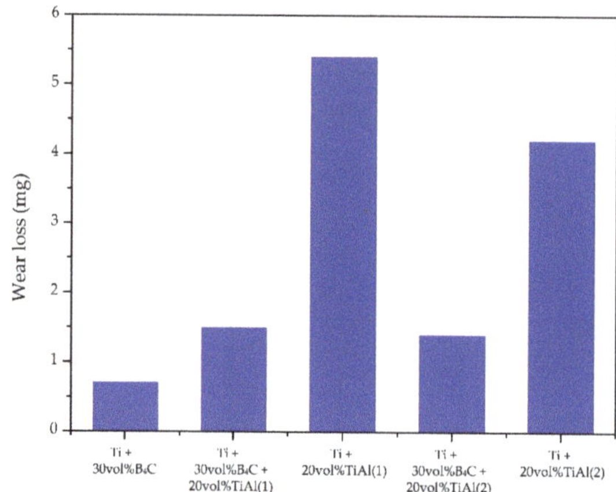

Figure 9. Wear loss vs. specimens.

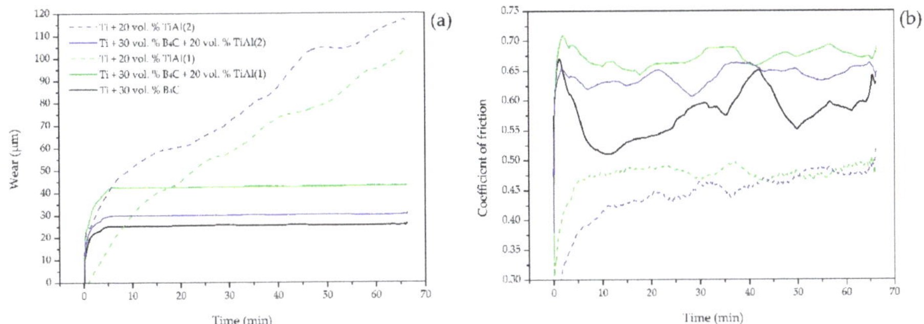

Figure 10. Tribology results: (**a**) wear vs. time and (**b**) coefficient of friction vs. time.

Figure 10 shows the wear and COF vs. time for all the specimens at the same wear conditions. In accordance with the commented above, the B_4C particles contributed to a lower penetration than in other specimens, as it can be appreciated in Figure 10a. In these specimens, the B_4C particles could participate in avoiding their pluck from the titanium matrix, maintaining the initial penetration corresponding to the applied load and speed. As expected, the absence of B_4C promoted the depth of the wear track, being the composite made from Ti-Al(2) the most affected. The COF vs. time is shown in Figure 10b. The coefficient values in the steady-state region were similar in specimens reinforced with B_4C. During the wear test, the particles and the hard precipitates in the titanium matrix could collide with the ball; therefore, COF was higher than in specimens without B_4C particles. Based upon the foregoing, when the intermetallic remained in the matrix, as in Ti-Al(1), there were also collisions between the ball and these precipitates, reflected in the COF values measured as seen in Figure 10b.

Optical micrographs in Figure 11 show representative worn track areas of the samples after the tribological characterisation. The composites with B_4C reinforcements seemed to be able to prevent the formation of severe worn surfaces, showing stronger behaviours. Moreover, in samples without ceramic reinforcements, Figure 11c,e, the material removal could be clearly appreciated with debris in the worn. The worn track surfaces were also significantly wider than in specimens with B_4C particles, being more noticeable when the intermetallic was added as an elementary blend.

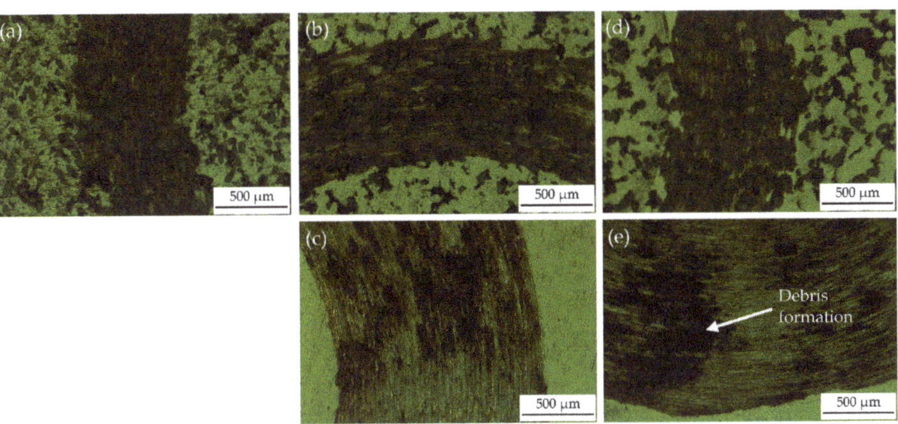

Figure 11. Worn surface: (**a**) Ti + 30 vol. % B_4C, (**b**) Ti + 30 vol. % B_4C + 20 vol. % Ti-Al(1), (**c**) Ti + 20 vol. % Ti-Al(1), (**d**) Ti + 30 vol. % B_4C + 20 vol. % Ti-Al(2) and (**e**) Ti + 20 vol. % Ti-Al(2).

4. Conclusions

The following conclusions were drawn from this research:

(1) The presence of intermetallic affected considerably the formation of the secondary phases TiC and TiB. If the aluminium was added as an elementary powder in the blend, it caused less formation of these secondary phases, in comparison to the effect of the prealloyed Ti-Al powder. The presence of Al in the matrix could block the C and B diffusion, leading to a weak reaction between these elements and the titanium.

(2) The reaction layers between B_4C and the matrix presented similar characteristics, regardless of the intermetallic powder in the blend. This reaction layer was narrower when only B_4C was used. When adding intermetallics, more TiB and TiC were accumulated around the ceramic particles due to the commented screen effect of Al in the matrix. The opposite occurred with the precipitates size of the secondary phases TiB and TiC being bigger when no intermetallic was added.

(3) In relation to the reaction layer of the prealloyed Ti_xAl_y with titanium, a decomposition occurred allowing the Al to be introduced into the matrix.

(4) Regarding the mechanical and tribological properties, composites with ceramic reinforcement showed excellent behaviour with high hardness values and good wear resistance. Without B_4C, the presence of prealloyed intermetallic in the matrix helped to a better performance than with the elementary Ti-Al blend.

Author Contributions: Authors have collaborated to obtain high-quality research work. Conceptualisation, E.N. and I.M.-M.; methodology, I.M.-M., E.M.P.S. and C.A.; formal analysis, I.M.-M.; investigation, C.A.; resources, M.K. and E.N.; data curation, E.M.P.S.; writing—original draft preparation, I.M.-M., E.M.P.S. and C.A.; writing—review and editing, C.A. and A.M.B.; visualisation, E.M.P.S. and funding acquisition, I.M.-M., E.M.P.S. and C.A. All authors have read and agreed to the published version of the manuscript.

Funding: The authors want to thank the Universidad de Sevilla for the use of experimental facilitates at CITIUS, Microscopy and X-Ray Laboratory Services (VI PPIT-2018-I.5 Eva María Pérez Soriano, VI PPIT-2019-I.5 Cristina Arévalo Mora and VI PPIT-2019-I.5 Isabel Montealegre Meléndez).

Conflicts of Interest: The authors declare no conflicts of interest.

References

1. Lütjering, G.; Williams, J.C. *Titanium*, 2nd ed.; Springer: Berlin, Germany, 2007.
2. Leyends, C.; Peters, M. *Titanium and Titanium Alloys: Fundamentals and Applications*, 1st ed.; Wiley-VCH Verlag GmbH & Co. KGaA: Weinheim, Germany, 2003.
3. Elias, C.N.; Lima, J.H.C.; Valiev, R.; Meyers, M.A. Biomedical applications of titanium and its alloys. *JOM* **2008**, *60*, 46–49. [CrossRef]
4. Leary, M. Design of titanium implants for additive manufacturing. *Titan. Med. Dent. Appl.* **2018**, 203–224. [CrossRef]
5. Tjong, S.C.; Mai, Y.-W. Processing-structure-property aspects of particulate- and whisker-reinforced titanium matrix composites. *Compos. Sci. Technol.* **2008**, *68*, 583–601. [CrossRef]
6. Ravi Chandran, K.S.; Panda, K.B.; Sahay, S.S. TiB_w-reinforced Ti composites: Processing, properties, application prospects, and research needs. *JOM* **2004**, *56*, 42–48. [CrossRef]
7. Zadra, M.; Girardini, L. High-performance, low-cost titanium metal matrix composites. *Mater. Sci. Eng. A* **2014**, *608*, 155–163. [CrossRef]
8. Neubauer, E.; Vály, L.; Kitzmantel, M.; Grech, D.; Rovira, A.; Montealegre-Meléndez, I.; Arévalo, C. Titanium Matrix Composites with High Specific Stiffness. *Key Eng. Mater.* **2016**, *704*, 38–43. [CrossRef]
9. Cao, Z.; Wang, X.; Li, J.; Wu, Y.; Zhang, H.; Guo, J.; Wang, S. Reinforcement with graphene nanoflakes in titanium matrix composites. *J. Alloys Compd.* **2017**, *696*, 498–502. [CrossRef]
10. Sabahi Namini, A.; Azadbeh, M.; Shahedi Asl, M. Effect of TiB_2 content on the characteristics of spark plasma sintered Ti–TiB_w composites. *Adv. Powder Technol.* **2017**, *28*, 1564–1572. [CrossRef]

11. Montealegre-Meléndez, I.; Neubauer, E.; Angerer, P.; Danninger, H.; Torralba, J.M. Influence of nano-reinforcements on the mechanical properties and microstructure of titanium matrix composites. *Compos. Sci. Technol.* **2011**, *71*, 1154–1162. [CrossRef]

12. Zhang, X.; He, M.; Yang, W.; Wu, K.; Zhan, Y.; Song, F. Structure and mechanical properties of in-situ titanium matrix composites with homogeneous Ti_5Si_3 equiaxial particle-reinforcements. *Mater. Sci. Eng. A* **2017**, *698*, 73–79. [CrossRef]

13. Ni, D.R.; Geng, L.; Zhang, J.; Zheng, Z.Z. Effect of B_4C particle size on microstructure of in situ titanium matrix composites prepared by reactive processing of Ti–B_4C system. *Scr. Mater.* **2006**, *55*, 429–432. [CrossRef]

14. Zhang, Y.; Sun, J.; Vilar, R. Characterization of (TiB+TiC)/TC4 in situ titanium matrix composites prepared by laser direct deposition. *J. Mater. Process. Technol.* **2011**, *211*, 597–601. [CrossRef]

15. Arévalo, C.; Kitzmantel, M.; Neubauer, E.; Montealegre-Meléndez, I. Development of Ti-MMCs by the use of different reinforcements via conventional Hot-Pressing. *Key Eng. Mater.* **2016**, *704*, 400–405. [CrossRef]

16. Wang, X.; Wang, L.; Luo, L.; Yan, H.; Li, X.; Chen, R.; Su, Y.; Guo, J.; Fu, H. High temperature deformation behavior of melt hydrogenated (TiB+TiC)/Ti-6Al-4V composites. *Mater. Des.* **2017**, *121*, 335–344. [CrossRef]

17. Jia, L.; Li, S.; Imai, H.; Chen, B.; Kondoh, K. Size effect of B_4C powders on metallurgical reaction and resulting tensile properties of Ti matrix composites by in-situ reaction from Ti–B_4C system under a relatively low temperature. *Mater. Sci. Eng. A* **2014**, *614*, 129–135. [CrossRef]

18. Radhakrishna Bhat, B.V.; Subramanyam, J.; Bhanu Prasad, V.V. Preparation of Ti-TiB-TiC & Ti-TiB composites by in-situ reaction hot pressing. *Mater. Sci. Eng. A* **2002**, *325*, 126–130. [CrossRef]

19. AlMangour, B.; Grzesiak, D.; Yang, J.-M. In-situ formation of novel TiC-particle-reinforced 316L stainless steel bulk-form composites by selective laser melting. *J. Alloys Compd.* **2017**, *706*, 409–418. [CrossRef]

20. Jia, L.; Wang, X.; Chen, B.; Imai, H.; Li, S.; Lu, Z.; Kondoh, K. Microstructural evolution and competitive reaction behavior of Ti-B_4C system under solid-state sintering. *J. Alloys Compd.* **2016**, *687*, 1004–1011. [CrossRef]

21. Li, S.; Kondoh, K.; Imai, H.; Chen, B.; Jia, L.; Umeda, J. Microstructure and mechanical properties of P/M titanium matrix composites reinforced by in-situ synthesized TiC–TiB. *Mater. Sci. Eng. A* **2015**, *628*, 75–83. [CrossRef]

22. Arévalo, C.; Montealegre-Meléndez, I.; Ariza, E.; Kitzmantel, M.; Rubio-Escudero, C.; Neubauer, E. Influence of Sintering Temperature on the Microstructure and Mechanical Properties of In Situ Reinforced Titanium Composites by Inductive Hot Pressing. *Materials* **2016**, *9*, 919. [CrossRef]

23. Ma, F.; Shi, Z.; Liu, P.; Li, W.; Liu, X.; Chen, X.; He, D.; Zhang, K.; Pan, D.; Zhang, D. Strengthening effect of in situ TiC particles in Ti matrix composite at temperature range for hot working. *Mater. Charact.* **2016**, *120*, 304–310. [CrossRef]

24. Ma, F.; Wang, T.; Liu, P.; Li, W.; Liu, X.; Chen, X.; Pan, D.; Lu, W. Mechanical properties and strengthening effects of in situ (TiB+TiC)/Ti-1100 composite at elevated temperatures. *Mater. Sci. Eng. A* **2016**, *654*, 352–358. [CrossRef]

25. Zhao, Q.; Liang, Y.; Zhang, Z.; Li, X.; Ren, L. Effect of Al content on impact resistance behavior of Al-Ti-B_4C composite fabricated under air atmosphere. *Micron* **2016**, *91*, 11–21. [CrossRef] [PubMed]

26. Zhang, J.; Lee, J.-M.; Cho, Y.-H.; Kim, S.-H.; Yu, H. Effect of the Ti/B_4C mole ratio on the reaction products and reaction mechanism in an Al–Ti–B_4C powder mixture. *Mater. Chem. Phys.* **2014**, *147*, 925–933. [CrossRef]

27. Schmidt, J.; Boehling, M.; Burkhardt, U. Preparation of titanium diboride TiB_2 by spark plasma sintering at slow heating rate. *Sci. Technol. Adv. Mater.* **2007**, *8*, 376–382. [CrossRef]

28. Kondoh, K. 16—Titanium metal matrix composites by powder metallurgy (PM) routes. In *Titanium Powder Metallurgy*; Qian, M., Froes, F.H., Eds.; Butterworth-Heinemann: Oxford, UK, 2015; pp. 277–297.

29. Montealegre-Meléndez, I.; Neubauer, E.; Arévalo, C.; Rovira, A.; Kitzmantel, M. Study of Titanium Metal Matrix Composites Reinforced by Boron Carbides and Amorphous Boron Particles Produced via Direct Hot Pressing. *Key Eng. Mater.* **2016**, *704*, 85–93. [CrossRef]

30. Montealegre-Meléndez, I.; Arévalo, C.; Perez-Soriano, E.M.; Kitzmantel, M.; Neubauer, E. Microstructural and XRD Analysis and Study of the Properties of the System Ti-TiAl-B_4C Processed under Different Operational Conditions. *Metals* **2018**, *8*, 367. [CrossRef]

31. ASM-International. *Nondestructive Evaluation and Quality Control*, 9th ed.; ASM-International: Materials Park, OH, USA, 1989.

32. ASTM C373-14. *Standard Test Method for Water Absorption, Bulk Density, Apparent Porosity, and Apparent Specific Gravity of Fired Whiteware Products, Ceramic Tiles, and Glass Tiles*; ASTM International: West Conshohocken, PA, USA, 2014.

33. Okulov, A.V.; Volegov, A.S.; Weissmüller, J.; Markmann, J.; Okulov, I.V. Dealloying-Based Metal-Polymer Composites for Biomedical Applications. *Scr. Mater.* **2018**, *146*, 290–294. [CrossRef]

34. Okulov, I.V.; Okulov, A.V.; Volegov, A.S.; Markmann, J. Tuning Microstructure and Mechanical Properties of Open Porous TiNb and TiFe Alloys by Optimization of Dealloying Parameters. *Scr. Mater.* **2018**, *154*, 68–72. [CrossRef]

35. Okulov, I.V.; Sarmanova, M.F.; Volegov, A.S.; Okulov, A.; Kühn, U.; Skrotzki, W.; Eckert, J. Effect of Boron on Microstructure and Mechanical Properties of Multicomponent Titanium Alloys. *Mater. Lett.* **2015**, *158*, 111–114. [CrossRef]

36. Okulov, I.V.; Bönisch, M.; Okulov, A.V.; Volegov, A.S.; Attar, H.; Ehtermam-Haghighi, S.; Calin, M.; Wang, Z.; Hohenwarter, A.; Kaban, I.; et al. Phase Formation, Microstructure and Deformation Behavior of Heavily Alloyed TiNb- and TiV-Based Titanium Alloys. *Mater. Sci. Eng. A* **2018**, *733*, 80–86. [CrossRef]

Article

Biocompatibility and Cellular Behavior of TiNbTa Alloy with Adapted Rigidity for the Replacement of Bone Tissue

Mercè Giner [1,*], Ernesto Chicardi [2,*], Alzenira de Fátima Costa [3], Laura Santana [2], María Ángeles Vázquez-Gámez [4], Cristina García-Garrido [5], Miguel Angel Colmenero [3], Francisco Jesús Olmo-Montes [3], Yadir Torres [2] and María José Montoya-García [4]

1   Departamento de Citología e Histología Normal y Patológica, Universidad de Sevilla, 41009 Sevilla, Spain
2   Departamento de Ingeniería y Ciencia de los Materiales y del Transporte, Escuela Politécnica Superior, Calle Virgen de África 7, 41011 Sevilla, Spain; lausanroc@alum.us.es (L.S.); ytorres@us.es (Y.T.)
3   Servicio de Medicina Interna, HUV Macarena, Avda Sánchez Pizjuán s/n, 41009 Sevilla, Spain; alzenira.costa@juntadeandalucia.es (A.d.F.C.); mangel.colmenero.sspa@juntadeandalucia.es (M.A.C.); franciscoj.olmo.sspa@juntadeandalucia.es (F.J.O.-M.)
4   Departamento de Medicina, Universidad de Sevilla, Avda. Dr. Fedriani s/n, 41009 Sevilla, Spain; mavazquez@us.es (M.Á.V.-G.); pmontoya@us.es (M.J.M.-G.)
5   Instituto Andaluz del Patrimonio Histórico (IAPH), Camino de los Descubrimientos s/n., 41092 Sevilla, Spain; cristina.g.garrido@juntadeandalucia.es
*   Correspondence: mginer@us.es (M.G.); echicardi@us.es (E.C.)

Abstract: In this work, the mechanical and bio-functional behavior of a TiNbTa alloy is evaluated as a potential prosthetic biomaterial used for cortical bone replacement. The results are compared with the reference Ti c.p. used as biomaterials for bone-replacement implants. The estimated mechanical behavior for TiNbTa foams was also compared with the experimental Ti c.p. foams fabricated by the authors in previous studies. A TiNbTa alloy with a 20–30% porosity could be a candidate for the replacement of cortical bone, while levels of 80% would allow the manufacture of implants for the replacement of trabecular bone tissue. Regarding biocompatibility, in vitro TiNbTa, cellular responses (osteoblast adhesion and proliferation) were compared with cell growth in Ti c.p. samples. Cell adhesion (presence of filopodia) and propagation were promoted. The TiNbTa samples had a bioactive response similar to that of Ti c.p. However, TiNbTa samples show a better balance of bio-functional behavior (promoting osseointegration) and biomechanical behavior (solving the stress-shielding phenomenon and guaranteeing mechanical resistance).

Keywords: prosthetic material; TiNbTa; osteoblastic cells; cellular adhesion; porous titanium; mechanical behavior

Citation: Giner, M.; Chicardi, E.; Costa, A.d.F.; Santana, L.; Vázquez-Gámez, M.Á.; García-Garrido, C.; Colmenero, M.A.; Olmo-Montes, F.J.; Torres, Y.; Montoya-García, M.J. Biocompatibility and Cellular Behavior of TiNbTa Alloy with Adapted Rigidity for the Replacement of Bone Tissue. *Metals* **2021**, *11*, 130. https://doi.org/10.3390/met11010130

Received: 11 December 2020
Accepted: 2 January 2021
Published: 11 January 2021

Publisher's Note: MDPI stays neutral with regard to jurisdictional claims in published maps and institutional affiliations.

1. Introduction

The increase in life expectancy and population aging generate increasing demand for the repair of worn or degenerated tissues, for the restoration of organ functions, and esthetic restoration. Among these, bone tissue is one of the most popular [1]. Replacement medical systems require materials with essential properties that enable interaction with bone. These materials must be resistant to corrosion and wear and have high hardness and ductility. The proper selection of the biomaterial is the key factor for the long-term success of the implant. Hitherto, titanium has been the material of choice for intraosseous applications due to its excellent osseointegration capacity, biocompatibility, high resistance to corrosion, a modulus of elasticity compatible with that of bone, and very good tolerance for soft tissues [2,3]. Its anchorage to bone tissue is possible thanks to the oxide layer formed in the material when passivated [4].

Pure commercial titanium (Ti c.p.) and the Ti-6Al-4 V alloy are currently largely used for bone implants. However, both elements present a high elastic modulus, E, (100–112 GPa) compared to the elastic modulus of cortical bone (15–25 GPa) and trabecular bone (less

203

than 1 GPa) [4,5] which produces the effect of Stress-Shielding [6]. This phenomenon arises since the stiffness of the bone-implant material is greater than the stiffness of the bone itself, which causes the whole load to fall on the bone implant. This imbalance in mechanical loads influences bone remodeling and increases bone formation in the heaviest areas and decreases it in areas with no gravitation.

The areas with fewer stimuli present lower bone density, which favors premature fractures and/or the loosening of the implant [7]. Currently, research is aimed at reducing the elastic modulus of the material. One approach to solve this issue involves the design and manufacture of porous metallic biomaterials that combine the low stiffness of porous structures [8,9]. It should also be borne in mind that the higher the porosity, the lower the mechanical strength and wear resistance of the porous structure. Another approach to address this specific issue in titanium alloys is to reduce the stiffness by designing alloys with low Young's modulus and the high strength of metallic alloys [10]. Two generations of Ti alloys have been developed for biomedical applications [3,11]. The first generation presents both α and (α + β) phases, while the second generation of Ti alloys is called the β-phase Ti-based alloys [10–12]. In the first, depending on the alloying elements, a cytotoxic effect must be taken into account [13]. For example, Ti, Zr, Ta, and Pd are elements of low cytotoxicity [13], while Al, V, Ni, and Co have been reported as cytotoxic [13]. On the other hand, the other generation (β-phase Ti-based alloys) have been preferred mainly due to their lower elastic modulus [14]. In these, the presence of β-phase is promoted by incorporating β-stabilizing elements, such as Pd, Ni, Mo, Ta, V, Mn, and Nb [15]. Several authors have identified the biological impact of a number of those elements on human health: Ti, Nb, Ta, Mn, Zr, and Ru are reported to have been found to exhibit the best bio-functional behavior [13,14], while the Ta, Nb, and Zr decrease the elastic modulus of this type of alloys and increase their mechanical strength [13–20]. Novel beta alloys based on this concept are the commercial TLM alloy (Ti-25Nb-3Zr-3Mo-2Sn) [21], the quaternary Ti-Nb-Sn-Cr [22], Ti-Nb-Zr-Cr [23] and Ti-Nb-Zr-Ta alloys systems [24], among others [3,25]. All of them showed elastic modulus still higher than the cortical bone tissue (minimum value reached around 45 GPa).

Specifically, Chicardi et al. [26,27] have recently fabricated a (β + γ)-TiNbTa material (bcc and innovative structures for Ti alloys, for α and β phases, respectively), with physico-chemical properties more similar to those of the bones and with Young's modulus more similar to cortical bone. In particular, this TiNbTa material has been shown to decrease the elastic modulus (49 + 3 GPa) and to maintain excellent yield strength (σy > 1860 MPa) with expected high biocompatibility [28], which places it among the ideal materials to prevent the *Stress-Shielding* phenomena by the two aforementioned synergic ways to solve this issue: (a) the formation of partial β (bcc) phase with low stiffness instead of the elimination of the α- phase, with high stiffness and (b) the introduction of porosity to further decrease its elastic modulus to approximate it to the cortical and/or trabecular bone without the inadmissible sudden drop in the yield strength.

However, the characterization of this material is still incomplete as regards its validation for use as a metallic biomaterial for bone grafts or implants, which constitutes the main objective of this work. It remains necessary to compare this material with the widely employed Ti c.p. and its foams by evaluating the potential mechanical properties of TiNbTa foams, the osseointegration capacity of this new TiNbTa-based biomaterial, and by determining its cytotoxicity, adhesion capacity, proliferation, and differentiation on osteoblastic cells.

2. Materials and Methods

2.1. Development and Mechanical Characterization of the TiNbTa Potential Prosthetic Biomaterial

The new potential TiNbTa material was developed from the elemental Ti, Nb, and Ta powder transition metals as raw materials using a Powder Metallurgy route, including the mechanical alloying (MA) and subsequent consolidation of the as-synthesized TiNbTa alloy by spark plasma sintering (SPS). A schema of the route used is displayed in Figure 1.

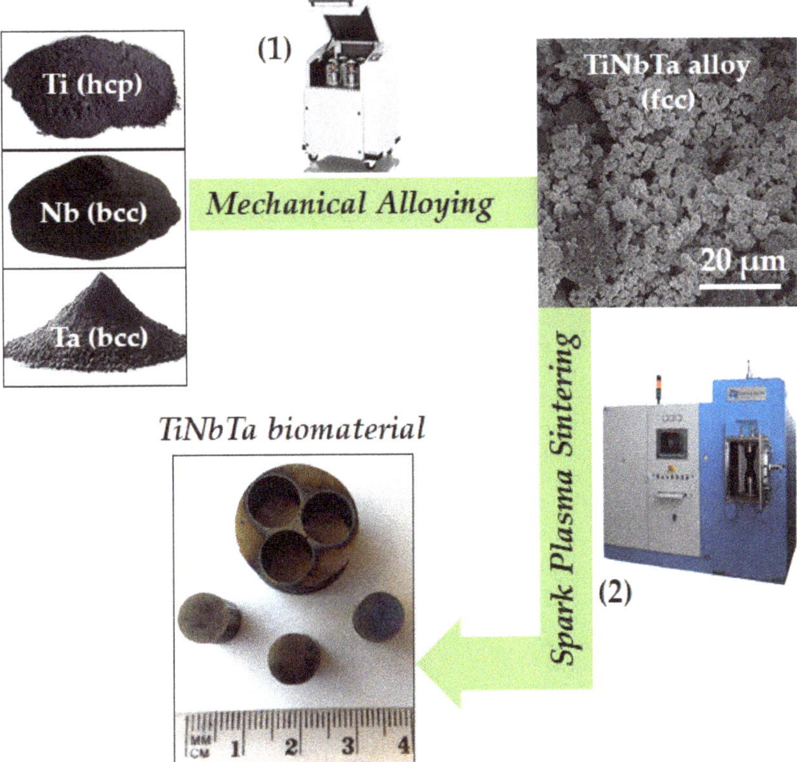

Figure 1. Schema of the powder metallurgy route used for the development of the new potential prosthetic TiNbTa material, showing the evolution of the material from raw elements (Ti, Nb, and Ta) to the synthesized TiNbTa alloy, and, finally, to the fabricated potential prosthetic TiNbTa biomaterial. (1) Image of the PM400 planetary ball mill, courtesy of Retsch. (2) Image of the HPD 25 spark plasma sintering (SPS) device, courtesy of FCT Systeme GmbH.

Specifically, the Ti, Nb, and Ta powders, with particle size < 325 mesh and purity > 99.6%, were supplied by NOAH tech, San Antonio, TX, USA. Therefore, 50 g of the mix of powders, with a nominal composition of 57% of Ti, 30% of Nb, and 13% of Ta, were milled in a PM400 planetary ball mill supplied by Retsch GmbH. The optimized milling conditions were a spinning rate of 250 rpm; a ball-to-powder ratio (BPR) of 10:1; yttria-stabilized zirconia (YSZ) balls and vial (500 mL in volume); an Ar atmosphere to prevent the oxidation of metals during milling; and 60 h of milling time.

The as-synthesized TiNbTa alloy was consolidated with cylindrical geometry (20 mm in diameter and 10 mm in height) by spark plasma sintering (SPS) using a commercial HPD 25 spark plasma sintering (SPS) device, fabricated by FCT Systeme GmbH. The sintering conditions included a dwell temperature of 1400 °C, a heating rate of 500 °C/min, no dwell time, and a free in-air cooling rate (calculated as approximately 700 °C/min). An in-depth description of the processing route can be found elsewhere [27].

The newly developed TiNbTa potential prosthetic material was microstructurally characterized [26], and the mechanical behavior, biocompatibility, and osseointegration were studied in this work. For mechanical behavior, two key parameters were used for a prosthetic biomaterial performance: the elastic modulus (E) and the yield strength (σy). These were determined from a stress–strain curve obtained in a uniaxial compression test. This test was carried out on five reproduced TiNbTa specimens in a universal mechanical testing machine (Model 6025, by Instron) and, in accordance with the ASTM E9-09 standard, with a displacement rate of 0.05 mm/min and a maximum load of 100 kN. The specimens were tested, of 8 mm in diameter and 10 mm in height, in accordance with the "short solid cylindrical specimens" in the ASTM E9-09, were machined from the aforementioned cylinders (20 mm in diameter and 10 mm in height, see Figure 1) by the wire electro-erosion process in electrical discharge machining (EDM).

On the other hand, the elastic modulus (E) and the yield strength (σy) for porous TiNbTa materials, with porosities from 10 up to 90 vol %, were estimated from the E and σy values corresponding to the fully dense TiNbTa presented and the empirical Mondal equation [29].

2.2. In Vitro Cell Experiments

MC3T3E1, a murine pre-osteoblast cell line (CRL-2593, from ATCC), was utilized to analyze the TiNbTa effect on cell metabolism and viability during the cell adhesion and proliferation process.

The different techniques employed to evaluate cellular response are described in Figure 2.

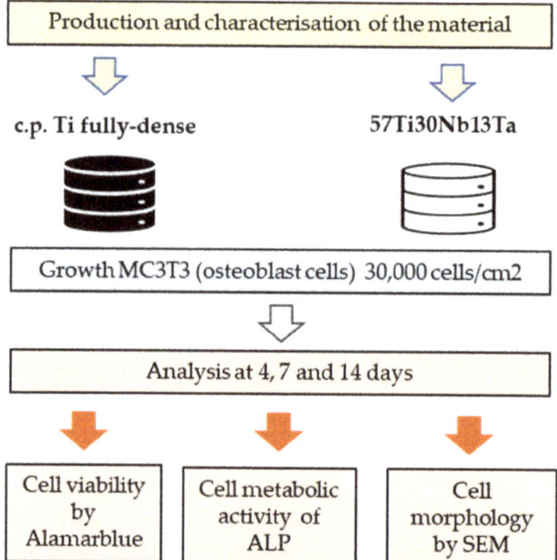

Figure 2. Diagram of in vitro experimental design. Osteoblast cell line (MC3T3E1) was used to analyze the biological response to fully dense Ti c.p. vs. TiNbTa samples.

2.3. Cell Culture

MC3T3E1, a murine pre-osteoblast cell line (CRL-2593, from ATCC), was employed to analyze the TiNbTa effect on cell metabolism and viability during the cell adhesion and proliferation process. Routine passaging of the cell line was performed on 25 cm^2 flasks with minimum essential medium (MEM), containing 10% fetal bovine serum plus antibiotics (100 U/mL penicillin and 100 mg/mL streptomycin sulfate) (Invitrogen). Sample discs were autoclaved at 121 °C for 30 min and were then placed onto a 24-well plate. Osteoblast cells were seeded at a cell density of 30,000 cells/cm^2 per sample and in 800 µL of prewarmed culture medium. Plates were kept at 37 °C in a humidified 5% CO^2 atmosphere, and as a control, triplicate blank TCP (tissue culture plastic) and fully dense pure titanium discs were employed as negative and positive controls on the same plate for each time of period.

At 48 h of osteoblast cultured on TiNbTa and fully dense Ti c.p. samples, the cell media was changed to osteogenic media (α-MEM medium) supplemented with 10 mM of ascorbic acid (Merck, Germany) and 50 µg/mL of β-glycerophosphate (StemCell Technologies,

Canada). The in vitro cell experiments were carried out at 4, 7, and 14 days of cell incubation in which the samples were transferred onto a new 24-well plate to prevent counting non-attached or attached cells on the well plate.

2.4. Cell Viability and Proliferation Assay

Cell proliferation and viability tests were evaluated using alamarBlue® reagent (Invitrogen). In accordance with the manufacturer's protocol, new fresh media (800 µL) and 80 µL of alamarBlue® reagent were added, and the plate was incubated for 1 h 30 min at 37 °C in dark conditions. The absorbance at 570 nm (TECAN, Infinity 200 Pro) was subsequently recorded.

2.5. Cell Differentiation by Alkaline Phosphatase (ALP) Evaluation

MC3T3 differentiation levels were evaluated through alkaline phosphatase (ALP) activity, using the alkaline phosphatase assay kit (Colorimetric) (Abcam ab83369, UK). The assay was performed on days 4, 7, and 14 in triplicate according to the manufacturer's protocol. The absorbance at 405 nm of 4-nitrophenol was measured in a 96-well microplate reader. Data are expressed as µmol/min/mL of pNPP.

2.6. Cell Morphology

Scanning electron microscopy (SEM) was employed to evaluate cell behavior at 4, 7, and 14 days. The samples were fixed in 10% formalin, which was followed by a dehydration step with ethanolic solutions and then coated by gold-coating using a sputter coater (Pelco 91000, Ted Pella, Redding, CA, USA). The images were obtained using a Jeol JSM-6330 F scanning electron microscope (JEOL, Tokyo, Japan) with an acceleration voltage of 10 kV.

2.7. Statistical Analysis

All experiments were performed in triplicate in order to ensure reproducibility. The results were expressed in terms of the mean and standard deviation to perform two-way ANOVA, followed by Tukey's posttest using SPSS v.22.0 for Windows. The significance level was considered at p values of $p < 0.05$ (*) and $p < 0.01$ (**).

3. Results and Discussion

The following section presents the results related to the mechanical and cellular behavior of the TiNbTa-based potential prosthetic biomaterial developed, comparing it with Ti c.p. (porous and fully dense). Therefore, the study of cellular response correlated to the mechanical properties of TiNbTa could be assessed.

3.1. Mechanical Behavior of the TiNbTa Potential Prosthetic Biomaterial and Comparison with Actual Biomaterials

The strain-stress curve obtained for the fully dense TiNbTa specimen from the compression tests gave elastic modulus (E) and yield strength (σy) values of 49 ± 3 GPa and 1860 ± 20 MPa, respectively (Table 1), as previously reported. This mechanical behavior is, a priori, interesting for the use of this material as a prosthetic biomaterial for bone tissue replacement. Not in vain, although the elastic modulus value obtained is similar, as an example, to another beta (bcc) Ti alloy already developed by Zhang et al. [30], concretely, the accepted as biomaterial Ti24Nb4Zr8Sn alloy (E = 53 GPa), the yield strength is much higher in comparison with this alloy (σy ~800 MPa). In addition, although the elastic modulus seems to remain elevated in comparison with the cortical (15–25 GPa) and cancellous (less than 1 GPa) bone tissues [31], the high yield strength value is shown grants this material the opportunity to adjust the elastic modulus value to both cortical and cancellous bone by the introduction of porosity without an impermissible decrease in yield strength. To this end, both key mechanical properties for metallic biomaterials for implants were estimated for hypothetical porous TiNbTa materials from the values corresponding to the fully dense TiNbTa presented and the empirical Mondal equation [29].

Furthermore, these determined values were compared with the various E and σy values for the titanium foams of previously published biomaterials that were fabricated by the space-holder powder metallurgy route and the fully dense Ti c.p. used as biomaterials for implants. Thus, both sets of serial data (E and σy for the TiNbTa and the fully dense Ti c.p. and foam materials) are compared in Figure 3. It can be observed how it is possible to reach Young's modulus corresponding to the cortical and even to the cancellous bone tissue through the introduction of a range of 20–30 vol % and more than 80 vol % of porosity, respectively, on the two materials, that is, for the new TiNbTa material and the currently used Ti c.p. (Figure 3a). However, regarding the damage to the yield strength for both materials, the introduction of porosity in the TiNbTa materials enables a higher value of σy to remain in comparison with the Ti c.p. (Figure 3b) and significantly higher than that of the cortical bone (227 MPa). This is a positive aspect that prevents or delays the plastic deformation of the TiNbTa in comparison with Ti c.p. for the same Young's modulus.

Table 1. Total volumetric porosity (p), elastic modulus (E), and yield strength (σy) for (a) the developed fully dense TiNbTa material and the corresponding TiNbTa foams, with volumetric porosities from 10 to 70 vol %, calculated by the Mondal equation [29]; (b) the Ti foams developed by the space-holder, also with volumetric porosities from 10 to 70 vol % and marked as "a" or "b" when NaCl or NH_4HCO_3 was used as the space-holder, respectively; (c) commercial Ti c.p., commercial Ti6Al4V [4]; (d) cortical and cancellous bone tissues [5,31].

Specimen	Total Porosity (p,%)	E (GPa)	σ_y (MPa)
TiNbTa_0	0	48.0	1860
TiNbTa_10	10	35.0	1357
TiNbTa_20	20	25.2	976
TiNbTa_30	30	17.7	685
TiNbTa_40	40	12.0	465
TiNbTa_50	50	7.7	300
TiNbTa_60	60	4.6	179
TiNbTa_70	70	2.4	95
TiNbTa_80	80	1.0	40
TiNbTa_90	90	0.2	9
Ti_60a	57	8.1	16
Ti_70a	64	3.5	42
Ti_30b	28	15.9	389
Ti_40b	38	5.8	272
Ti_50b	45	8.5	192
Ti_60b	55	3.7	112
Ti_70b	63	4.6	57
Ti c.p.	0	105	480
Ti6Al4V	0	110	870
Cortical bone tissue	~0	[15–25]	[80–200] *
Cancellous bone tissue	>30	<1	[1–9] *

* ultimate tensile strength (σUTS).

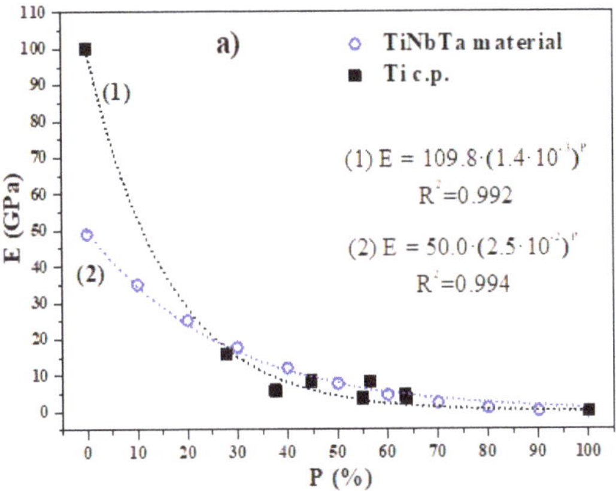

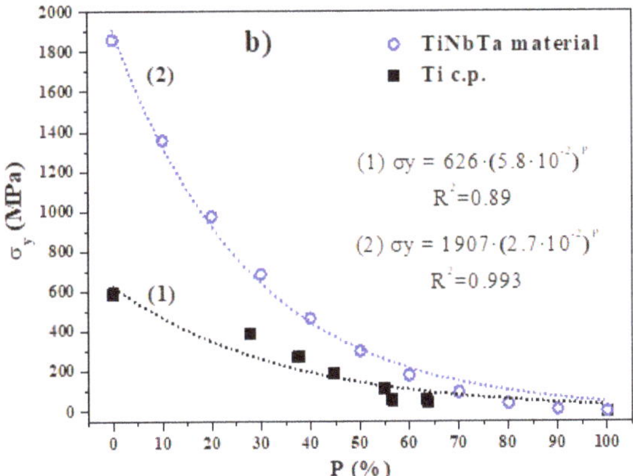

Figure 3. (**a**) Elastic modulus (E) and (**b**) yield strength (σy) for both TiNbTa material (○) and Ti c.p. (v) foams in terms of the porosity introduced in the specimen. Note that the values for TiNbTa are determined by the Mondal equation, except the experimental values obtained from the compression test for fully dense TiNbTa, and that the values for Ti c.p. are experimentally obtained in previous work. The equations displayed correspond to the best fit (exponential curves) to the serial data.

Subsequently, Figure 4 shows the double logarithmic E-σy curves for the Ti c.p. and TiNbTa fully dense and foam materials. Moreover, the same values for the Ti6Al4V biomaterial and the interval values for the cortical and cancellous bone tissues were introduced for comparison purposes. It should be borne in mind that the main objective of this linearized figure is to maximize the index $\sigma y/E$, that is, minimize the E and, consequently, diminish the potential detrimental "Stress-Shielding" phenomena of a possible implant fabricated with these materials, but with a maximization of the yield strength (σy), thereby avoiding the plastic deformation of the implant. Thus, after linearization, the index $\sigma y/E = C$ evolves to the line equation log σy = log (E) + log C. Using a line with a slope equal to 1 (marked in Figure 4 as a dotted line), it is possible to maximize the ordinate at the origin (log (C)), and, therefore the index $C = \sigma y/E$, as desired. With this reasoning, the specimens that optimize this index $\sigma y/E$ for the cortical bone are those named either TiNbTa_20 or TiNbTa_30, that is, the TiNbTa foams with 20 and 30 vol % of porosity, respectively. This assertion is in concordance with the optimized specimens selected attending to the relationship porosity-E and porosity-(σy) (TiNbTa_20 and TiNbTa_30) as determined in

Figure 3. However, a higher amount of porosity (10% more) could be introduced for TiNbTa_30, thereby enabling improvements in the cells' ingrowth to the implant without any detrimental effect on mechanical behavior; these last specimens could be accepted as the optimized materials for cortical bone tissue replacement. Furthermore, for both specimens, the yield strength obtained is sufficient in comparison with the maximum yield strength of the cortical bone tissue. It should be taken into consideration that, although the Ti_40b seems to have a higher $\sigma_y/E = C$ than the cortical bone, its lower E value in comparison with the cortical bone could affect the mechanical behavior of the implant with a non-acceptable elastic deformation.

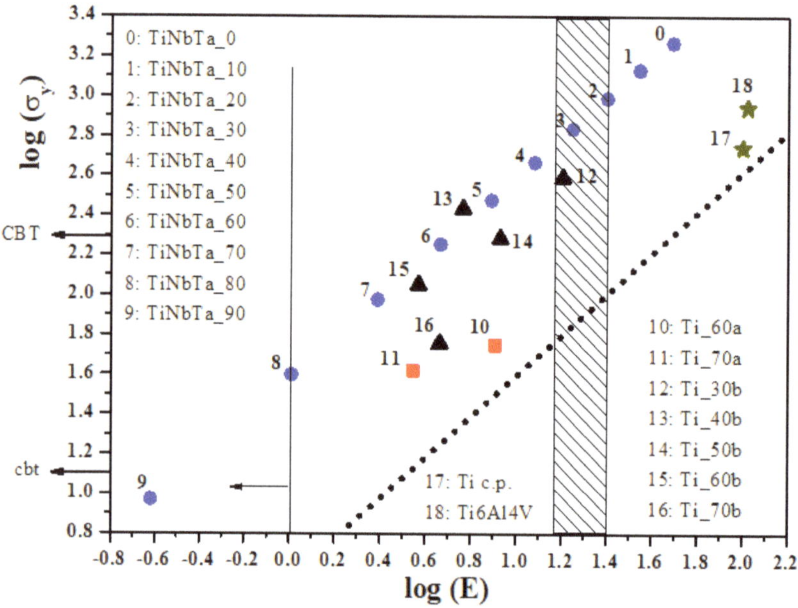

Figure 4. Comparison between elastic modulus (E) and yield strength (σ_y) for: (**a**) the developed fully dense TiNbTa material and the corresponding porous TiNbTa (•); (**b**) Ti foams (fabricated by the space-holder, using sodium chloride (NaCl) (v), and ammonium bicarbonate (NH_4HCO_3) (▲); (**c**) commercial Ti c.p. (number 17) and commercial Ti6Al4V (number 18). The striped column marks the limits between the interval E values for cortical bone tissues. In turn, the E values for cancellous bones (less than 1 GPa) are marked with the continuous line and the arrow to the left. CBT and cbt: maximum value of yield strength for cortical (CBT) and cancellous (cbt) bone tissues, respectively. In addition, for comparison purposes, the dotted line marks the slope line equal to 1 to observe the specimens that maximize the index σ_y/E.

On the other hand, related to the selection of an optimal material to replace the cancellous bone tissue, only specimen TiNbTa_80 accomplishes the requirements (<1 GPa and 9 MPa for E and σ_y for the cancellous bone tissue). Specifically, Young's modulus is equal to 1 GPa, and 40 MPa of yield strength was calculated (see Table 1). However, the high level of porosity (80 vol %) could complicate the manufacturing of this potential biomaterial to replace the cancellous bone by means of powder metallurgy routes.

3.2. Biocompatibility of the TiNbTa Potential Prosthetic Biomaterial

MC3T3 cells were seeded on TiNbTa and Ti c.p. substrates to evaluate proliferation, metabolism cells, and cell adhesion for 4, 7, and 14 days. Cell cultures in both samples had similar cell growth. At 7 days of growth, the cells present a columnar and basophilic morphology compatible with the onset of differentiation.

Figure 5 shows the viability of the cell line as a function of the cell growth time (4, 7, and 14 days) on the Ti c.p. and TiNbTa. At 4 and 7 days of culture, the cell viability in the TiNbTa discs is similar to that of the Ti c.p., while at 14 days, a slight decrease in viability in the TiNbTa samples and an increase in the Ti c.p. can be observed, although without significant differences. Significant differences appear between days 7 and 14 of the cells grown on the surface of TiNbTa, in which there is a decrease in proliferation at 14 days. This decrease is to be expected since the differentiation begins on osteoblastic cells, and when differentiating, they lose their proliferation capacity.

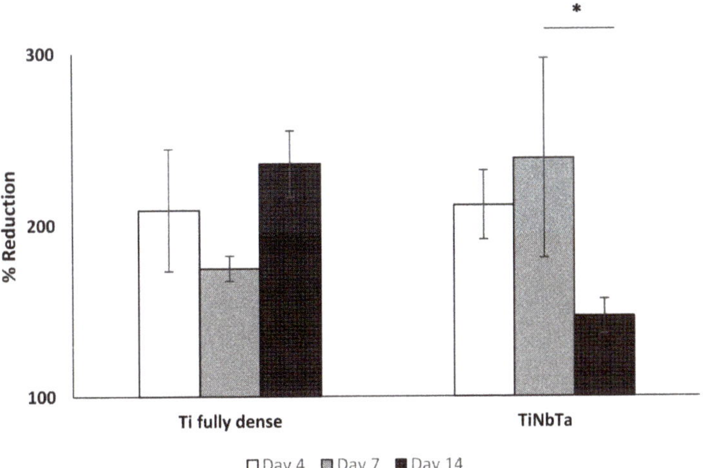

Figure 5. Cell viability of cultured cells from fully dense Ti or TiNbTa at 4, 7, and 14 days of cell growth. The results are presented as mean ± standard deviation, * $p < 0.05$.

Currently, titanium constitutes the most used material in biomedical implants [19], mainly commercially pure titanium and the alloy with titanium, aluminum, and vanadium (Ti-6Al-4 V). Despite its good biocompatibility and resistance to corrosion, there are still certain limitations from the clinical point of view due to complications, such as the toxicity of vanadium and aluminum and the stress-shielding effect [12,20].

Previous studies carried out in fibroblasts demonstrated that the TiNbZrTa alloy does not have a direct or indirect cytotoxic effect on these cells [32], which indicates that it is a safe alloy for use as a biomaterial. It has been shown that Nb and Ta, in addition to not being toxic substances, also inhibit the release of metal ions through the formation of an oxide layer formed by TiO_2 and Nb_2O_5 on the surface of the material [33–35], and this increases their resistance to corrosion [26]. These studies indicate that the use of the TiNbZrTa alloy in the biomedical area is highly recommended due to the low release of metals in vitro and a low elastic modulus, which is very beneficial, especially for long-term implants [36].

Several preliminary studies have shown that the physical and chemical characteristics of the biomaterial surface directly or indirectly affect the adhesion, proliferation, differentiation, and interaction with the cell and consequently the cell environment in the various stages of differentiation in response to the surface of the material [37]. Our assays demonstrated cell viability and proliferation on every surface studied. No differences in cell viability were observed between the two types of the disc at four days of culture. However, in the samples cultured on the TiNbTa discs, a slight increase in cell viability at seven days was observed, but without significant differences (Figure 5). These results are similar to those previously published by Qian Wang et al. [38] in studies carried out on osteoblasts cultured on porous Ta, and they also agree with the results published by Donato et al. [32] in studies carried out on fibroblast cultures.

In all cases, the viability percentage is always greater than 150%. A toxic effect is considered when the viability is less than 75%; therefore, with the results obtained, it can be deduced that cell culture is viable throughout the entire duration of the culture in all conditions, fully agreeing with previous results published by Qian Wang et al. [38] and Shimko et al. [39].

Another important consideration involves the levels of alkaline phosphatase (ALP) secreted by osteoblasts during the osseointegration process. ALP is a glycoprotein abundantly expressed in the extracellular matrix and is considered a marker in the initial stage of osteogenic differentiation [40,41]. Several studies relate the activity of alkaline phosphatase with the process of cell differentiation, together with other specific proteins, such as osteocalcin, osteopontin, osteonectin, and type I collagen [42,43].

Figure 6 represents the alkaline phosphatase (ALP) activity values in MC3T3 cell cultures at 4, 7, and 14 days. In TiNbTa discs, a decrease in activity is also observed with the culture time, and there is a significant difference between days 4 and 7, and also between days 7 and 14 of culture ($p < 0.001$). In Ti c.p. discs, the results obtained show a decrease in enzyme activity, whereby the difference found between days 4 and 7 are statistically significant ($p < 0.05$). In both conditions, at 4 days, the maximum enzymatic activity is obtained since there is a greater proliferation and cell growth as the viability values corroborate. From day 7, the cells present a more differentiated phenotype as can be observed in the SEM images, and the ALP activity decreases.

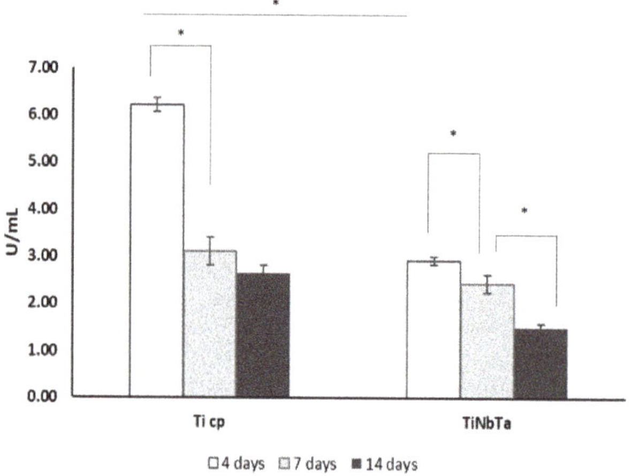

Figure 6. Alkaline phosphatase activity in cell growth on discs of fully dense Ti or TiNbTa at 4, 7, and 14 days. * $p < 0.05$.

According to these results, both discs stimulate cell differentiation. The maximum levels of ALP activity observed at four days of culture suggest a high level of cell differentiation. The decrease in the levels of this protein observed after seven days are compatible with the beginning of the cell differentiation process to a mineralizing phenotype, which supports the fact that ALP is indeed an early marker of cell differentiation. Similar results were observed in other studies in which alkaline phosphatase activity was analyzed in osteoblast cell culture. Boyan et al. [44] also observed a decrease in alkaline phosphatase activity in cells grown on rough surfaces on analyzing Saos2 (Osteogenic Sarcoma) cells, a cell line similar to human osteoblasts.

A preliminary study was conducted to study cell morphology and adhesion on TiNbTa study material compared to growth in Ti c.p. To this end, the samples of both types of discs at 4, 7, and 14 days of culture were visualized. At four days, small cell clusters were observed on both types of surface and of similar density. At seven days, the images demonstrated a monolayer growth on the entire surface of the disc; these results were similar for both materials. In addition, we can begin to observe small morphological changes in the cells grown on Ti cp; they are more stellate cells, while in the alloy, they maintain the fibroblastic shape. Cell-cell and cell-biomaterial junctions were also observed. The cells adhere through filopodia (fine cell projections, yellow arrow) and lamellipodia (wider extensions, yellow asterisk), thus demonstrating the connection of the cells with the biomaterial. At 14 days, the presence of small hexagonal-structure vesicles (red asterisk) can already be observed, which suggests possible nucleation of hydroxyapatite on the cell surface, as an indication of the start of the development of the mineralization process, and a clear difference is seen between the morphology of the cells grown on fully dense Ti c.p. and TiNbTa: in the alloy, the cell shows a more differentiated phenotype with extracellular matrix secretion (Figure 7).

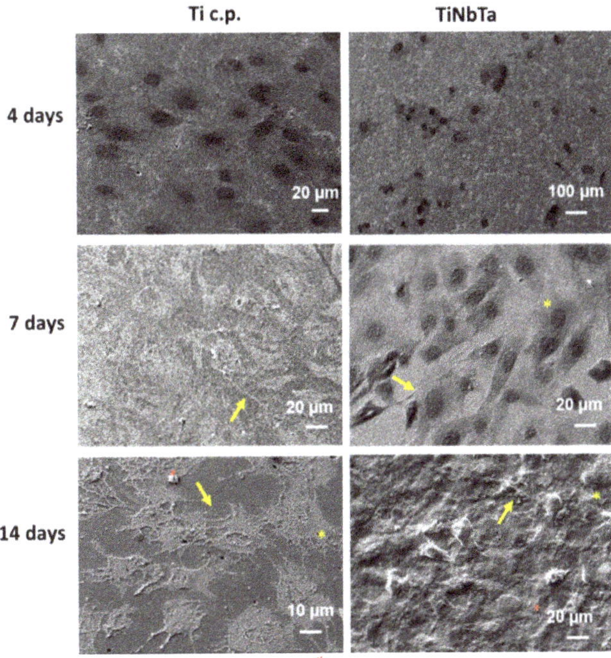

Figure 7. SEM micrographs at 4, 7, and 14 days of osteoblast culture growing on fully dense Ti c.p. or TiNbTa surface. Cell morphology and osteoblast proliferation are observed with filopodia marked by yellow arrows and lamellipodia indicated by yellow asterisks (flattened extension of a cell, by which it adheres to a surface). The small hexagonal-structure vesicles are indicated by a red asterisk, which suggests hydroxyapatite on the cell surface.

From day 7, the cells present a more differentiated phenotype as observed in the SEM images, and the ALP activity decreases. These results confirm those published by Lin-Jie Li et al. in which they observed that, on smooth surfaces, cells exhibited a high proliferation rate; however, the levels of alkaline phosphatase and osteocalcin were low, which suggests a phenotypic change in differentiated osteoblasts [40,44]. In another interesting study, Sungsin Jo et al. [45] observed a delay in the mineralization and differentiation of osteoblastic cells following the elimination of the ALP gene, thereby demonstrating that higher levels of the enzyme are necessary to promote the initiation of differentiation in osteoblasts.

In summary, these results show that the TiNbTa alloy has a good growth capacity and a low elastic modulus, and therefore this material should be considered as a good candidate for use in the biomedical area. However, complementary and in vivo, biological tests remain necessary to fully understand how the TiNbTa alloy affects the osseointegration process and how its use in biomedical applications could be improved.

4. Conclusions

A TiNbTa potential prosthetic biomaterial is developed and evaluated, and its biocompatibility is compared with the current Ti c.p. used as a bone-replacement implant material. The main conclusions obtained can be set out as follows:

- It is possible to obtain a similar Young's modulus and higher yield strength than those of the cortical bone tissue through the development of TiNbTa foams and Ti c.p. foams. These interesting properties are even better for the TiNbTa foams in comparison with the Ti c.p. foams;

- In particular, regarding the $\sigma y/E$ index, the TiNbTa_20 and TiNbTa_30, corresponding to TiNbTa foams with volumetric porosity values of 20 and 30%, respectively, appear to present the optimal potential prosthetic biomaterials for cortical bone replacements (higher index). They also displayed interesting values of E = 25.5 GPa and 17.7 GPa, and σy = 976 Mpa and 685 MPa, respectively. In addition, the high yield strengths help to prevent any plastic deformation for a hypothetical implant fabricated by this material;

- In turn, the mechanical behavior for cancellous bone tissue (E less than 1 GPa and σy = 9 MPa) could also be reached by the TiNbTa foams with an 80 volumetric percentage of porosity (E = 1 GPa and σy = 40 MPa), However, the high porosity makes its manufacture less feasible;

- In vitro tests show the successful adhesion, proliferation, and differentiation of osteoblasts (MC3T3 cell line), thereby confirming the biocompatibility and non-toxicity of TiNbTa samples;

- On the other hand, osteoblasts cultured in the TiNbTa samples showed the same metabolic and alkaline phosphate activity as those cultured in the Ti c.p. samples. Therefore, TiNbTa samples show a bio-functional behavior balance, which promotes in vitro osseointegration and solves the stress-shielding phenomenon.

Author Contributions: Conceptualization, project administration, supervision, methodology, M.G., M.J.M.-G., E.C. and Y.T.; investigation A.d.F.C., L.S., M.Á.V.-G. and C.G.-G.; formal analysis, M.A.C. and F.J.O.-M. validation, M.G. and E.C.; discussion and writing—original draft preparation, all the authors. All authors have read and agreed to the published version of the manuscript.

Funding: This work was supported by the Ministry of Science and Innovation of Spain under the grant PID2019-109371GB-I00 and by the Junta de Andalucía—FEDER (Spain) through the Project Ref. US-1259771.

Institutional Review Board Statement: Not applicable.

Informed Consent Statement: Not applicable.

Data Availability Statement: The data presented in this study are available on request from the corresponding author. The data are not publicly available due to privacy.

Conflicts of Interest: Authors declare no conflict of interest.

References

1. Holzapfel, B.M.; Reichert, J.C.; Schantz, J.-T.; Gbureck, U.; Rackwitz, L.; Nöth, U.; Jakob, F.; Rudert, M.; Groll, J.; Hutmacher, D.W. How smart do biomaterials need to be? A translational science and clinical point of view. *Adv. Drug Deliv. Rev.* **2013**, *65*, 581–603. [CrossRef] [PubMed]
2. Li, Y.; Yang, C.; Zhao, H.; Qu, S.; Li, X.; Li, Y. New developments of ti-based alloys for biomedical applications. *Materials* **2014**, *7*, 1709–1800. [CrossRef] [PubMed]
3. Geetha, M.; Singh, A.K.; Asokamani, R.; Gogia, A.K. Ti based biomaterials, the ultimate choice for orthopaedic implants—A review. *Prog. Mater. Sci.* **2009**, *54*, 397–425. [CrossRef]
4. Zhang, L.C.; Chen, L.Y. A Review on Biomedical Titanium Alloys: Recent Progress and Prospect. *Adv. Eng. Mater.* **2019**, *21*, 1801215. [CrossRef]

5. Rho, J.Y.; Ashman, R.B.; Turner, C.H. Young's modulus of trabecular and cortical bone material: Ultrasonic and microtensile measurements. *J. Biomech.* **1993**, *26*, 111–119. [CrossRef]
6. Muñoz, S.; Castillo, S.M.; Torres, Y. Different models for simulation of mechanical behaviour of porous materials. *J. Mech. Behav. Biomed. Mater.* **2018**, *80*, 88–96. [CrossRef]
7. Kourtis, S.; Damanaki, M.; Kaitatzidou, S.; Kaitatzidou, A.; Roussou, V. Loosening of the fixing screw in single implant crowns: Predisposing factors, prevention and treatment options. *J. Esthet. Restor. Dent.* **2017**, *29*, 233–246. [CrossRef]
8. Trueba, P.; Chicardi, E.; Rodríguez-Ortiz, J.A.; Torres, Y. Development and implementation of a sequential compaction device to obtain radial graded porosity cylinders. *J. Manuf. Process.* **2020**, *50*, 142–153. [CrossRef]
9. Wang, X.; Xu, S.; Zhou, S.; Xu, W.; Leary, M.; Choong, P.; Qian, M.; Brandt, M.; Xie, Y.M. Topological design and additive manufacturing of porous metals for bone scaffolds and orthopaedic implants: A review. *Biomaterials* **2016**, *83*, 127–141. [CrossRef]
10. Zhou, L.; Yuan, T.; Tang, J.; He, J.; Li, R. Mechanical and corrosion behavior of titanium alloys additively manufactured by selective laser melting—A comparison between nearly β titanium, α titanium and α + β titanium. *Opt. Laser Technol.* **2019**, *119*, 105625. [CrossRef]
11. Pilliar, R.M. Metallic biomaterials. In *Biomedical Materials*; Springer: Boston, MA, USA, 2009; pp. 41–81.
12. Abdel-Hady Gepreel, M.; Niinomi, M. Biocompatibility of Ti-alloys for long-term implantation. *J. Mech. Behav. Biomed. Mater.* **2013**, *20*, 407–415. [CrossRef] [PubMed]
13. Biesiekierski, A.; Wang, J.; Abdel-Hady Gepreel, M.; Wen, C. A new look at biomedical Ti-based shape memory alloys. *Acta Biomater.* **2012**, *8*, 1661–1669. [CrossRef]
14. Cui, C.; Hu, B.M.; Zhao, L.; Liu, S. Titanium alloy production technology, market prospects and industry development. *Mater. Des.* **2010**, *32*, 1684–1691. [CrossRef]
15. Kim, E.S.; Jeong, Y.H.; Choe, H.C.; Brantley, W.A. Formation of titanium dioxide nanotubes on Ti-30Nb-xTa alloys by anodizing. *Thin Solid Film.* **2013**, *549*, 141–146. [CrossRef]
16. Schmidt, R.; Pilz, S.; Lindemann, I.; Damm, C.; Hufenbach, J.; Helth, A.; Geissler, D.; Henss, A.; Rohnke, M.; Calin, M.; et al. Powder metallurgical processing of low modulus β-type Ti-45Nb to bulk and macro-porous compacts. *Powder Technol.* **2017**, *322*, 393–401. [CrossRef]
17. Chicardi, E.; Aguilar, C.; Sayagués, M.J.; García-Garrido, C. Influence of the Mn content on the TiNbxMn alloys with a novel fcc structure. *J. Alloys Compd.* **2018**, *746*, 601–610. [CrossRef]
18. Shen, X.; Zhang, Y.; Jiang, Y.; Zhou, R. Corrosion behavior of Ti-35Nb-7Zr-5Ta alloy prepared by spark plasma sintering in Hank's artificial body fluid. *Corros. Sci. Prot. Technol.* **2016**, *28*, 543–548.
19. Stráský, J.; Harcuba, P.; Václavová, K.; Horváth, K.; Landa, M.; Srba, O.; Janeček, M. Increasing strength of a biomedical Ti-Nb-Ta-Zr alloy by alloying with Fe, Si and O. *J. Mech. Behav. Biomed. Mater.* **2017**, *71*, 329–336. [CrossRef]
20. Liu, Q.; Meng, Q.; Guo, S.; Zhao, X. α′ Type Ti–Nb–Zr alloys with ultra-low Young's modulus and high strength. *Prog. Nat. Sci. Mater. Int.* **2013**, *23*, 562–565. [CrossRef]
21. Zhentao, Y.; Lian, Z. Influence of martensitic transformation on mechanical compatibility of biomedical β type titanium alloy TLM. *Mater. Sci. Eng. A* **2006**, *438–440*, 391–394. [CrossRef]
22. Jawed, S.F.; Rabadia, C.D.; Liu, Y.J.; Wang, L.Q.; Li, Y.H.; Zhang, X.H.; Zhang, L.C. Mechanical characterization and deformation behavior of β-stabilized Ti-Nb-Sn-Cr alloys. *J. Alloys Compd.* **2019**, *792*, 684–693. [CrossRef]
23. Jawed, S.F.; Rabadia, C.D.; Liu, Y.J.; Wang, L.Q.; Li, Y.H.; Zhang, X.H.; Zhang, L.C. Beta-type Ti-Nb-Zr-Cr alloys with large plasticity and significant strain hardening. *Mater. Des.* **2019**, *181*, 108064. [CrossRef]
24. Samuel, S.; Nag, S.; Nasrazadani, S.; Ukirde, V.; El Bouanani, M.; Mohandas, A.; Nguyen, K.; Banerjee, R. Corrosion resistance and in vitro response of laser-deposited Ti-Nb-Zr-Ta alloys for orthopedic implant applications. *J. Biomed. Mater. Res.* **2010**, *94*, 1251–1256. [CrossRef] [PubMed]
25. Kaur, M.; Singh, K. Review on titanium and titanium-based alloys as biomaterials for orthopaedic applications. *Mater. Sci. Eng. C* **2019**, *102*, 844–862. [CrossRef] [PubMed]
26. Chicardi, E.; Gutiérrez-González, C.F.; Sayagués, M.J.; García-Garrido, C. Development of a novel TiNbTa material potentially suitable for bone replacement implants. *Mater. Des.* **2018**, *145*, 88–96. [CrossRef]
27. García-Garrido, C.; Gutiérrez-González, C.; Torrecillas, R.; Pérez-Pozo, L.; Salvo, C.; Chicardi, E. Manufacturing optimisation of an original nanostructured (beta + gamma)-TiNbTa material. *J. Mater. Res. Technol.* **2019**, *8*, 2573–2585. [CrossRef]
28. Hussein, A.H.; Gepreel, M.A.H.; Gouda, M.K.; Hefnawy, A.M.; Kandil, S.H. Biocompatibility of new Ti-Nb-Ta base alloys. *Mater. Sci. Eng. C* **2016**, *61*, 574–578. [CrossRef]
29. Mondal, D.P.; Ramakrishnan, N.; Suresh, K.S.; Das, S. On the moduli of closed-cell aluminum foam. *Scr. Mater.* **2007**, *57*, 929–932. [CrossRef]
30. Zhang, L.C.; Klemm, D.; Eckert, J.; Hao, Y.L.; Sercombe, T.B. Manufacture by selective laser melting and mechanical behavior of a biomedical Ti-24Nb-4Zr-8Sn alloy. *Scr. Mater.* **2011**, *65*, 21–24. [CrossRef]
31. Currey, J.D. Bone and Natural Composites: Properties. In *Encyclopedia of Materials: Science and Technology*; Elsevier: Amsterdam, The Netherlands, 2001; pp. 776–781.
32. Donato, T.A.; de Almeida, L.H.; Nogueira, R.A.; Niemeyer, T.C.; Grandini, C.R.; Caram, R.; Schneider, S.G.; Santos, A.R., Jr. Cytotoxicity study of some Ti alloys used as biomaterial. *Mater. Sci. Eng. C* **2009**, *29*, 1365–1369. [CrossRef]

33. Cremasco, A.; Messias, A.D.; Esposito, A.R.; Duek, E.A.D.R.; Caram, R. Effects of alloying elements on the cytotoxic response of titanium alloys. *Mater. Sci. Eng. C* **2011**, *31*, 833–839. [CrossRef]
34. Wang, Y.B.; Zheng, Y.F. Corrosion behaviour and biocompatibility evaluation of low modulus Ti-16Nb shape memory alloy as potential biomaterial. *Mater. Lett.* **2009**, *63*, 1293–1295. [CrossRef]
35. Li, Y.; Wong, C.; Xiong, J.; Hodgson, P.; Wen, C. Cytotoxicity of titanium and titanium alloying elements. *J. Dent. Res.* **2010**, *89*, 493–497. [CrossRef] [PubMed]
36. Okazaki, Y.; Gotoh, E. Comparison of metal release from various metallic biomaterials in vitro. *Biomaterials* **2005**, *26*, 11–21. [CrossRef]
37. Lamolle, S.F.; Monjo, M.; Lyngstadaas, S.P.; Ellingsen, J.E.; Haugen, H.J. Titanium implant surface modification by cathodic reduction in hydrofluoric acid: Surface characterization and in vivo performance. *J. Biomed. Mater. Res.-Part A* **2009**, *88*, 581–588. [CrossRef]
38. Wang, Q.; Zhang, H.; Li, Q.; Ye, L.; Gan, H.; Liu, Y.; Wang, H.; Wang, Z. Biocompatibility and osteogenic properties of porous tantalum. *Exp. Ther. Med.* **2015**, *9*, 780–786. [CrossRef]
39. Shimko, D.A.; Shimko, V.F.; Sander, E.A.; Dickson, K.F.; Nauman, E.A. Effect of porosity on the fluid flow characteristics and mechanical properties of tantalum scaffolds. *J. Biomed. Mater. Res.-Part B Appl. Biomater.* **2005**, *73*, 315–324. [CrossRef]
40. Li, L.J.; Kim, S.N.; Cho, S.A. Comparison of alkaline phosphatase activity of MC3T3-E1 cells cultured on different Ti surfaces: Modified sandblasted with large grit and acid-etched (MSLA), laser-treated, and laser and acid-treated Ti surfaces. *J. Adv. Prosthodont.* **2016**, *8*, 235–240. [CrossRef]
41. Owen, T.A.; Aronow, M.; Shalhoub, V.; Barone, L.M.; Wilming, L.; Tassinari, M.S.; Kennedy, M.B.; Pockwinse, S.; Lian, J.B.; Stein, G.S. Progressive development of the rat osteoblast phenotype in vitro: Reciprocal relationships in expression of genes associated with osteoblast proliferation and differentiation during formation of the bone extracellular matrix. *J. Cell. Physiol.* **1990**, *143*, 420–430. [CrossRef]
42. Siddiqui, J.A.; Swarnkar, G.; Sharan, K.; Chakravarti, B.; Sharma, G.; Rawat, P.; Kumar, M.; Khan, F.M.; Pierroz, D.; Maurya, R.; et al. 8,8″-Biapigeninyl stimulates osteoblast functions and inhibits osteoclast and adipocyte functions: Osteoprotective action of 8,8″-biapigeninyl in ovariectomized mice. *Mol. Cell. Endocrinol.* **2010**, *323*, 256–267. [CrossRef] [PubMed]
43. Martín González, A.; Allo Miguel, G.; Aramendi Ramos, M.; Librizzi, S.; Jiménez, C.; Hawkins, F.; Martínez Díaz-Guerra, G. Different development of serum sclerostin compared to other bone remodeling markers in the first year after a liver transplant. *Rev. Osteoporos. Metab. Miner.* **2019**, *11*, 25–29. [CrossRef]
44. Boyan, B.D.; Hummert, T.W.; Dean, D.D.; Schwartz, Z. Role of material surfaces in regulating bone and cartilage cell response. *Biomaterials* **1996**, *17*, 137–146. [CrossRef]
45. Jo, S.; Han, J.; Lee, Y.L.; Yoon, S.; Lee, J.; Wang, S.E.; Kim, T.H. Regulation of osteoblasts by alkaline phosphatase in ankylosing spondylitis. *Int. J. Rheum. Dis.* **2019**, *22*, 252–261. [CrossRef] [PubMed]

Article

Reconstruction of Complex Zygomatic Bone Defects Using Mirroring Coupled with EBM Fabrication of Titanium Implant

Khaja Moiduddin [1,*], Syed Hammad Mian [1], Usama Umer [1], Naveed Ahmed [1,2], Hisham Alkhalefah [1] and Wadea Ameen [1]

[1] Advanced Manufacturing Institute, King Saud University, Riyadh 11421, Saudi Arabia; syedhammad68@yahoo.co.in (S.H.M.); uumer@ksu.edu.sa (U.U.); anaveed@ksu.edu.sa (N.A.); halkhalefah@ksu.edu.sa (H.A.); wqaid@ksu.edu.sa (W.A.)

[2] Department of Industrial and Manufacturing Engineering, University of Engineering and Technology, Lahore 54000, Pakistan

* Correspondence: khussain1@ksu.edu.sa; Tel.: +966-11-469-7372

Received: 17 October 2019; Accepted: 18 November 2019; Published: 22 November 2019

Abstract: Reconstruction of zygomatic complex defects is a surgical challenge, owing to the accurate restoration of structural symmetry as well as facial projection. Generally, there are many available techniques for zygomatic reconstruction, but they hardly achieve aesthetic and functional properties. To our knowledge, there is no such study on zygomatic titanium bone reconstruction, which involves the complete steps from patient computed tomography scan to the fabrication of titanium zygomatic implant and evaluation of implant accuracy. The objective of this study is to propose an integrated system methodology for the reconstruction of complex zygomatic bony defects using titanium comprising several steps, right from the patient scan to implant fabrication while maintaining proper aesthetic and facial symmetry. The integrated system methodology involves computer-assisted implant design based on the patient computed tomography data, the implant fitting accuracy using three-dimensional comparison techniques, finite element analysis to investigate the biomechanical behavior under loading conditions, and finally titanium fabrication of the zygomatic implant using state-of-the-art electron beam melting technology. The resulting titanium implant has a superior aesthetic appearance and preferable biocompatibility. The customized mirrored implant accurately fit on the defective area and restored the tumor region with inconsequential inconsistency. Moreover, the outcome from the two-dimensional analysis provided a good accuracy within 2 mm as established through physical prototyping. Thus, the designed implant produced faultless fitting, favorable symmetry, and satisfying aesthetics. The simulation results also demonstrated the load resistant ability of the implant with max stress within 1.76 MPa. Certainly, the mirrored and electron beam melted titanium implant can be considered as the practical alternative for a bone substitute of complex zygomatic reconstruction.

Keywords: zygomatic bone; customized reconstruction; electron beam melting; 3D comparison; titanium alloy; finite element analysis

---

1. Introduction

Advancements in the field of biomaterials, fabrication techniques, and computer-assisted technologies have been taking place owing to the huge demand for medical implants [1]. Biomaterials are natural or artificial materials that are used to enhance or replace any tissue, organ, or biological structure in order to improve the quality of human life [2]. Titanium and its alloys have a unique combination of high strength to low weight ratio, low density and better corrosion resistance, are biocompatible, and have non-magnetic properties. They are one of the few biomaterials that naturally match the requirement of bone tissue replacement in the human body. Among them, commercial pure titanium (grade 2), Ti-6Al-4V (grade 5), and Ti-6Al-4V ELI (Extra low interstitial) (grade 23) are widely used biomaterial in medical application [3]. In addition, a significant rise in the fabrication of custom-built titanium implants with better design and biomechanical properties has been realized in clinical applications. The facilitation and application of tailor-made effective implants are crucial to improve the quality of the patient's life. Moreover, it is of utmost importance to study the implant concerning its design, fitting accuracy, biomechanical properties, and fabrication method.

Zygomatic bone reconstruction is indeed a challenging procedure for maxillofacial surgeons due to its unique position which is close to the orbital rim that houses the eyeball [4]. The integrity of the zygomatic reconstruction can be achieved through the maintenance of the person's aesthetic appearance, its functionality, and the prominence of the cheek [5]. The zygomatic bone is highly robust and variable in shape. It is located at the crucial junction between the zygomatic arc, orbital wall, and tooth bearing snout. It consists of multiple bone tissues types, namely the cancellous bone which is surrounded by cortical bone shell. The cortical bone was found to be thickest (5.0 mm) in the upper zygomatic bone and thinnest (1.1 mm) at the anterior of maxillary sinus region [6]. In addition, the cortical bone has greater volume fraction in the zygomatic bone than the cancellous bone [7]. Studies indicate that the young modulus of zygomatic bone is in the range from 10.4 GPa to 19.6 GPa, respectively [8]. Titanium implants are commonly used for zygomatic bony reconstruction because they provide reliable support for the orbital contents [9]. The first zygomatic implant reconstruction was developed by Branemark without grafting procedures for maxillectomized patients [10]. Different surgical approaches and treatments have been employed since then, for successful zygomatic complex reconstruction, including osteotomy, autologous bone graft, and synthetic implants [11]. Among all, the autologous bone graft is considered gold standards; however, limited bone availability, volatile resorption rate, and deformities remain serious challenges. Hence, various alloplastic implants, including polymers [12], silicone [13], metals [14], and hydroxyapatite based products [15] were also used to replace the autologous bone graft.

Long-term success and effectiveness of the implants depend on several factors, including the design, implant material, accuracy, biomechanical study, fabrication process, and, of course, the skills of the surgeon [4]. In all, the mirror reconstruction technique is one of the widely used implant design techniques for medical applications [16,17]. It can replace the defective bony region with a healthy region, thus maintaining the symmetry and ideal anatomical structure [18]. Titanium alloy (Ti-6Al-4V ELI) is one of the most favored biocompatible materials owing to its high strength to weight ratio, excellent corrosion resistance, and mechanical properties [19]. Titanium and its alloys have excellent biocompatibility when compared to other metallic biomaterials such as stainless steel and cobalt chromium alloys. The most widely used titanium alloy (Ti-6Al-4V ELI) used for bone replacement has a modulus of elasticity (110 GPa) almost half that of stainless steel (200 GPa) and Cobalt chromium alloy (210 GPa) [20]. Development of new titanium alloys with shape memory is a growing attraction for biomedical application with superior elasticity and properties closer to that of bone [21]. One of the important properties that differentiate titanium and its alloys from other biomaterials is the natural formation of thin oxide film on its surface which is responsible for good chemical stability and biocompatibility [22]. The biomechanical study involving stress analysis on the implant and its surrounding bone is also critical to study the orthopedic mechanical failure [23]. It helps to evaluate different designs virtually during functional loading, thus reducing material usage, physical prototyping, and testing methods [24]. Implant accuracy evaluation is also a mandatory procedure to analyze its fitting accuracy. The implant–bone interface contact is necessary for immediate restoration as well as to avoid any damage to the vital structures [25].

Fabrication of an implant from the patient computer tomography (CT) scan involves several steps encompassing data collection and processing, design template, virtual simulation, and fabrication utilizing three-dimensional (3D) printing. 3D printing also known as rapid prototyping and additive manufacturing is a technique of producing physical 3D objects from computer-aided-design (CAD) files in a successive material layering. Electron beam melting (EBM) is one of the widely used metal 3D printing processes with a US Federal Drug Administration (FDA) approval [26]. It has been widely adopted by surgeons at an impressive rate and is used in a large variety of medical applications [27,28]. The 3D printed models can also be used for mock surgeries, pre-operative planning, surgical guides, and educational training purposes [29].

It is indeed very difficult to reconstruct a complex anatomical structure with proper aesthetic and facial symmetry [30]. Any deviation or aberration in structural alignment between the implant and bone contours may most likely lead to functional disturbance and implant failure. In this study, an integrated methodology involving custom design using the mirror reconstruction technique has been proposed. It also consists of biomechanical evaluation for the custom built implant, implant accuracy assessment using 3D comparison technique, and, finally, the three-dimensional (3D) printing of the titanium implant using state-of-the-art EBM technology. Certainly, the objective is to produce a zygomatic titanium implant, which is reliable based on its functionality, appearance, and mechanical strength. The methodology adopted in this work involves multidisciplinary fields including custom-built design, biomechanical analysis, implant accuracy, and titanium-based EBM fabrication. Most often, the clinicians or engineers, owing to the vastness or intricacies of the field, overlook some important step or assessment test. Henceforth, in this work, the different steps that are pertinent in zygomatic rehabilitation as well as implant realization are comprehensively described.

2. Materials and Methods

The methodology as presented in Figure 1 involved the integration of several technologies to realize the customized zygomatic titanium implant. This approach also comprised of the formal meeting and communication between the engineering department and the medical field in each step, to evaluate and verify the design model for the enhanced aesthetic outcome, and to reduce implant revision and failure. Indeed, a multidisciplinary approach involving biomedical engineers, surgeons, and researchers are vital for the successful fabrication and implementation of implants [31]. Studying implant failure would help the clinicians and engineers in understanding the failure factors and to develop better prosthesis. In this study, a zygomatic implant was first produced using conventional techniques as illustrated in Figure 1a. Later, a new computer-aided custom design implant was produced as shown in Figure 1b and compared with the conventionally produced implant. Some of the drawbacks of conventional techniques in implant design are the excessive time consumption, inaccurate fitting of implant, diagnostic limitations, and lack of surgical planning. From the past decade, the development of computer-assisted implant design and surgical planning has become extremely important in orthopedic surgeries.

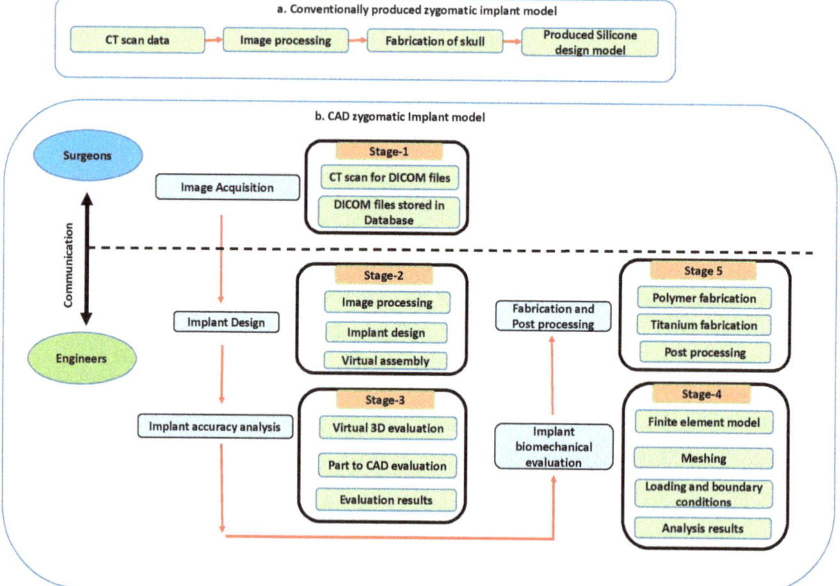

Figure 1. Methodology employed in the fabrication of (**a**) conventionally produced zygomatic implant model and (**b**) computer-aided-design zygomatic implant model.

2.1. Implant Customization and Fabrication

A cone-beam computed tomography (CBCT) scan was performed using Promax 3D (Planmeca, Helsinki, Finland) on a 34-year-old male patient who was suffering from painful cheek swelling. The series of two-dimensional (2D) images obtained from the CT scan were stored as a Digital Imaging and Communications in Medicine (DICOM) file in a database. The DICOM file was then imported into Mimics® (version, Materialise, Leuven, Belgium), a medical modeling software, which distinctly converted the series of 2D images into a 3D model as shown in Figure 2. The thresholding function from Mimics® with a thresholding value of 226 to 3071 Hounsfield Units (HU) was used to segregate the hard and soft tissues. The HU is a universally acceptable dimensionless unit that is used to express CT numbers. The Segmentation and region growing functions using Mimics® were used to divide and subdivide the image into several regions until the region of interest–Skull (green) was obtained. The obtained 3D skull model was then saved as a Standard Translation Language (STL) file and imported into a polymer 3D printer-fused deposition modeling (FDM) for fabrication.

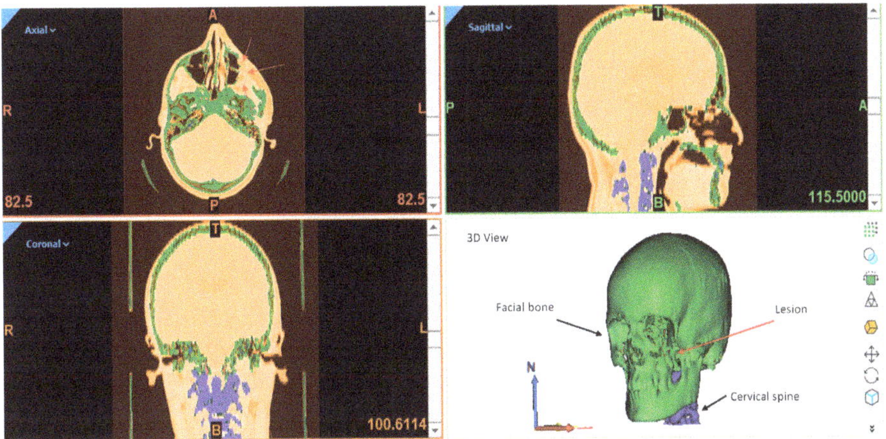

Figure 2. 3D model of the patient skull anatomy revealing the tumor location.

The FDM fabricated skull 3D model was used for designing the zygomatic implant. A silicone impression material was placed over the defective area and gently molded to obtain the shape of the zygomatic bone as shown in Figure 3a. As the process was done manually, a lot of contours and rough surfaces could be observed on the posterior end of the implant (Figure 3c). The conventionally designed implant (Figure 3b) did not adapt to the recipient defect perfectly, and hence it was rejected based on the formal meeting with the medical clinicians. On further discussion with the surgeons, it was decided to have a new CAD implant model using reconstruction techniques.

There are two types of implant reconstruction techniques widely used in custom designs [32]. The first is a mirror reconstruction technique and the other is an anatomical reconstruction design based on a curve based and refinement approach [4]. The mirroring technique provides better facial symmetry and is more accurate for medium and large complex tumors, whereas the anatomical reconstruction technique is a lengthy process, requires lots of human expertise, and provides good results for smaller tumors [33,34]. One of the major differences between these two techniques is that mirroring techniques can be applied only in symmetrical regions, whereas anatomical reconstruction can be applied to both symmetrical and asymmetrical parts.

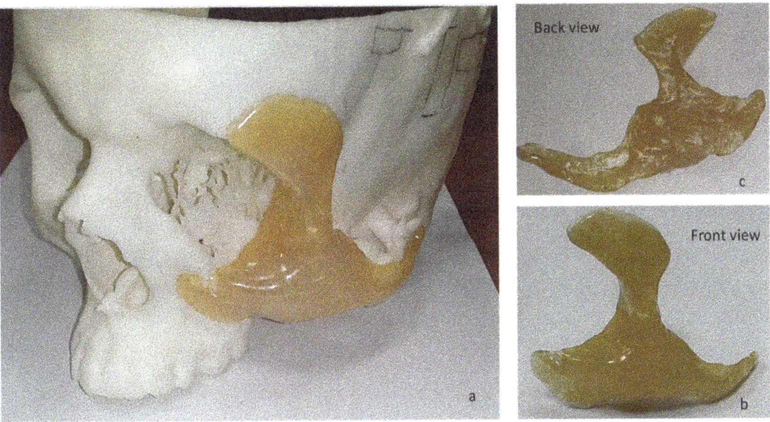

Figure 3. (**a**) Fused deposition Modeling fabricated skull model with the attached silicone implant; (**b**) silicone implant front view; and (**c**) silicone implant back view.

Based on the inputs from the medical clinicians, an implant mirror reconstruction technique was utilized to replace the defective region with the healthier bony region. The steps involved in the mirroring reconstruction technique are illustrated in Figure 4. Primarily, the STL defective model (Figure 4a) was first resected into two halves (Figure 4b) using Magics® (Materialise, Leuven, Belgium). The defective side was removed (Figure 4c) and replaced by the healthy right side (Figure 4d) using the mirror reconstruction technique. The next step was to join the two error-free sides through merging operation (Figure 4e). Any voids or gaps were removed by wrapping operations (Figure 4f). Boolean subtraction process (Figure 4g) was performed between the error-free model (Figure 4f) and the tumor model (Figure 4a) to obtain the zygomatic implant template (Figure 4h), which was then saved as an STL file for subsequent fabrication.

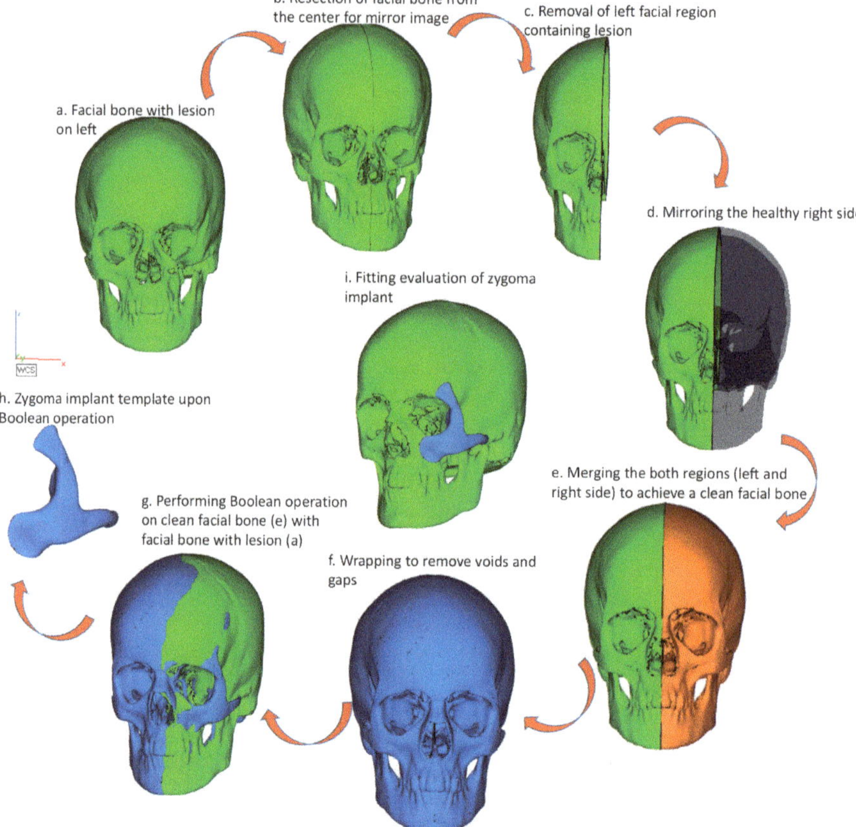

Figure 4. Computer-aided customized zygomatic implant process cycle. (**a**) facial bone with lesion; (**b**) resection of facial bone for mirror image; (**c**) removal of left facial region with lesion; (**d**) mirroring the healthy right side; (**e**) merging both left and right regions to obtain clean facial bone; (**f**) wrapping operation to remove gaps; (**g**) boolean operation between clean facial bone and facial bone with lesion; (**h**) obtained zygomatic bone template upon Boolean operation; (**i**) fitting evaluation of zygoma implant.

Figure 5 illustrates the FDM fabricated zygomatic implant designed using a mirror reconstruction technique as well as its counterpart that is a conventionally produced implant model. On visual observation, it can be clearly seen that the manually produced implant model has lots of curves and rough surfaces when compared to the mirror implant model produced through FDM. Figure 6 demonstrates the precise fitting of the mirror implant model onto the skull model. The polymer produced FDM models also assisted the medical clinicians in comprehensive surgical planning and rehearsal in addition to precise drilling of screw holes.

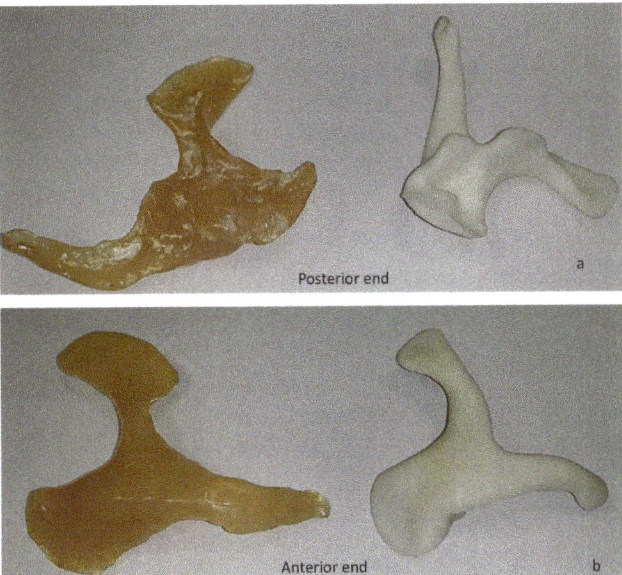

Figure 5. Silicone based zygomatic implant produced manually without CAD technique (**a**) and FDM fabricated zygomatic implant by CAD mirroring technique (**b**).

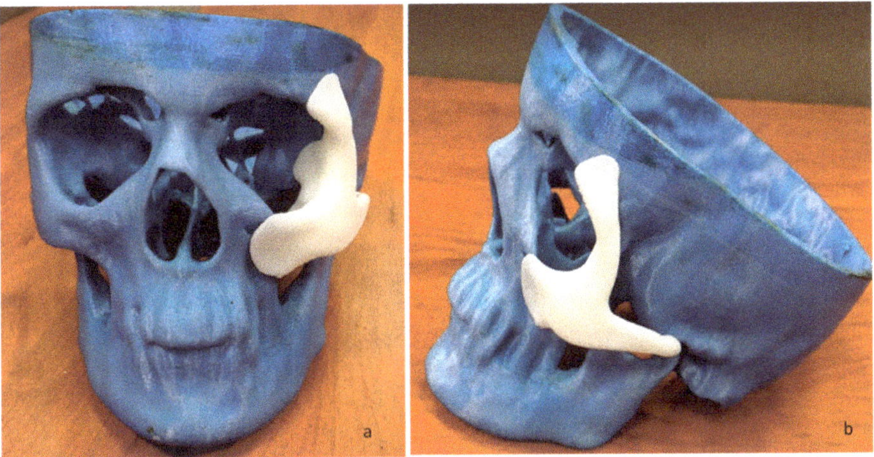

Figure 6. (**a**) FDM fabricated customized zygomatic implant fitted onto the skull model (front view); (**b**) side view illustration.

After the successful rehearsal and fitting operation of the FDM produced a zygomatic implant onto the skull model and based on the approval from the clinicians, the mirror implant model was fabricated using EBM technology. Arcam's EBM technology offers a new method of rapid manufacturing of near net shaped titanium products, thus eliminating the time, cost and challenges in machining and casting [35]. Arcam EBM A2 machine (Arcam AB, Mölndal, Sweden) as illustrated in Figure 7b was used for the fabrication of titanium zygomatic implant. The accuracy of the EBM machine is in the range of 0.13 mm to 0.20 mm for shorter to long range build parts and the highest resolution is 50 μm [36,37]. The EBM process cycle is illustrated in Figure 7c where the filament (1), when heated to a temperature of above 2500 °C, accelerates a beam of electron (5) through the series of magnetic lenses including astigmatism lens, focus lens, and deflection lens before hitting the titanium powder (Ti-6Al-4V ELI) (8). The first magnetic lens (2) generates a circular beam of electrons and corrects astigmatism while the second magnetic lens (3) focuses the electron beam to the desired diameter, and the third magnetic lens (4) deflects the focused beam to the desired position. The powder hoppers (6) continuously feed the titanium powder (8) onto the start plate (10) inside the build platform. A mechanical raking blade (7) spreads the titanium powder evenly onto the build platform (9). Initially, a high-speed beam of electrons scans the titanium powder to preheat the powder to a sintered state. After preheating, the melting of powder takes place at slower beam scans. On completion of each melting cycle, the build platform is lowered by one-layer thickness. The entire EBM build process takes place under vacuum and under elevated temperature. This is done to prevent reactions between reactive metals (titanium) with oxygen and to prevent residual stresses [38].

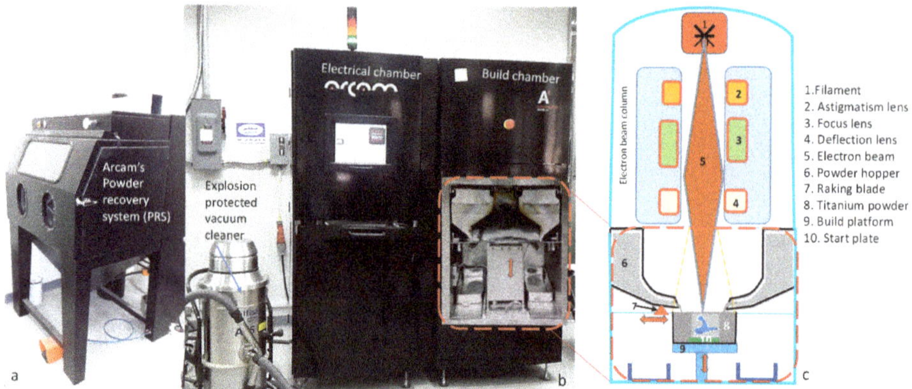

Figure 7. (**a**) ARCAM powder recovery system unit to remove the sintered powder attached to the build part, (**b**) ARCAM A2 electron beam melting machine, and (**c**) the schematic working diagram of electron beam melting machine.

Figure 8a illustrates the fabricated titanium zygomatic implant with support structures. The titanium implant after fabrication was placed inside a powder recovery system (PRS) to remove the semi-sintered powder attached to the implant. The support structures were removed manually using pliers. The titanium zygomatic implant was then fixed onto the polymer skull model for fitting evaluation (Figure 8b).

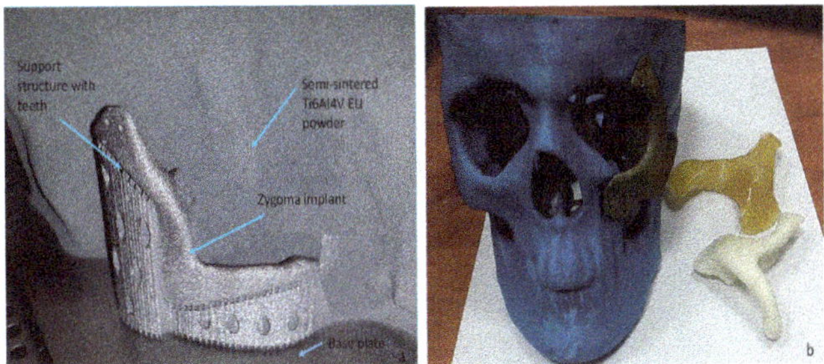

Figure 8. (**a**) EBM fabricated titanium zygomatic implant with support structures inside the powder recovery system and (**b**) titanium zygomatic implant fitted to the skull model after support removal.

2.2. Evaluation

To study the accurate fitting of the implant onto the skull region, an accuracy analysis of the implant was executed in Geomagics Control® (3D systems, Rock Hill, SC, USA). It is one of the most dynamic and comprehensive techniques to estimate the deviation between the test and reference CAD object [39]. It was carried out to quantify the error between the implant and the face model. The implant accuracy analysis was performed in two stages as depicted in Figure 9.

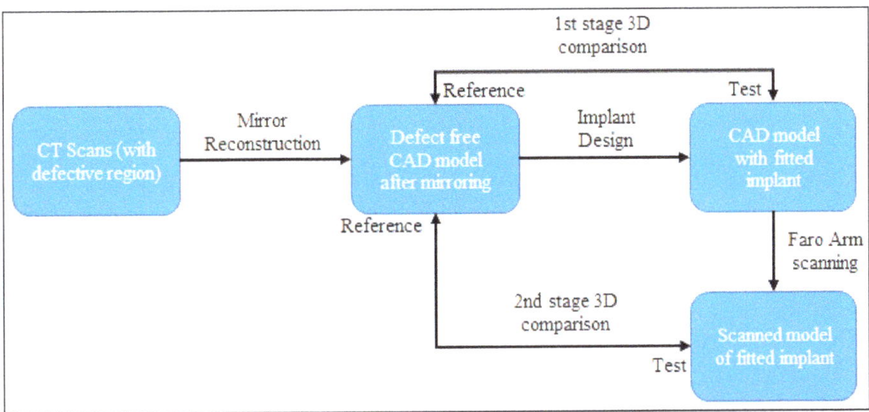

Figure 9. Evaluation of designed implant using comparison approach in Geomagics®.

- First stage comparison before fabrication. In this stage, the tumor-free model after mirroring was taken as a reference and the skull model with the implant was considered as a test file. It was done to quantify the error from the mirroring stage.
- The second stage after fabrication. The 3D model obtained using the Faro platinum arm scanner (FARO, Lake Mary, FL, USA) was used to inspect the implant with the skull model. The 3D model obtained from a Faro arm scanner was considered as a test file and the tumor-free model after mirroring was fixed as the reference file. The test was aligned and superimposed on the error-free mirror model to analyze the deviation error.

The Finite Element Analysis (FEA) was also conducted to investigate the biomechanical behavior of the zygomatic implant and its supporting bone. Typical occlusal loads were applied and stresses on the implants and its surrounding bones were examined. The STL models of Skull and zygomatic implant were first transformed to Solid B-Rep models before analysis. The finite element model (FEM) was developed using ABAQUS/CAE (Version 6.14, Dassault Systemes, Veliez-Villacublai, France).

ABAQUS/CAE is an interactive, graphical environment for Abaqus software. It allows models to be created quickly and easily by producing or importing the geometry of the structure to be analyzed and decomposing the geometry into meshable regions. Once the model is complete, ABAQUS/CAE can submit, monitor, and control the analysis jobs. The Visualization module can then be used to interpret the results. The attributes of cortical bone were assigned to the skull model, and titanium alloy (Ti-6Al-4V ELI) was assigned to the zygomatic implant [40,41]. The material properties designated in the FEA model are presented in Table 1.

Table 1. Material properties utilized in the finite element analysis model [40,41].

Materials	Young's Modulus (GPa)	Poisson's Ratio	Yield Strength (MPa)
Cortical bone	13.7	0.3	122
Zygomatic implant (Ti-6Al-4V ELI)	120	0.3	930

The FEM with load and boundary conditions is shown in Figure 10. The skull was fixed around the neck area and a force of 50 Newton was applied to the zygomatic implant over an area of 500 mm^2. In previous studies, researchers have also applied similar loading conditions of 5.5 kg on zygomatic under the mastication process [42]. The joints between the skull and zygomatic implant were modeled using mesh independent fastener available in ABAQUS/STANDARD. ABAQUS/STANDARD (Version 6.14, Dassault Systemes, Veliez-Villacublai, France) is a general-purpose analysis product that can solve a wide range of linear and nonlinear problems involving the static, dynamic, thermal, electrical, and electromagnetic response of components. Abaqus/Standard solves a system of equations implicitly at each solution increment. Mesh independent fasteners provide a point-based connection between surfaces similar to spot-weld connections. A total of 12 fasteners were utilized across the three joint areas of the zygomatic implant (red-colored) as shown in Figure 10.

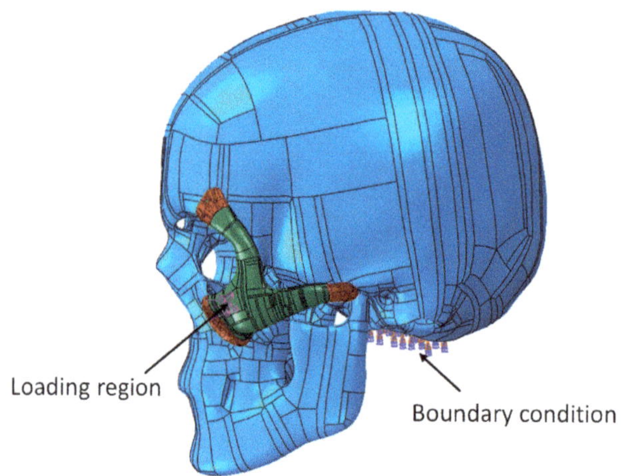

Figure 10. Loading and boundary conditions on the skull and the zygomatic implant.

3D stress quadratic tetrahedron elements (C3D10) with 10 nodes were selected for the FEM as shown in Figure 11. To save computational time, an optimum size mesh is selected that resulted in 149,966 elements in the whole model. To model interaction between implant and skull surfaces, a general contact algorithm has been selected along with frictionless tangential behavior.

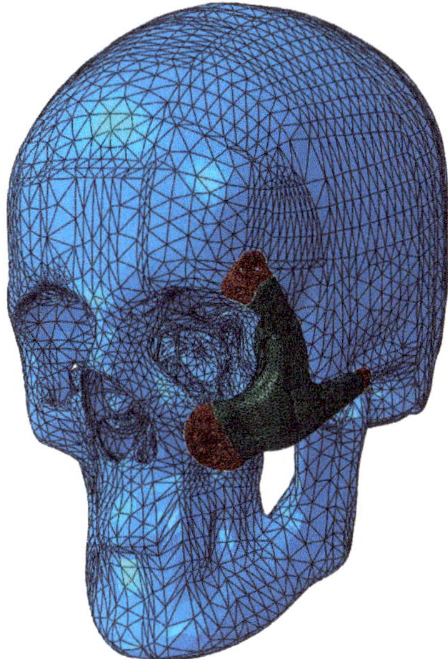

Figure 11. Finite element mesh of the skull and zygomatic implant.

3. Results and Discussion

The tumor-free skull model obtained from the mirror reconstruction technique (Figures 4f and 12b) was used as a reference and the skull model with the attached implant (Figure 12c) was taken as a test file. The test file was aligned and superimposed onto the reference file to obtain the deviation error, which reflected both the mirroring as well as implant shape effect as shown in Figure 12a. In addition to 3D analysis, the 2D comparison was also conducted which quantified the implant shape effect. As it can be observed, there is an inappreciable deviation onto the zygomatic implant region, which states that the obtained zygomatic implant model fitted precisely and replaced the tumor region without much deviation.

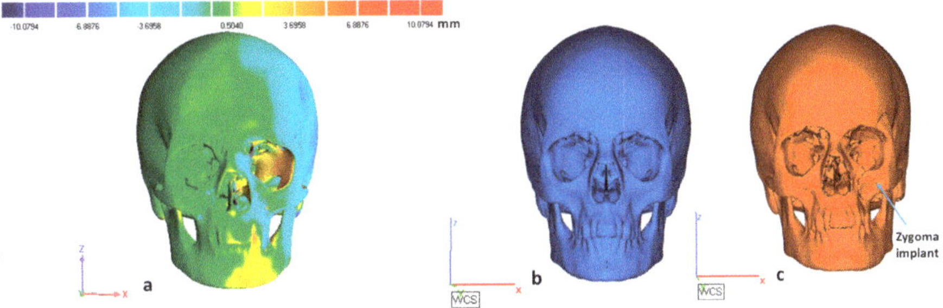

Figure 12. (**a**) 3D inspection results of model deviation between the (**b**) mirror reconstruction model and (**c**) zygomatic implant model.

The 2D comparison results between the reference and the test model are illustrated in Figure 13. In 2D comparison, a cross-sectional plane was created onto the test and reference model in the zygomatic region. Figure 13a–c illustrate the different cross-sectional views of the superimposition of the test file (Figure 13e) onto the reference file (Figure 13d). The color spectrum in the whisker deviation (Figure 13f) provides the deviation error between them. The 3D and 2D comparison results obtained through virtual model inspection in 1st stage are provided in Table 2.

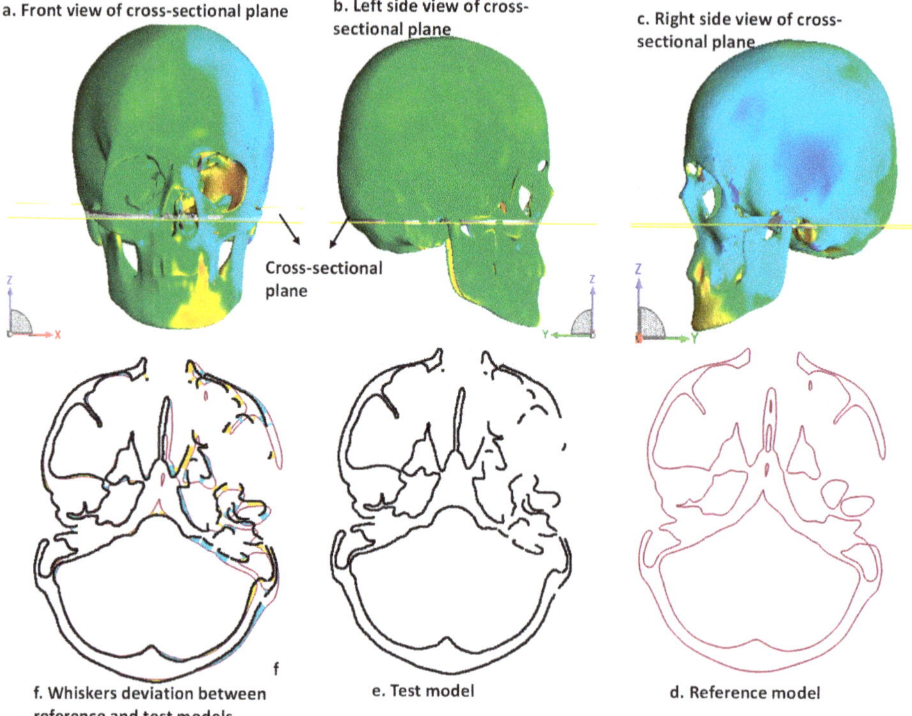

Figure 13. Different views (**a–c**) of cross-sectional plane on the implant region. The 2D comparison results (**f**) illustrating the deviation between the (**d**) reference model and the (**e**) test model.

The statistics used to investigate the implant accuracy were the average (AVG) deviation in the positive and negative directions and the root mean square error (RMSE) as illustrated in Equation (1). The average deviation was used as they approximate the differences between the test and reference files in the inward and outward direction. The RMSE represents the average magnitude of the error between two data sets or models. It also quantifies the overall accuracy of the models:

$$\text{RMSE} = \frac{1}{\sqrt{n}} \sqrt{\sum_{i=1}^{n} (X_{1,i} - X_{2,i})^2},$$

(1)

where $X_{1,i}$ is the measurement point of i in the reference data, $X_{2,i}$ is the measurement point of i in the test data, and n is the number of measuring points.

Similarly, the Faro arm scanning and its deviation analysis were carried out on the physical models as shown in Figure 14. The faro arm scanner (Figure 14a) was used to scan the physical facial model attached with an EBM fabricated zygomatic implant. The scanned model was taken as a test file, whereas the mirror reconstruction CAD model was taken as a reference file as illustrated in Figure 14b. The results obtained on the superimposition of test file over a reference file in the 3D comparison technique are shown in Table 2.

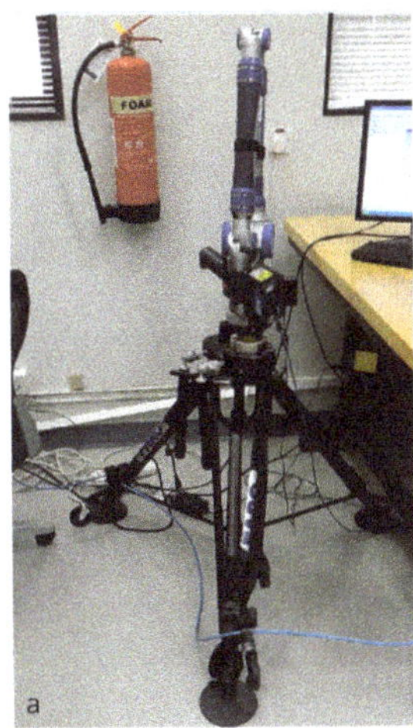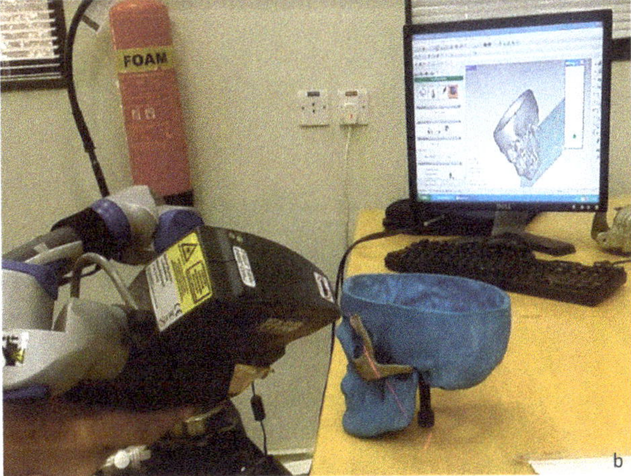

Figure 14. 3D measurement (**a**) Faro platinum arm with scanner; (**b**) scanning of the skull model fixed with titanium zygomatic implant.

Table 2. 3D and 2D comparison results for implant accuracy evaluation.

Comparison	Models	AVG Deviation (mm)	Root Mean Square Error (RMSE) (mm)
1st Stage (virtual evaluation using Geomagics®)	Zygomatic mirror and zygomatic with implant (3D comparison)	1.65/−1.55	2.38
	Zygomatic mirror and zygomatic with implant (2D comparison on Implant region)	0.86/−0.97	1.28
2nd Stage (physical model evaluation using Faro arm scanner)	Zygomatic mirror and zygomatic with implant (3D comparison)	1.82/−3.86	4.25
	Zygomatic mirror and zygomatic with implant (2D comparison on Implant region)	1.40/−1.79	1.96

The results from the 1st stage accuracy analysis, where the remodeled or reconstructed face was analyzed with the mirrored face virtually, the average deviation was in the range of 1.65 mm in the outside direction and −1.55 mm in the inward of the face with RMSE as 2.38 mm. In the second stage, when the remodeled physical face model was analyzed with the mirrored face, the average deviation exhibits in the range of 1.82 mm in the outside direction and −3.86 mm in the inward of the face along with RMSE of 4.25 mm. The deviation (or error) was slightly higher in the physical evaluation when compared to virtual evaluation due to the inclusion of several uncertainties from fabrication as well as the scanning procedure performed by the Faro Arm scanner (FARO, Lake Mary, FL, USA). The lesser RSME in 2D comparison of virtual and physical evaluation further validated the implant accuracy for fitting and aesthetic appearance. The 2D comparison was implemented to evaluate the test model at the defect region where the implant was fixed. It did not consider the entire model, but only the region of interest for comparison. Notice that, at this point, it is difficult to establish or compare these results due to the unavailability of similar studies.

This study is the first of its kind where authors have made an effort to quantify the fitting accuracy of the zygomatic implant on the human face. However, the authors along with this analysis also physically anchored the part onto the face to visualize the aesthetics as well as the facial symmetry. The results were satisfactory and acceptable. In addition, the 2D average deviation of physical model with 1.40 mm in the outward direction and −1.79 mm in the inward direction also justify the results—that there is not much deviation at the zygomatic implant. In addition, the authors will carry out similar studies in the future with zygomatic defects to prepare the data for fitting accuracy and determine their acceptable values. As part of the future studies, an in vivo study will also be performed with the zygomatic implant to confirm its fitting performance.

The FEA outcomes can be realized in Figure 15. The Von Mises stress distribution in the model is shown in Figure 15a. Highly stressed regions can be visualized around the joint and contact areas between the skull and the implant. An inverted image of the zygomatic implant is shown in Figure 15b, which identifies different stress regions. Maximum von Mises stress is found to be 1.76 MPa at one of the fastener positions that is well below the yield strength of implant material—thus ensuring the absence of any plastic deformation. Figure 15c shows the maximum principal strain contours for the zygomatic implant. It is clear that high strained regions comprise interaction and fasteners' areas. Nonetheless, these strains are very low and the maximum value approaches 1.34×10^{-5}. The total displacement pattern on the zygomatic implant due to the applied load is shown in Figure 15d. As depicted, maximum displacement occurs at the load-bearing area. However, maximum deformation has been around 2 microns, which again confirms high stiffness of the implant design.

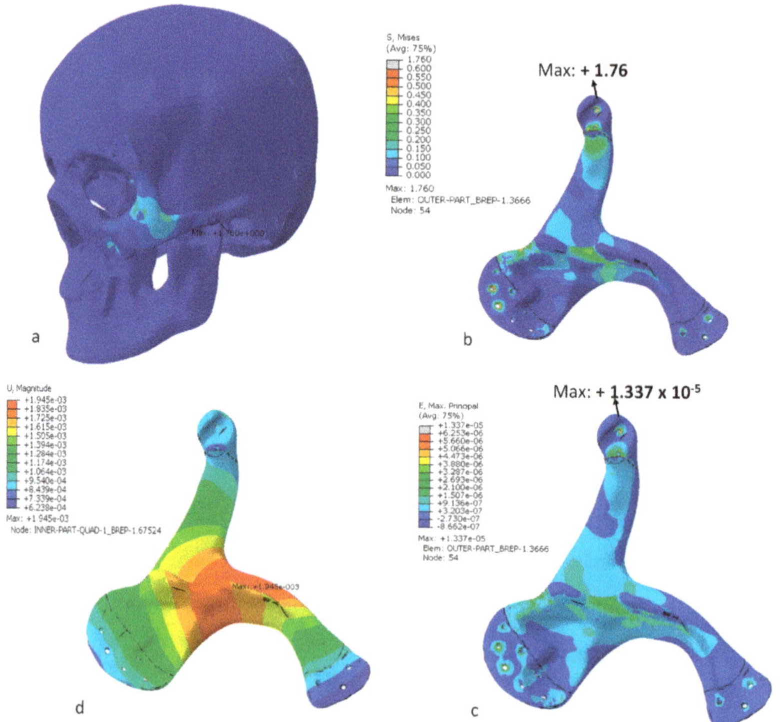

Figure 15. (**a**) Mises stress distributions for the zygomatic implant (MPa), (**b**) Mises stress distributions for the zygomatic implant (MPa), (**c**) maximum principal strain for the zygomatic implant and (**d**) total displacement contour for the zygomatic implant (mm).

4. Conclusions

In this study, a customized zygomatic titanium implant reconstructed using the mirror reconstruction technique was investigated based on the design, biomechanical study, and implant fitting accuracy. The work is particularly important from earlier studies as the authors have identified a technique (2D and 3D comparison) to quantify the implant fitting accuracy or fitting error. Initially, the zygomatic implant was designed manually using silicone, which could not adapt to the bone contours perfectly. Later, a computer-aided custom design implant was constructed using a computer-aided mirroring technique and fabricated using titanium based on state-of-the-art electron beam melting technology. The custom design mirror reconstruction implant precisely fits on the facial region with a maximum deviation error (RMSE) of 2.38 mm in the virtual assembly and 4.25 mm in the physical assembly. Furthermore, the outcomes from the 2D analysis revealed average deviations within 2 mm at the implant region. The implant when practiced on a physical prototype provided a flawless fitting, good symmetry, and pleasant aesthetics. This fitting evaluation study provides crucial information about the implant actualization on the patient's face. This outcome also provides indispensable knowledge for future in vivo or cadaveric implantation studies to further establish the implant's superior performance. The designed implant also successfully withstands the load, with max stress found to be of 1.76 MPa, which is well below the yield strength of implant material (titanium). This proves that the fabricated titanium zygomatic implant possesses the required mechanical strength. Finally, the authors conclude that the EBM fabricated mirror designed titanium implant satisfies the aesthetic, functional, and mechanical properties for efficient zygomatic bone reconstruction. The proposed design methodology can also be applied for other bone reconstruction surgeries. An in vivo or cadaveric based study of the zygomatic titanium implant will be performed as part of future investigation to confirm its fitting performance.

Author Contributions: K.M.: Conceptualization and experiments; K.M. and S.H.M.: Methodology and draft preparation; U.U.: Analysis and Investigation; N.A.: Data curation, resources, and investigation; H.A.: Project administration, revision and funding acquisition, W.A.: Revision and validation.

Funding: This research was financially supported by the Deanship of Scientific Research, King Saud University: Research group No: RG-1440-034.

Acknowledgments: The authors extend their appreciation to the Deanship of Scientific Research at King Saud University for funding this work through Research Group no. RG-1440-034.

References

1. Di, M.P.; Coburn, J.; Hwang, D.; Kelly, J.; Khairuzzaman, A.; Ricles, L. Additively manufactured medical products-the FDA perspective. *3D Print Med.* **2016**, *2*, 1–6.
2. de Viteri, V.S.; Fuentes, E. Titanium and Titanium Alloys as Biomaterials. *Tribol. Fundam. Adv. InTech.* **2013**, 155–181.
3. Balazic, M.; Kopac, J.; Jackson, M.J.; Ahmed, W. Review: Titanium and titanium alloy applications in medicine. *IJNBM* **2007**, *1*, 3. [CrossRef]
4. Parthasarathy, J. 3D modeling, custom implants and its future perspectives in craniofacial surgery. *Ann. Maxillofac. Surg.* **2013**, *4*, 9. [CrossRef]
5. Starch-Jensen, T.; Linnebjerg, L.B.; Jensen, J.D. Treatment of Zygomatic Complex Fractures with Surgical or Nonsurgical Intervention: A Retrospective Study. *Int. J. Oral Maxillofac. Surg.* **2018**, *12*, 377–387. [CrossRef]
6. Lee, H.-S.; Choi, H.-M.; Choi, D.-S.; Jang, I.; Cha, B.-K. Bone thickness of the infrazygomatic crest area in skeletal Class III growing patients: A computed tomographic study. *Imaging Sci. Dent.* **2013**, *43*, 261–266. [CrossRef]
7. Pryor McIntosh, L.; Strait, D.S.; Ledogar, J.A.; Smith, A.L.; Ross, C.F.; Wang, Q.; Opperman, L.A.; Dechow, P.C. Internal Bone Architecture in the Zygoma of Human and Pan. *Anat. Rec.* **2016**, *299*, 1704–1717. [CrossRef]

8. Milne, N.; Fitton, L.C.; Kupczik KFagan, M.J.; O'Higgins, P. The role of the zygomaticomaxillary suture in modulating strain distribution within the skull of Macaca fascicularis. *Homo J. Comp. Hum. Biol.* **2009**, *60*, 281.

9. Foletti, J.M.; Martinez, V.; Haen, P.; Godio-Raboutet, Y.; Guyot LThollon, L. Finite element analysis of the human orbit. Behavior of titanium mesh for orbital floor reconstruction in case of trauma recurrence. *J. Stomatol. Oral Maxillofac. Surg.* **2019**, *120*, 91–94. [CrossRef]

10. Parel, S.M.; Brånemark, P.I.; Ohrnell, L.O.; Svensson, B. Remote implant anchorage for the rehabilitation of maxillary defects. *J. Prosthet. Dent.* **2001**, *86*, 377–381. [CrossRef]

11. Quatela, V.C.; Chow, J. Synthetic facial implants. *Facial Plast. Surg. Clin. N. Am.* **2008**, *16*, 1–10. [CrossRef] [PubMed]

12. Scolozzi, P. Maxillofacial reconstruction using polyetheretherketone patient-specific implants by 'mirroring' computational planning. *Aesthetic Plast. Surg.* **2012**, *36*, 660–665. [CrossRef] [PubMed]

13. Ivy, E.J.; Lorenc, Z.P.; Aston, S.J. Malar augmentation with silicone implants. *Plast. Reconstr. Surg.* **1995**, *96*, 63–68. [CrossRef] [PubMed]

14. El-Khayat, B.; Eley, K.A.; Shah, K.A.; Watt-Smith, S.R. Ewings sarcoma of the zygoma reconstructed with a gold prosthesis: A rare tumor and unique reconstruction. *Oral Surg. Oral Med. Oral Pathol. Oral Radiol. Endod.* **2010**, *109*, e5–e10. [CrossRef] [PubMed]

15. Hoffmann, J.; Cornelius, C.P.; Groten, M.; Pröbster, L.; Pfannenberg, C.; Schwenzer, N. Orbital reconstruction with individually copy-milled ceramic implants. *Plast. Reconstr. Surg.* **1998**, *101*, 604–612. [CrossRef] [PubMed]

16. Zhang, Y. Orbital Defect Repair and Secondary Reconstruction of Enophthalmos with Mirror-Technique Fabricated Titanium Mesh. *J. Oral Maxillofac. Surg.* **2008**, *66*, 19–20. [CrossRef]

17. Moiduddin, K.; Mian, S.H.; Umer, U.; Alkhalefah, H. Fabrication and Analysis of a Ti6Al4V Implant for Cranial Restoration. *Appl. Sci.* **2019**, *9*, 2513. [CrossRef]

18. Liu, Y.; Xu, L.; Zhu, H.; Liu, S.S.-Y. Technical procedures for template-guided surgery for mandibular reconstruction based on digital design and manufacturing. *Biomed Eng. Online* **2014**, *13*, 63. [CrossRef]

19. Christensen, A.; Kircher, R.; Lippincott, A. Qualification of electron beam melted (EBM) Ti6Al4V-ELI for orthopaedic implant applications. In *Medical Device Materials IV: Proceedings of the Materials and Processes for Medical Devices Conference*; Jeremy, G., Ed.; ASM International: Cleveland, OH, USA, 2007; Volume 6, pp. 48–53.

20. Niinomi, M. Recent research and development in titanium alloys for biomedical applications and healthcare goods. *Sci. Technol. Adv. Mater.* **2003**, *4*, 445. [CrossRef]

21. Ehtemam-Haghighi, S.; Prashanth, K.G.; Attar, H.; Chaubey, A.K.; Cao, G.H.; Zhang, L.C. Evaluation of mechanical and wear properties of Ti xNb 7Fe alloys designed for biomedical applications. *Mater. Des.* **2016**, *111*, 592–599. [CrossRef]

22. Huynh, V.; Ngo, N.K.; Golden, T.D. Surface Activation and Pretreatments for Biocompatible Metals and Alloys Used in Biomedical Applications. *Int. J. Biomater.* **2019**, *2019*, 21. [CrossRef] [PubMed]

23. Coquim, J.; Clemenzi, J.; Salahi, M.; Sherif, A.; Tavakkoli Avval, P.; Shah, S.; Schemitsch, E.H.; Shaghayegh Bagheri, Z.; Bougherara, H.; Zdero, R. Biomechanical Analysis Using FEA and Experiments of Metal Plate and Bone Strut Repair of a Femur Midshaft Segmental Defect. *Biomed. Res. Int.* **2018**, *2018*, 11. [CrossRef] [PubMed]

24. Lemu, H.G. Virtual engineering in design and manufacturing. *Adv. Manuf.* **2014**, *2*, 289–294. [CrossRef]

25. Shah, F.A.; Thomsen, P.; Palmquist, A. Osseointegration and current interpretations of the bone-implant interface. *Acta Biomater.* **2019**, *84*, 1–15. [CrossRef] [PubMed]

26. Wysocki, B.; Maj, P.; Sitek, R.; Buhagiar, J.; Kurzydłowski, K.; Święszkowski, W. Laser and Electron Beam Additive Manufacturing Methods of Fabricating Titanium Bone Implants. *Appl. Sci.* **2017**, *7*, 657. [CrossRef]

27. Li, X.; Wang, C.; Zhang, W.; Li, Y. Fabrication and characterization of porous Ti6Al4V parts for biomedical applications using electron beam melting process. *Mater. Lett.* **2009**, *63*, 403–405. [CrossRef]

28. Moiduddin, K.; Hammad Mian, S.; Alkindi, M.; Ramalingam, S.; Alkhalefah, H.; Alghamdi, O. An in vivo Evaluation of Biocompatibility and Implant Accuracy of the Electron Beam Melting and Commercial Reconstruction Plates. *Metals* **2019**, *9*, 1065. [CrossRef]

29. Hoang, D.; Perrault, D.; Stevanovic, M.; Ghiassi, A. Surgical applications of three-dimensional printing: A review of the current literature & how to get started. *Ann. Transl. Med.* **2016**, *4*, 456.

30. Lee, J.-W.; Fang, J.-J.; Chang, L.-R.; Yu, C.-K. Mandibular Defect Reconstruction with the Help of Mirror Imaging Coupled with Laser Stereolithographic Modeling Technique. *J. Formos. Med. Assoc.* **2007**, *106*, 244–250. [CrossRef]

31. Kashi, A.; Saha, S. Mechanisms of failure of medical implants during long-term use. In *Biointegration of Medical Implant Materials*; Sharma, C.P., Ed.; Woodhead Publishing: Cambridge, UK, 2010; pp. 326–348.

32. Singare, S.; Shenggui, C.; Sheng, L. The use of 3D printing technology in human defect reconstruction-a review of cases study. *Med. Res. Innov.* **2017**, *1*, 1–4. [CrossRef]

33. Moiduddin, K.; Al-Ahmari, A.; Nasr, E.S.A.; Mian, S.H.; Al Kindi, M. A comparison study on the design of mirror and anatomy reconstruction technique in maxillofacial region. *Technol. Health Care* **2016**, *24*, 377–389. [CrossRef] [PubMed]

34. Jardini, A.L.; Larosa, M.A.; Maciel Filho, R.; de Carvalho Zavaglia, C.A.; Bernardes, L.F.; Lambert, C.S.; Calderoni, D.R.; Kharmandayan, P. Cranial reconstruction: 3D biomodel and custom-built implant created using additive manufacturing. *J. Cranio-Maxillofac. Surg.* **2014**, *42*, 1877–1884. [CrossRef] [PubMed]

35. Lütjering, G.; Williams, J.C. *Titanium*; Springer Science & Business Media: Berlin, Germany, 2007.

36. Arcam, A. Arcam A2 Technical Specification. ARCAM A2 TECHNICAL DATA. 2019. Available online: http://www.arcam.com/wp-content/uploads/Arcam-A2.pdf (accessed on 23 May 2019).

37. New! 50 μm Process for High Resolution and Surface Finish. 21 May 2012. Available online: http://www.arcam.com/new-50-um-process-for-high-resolution-and-surface-finish (accessed on 14 September 2019).

38. Umer, U.; Ameen, W.; Abidi, M.H.; Moiduddin, K.; Alkhalefah, H.; Alkahtani, M.; Al-Ahmari, A. Modeling the Effect of Different Support Structures in Electron Beam Melting of Titanium Alloy Using Finite Element Models. *Metals* **2019**, *9*, 806. [CrossRef]

39. Mian, S.H.; Mannan, M.A.; Al-Ahmari, A.M. The influence of surface topology on the quality of the point cloud data acquired with laser line scanning probe. *Sens. Rev.* **2014**, *34*, 255–265. [CrossRef]

40. El-Anwar, M.I.; Mohammed, M.S. Comparison between two low profile attachments for implant mandibular overdentures. *J. Genet. Eng. Biotechnol.* **2014**, *12*, 45–53. [CrossRef]

41. Arcam, Ti6Al4V ELI Titanium Alloy, Ti6Al4V ELI Titanium Alloy. Available online: http://www.arcam.com/wp-content/uploads/Arcam-Ti6Al4V-ELI-Titanium-Alloy.pdf (accessed on 23 September 2019).

42. Nagasao, M.; Nagasao, T.; Imanishi, Y.; Tomita, T.; Tamaki, T.; Ogawa, K. Experimental evaluation of relapse-risks in operated zygoma fractures. *Auris. Nasus. Larynx.* **2009**, *36*, 168–175. [CrossRef]

 *metals*

Article

Structural Aspects and Characterization of Structure in the Processing of Titanium Grade4 Different Chips

Krzysztof Topolski *, Jakub Jaroszewicz and Halina Garbacz

Faculty of Materials Science and Engineering, Warsaw University of Technology, Wołoska 141,
02-507 Warsaw, Poland; jakub.jaroszewicz@pw.edu.pl (J.J.); halina.garbacz@pw.edu.pl (H.G.)
* Correspondence: kt.topolski@gmail.com; Tel.: +48-22-2348740

Abstract: This study presents the structural aspects of the solid-state processing of various titanium chips. The structural characterization of: (1) commercial pure Ti in the as-received state, (2) manufactured chips, and (3) products of the chip processing are presented. Pure single-phase titanium Grade4 (Ti Gr4) was processed which, among all grades of pure titanium, is characterized by the lowest possible purity and the highest possible strength at the same time. Four geometries of chips were processed, i.e., chips after turning (thin and coarse), and chips after milling (thin and coarse). An unconventional plastic working method was applied to transform a dispersed form (chips) into solid, bulk metal in the form of rods without re-melting. The rods with a diameter of Ø8 mm and a length of about 500 mm were manufactured. Based on computer tomography and Archimedes measurements, it was found that the manufactured rods were consolidated and near fully dense. In turn, microscopy investigations proved that conventional, polycrystalline, grained structures were obtained. Only an insignificantly small number of internal defects were revealed, meaning that the obtained rods exhibited a proper structure typical for commercial titanium. Obtained materials, except of small surface inclusions, were fee of impurities. Whereas the results of the compression tests proved that the manufactured rods are characterized by new interatomic bonds, cohesion and plasticity analogous to those of titanium in the as-received state.

Keywords: titanium; chips; structural properties; structural characterization; recycling

Citation: Topolski, K.; Jaroszewicz, J.; Garbacz, H. Structural Aspects and Characterization of Structure in the Processing of Titanium Grade4 Different Chips. *Metals* **2021**, *11*, 101. https://doi.org/10.3390/met11010101

Received: 7 December 2020
Accepted: 30 December 2020
Published: 6 January 2021

Publisher's Note: MDPI stays neutral with regard to jurisdictional claims in published maps and institutional affiliations.

1. Introduction

For many years, the recycling of metals has been an important and developing field of industry. One area of this industry is the processing of chips and its transformation to useful semi-products. A basic technique of metal chips recycling is remelting. However, remelting is characterized by relatively high energy consumption, emissions of harmful gases, and solid waste. In addition, in the case of titanium and its alloys, protection against atmospheric air is necessary.

An alternative technique for chips recycling is processing without remelting, i.e., solid state processing. This is carried out using plastic working technology at an elevated temperature. For this purpose, (1) conventional methods of plastic working (direct extrusion [1–3], rolling [1,4]) as well as relatively new, (2) unconventional methods (Severe Plastic Deformation (SPD) [5–9], the KOBO method [10,11], friction stir extrusion [12] or friction back extrusion [13]) are used. In recent years, in this field, a significant increase in the number of publications on the subject has been observed.

In the context of solid state chips treatment, the most commonly processed metals are aluminum and its alloys [1,2,5,11,13–15], followed by magnesium and its alloys [3,5,12,16,17], whereas there is relatively little data in the literature on the processing of chips of pure titanium and its alloys [4,7–10,18].

In the context of chip processing-recycling, the structure of the material obtained is an important issue. The final quality and properties of the manufactured recycled products depend on: the amount and type of internal defects, impurities, chip bonding, grain shape, and grain size. Especially in the case of plastic working obtaining a proper structure (transformed from a dispersed form into a bulk and solid state without remelting) is a challenge. Hence, in order to consolidate chips into a new solid structure, the following conditions must be fulfilled: (1) sufficiently high stress, (2) sufficiently high temperature, (3) sufficiently high plastic deformation, and no external contamination [1,2,14].

Chips are dispersed macroscopic forms that have their own microstructure. After initial consolidation, the chips form a briquette, which is a consolidated macro-element having their own macrostructure consisting of chips. It should be noted that these chips have still their own microstructure. Therefore, in the context of chip processing, the material structure during, and after recycling, can be considered as both macrostructure and microstructure. The properties of the material structure after processing are important, because they determine whether the chip recycling process has been carried out correctly or incorrectly. A proper and defect-free structure ensures that the product possesses the quality and properties needed for it to be useful.

In the area of the processing of chips of titanium and its alloys, some studies in the literature regarding structural investigations are worthy of mention. For example, in the work [4] chips of pure titanium VT1-0 grade were processed by hot rolling, and this way strips were produced. That work presents the metallographic microstructures immediately after hot rolling, as well after annealing and recrystallization. In turn, in work [7] the ECAP method with back pressure was applied to recycle pure titanium Grade2 chips after milling. As a result, dense and consolidated products in the form of rods were manufactured. In order to characterize the obtained microstructures, high resolution electron backscatter diffraction (EBSD), transmission electron microscopy (TEM) and light microscopy were employed. Whereas in work [8], various types of Ti6Al4V chips were recycled using ECAP with back pressure and hot forging to produce the desired final microstructure. In that work, scanning electron microscopy (SEM) was used to observe the microstructures and chip boundaries. Another example of Ti6Al4V chips processed by ECAP with back pressure was presented in work [9]. In that study, structural effects, including chips interlocking, were analyzed using SEM and light microscopy, while transmission electron microscopy (TEM) was used to analyze the banded structure.

The novelty of our work is: (1) the material processed (tough pure mono-phase α titanium Grade4 with the highest possible hardness and strength), (2) the geometry of the materials (various shapes and thicknesses of chips), and (3) the unconventional plastic working technique, which makes it possible to process chips in the solid state, without remelting (direct extrusion with cyclic die rotation realized in two directions). The processing of titanium chips like these without remelting has not been the subject of any previous research, particularly in the context of structural analyses using such techniques of investigations.

The aim of this study is to present the structural aspects and structural transformation resulting from the solid state processing of various titanium chips. Four types of titanium chips were processed into four individual rods. These materials were characterized by means of a complex structural investigation. The effect of chip type on the results obtained is presented. Moreover, the relationship between the processing and the structures thus obtained is analyzed.

2. Materials Investigated

In this work, the material processed and investigated was pure mono-phase α titanium Grade4 (Ti Gr4) representing (among all grades of pure titanium) the lowest possible purity as well as the highest possible yield stress, tensile strength and hardness. Its chemical composition is given in Table 1. Generally, Ti Gr4 exhibits significantly higher strength and hardness than other grades of Ti and is more contaminated and less ductile. Therefore Ti Gr4 was hard to deform and more difficult to process. For example, the hardness of this material in the as-received state was 190 HV10, whereas the yield stresses (evaluated by compression tests) was 492 MPa.

In the as-received state, this investigated material was in the form of rod. The rod was subjected to machining into specific chips in order to obtain various and required types of chips (variants V1–V4). Ti Gr4 rod was subjected to turning (variants V1, V2) and milling (variants V3, V4) for two thickness of chips each time. The machining processes were carried out under conditions without emulsion cooling. This approach reduced the contamination of the chips. After the machining, the chips thus obtained were cleaned using acetone and an ultrasonic washing machine.

Next step of the processing was an initial consolidation of the chips into briquettes, heating, and finally, the extrusion of the briquettes into the final products in the form of rods.

Table 1. Chemical composition of investigated titanium Grade4.

Element	Fe	O	N	H	C	Al	V	Ti
[%] weight	0.314	0.26	0.012	0.001	0.02	-	-	Base

3. Investigation Methods

The structural characterization of the manufactured materials was conducted using: (1) computer tomography, (2) density measurements, (3) fluorescence spectrometry, (4) energy dispersive spectroscopy, (5) light microscopy, and (6) scanning electron microscopy.

To investigate and illustrate the transformation effect, X-ray computer tomography was used (Xradia Micro XCT 400 tomograph). Structural research using this method was carried out at a resolution of 25 μm (the "detection threshold"), meaning that any structural elements (e.g., pores) smaller than 25 μm were not detected. The samples intended for computer tomography investigation were 8 mm in diameter and 50 mm long, thus they were volumetric and bulk materials. They were cut out from the middle of the rods and then cleaned using acetone and an ultrasonic washing machine.

The density measurements were performed using the Archimedes method and an analytical balance with an accuracy of 0.01 g/cm^3. Exactly the same samples as for the computer tomography were used for the density measurements. The tests were carried out on the samples in as-received state (reference) and on all variants after processing. Three measurements were made for each sample.

In order to evaluate the potential contamination of the manufactured rods, chemical composition measurements using X-ray fluorescence spectrometry were realized. An Olympus DELTA XRF spectrometer was used. This device makes it possible to measure a surface in area of diameter of 1.5 mm and a depth of up to 100 μm. The samples were tested: (1) on the external cylindrical surface, (2) in transverse section, and (3) in longitudinal section. For each sample and area, two measurements were taken at random locations. The tests were carried out on the samples in as-received state (reference) and for all variants after processing.

To obtain additional information about the chemical composition and chemical purity of the manufactured rods, the chemical microanalysis using scanning electron microscopy with energy dispersive spectroscopy attachment (SEM-EDS) was also used. Samples from each variant (V1–V4) were tested. Each variant was examined using one sample presenting transverse section and one sample presenting longitudinal sections. Moreover, two analyzes were performed for each sample: (1) on the circuit of the sample-rod, (2) in the area of the axis-center of the sample-rod. This gave sixteen measurements in total. The samples were subjected to elemental qualitative and quantitative analysis by taking a spectrum from their surface. Square areas with dimensions of 1 mm $\times$ 1 mm were analyzed.

In order to reveal the microstructures of the processed materials, the samples were ground, polished and etched. In this way special metallographic samples were prepared. On the transverse section, for each variant the entire area of the sample was etched and analyzed, i.e., the area of a circle with a diameter of 8 mm. Similarly, in the longitudinal section, the entire area of the sample section was etched and analyzed. In the longitudinal section, the tested surface was rectangular, with sides of 8 mm (height) $\times$ 10 mm (length).

The revealed microstructures were analyzed using light microscopy. For the purpose of estimating grain size and grain shape, image analysis software was used [19]. The grains were characterized based on the following stereological parameters: (1) average grain size $E(d_2)$ (i.e., the average value (E) of the average equivalent grain diameters (d_2)), (2) SD—standard deviation, (3) CV—coefficient of variation (where CV = SD/average grain size), (4) "α" shape factor (defined as: $\alpha = d_{max}/d_2$, where d_{max} is the maximum chord of a grain), and (5) "β" shape factors (defined as: $\beta = p/(\pi^* d_2)$, where "p" is the perimeter of a grain). It should be clarified that "α" shape factors determine the deviation of the shape of a grain from that of a circle, and characterize the elongation of the grain, while "β" shape factors determine the degree of development of the grain boundary surface. For each case, approximately one hundred grains were used for the stereological analysis.

Some of the structural examinations were also performed in a scanning electron microscope (SEM, Hitachi S3500) using the secondary electron (SE) mode. Observation was performed on the transverse and longitudinal sections of the samples.

The compression tests were conducted at room temperature using a universal vertical MTS 858 hydraulic testing machine. The samples had the standard cylindrical shape and were 7 mm in diameter and 10.5 mm high (height to diameter ratio equal 1.5). Two tests were conducted for each state of material.

4. Results and Discussion

The processing procedure consisted of the following stages: (1) characterization of titanium in the as received state, (2) the machining of the Ti in the as-received state into the specific chips required, (3) the preliminary consolidation of the chips into the form of "briquettes", (4) the heating of the briquettes before extrusion, and (5) extrusion of the briquettes into the form of solid rods using the KOBO method.

4.1. Titanium in the As-Received State

Light microscopy observations of the etched samples revealed that the microstructures of Ti Gr4 in the as-received state were the same in the transverse section (Figure 1a) and in the longitudinal section (Figure 1b). In both these cases, the microstructure was characterized by equiaxed grains whose size ranged from 10–50 μm. For the transverse section, the average grain size of the initial Ti Gr4 was 32 μm (with SD = 9.71, CV = 0.30), while for the longitudinal section 28 μm (with SD = 8.34, CV = 0.30; also see Table 8).

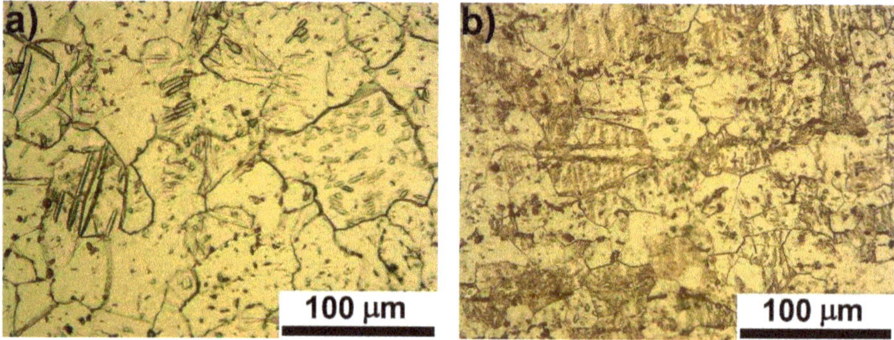

Figure 1. Samples of microstructures revealed by etching, Ti Gr4 as-received state, light microscopy: (**a**) transverse section, (**b**) longitudinal section.

The average density of the Ti Gr4 in the as-received state was 4.51 [g/cm^3]. This value, measured by the Archimedes method, was consistent with the theoretical value of titanium (4.51 [g/cm^3]). As mentioned previously, among others titanium grades, Ti Gr4 is tough grade. For example, the hardness of the processed Ti in the as-received state was 190 HV10, whereas the yield stresses (evaluated by compression tests) was 492 MPa.

4.2. Chips Manufactured by Machining

In this step of processing, Ti Gr4 rod in the as-received state was subjected to turning and milling without any cooling with the aim to produce the various, specific chips. The chips were prepared by (1) turning and (2) milling. Beside, for each type of these chip, two thicknesses were prepared: (1) thin (~0.1 mm) and (2) coarse (~0.3–0.4 mm), thus making four variants of the experiments. A detailed description of these variants is provided in Table 2.

Table 2. Detailed description of the four variants of TiGr4 used in the experiments.

Variant No.	Experiment–Samples			Variant Description
	Material	Machining	Chip Thickness	
1		turning	thin (~0.1 mm)	V1
2	Ti Gr4		coarse (~0.3–0.4 mm)	V2
3		milling	thin (~0.1 mm)	V3
4			coarse (~0.3–0.4 mm)	V4

Examples views of these chips are presented in Figure 2a (variant V2) and Figure 2b (variant V4). Generally, the chips after turning (V1,V2) were long, twisted, ribbon-like and tangled, whereas the chips after milling (V3,V4) were short, flaky and loose. During the machining process the cooling by emulsion was not applied. Thus, titanium chips were not contaminated chemically by coolant. Moreover, the obtained chips were not significantly oxidized because their color remained typical for titanium, i.e., silvery-white. In contrast, the surface of the manufactured rods was significantly oxidized and, for this reason, these surfaces exhibited other, various colors.

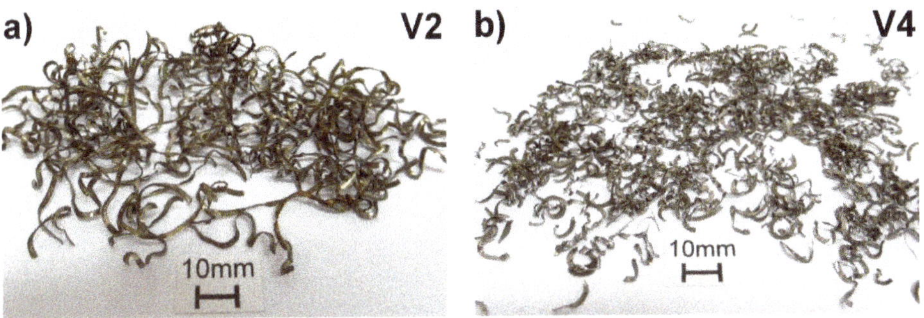

Figure 2. Macroscopic images of titanium Gr4 chips manufactured by machining: (**a**) coarse chips obtained by turning—variant V2, (**b**) coarse chips obtained by milling—variant V4.

In the context of chip microstructure, it should be explained that, during the applied machining, a very large plastic strain was introduced into each chip. This extremely high deformation changed the structures of the titanium significantly. Comparing the microstructure of the chips with those in the as-received state there were no equiaxial grains, and the grain boundaries were poorly visible. Instead, texture and deformation bands were visible, with only a few elongated grains. Some sample views of the microstructure of the chips are presented in Figure 3.

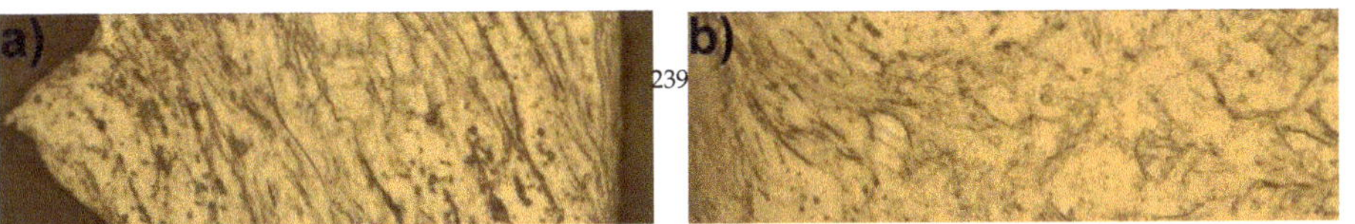

4.3. Briquettes Obtained after Preliminary Consolidation of Chips

In this step, the chips thus obtained were subjected to a preliminary consolidation using the upsetting method. The compaction process was carried out in a vertical position using special tools and a conventional hydraulic press. The applied stamp load was increased up to 30 tons. Each process was conducted in multiple strokes. In this way, cylindrical billets with a diameter of about 38 mm, called "briquettes", were obtained. Each of the four types of chips was compacted separately, thus four individual briquettes were prepared.

The briquettes were characterized by a porous macrostructure consisting of deformed, tangled, bent and compacted chips. In this way, special, characteristic macrostructures consisting of chips were formed, where the chips had their own microstructure. For each type of chip, one separate briquette was obtained. Sample macroscopic views of the manufactured briquettes are presented in Figure 4a (variant V2) and Figure 4b (variant V4). As can be seen, the briquettes were solid and they formed the agglomerate, with relatively strong mechanical connections between particular macro-elements (chips).

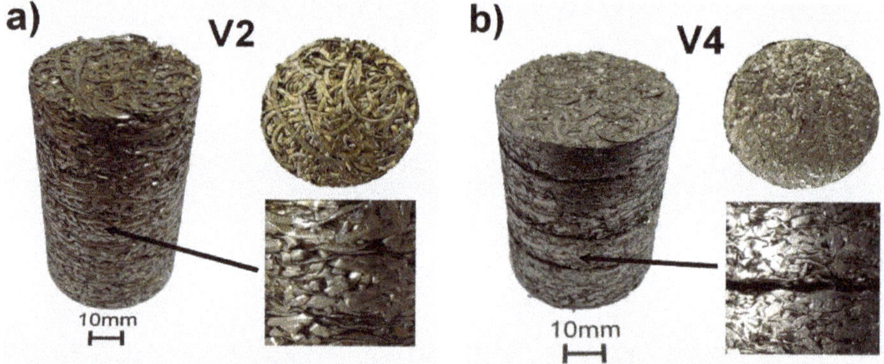

Figure 4. Sample views of titanium Gr4 briquettes obtained by preliminary consolidation: (**a**) briquette consisting of coarse chips after turning—variant V2, (**b**) briquette consisting of coarse chips after milling—variant V4.

The briquettes obtained were subjected to the density measurements. Density was estimated based on the weight and dimensions of the briquettes (Table 3). It was observed that the densities of all of the briquettes were significantly lower than the theoretical density of solid commercial titanium (4.51 g/cm^3) and that of the rods produced by extrusion (about 4.50 g/cm^3, also see Table 7). This confirms that the briquettes were porous, and that there were some voids between the individual chips. The densities of the briquettes obtained from the chips after turning were lower than those of the briquettes obtained from the chips after milling, which is also confirmed by the macroscopic photos of the briquettes.

Table 3. Briquette density evaluated on the basis of mass and dimension measurements.

Sample	V1	V2	V3	V4
Density [g/cm^3]	2.63	2.52	2.77	2.70

4.4. Heating of the Briquettes before Extrusion

Before extrusion, each briquette was separately placed in a press chamber and was heated (through heating the chamber, together with the chamber), to a temperature of 350 °C. Annealing at that temperature was then applied for 20 min, after which the briquettes remained in the press chamber, which was not cooled or subjected to any other thermal process before extrusion. The heating was ongoing during the extrusion process.

Thus, during the short time before the extrusion was initiated, the temperatures of the briquette and press chamber did not decrease.

The purpose of the annealing was: (1) to improve the plasticity of the chip-briquettes, (2) to reduce the extrusion force, (3) to facilitate the phenomenon of chip bonding, which is an important factor in the context of consolidating the chips and transforming them into solid form, and (4) to further facilitate, as an important part in the heat balance, the phenomenon of dynamic recrystallization.

During the extrusion process, due to friction and plastic deformation phenomena, the temperature of the processed titanium increased additionally probably by about 200–300 °C, which further improved the ductility of the titanium and promoted bonding between the chips. On the other hand, both the rod-product and the die were cooled with water in the area of the die exit. Finally, in short, it can be assumed that the real-total extrusion temperature in the zone of deformation and die entry was about 600 °C. This temperature also resulted in the desired phenomenon of dynamic recrystallization.

It is important to mention that, due to the initial annealing of the briquettes, the microstructure of the chips did not change essentially, and this treatment had no significant impact on the initial chip microstructure. Thus, the microstructure after annealing remained the same, and annealing was not a separate and additional step in the microstructure evolution.

4.5. Rods Manufactured by Extrusion

The final and key operation was direct extrusion with cyclic die rotation, realized alternately in both directions. Applied twist angle of die was ±6°, whereas oscillation frequency of die was 5 Hz. This unconventional plastic working process is commonly known as the KOBO method. Detailed information on this method was presented in the work [10]. The dies for extrusion of titanium chips were made of a special Inconel 718 alloy after solution and ageing. The chemical composition of the applied Inconel 718 was multi-element. However, the major elements were as follows (%weight): 53.5 Ni; 18.61 Fe; 17.61 Cr; 5.05 Nb; 3.00 Mo. This chemical composition of the applied dies is important information in the context of the subsequent results of chemical composition tests of the rods-products.

The processes of extrusion (V1–V4) resulted in obtaining four separate volumetric products in the form of solid, bulk, consolidated rods. Their diameter was 8 mm, and their length ranged from 400–600 mm (see example photos: Figure 5a—variant V2 and Figure 5b—variant V3).

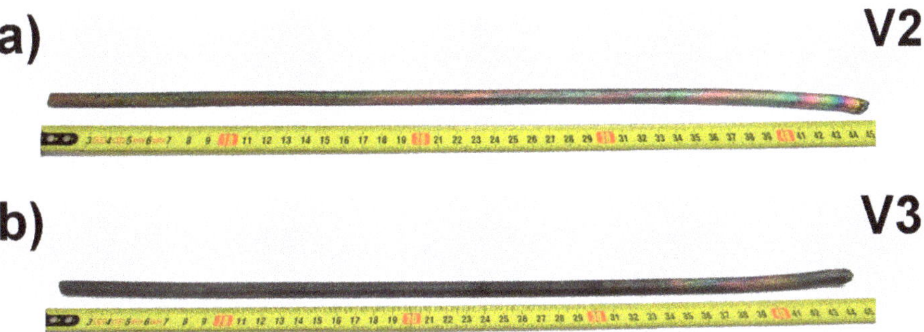

Figure 5. Example photos of final rods-products obtained by KOBO extrusion of briquettes: (**a**) variant V2 (Ti Gr4, coarse chips after turning), (**b**) variant V3 (Ti Gr4, thin chips after milling).

4.6. X-ray Computer Tomography Examinations of Rods

The rods obtained by processing the chips were subjected to structural investigations by X-ray computer tomography. This method made it possible to detect and visualize all spaces with a density other than that of titanium. These spaces then had a different color that creates a contrast. Thus, this technique reveals any matter different than titanium, i.e., (1) possible internal defects such as: pores, voids, delaminations or cracks, and (2) any possible metallic impurities resulting from the multi-stage processing. These non-destructive investigations were performed using the volumetric samples, so it was possible to precisely determine (in the all sample volume) the location of these areas and as well as to estimate their quantity. It should be emphasized that this technique has not previously been reported in the literature data regarded the recycling of chips. Using this computer tomography method, a special approach was taken, that made it possible to examine all the large volumes of the bulk, volumetric samples (length 50 mm and diameter 8 mm).

On the basis of the set of photos obtained during the tests, three-dimensional models were generated that illustrated the internal structure of the tested samples. These models are semi-transparent. The results obtained are presented in Figure 6 (variants V1, V2, V3, V4). In order to illustrate the geometry and arrangement of individual areas, two views are presented: longitudinal and transverse. For each sample, two separate models were generated: one illustrating spaces with a density less than titanium (material ρ < Ti ρ), and second illustrating spaces with a density greater than titanium (material ρ > Ti ρ). Hence, the following colored spaces are visible in the models:

(a) semi-transparent, gray color—titanium,

(b) blue color- spaces less dense than titanium (material ρ < Ti ρ). These spaces should be identified and interpreted as: closed pores, voids and/or cracks, or delamination. Thus this covers a wide group of defects that could have resulted from completely different causes. However, all of them are less dense than titanium. They appeared together and simultaneously, but can be distinguished on the basis of geometrical analysis and other microscopic research techniques,

(c) orange color—material with a density greater than titanium (material ρ > Ti ρ). These spaces should be identified and interpreted as metallic impurities-inclusions coming from the die. These materials-particles were transferred during extrusion as a result of tribological phenomena.

Three-dimensional image reconstruction made it possible to conduct a quantitative morphometric analysis of the samples. The total percentage share of material whose density was less than that of titanium and the total percentage share of material whose density was greater than that of titanium were calculated. The results of these calculations are presented in Table 4. The resolution of this investigation was 25 µm (detection threshold) meaning, that any structural elements smaller than 25 µm were not detected.

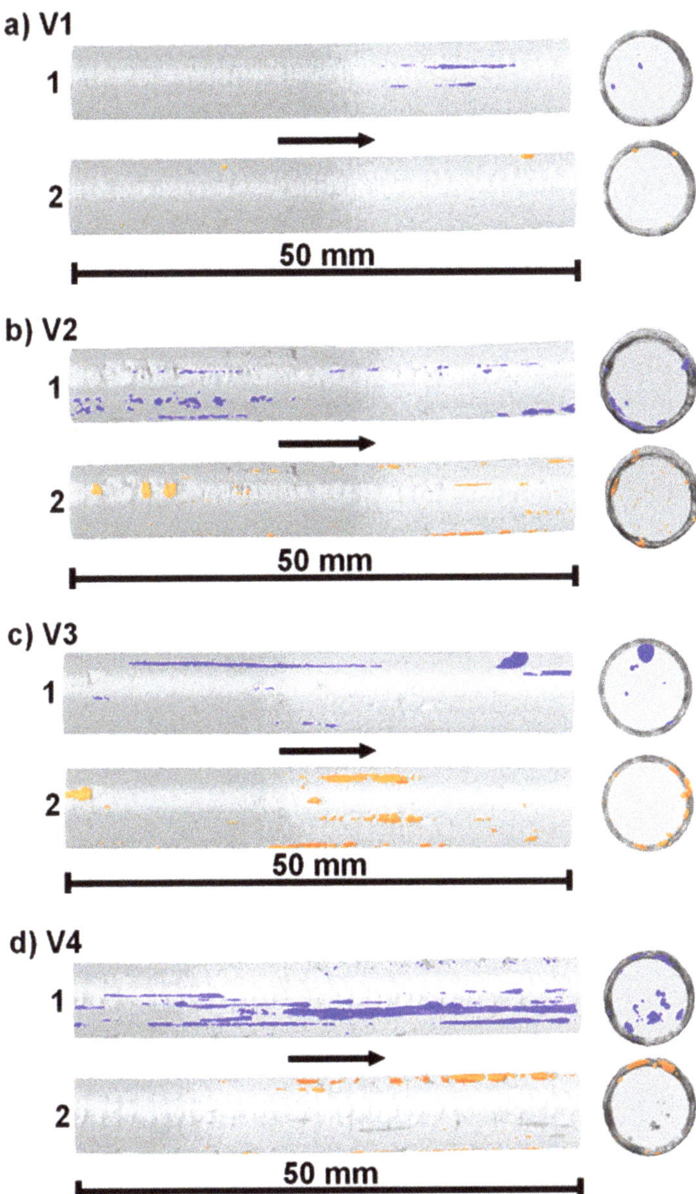

Figure 6. Results of X-ray computer tomography obtained for TiGr4; (1) visualization of detected material having a density less than that of titanium, (2) visualization of detected material having a density higher than that of titanium; (**a**) V1 turning, thin chips, (**b**) V2 turning, coarse chips, (**c**) V3 milling, thin chips, (**d**) V4 milling, coarse chips. The arrow indicates the direction of extrusion.

Table 4. Results of the quantitative calculation of the 3D models referring the total percentage share of material with density different than those of Ti. Analysis for spaces with a density less than that of titanium (material ρ < Ti ρ, marked in blue on the 3D models) and for a density greater than that of titanium (material ρ > Ti ρ, marked in orange on the 3D models). The applied resolution was 25 μm (detection threshold), thus the calculation covers structural elements larger than 25 μm only.

Samples TiGr4	Material ρ < Ti ρ [%]	Material ρ > Ti ρ [%]
V1	0.02	0.01
V2	0.11	0.10
V3	0.19	0.20
V4	0.54	0.24
Average	0.21	0.14

The total percentage share of materials with a density less than that of titanium (material ρ < Ti ρ) was calculated as the ratio of the sum of the volumes of all such materials ($V_{\rho<Ti}$) to the volume of the entire sample (V_{sample}), i.e., ($V_{\rho<Ti}/V_{sample}$) * 100%. Similarly, the total percentage share of materials with a density greater than that of titanium (material ρ > Ti ρ) was calculated as the ratio of the sum of the volumes of all such materials ($V_{\rho>Ti}$) to the volume of the entire sample (V_{sample}), i.e., ($V_{\rho>Ti}/V_{sample}$) * 100%.

All of the spaces (defects) with a density less than that of titanium were elongated and dragged along the rods in the direction of extrusion, which is typical for extrusion processes. The spaces were continuous or intermittent. On the transverse section, these spaces were characterized by small dimensions and were in the shape of a circle or ellipse. The geometry of these defects suggests that they are pores (voids), i.e., closed spaces filled with atmospheric air. Such structural elements may form when, during extrusion, air present in a briquette becomes trapped inside (blocked and closed). The porosity detected occurred in various locations distributed randomly.

Considering the samples of Ti Grade4 (V1–V4), the porosity was in the range of from 0.02–0.54%, and the average was only 0.21%, what is a favorable result. In these cases, good consolidated and compacted materials were obtained whose porosity was relatively small.

Despite the fact that, the densities of the briquettes obtained from the chips after turning were lower than those of the briquettes obtained from the chips after milling, it was observed that, the rods obtained from chips after turning had a slightly lower porosity than those obtained from chips after milling. The possible reason for this was that, during extrusion of the V3–V4 chips, air present in a briquette was more blocked. Besides, the V3–V4 chips were short, flaky and fine. This could have caused them to adhere more to the die. Thus the processing of V3–V4 chips may have been disrupted. In addition, the extrusion of coarse chips resulted in higher rod porosity than the extrusion of thin chips. This could be due to the greater stiffness of coarse chips and their higher resistance to plastic forming. This was probably due to the purity of titanium, and specifically to their relatively high yield point (~500 MPa). With this high yield point, the thickness of the chips probably influence on the extrusion.

Nevertheless, it should be emphasized that the differences in porosity obtained for some individual variants were relatively small. In addition, the specific nature of such plastic forming is that, the porosity in different areas of the rods can differ slightly. There-fore, the most reasonable conclusion is that, regardless of chip type, the porosity of the rods obtained using the KOBO method is in a narrow range and does not exceed a certain, relatively small limit. The porosity of the Ti Gr4 rods did not exceed 0.6%. Hence, it can be concluded that processing titanium chips by preliminary upsetting followed by KOBO extrusion makes it possible to obtain rod-products whose porosity is less than 0.6%.

The defects with a density greater than that of titanium (metallic impurities-inclusions, material ρ > Ti ρ, orange color on the 3D models) did not occur randomly. They were revealed on the perimeter, i.e., on the external surface of the rods. Exception could be sample V2, which exhibited some insignificant number of such defects inside the rod also.

Defects with a density greater than Ti were elongated and dragged along the rods, i.e., in the direction of extrusion, which is typical for such processes. It was observed that the sample V1 exhibited a quantity of metallic defects close to zero. In this case, porosity was close to zero also, thus very high quality product was obtained, which level of defects was insignificantly low. In the remaining cases (V2, V3, V4), the amount of impurity was higher, but still relatively low.

Taking into account all of the samples of Ti Gr4 (V1–V4), the impurities ranged from 0.01–0.24% (average 0.14%). These results are satisfactory and (despite the difficult tribological conditions associated with extrusion) confirm the high purity of the products obtained. Generally, the extrusion of Ti Gr4 resulted in some small amount of impurities along the rod's surface caused by the metals being transferred from the dies. This was because tribological wear of the dies and, thus, transfer of metals from the dies to the surface of the rods. Generally, it was found that, regardless of the type of chips, the impurities in the rods obtained by the KOBO method were in a narrow range and did not exceed a relatively small limit of 0.3%.

As stated previously, the rods obtained from chips after turning had a slightly lower porosity than those obtained from chips after milling. Similarly, it was observed, that these rods obtained from chips after turning had also a smaller amount of impurities than those obtained from chips after milling. The possible reason for this was that, in the V3–V4 case, thin and finest chips more adhered to the die. Thus, there was a greater wear of the front surface of the dies and greater transferring of particles from the dies.

In addition, analogous to the results of porosity, the extrusion of the thin chips resulted in slightly less impurity than the extrusion of the coarse chips. It results from greater difficulty in deforming coarse chips. Nevertheless, it should be emphasized that the differences obtained for individual variants were relatively small. In addition, a characteristic feature of such processing is that the share of defects in different areas of a rod can differ slightly.

Some relationship between porosity and defects having a density greater than that of Ti has been observed. The samples with higher porosity also had a higher content of non-Ti metals. This suggests the influence of die wear on the quality of the rod obtained.

Summing up, a dispersed form (chips) was successfully transformed into new, consolidated, near fully dense and compacted solid state structures. Based on the computer tomography examination, in the manufactured rods, no initial chip boundaries were revealed. Moreover, the voids present between the chips in the briquettes were not detected too. In other words, the macrostructures of the briquettes (consisting of chips) were fully processed.

4.7. Chemical Analysis of the Surfaces of the Rods by XRF

The computer tomography research revealed the presence on the rod's surface of some particles characterized by a density higher than that of titanium (material $\rho >$ Ti ρ). To identify and verify these materials, an analysis was performed using X-ray fluorescence spectrometry. The results of the chemical analysis of the surfaces of the manufactured rods are presented in Table 5. The applied XRF technique allowed identification of elements ranging from magnesium to uranium, inclusive. This means that, using XRF, it was impossible to detect light elements such as oxygen, hydrogen, carbon and nitrogen present in Ti Gr4. Therefore, the XFR studies did not include these elements.

Based on the chemical tests conducted, additional, foreign metallic elements—Ni, Nb, Mo and Fe were found on the external cylindrical surfaces of the rods (note, Fe is part of the chemical composition of Ti Gr4 simultaneously). A chemical analysis confirmed that these particles-structural components on the rod's surface were metallic impurities-inclusions. According to the results obtained using computer tomography, each of these elements had a density greater than that of titanium. These were defects, and should be interpreted as impurities that came from the die. The dies for the KOBO extrusion were made of a special Inconel 718 alloy containing exactly such metallic elements. Hence, these metals were transferred from the die during the extrusion process as an effect of tribological phenomena.

These multicomponent (multimetallic) particles were formed, in some places on the surface of the rods, as a result of mixing, bonding, welding and grafting of the titanium with the Inconel alloy.

Nevertheless, it should be emphasized that the chemical analysis confirmed the results obtained by tomography, that quantitative content of these impurities was relatively small and insignificant. Further, the small amounts of these foreign metals occurred on the surface of the rods mainly, and could easily be removed by surface treatment. Our examinations by XRF of the transverse and longitudinal sections of the rods did not reveal the presence of such impurities inside. The only exception was the sample V2 on the transverse section (detected Cr = 0.42 ± 0.12%wt.) and the sample V3 on the transverse section (detected Cr = 0.43 ± 0.12%wt.; Mo = 0.02%wt.) where the some insignificant amount of foreign elements were transferred inside the rod also.

Table 5. Results of chemical analysis performed on the external cylindrical surfaces of the manufactured rods. Results obtained by X-ray fluorescence spectrometry. Description: [%wt.]—content in weight percent, Av—average value. Statistic—number of detected cases per two measurements (two surface measurements were made for each variant). Error of measurements: Ti ±0.05; Fe ±0.04.

Additional Element		Fe	Ni	Mo	Nb	Cr	Ti (Base)
Range [%wt.] (average)		0.28–3.42 (Av = 1.181)	0.00–0.24 (Av = 0.120)	0.00–0.06 (Av = 0.015)	0.00–0.02 (Av = 0.007)	0.00 (Av = 0.00)	96.26–99.72 (Av = 98.677)
Statistics	V1	2/2	2/2	0/2	0/2	0/2	2/2
	V2	2/2	2/2	1/2	1/2	0/2	2/2
	V3	2/2	2/2	1/2	1/2	0/2	2/2
	V4	2/2	1/2	2/2	1/2	0/2	2/2

4.8. Chemical Analysis of the Rods by SEM-EDS

To extend the analysis of chemical purity/contamination of the manufactured rods, the chemical microanalysis using SEM-EDS was also used (Table 6). The results showed, that the tested areas selected from manufactured rods, in a total of fourteen out of sixteen cases, consist entirely of titanium and are not contaminated with other elements. Although insignificant amount of Si was detected in several cases, it should not be taken into account. Silicon occurs only superficial. Its presence is the result of sample preparation, i.e., it is a polishing residue. In only two out of sixteen cases, some amount of Ni and Cr were detected. These are the only contaminants found. Foreign elements Ni and Cr come from the dies and were detected only on the circuit of the samples (rods).

It should be explained that, according to the Ti Grade4 standard, this grade contains about 0.8% wt. solid solutes elements other than Ti, such as: Fe, and light elements: O, N, H, C (see Table 1). However, in this case, their presence is not detectable by SEM-EDS. Based on the SEM-EDS investigation, it was theoretically possible to detect light elements such as O, H, N and C. However, they were not detected (the content was evaluated as zero) because their quantity and concentration were apparently too low. Thus, this content had to be relatively small, and, at the same time, similar to the Ti in the as received state. In the context of: contamination, content of light metal and oxidation, it is worth explaining, that the microscopic observations did not reveal the presence of the oxide lattice in the manufactured materials.

Generally, the results of the chemical composition tests obtained by the SEM-EDS method are consistent with the results obtained using computer tomography and the XRF method. However, some differences resulted from a different testing method and different samples (external cylindrical surfaces versus metallographic specimen from the sections).

Table 6. Results of the chemical composition obtained using SEM-EDS technique. Samples V1–V4 analyzed on the transverse section and on the longitudinal section in different areas (circuit, center). Error of measurements: ±0.66% wag.

Element [%] Weight	V1 Transverse		V2 Transverse		V3 Transverse		V4 Transverse	
	Circuit	Center	Circuit	Center	Circuit	Center	Circuit	Center
Ti	100	100	99.77	99.72	100	99.77	100	100
Ni	-	-	-	-	-	-	-	-
Fe	-	-	-	-	-	-	-	-
Cr	-	-	-	-	-	-	-	-
Nb	-	-	-	-	-	-	-	-
Mo	-	-	-	-	-	-	-	-
Si	-	-	0.23	0.28	-	0.23	-	-
Other total	-	-	-	-	-	-	-	-

Element [%] Weight	V1 Longitudinal		V2 Longitudinal		V3 Longitudinal		V4 Longitudinal	
	Circuit	Center	Circuit	Center	Circuit	Center	Circuit	Center
Ti	100	100	93.59	100	90.41	100	99.68	99.71
Ni	-	-	4.97	-	7.50	-	-	-
Fe	-	-	-	-	-	-	-	-
Cr	-	-	1.44	-	2.09	-	-	-
Nb	-	-	-	-	-	-	-	-
Mo	-	-	-	-	-	-	-	-
Si	-	-	-	-	-	-	0.32	0.29
Other total	-	-	-	-	-	-	-	-

4.9. Density Measurements

The manufactured materials were also subjected to density measurements in order to evaluate their structural quality and the consolidation effect. Ti Gr4 sample in the as-received state was also tested as references for the materials after processing. The density result obtained for titanium in the as-received state was designated as 100%, and then the density results for titanium after processing (V1–V4) were calculated in relation to the density of Ti in the as-received state, described as percentages, and defined as relative density.

The results show that all the rods obtained (V1–V4) were characterized by the typical, correct density for solid titanium (Table 7). The theoretical-nominal density of solid commercial Ti is 4.51 g/cm^3 (4.507 g/cm^3), and may differ slightly depending on purity and temperature. The values of densities obtained were only slightly and insignificantly lower (0.01 g/cm^3) than the theoretical density of solid titanium. This could be the effect of the minimal porosity revealed in the computer tomography investigations. The density of the Ti Gr4 rods remained the same, at 4.50 g/cm^3 for all V1–V4 cases. The densities of the V1–V4 rods were significantly higher than those of the briquettes (about 3.00 g/cm^3, see Table 3), which is another evidence of the transformation of the chips.

Although the plastic shaping and deformation of Ti Gr4 is more difficult than other grades of Ti, surprisingly, the Ti Gr4 had practically a correct density. In addition, the density of the Ti Gr4 rods was correct, despite the fact that the porosity of these rods was higher than 0.00%. These inconsistent differences in the porosity-density results may be explained by the fact, that these results are influenced by surface contamination of Ti Gr4 with metals-particles heavier than Ti (which inflate the density results). No clear and significant effect of: (1) chips type (turning, milling) or (2) chips thickness on the density results was observed.

Referring the obtained results to solid Ti in the as-received state, the Ti Gr4 rods (V1–V4) exhibited a very high relative density equal 99.78%. Comparing these results to the literature, for example, in work [9], Ti6Al4V chips were processed by ECAP, and relative densities of from 92–99.9% were obtained, depending on: the number of ECAP passes, the value of back pressure, and the processing temperature. Whereas Ti Gr2 chips obtained by milling and processed by ECAP resulted in products with a density ranging from 4.50 to 4.53 g/cm^3 [7].

In summary, the results of the density measurements confirmed that the applied extrusion made it possible to transform the briquettes (having a porous structure) into solid, volumetric, near fully dense and condensed materials. The densities of the manufactured rods were very close to that of commercial titanium available in the trade, and quite satisfactory.

Table 7. Results of density measurements [g/cm^3] obtained for Ti Gr4 and Ti hp 99.99 performed using the Archimedes method.

Sample Ti Gr4	As-Received	V1	V2	V3	V4
Measurement No.1	4.52	4.50	4.50	4.50	4.49
Measurement No.2	4.51	4.49	4.50	4.50	4.50
Measurement No.3	4.51	4.50	4.50	4.50	4.50
Average density	4.51	4.50	4.50	4.50	4.50
Relative density [%]	100.00	99.78	99.78	99.78	99.78

4.10. Structural Investigations of Rods Using Light Microscopy

Our research of the manufactured rods showed that in all variants (V1–V4) a grained, solid state structure typical of polycrystalline metals was obtained. Exemplary and representative photos of rod's microstructures are presented in Figure 7 (transverse section) and in Figure 8 (longitudinal section). It should be emphasized, that the grained microstructures obtained are typical of commercial titanium, and were present in the as-received Ti as well. For each variant, after the metallographic observations the grain size and the grain shape were calculated. The results of the stereological analysis of these materials are summarized in Table 8.

Generally, for each rod (V1–V4), a very homogeneous microstructure with equiaxed grains was obtained. The microstructures of all the rods were characterized by similar grains. On the entire tested surface of each sample, there were grains of very similar size and shape, which was confirmed by the small values of the SD and CV coefficients, and similar values of α parameters and β parameters that were close to "1". For each variant, the entire surface of the sample exhibited the same, homogeneous microstructure. In other words, there were no differences between the various regions of the samples.

For each variant, transverse and longitudinal sections were tested. Comparing both these sections of a given variant, it was found that in each case the microstructure was the same. On the longitudinal sections no effects of texture or elongated grains were observed such as often occur in products such as rods.

A characteristic feature of all revealed microstructures (V1–V4) was that they had a very small grain size, significantly smaller than the grains of the titanium in the as-received state. This was the effect of a dynamic recrystallization that took place during the extrusion process, and of a lack of subsequent grain growth. Grain growth was deliberately limited by the: (1) short extrusion time, (2) cooling of the rods at the die exit, and (3) the optimal temperature of the briquettes. In machining, the chip formation occurs by shear deformation, which is concentrated in a narrow deformation zone inside the treated material. Hence, during each single machining pass, a very large plastic strain is introduced into each chip [10,20,21]. As a result of this extremely high deformation, the structure of the material changes significantly, and becomes different than in the as-received state before machining (see Figure 3). In chip microstructures, there are no equiaxed grains, and the grain boundaries are often invisible. Instead, there are texture and deformation bands. Since the equiaxial grains visible in the rods did not come from the chips, they had to be the result of dynamic recrystallization, which must have occurred during the extrusion of the briquettes. The chip structure is in a metastable state, the chips contain some stored energy and structural defects. In the next stage, when the chips were extruded using appropriate parameters (temperature, strain), this made the initiation of dynamic recovery and the recrystallization of the titanium.

Another reason for a hypothesis on recrystallization is the presence of the same equiaxed grain shape in both sections. In the longitudinal section, the grains are not elongated in the direction of extrusion. Extruding at a lower temperature without recrystallization leads to crystallographic texturing with strongly deformed, stretched grains, such were not present in the recycled titanium. Thus, it can be assumed that a dynamic recrystallization occurred during the extrusion process. The same conclusion was reached in work [2], where aluminum chips were recycled.

In the context of the grain microstructure obtained by chip recycling, the literature reports, for example, on the ECAP method applied to recycle pure Ti Gr2 chips after milling [7]. There, as well as in our work, it was observed that very small grains (with an average size of ~0.8 μm) were obtained at an extrusion temperature of 450 °C, while grains of about 3.9 μm were obtained at a higher temperature of 590 °C. It was also found that dynamic recovery and recrystallization processes may have occurred.

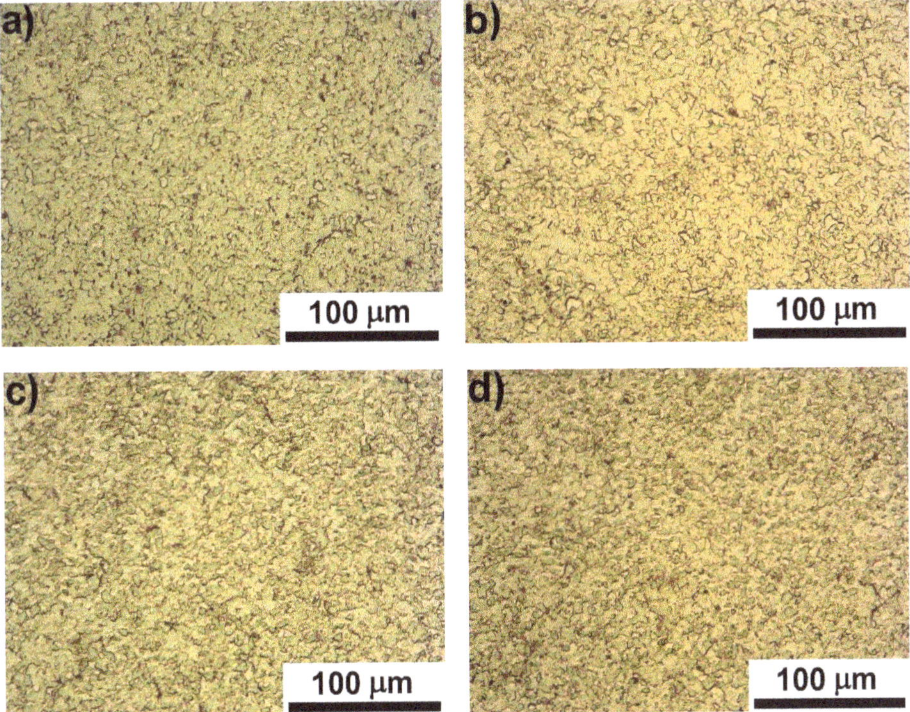

Figure 7. Microstructures of manufactured Ti Gr4 rods revealed by etching, transverse section, light microscopy: (**a**) variant V1, (**b**) variant V2, (**c**) variant V3, (**d**) variant V4.

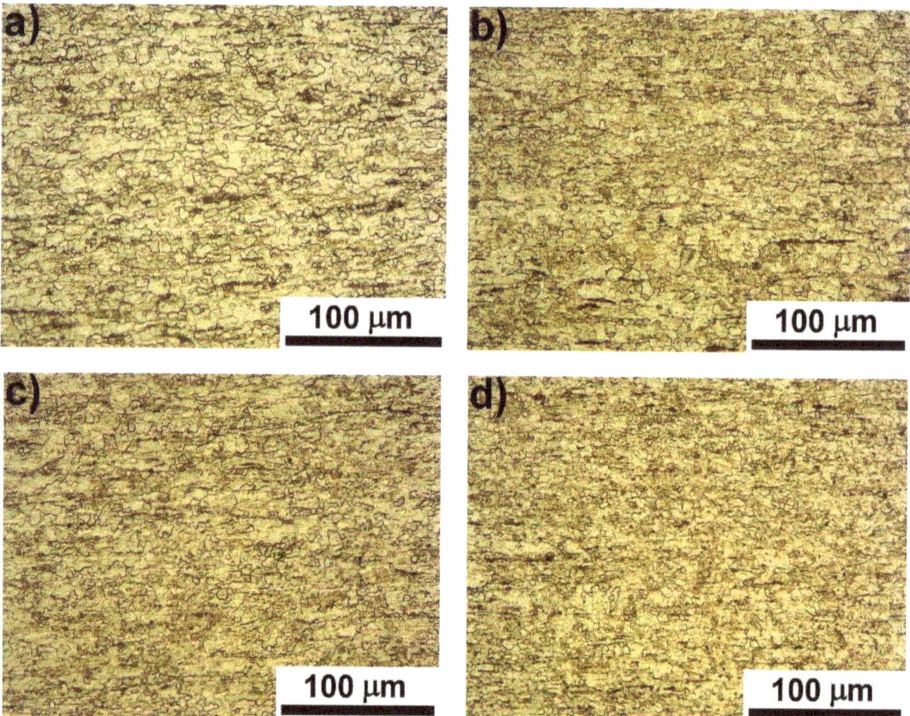

Figure 8. Microstructures of manufactured Ti Gr4 rods revealed by etching, longitudinal section, light microscopy: (**a**) variant V1, (**b**) variant V2, (**c**) variant V3, (**d**) variant V4.

In the case of Ti Gr4 (V1–V4), by comparing the four individual variants with each other it was found that the type of chips (turning, milling) and their thickness (thin, coarse) did not affect the microstructures of the obtained rods, since all four microstructures were characterized by practically the same grain size and grain shape (Table 8).

Table 8. Results of the stereological analysis of the microstructure of Ti Gr4, where: $E(d_2)$—average grain size, SD—standard deviation, CV—coefficient of variation, α and β—shape factors (a description of the stereological parameters is included in chapter "Investigation methods").

Ti Gr4 Samples (Chips)—Section	$E(d_2)$	$SD(d_2)$	$CV(d_2)$	α	β
As-received—transverse	32.09	9.71	0.30	1.32	1.20
As-received—longitudinal	27.78	8.34	0.30	1.29	1.17
V1 (turning, thin)—transverse	4.70	0.95	0.20	1.28	1.16
V1 (turning, thin)—longitudinal	4.06	1.01	0.25	1.27	1.15
V2 (turning, coarse)—transverse	4.26	0.86	0.20	1.25	1.14
V2 (turning, coarse)—longitudinal	3.92	1.12	0.29	1.26	1.14
V3 (milling, thin)—transverse	3.44	0.84	0.25	1.24	1.13
V3 (milling, thin)—longitudinal	3.43	0.91	0.27	1.26	1.14
V4 (milling, coarse)—transverse	3.72	0.71	0.19	1.28	1.14
V4 (milling, coarse)—longitudinal	3.10	0.72	0.23	1.28	1.15

Since in all of the variants similar final microstructures were obtained, it can be stated that the most important factors determining the final microstructures were the parameters of the extrusion process. It should be emphasized that, in our experiments, the extrusion parameters were the same for all four variants.

The relatively small magnification applied in Figures 7 and 8 made it possible to show the homogeneity of the V1–V4 microstructures, because the area of the image (where the average grain size was only about 4 μm) contained a relatively large number of grains. This magnification was also adjusted to the microstructure of the Ti in as-received state, which was characterized by a significantly larger grain size (averaging about 30 μm, Figure 1). Thus, it was possible to easily compare the as-received microstructures with those of the manufactured rods. However, when applied to Figures 7 and 8, this magnification did not permit an accurate analysis of the individual grains because the structures V1–V4 were too fine-grained. For this reason, observations were also made at higher magnifications (Figure 9). Figure 9 shows some sample microstructures (V2, V4) presented at a higher magnification where individual, single grains are visible.

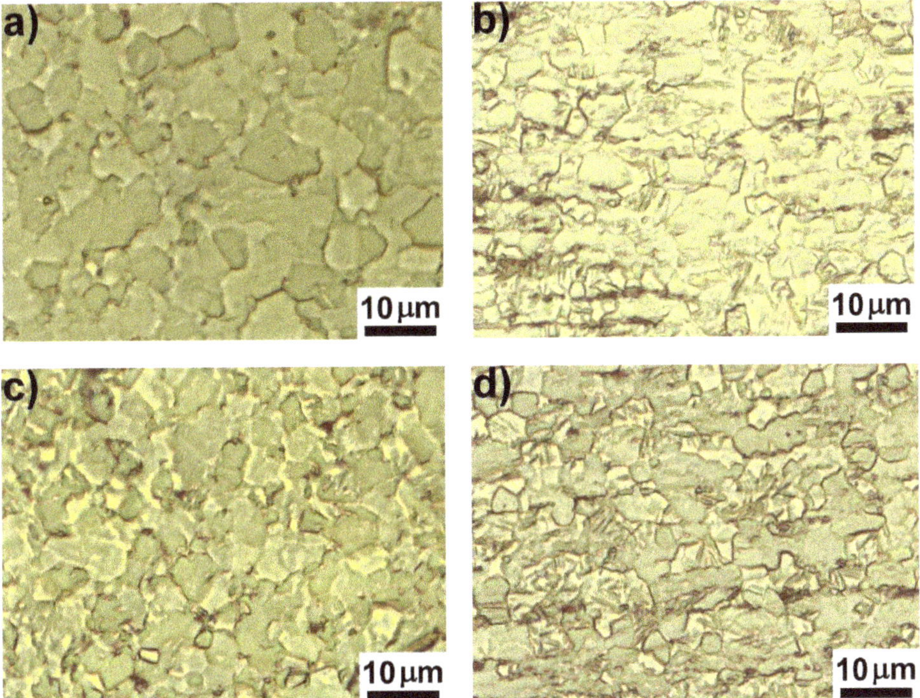

Figure 9. Microstructures of manufactured Ti Gr4 rods revealed by etching, visible under higher magnification: (**a**) variant V2, transverse section, (**b**) variant V2, longitudinal section, (**c**) variant V4, transverse section, (**d**) variant V4, longitudinal section.

Summarizing this chapter, in all of the manufactured rods new solid, grained, monolithic structures without the initial chips boundaries were obtained. This confirms that atomic (metallic) bonds between the chips were formed, and the chips became bonded. Due to adhesion and cohesion, consolidation phenomena took place.

4.11. Structural Investigations of Rods by Scanning Electron Microscopy (SEM)

The SEM investigations were conducted in order to confirm or/and obtain other information about the manufactured structures. This technique makes it possible to analyze recycled structures precisely, what has been also demonstrated by others authors who investigated pores and bonds [17,18].

The SEM investigations (conducted at different magnification) confirmed that all of the materials obtained (V1–V4) were near fully consolidated, as well as near fully dense. The dispersed form of the chips was transformed into bulk, solid materials. Sample results obtained using SEM are presented in Figure 10 (V1 variant). The monolithic, compact and homogeneous materials with no visible initial chip boundaries were observed. No voids were observed as had been present between the chips in the briquettes. In other words, no macrostructure consisting of chips, as was the case in the briquettes, was observed. As can be seen from the SEM pictures, the chips were fully processed and the briquettes were transformed into a new, solid structure with a new bonding on the atomic scale. Moreover, no delamination effect caused by an oxide layer was observed. Such defects can occur because at elevated temperatures titanium chips are subjected to oxygenation phenomena due to the machining and extrusion. It should be emphasized that the plastic working process applied (conducted at an elevated temperature, i.e., without remelting) considerably reduced chip oxidation and its negative effect on structure. Similarly, as in previous investigations, it can be concluded that adhesion and bonding between the initial chip surfaces occurred.

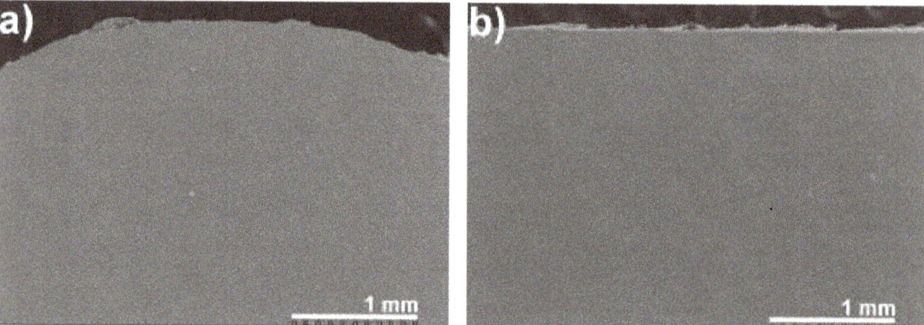

Figure 10. The results obtained for the polished samples using the SEM technique. Variant V1: (**a**) transverse section, (**b**) longitudinal section.

However, the obtained materials were also characterized by not numerous voids. The SEM structural observations made it possible to analyze the voids detected during the computer tomography examination. It was observed that these structural defects were three-dimensional. Some of them occurred in groups, and some separately. The voids in the transverse section of the samples were relatively equiaxial, whereas in the longitudinal section they were elongated and arranged in lines according to the direction of extrusion. The voids were surrounded by smooth, rounded surfaces. They were not fracture surfaces, delaminations or fragments of not-bonded chips. It was found that they were the micro-pores, i.e., enclosed spaces inside the material filled with atmospheric air. This air came from the briquette voids, and has been blocked-trapped in the material during extrusion. Sample results of these structural defects are presented in Figure 11.

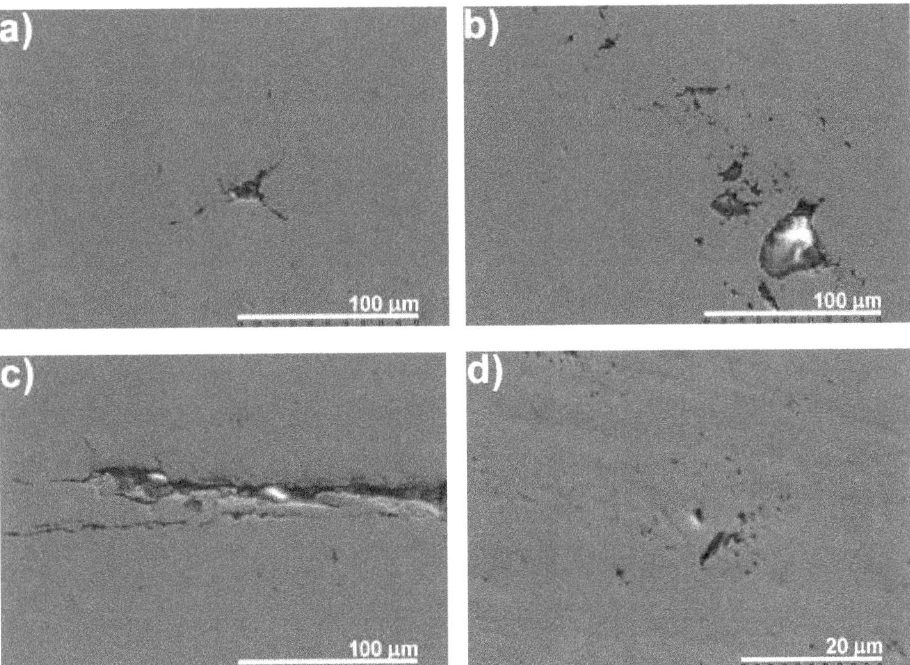

Figure 11. Example results of structural defects visible using SEM: (**a**) V2 transverse section, (**b**) V3 transverse section, (**c**) V2 longitudinal section, (**d**) another area of V2 transverse section visible in higher magnification.

As can be seen in the SEM pictures, the size of the pores is different. Including computer tomography results the size of the pores is in a very wide range (from a few micrometers to several millimeters). When using the computer tomography technique, pores smaller than 25 μm were not revealed (detection threshold), because that research concerning examining the large volumes of the bulk, volumetric samples (Ø8 mm × 50 mm). However, using the SEM technique, precise investigations were conducted at high magnification in the secondary electron mode (SE). As a result, very small voids, just several micrometers in size, were also detected (Figure 11).

4.12. Mechanical Properties Evaluated by Compression Tests and Hardness

The compression and hardness tests were selected because they provide the suitable information in the context of the presented structural research.

Uniaxial compression tests were applied in order to investigate the coherence and plasticity of volumetric, massive samples under the acting of high value of stress and plastic deformation. These tests provided information about the coherence and mechanical properties of the products obtained. The yield stress was determined, and the plasticity of the rods was tested. The samples were deformed within a range that significantly exceeded the yield stress. During the tests, the intentional, final value of plastic deformation was relatively high ($\varepsilon = 50\%$), which resulted in a reduction of the initial height of the sample by half (i.e., from 10.5 mm to 5.25 mm). This value of deformation is unattainable in the case of the tensile tests.

All the samples V1–V4 were selected for the research. For comparison, the material in as-received state was also tested. Based on the results of computer tomography it was assumed, that the samples of V4 variant had the greatest amount of defects (porosity), therefore it was most vulnerable to destruction. Thus, the V4 samples were characterized by the highest risk of cracks and decohesion. In sample V4, the absence of failure and the possibility of plastic deformation would be the evidence of high quality of manufactured materials.

The strain-stress curves obtained based on the compression tests are presented in Figure 12. As can be seen, in all the samples V1–V4 the deformation phenomena proceeded properly and similarly. The courses of compression for the samples V1–V4 were analogous to those for the material in the as received state, what is a particularly important and favorable result. All the samples examined reacted to the compressive stress in the same way. For all of them, the stage of elastic deformation was followed by a stage of plastic deformation. The graphs show a gradual strengthening of the materials. During the application of the stress, there were no clear and significant stress fluctuations or jumps. This proves that no significant cracking or fracturing occurred during compression, and that the materials did not separate. Throughout the broad scope of deformation applied, the materials deformed plastically and remained coherent. The course of the compression curves confirms the results of the macroscopic observations that only small, insignificant, local surface cracks and local surface delaminations occurred. This applies to all samples and is a particularly favorable result for the samples after chip consolidation.

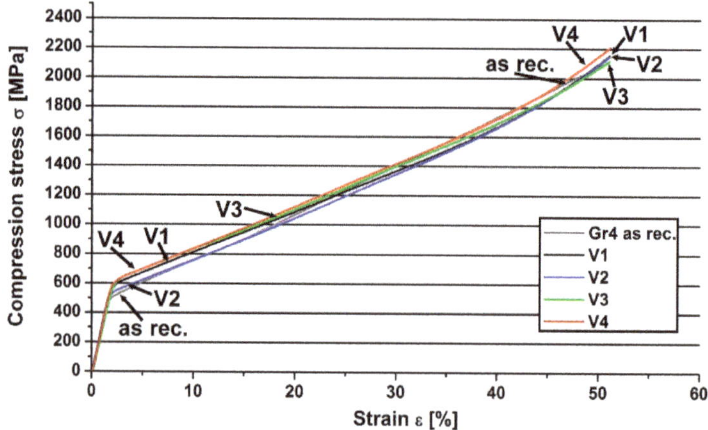

Figure 12. Strain-stress curves obtained from the compression tests.

Based on the strain-stress records, it was possible to determine the yield stress (Table 9). Comparing the values, it was found, that the titanium after chips processing were characterized by higher values of yield stress than the titanium in as-received state. Analyzing the structure and the mechanical results together, an additional strengthening mechanism could be observed. This was strengthening by grain boundaries caused by the grain refinement, because, in the extruded rods, significantly smaller grains were present. This phenomenon was widely discussed in the previous chapters. The results of yield strength obtained for samples after milling are slightly higher than those obtained for samples after turning. However, the differences are insignificant and the results are similar. Hence, no clear influence of the chip geometry on the yield stress was stated. Regardless of the values of the yield stress, the obtained results were within the range typical for titanium Gr4. These results are particularly important and beneficial in the context of titanium obtained by processing chips.

Table 9. Mechanical properties obtained for as-received Ti and processed Ti evaluated from the uniaxial compression tests.

Sample Ti Gr4		As Received	V1	V2	V3	V4
Yield stress $\sigma_{0.2}$ [MPa]	test No.1	493	563	521	589	588
	test No.2	490	565	525	594	593

It should be emphasized that all the tested samples were characterized by very good plasticity. In each case, the large deformation value ($\varepsilon = 50\%$) planned was achieved. These results are important and encouraging, especially in the cases of samples V1–V4, which were manufactured from chips, i.e., from a dispersed form. None of the samples was destroyed (defragmented) into separate parts, and so the compressive strength of the materials was not calculated.

The results of the macroscopic observations of the samples after the compression tests are shown in Figure 13. They are consistent with the recorded compression curves. The macroscopic examination proved that all of the samples remained coherent, which is especially important in the cases of samples V1–V4, which were obtained from chips. The samples did not crack into separate parts, nor was any defragmentation observed. In some cases, local, relatively small surface defects were found. However, these were only local delaminations on the circuit of the samples. They resulted from the local porosity and maximum tensile stresses acting on the circumference.

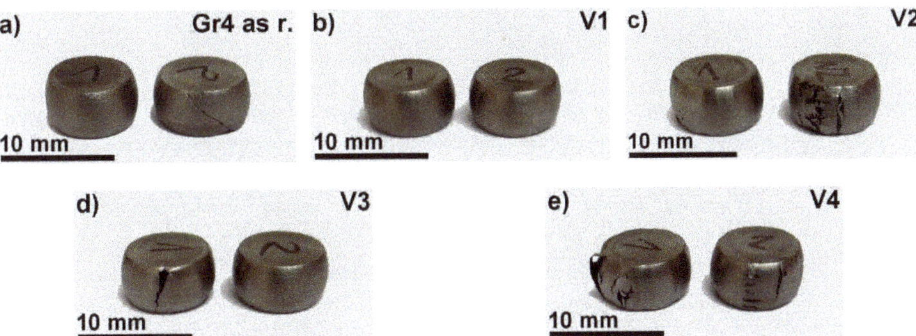

Figure 13. Macroscopic view of the samples after the uniaxial compression tests: (**a**) Ti Gr4, as received state, (**b**) variant V1, (**c**) variant V2, (**d**) variant V3, (**e**) variant V4.

The results of compression tests are another evidence of the chips transformation and of the forming of new, high-strength interatomic bonding.

The results of the hardness measurements (HV10) revealed that all of the obtained materials V1–V4 were characterized by the hardness in the range of 196–211 HV10. These values were very close to the hardness of Ti Gr4 in the as received state (190 HV10). This means that hardness of the manufactured rods is proper and typical for the hardness of commercial TiGr4. Therefore, the results of the HV10 measurements are another proof that the structure of the rods obtained is correct. Moreover, in the context of the presented structural research, the measurements of hardness also provided other important information. Namely, the measurements were made under a relatively high load of 10 kg (~100 N). Despite this, during the tests, all the samples V1–V4 remained coherent and none of them has been damaged by cracks or decohesion.

5. Summary and Conclusions

In this work, four variants of pure titanium chips were processed in order to consolidate and transform this dispersed form into solid, bulk materials. These four variants were Ti Gr4 chips after: turning (thin, coarse) and after milling (thin, coarse). The applied plastic working technology consisted of preliminary upsetting followed by direct extrusion at an elevated temperature using a rotating die (the KOBO method). A characteristic feature of the processing applied was that the deformed chips always remained in the solid state. In addition to the above, the following conclusions have been formulated:

1. The technology applied resulted in processing the chips into bulk, volumetric products in the form of rods Ø8 mm in diameter and with a length in a range of from 400–600 mm.
2. A dispersed form (chips) was successfully transformed into new, consolidated, solid state structures.
3. The manufactured rods were characterized by structures with a value of porosity between 0.02–0.54% and a density equal 4.50 g/cm^3.
4. The new structures of the manufactured rods were typical of commercial titanium. In this work, polycrystalline, grained microstructures were obtained, having equiaxial grains with an average size in the range 3–5 μm.
5. In all the rods, new, monolithic structures without initial chip boundaries were formed, which confirms the presence of new atomic bonds in the materials.
6. Within the compression tests, it was found that the manufactured rods exhibited proper cohesion and were not brittle. The rods were characterized by new interatomic bonds, cohesive forces and plasticity analogous to those of titanium in the as-received state.

Author Contributions: Conceptualization, K.T.; methodology, K.T.; formal analysis, K.T. and H.G.; investigation, K.T. and J.J.; writing—original draft preparation, K.T.; writing—review and editing, K.T.; supervision, K.T.; project administration, K.T.; funding acquisition, K.T. All authors have read and agreed to the published version of the manuscript.

Funding: This research was funded by National Science Center, POLAND (Narodowe Centrum Nauki, POLSKA), grant number UMO-2017/26/D/ST8/00051.

Institutional Review Board Statement: Not applicable.

Informed Consent Statement: Not applicable.

Data Availability Statement: Not applicable.

Conflicts of Interest: The authors declare no conflict of interest.

References

1. Shamsudin, S.; Lajis, M.A.; Zhong, Z.W. Evolutionary in Solid State Recycling Techniques of Aluminium: A review. *Procedia CIRP* **2016**, *40*, 256–261. [CrossRef]
2. Tekkaya, A.E.; Schikorra, M.; Becker, D.; Biermann, D.; Hammer, N.; Pantke, K. Hot profile extrusion of AA-6060 aluminum chips. *J. Mater. Process. Technol.* **2009**, *209*, 3343–3350. [CrossRef]
3. Wen, L.; Ji, Z.; Li, X.; Xin, M. Effect of Heat Treatment on Microstructure and Mechanical Properties of ZM6 Alloy Prepared by Solid Recycling Process. *J. Mater. Eng. Perform.* **2010**, *19*, 107. [CrossRef]
4. Mal'kov, A.V.; Shevchenko, V.V.; Nizkin, I.D.; Luk'yanova, E.V. Structure and Properties of Hot-Rolled Strips Obtained from Titanium Chips without Chip Remelting. *Russ. J. Non Ferrous Met.* **2008**, *49*, 258–260. [CrossRef]
5. Lapovok, R.; Qi, Y.; Ng, H.P.; Maier, V.; Estrin, Y. Multicomponent materials from machining chips compacted by equal-channel angular pressing. *J. Mater. Sci.* **2014**, *49*, 1193–1204. [CrossRef]
6. Zhilyaev, A.P.; Gimazov, A.A.; Raab, G.I.; Langdon, T.G. Using high-pressure torsion for the cold-consolidation of copper chips produced by machining. *Mater. Sci. Eng. A* **2008**, *486*, 123–126. [CrossRef]
7. Luo, P.; McDonald, D.T.; Palanisamy, S.; Dargusch, M.S.; Xia, K. Ultrafine-grained pure Ti recycled by equal channel angular pressing with high strength and good ductility. *J. Mater. Process. Technol.* **2013**, *213*, 469–476. [CrossRef]
8. Lui, E.W.; Palanisamy, S.; Dargusch, M.S.; Xia, K. Effects of chip conditions on the solid state recycling of Ti-6Al-4V machining chips. *J. Mater. Process. Technol.* **2016**, *238*, 297–304. [CrossRef]

9. Shi, Q.; Tse, Y.Y.; Higginson, R.L. Effects of processing parameters on relative density, microhardness and microstructure of recycled Ti–6Al–4V from machining chips produced by equal channel angular pressing. *Mater. Sci. Eng. A* **2016**, *651*, 248–258. [CrossRef]

10. Topolski, K.; Bochniak, W.; Łagoda, M.; Ostachowski, P.; Garbacz, H. Structure and properties of titanium produced by a new method of chip recycling. *J. Mater. Process. Technol.* **2017**, *248*, 80–91. [CrossRef]

11. Korbel, A.; Bochniak, W.; Śliwa, R.; Ostachowski, P.; Łagoda, M.; Kusion, Z.; Trzebuniak, B. Low-temperature consolidation of machining chips from hardly-deformable aluminum alloys. *Met. Form.* **2016**, *27*, 133–152.

12. Baffari, D.; Buffa, G.; Campanella, D.; Fratini, L.; Reynolds, A.P. Process mechanics in Friction Stir Extrusion of magnesium alloys chips through experiments and numerical simulation. *J. Manuf. Process.* **2017**, *29*, 41–49. [CrossRef]

13. Tang, W.; Reynolds, A.P. Production of wire via friction extrusion of aluminum alloy machining chips. *J. Mater. Process. Technol.* **2010**, *210*, 2231–2237. [CrossRef]

14. Gronostajski, J.; Matuszak, A. The recycling of metals by plastic deformation: An example of recycling of aluminium and its alloys chips. *J. Mater. Process. Technol.* **1999**, *92–93*, 35–41. [CrossRef]

15. Wan, B.; Chen, W.; Lu, T.; Liu, F.; Jiang, Z.; Mao, M. Review of solid state recycling of aluminum chips. *Resour. Conserv. Recycl.* **2017**, *125*, 37–47. [CrossRef]

16. Hu, M.L.; Ji, Z.S.; Chen, X.Y.; Wang, Q.D.; Ding, W.J. Solid-state recycling of AZ91D magnesium alloy chips. *Trans. Nonferrous Met. Soc. China* **2012**, *22*, 68–73. [CrossRef]

17. Peng, T.; Wang, Q.D.; Han, Y.K.; Zheng, J.; Guo, W. Consolidation behavior of Mg–10Gd–2Y–0.5Zr chips during solid-state recycling. *J. Alloys Compd.* **2010**, *503*, 253–259. [CrossRef]

18. Qi, Y.; Timokhina, I.B.; Shekhter, A.; Sharp, K.; Lapovok, R. Optimization of upcycling of Ti-6Al-4V swarf. *J. Mater. Process. Technol.* **2018**, *255*, 853–864. [CrossRef]

19. Wejrzanowski, T.; Spychalski, W.L.; Rożniatowski, K.; Kurzydłowski, K.J. Image based analysis of complex microstructures of engineering materials. *Int. J. Appl. Math. Comput. Sci.* **2008**, *18*, 33–39. [CrossRef]

20. Nouari, M.; Makich, H. On the Physics of Machining Titanium Alloys: Interactions between Cutting Parameters, Microstructure and Tool Wear. *Metals* **2014**, *4*, 335–358. [CrossRef]

21. Shunmugavel, M.; Goldberg, M.; Polishetty, A.; Nomani, J.; Sun, S.; Littlefair, G. Chip formation characteristics of selective laser melted Ti–6Al–4V. *Aust. J. Mech. Eng.* **2019**, *17*, 109–126. [CrossRef]

Article

Ambivalent Role of Annealing in Tensile Properties of Step-Rolled Ti-6Al-4V with Ultrafine-Grained Structure

Geonhyeong Kim [1], Taekyung Lee [2,*], Yongmoon Lee [3], Jae Nam Kim [1], Seong Woo Choi [4], Jae Keun Hong [4] and Chong Soo Lee [1,*]

1   Graduate Institute of Ferrous Technology (GIFT), Pohang University of Science and Technology (POSTECH), Pohang 37673, Korea; kkh285@postech.ac.kr (G.K.); jnkims@postech.ac.kr (J.N.K.)
2   School of Mechanical Engineering, Pusan National University, Busan 46241, Korea
3   Center for Advanced Aerospace Materials, Pohang University of Science and Technology (POSTECH), Pohang 37673, Korea; ymlee0725@postech.ac.kr
4   Advanced Metals Division, Korea Institute of Materials Science, Changwon 51508, Korea; rhkdgh@kims.re.kr (S.W.C.); jkhong@kims.re.kr (J.K.H.)
*   Correspondence: taeklee@pnu.ac.kr (T.L.); cslee@postech.ac.kr (C.S.L.); Tel.: +82-51-510-2985 (T.L.); +82-54-279-9009 (C.S.L.)

Received: 23 April 2020; Accepted: 21 May 2020; Published: 22 May 2020

Abstract: Step rolling can be used to mass-produce ultrafine-grained (UFG) Ti-6Al-4V sheets. This study clarified the effect of subsequent annealing on the tensile properties of step-rolled Ti-6Al-4V at room temperature (RT) and elevated temperature. The step-rolled alloy retained its UFG structure after subsequent annealing at 500–600 °C. The RT ductility of the step-rolled alloy increased regardless of annealing temperature, but strengthening was only attained by annealing at 500 °C. In contrast, subsequent annealing rarely improved the elevated-temperature tensile properties. The step-rolled Ti-6Al-4V alloy without the annealing showed the highest elongation to failure of 960% at 700 °C and a strain rate of 10^{-3} s^{-1}. The ambivalent effect of annealing on RT and elevated-temperature tensile properties is a result of microstructural features, such as dislocation tangles, subgrains, phases, and continuous dynamic recrystallization.

Keywords: Ti-6Al-4V; step rolling; grain refinement; superplasticity; continuous dynamic recrystallization

---

1. Introduction

Ti-6Al-4V alloy is widely used in aerospace industries due to its high specific strength, excellent corrosion resistance, and high service temperature [1]. This alloy also has superior superplasticity and consequent high formability. Accordingly, many complicated aircraft structures are manufactured from Ti-6Al-4V alloys, using superplastic forming to reduce the material's weight for cost saving [2]. This process is typically performed at relatively high temperatures of over 850 °C and low strain rates of less than 10^{-3} s^{-1}; these processes are expensive, so they increase production cost. Thus, research has been conducted to find ways to reduce the temperature or increase the strain rate (or both) at which superplastic forming is conducted. Grain refinement may be a solution to this challenge [3,4].

An increased fraction of grain boundaries assists grain-boundary sliding (GBS), which is the main mechanism of superplasticity. GBS by dislocation processes or diffusional flow can accommodate a large amount of deformation. Grain refinement can shift region II (i.e., the region of the highest strain rate sensitivity) to faster strain rates [5]. Therefore, grain refinement can achieve superplasticity at low temperatures or fast strain rates or both. Grain size has been reduced using a variety of severe plastic deformation (SPD) processes, such as equal-channel angular pressing (ECAP) [6], high-pressure torsion (HPT) [7], and multiaxial forging [8]. The grains of Ti-6Al-4V alloy have been reduced to a size of ~0.4 μm, which has enabled superplasticity at 550 °C and a strain rate of 2×10^{-4} s^{-1} [4]. However, SPD cannot easily produce an appreciable size of product, so the method is not useful for industrial production. Multi-pass caliber rolling has been proposed as a solution [9–11] but is only applicable to fabricating metallic rods rather than plates.

Step rolling [12] has been developed as a grain-refining process to fabricate relatively large Ti plates. This method can fabricate ultrafine-grained (UFG) bulk Ti-6Al-4V plate (i.e., 300 mm × 70 mm × 4 mm, with grain size $d = 0.5$ μm) with only a small cumulative strain of 1.6. Step rolling is performed by decreasing the rolling temperature step by step. Initial deformation is imposed at a high temperature to effectively fragment the martensitic laths. This fragmentation increases the formability of the material so that the plate can be rolled at progressively lower temperatures in subsequent rolling passes. This low-temperature rolling suppresses grain growth and thereby achieves significant grain refinement with a relatively low amount of plastic strain [12].

The superplasticity of Ti-6Al-4V obtained by step rolling and subsequent annealing was recently reported [13], but the study used only one deformation condition (i.e., a deformation temperature of 750 °C and a strain rate of 10^{-3} s^{-1}). However, superplastic behavior drastically varies with deformation temperature and $\dot{\varepsilon}$, so to exploit the advantage of step rolling, the high-temperature deformation behavior must be quantified at a wide range of these parameters. Therefore, the present study is a systematic investigation of the deformation behavior of step-rolled Ti-6Al-4V alloys at room temperature (RT) and elevated temperatures under a range of $\dot{\varepsilon}$.

2. Materials and Methods

The Ti-6Al-4V alloy used in this study (supplied by ATI, USA) was a 20 mm-thick plate with a chemical composition (mass %) of 6.28 Al, 4.00 V, 0.18 Fe, 0.178 O, and 0.012 C, in balance with Ti. Thermomechanical processing was achieved using step rolling (Figure 1). The as-received plate was solution-treated at 1050 °C for 30 min, then quenched in water to induce the formation of a full martensitic structure. The plate was soaked at 800 °C for 1 h, then step-rolled to a thickness of 3 mm. The step rolling was composed of eight passes with decreasing temperatures: 800 °C for the first pass, 725 °C for the second and third passes, and 650 °C for the fourth to eighth passes. The sample was then cooled in air to RT. The step-rolled sample without additional heat treatment is denoted as STEP-0. Sections of the STEP-0 sample were further annealed at 500 °C (sample STEP-5) or 600 °C (STEP-6) for 150 min in an electric furnace (AWF13, Lenton, Hope Valley, UK), then cooled in air to RT.

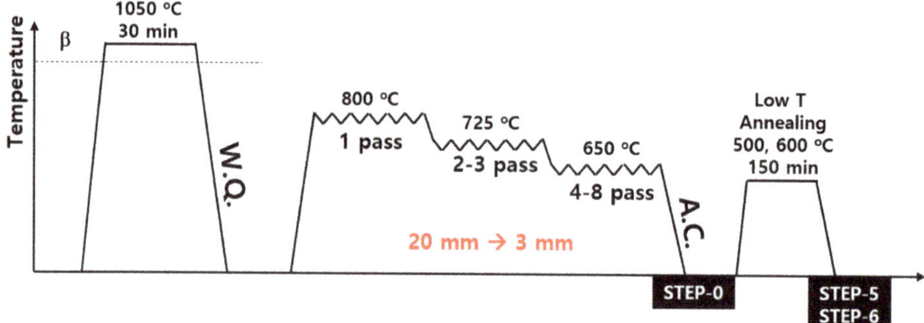

Figure 1. Schematic illustration of the step rolling and subsequent annealing employed in this work.

An RT tensile test was performed at a strain rate of 5×10^{-3} s^{-1} using a universal mechanical testing machine (8801, INSTRON, Norwood, MA, USA) with an extensometer (3542-025M-100-ST, Epsilon, Jackson, WY, USA), based on the procedure of the ASTM-E8 standard. Plate-type tensile specimens with a gage length of 25 mm, width of 6 mm, and thickness of 3 mm were machined along the rolling direction. An elevated-temperature tensile test was performed using a universal mechanical testing machine (8862, INSTRON, Norwood, MA, USA) attached to a halogen furnace (DF-60HG, Dae Heung Scientific Company, Incheon, Korea). For this experiment, specimens were prepared with a gauge length of 5 mm, width of 5 mm, and thickness of 2 mm. The elevated-temperature tensile tests were performed at all combinations of the three temperatures (650 °C, 700 °C, 750 °C) and the three initial strain rates (10^{-2} s^{-1}, 10^{-3} s^{-1}, 2×10^{-4} s^{-1}). Samples were held at a given testing temperature for 10 min before the tensile test was started.

For microstructural observation, samples were electro-polished (LectroPol-5, STRUERS, Denmark) in a solution of 410 mL of methanol, 245 mL of 2-butoxy ethyl alcohol, and 40 mL of HClO$_4$ at 22 V for 30 s. The plane normal to the transverse direction was examined using electron backscatter diffraction (EBSD, FEI QUANTA 3D FEG, Hillsboro, OR, USA) analysis with a step size of 0.1 μm. Quantitative microstructural characterizations were obtained using TSL-OIM Ver. 7.3 software (EDAX, Mahwah, NJ, USA); for data reliability, the scans were only analyzed for points that had a confidence index > 0.07. More than 1000 validated grains per condition were used for the analysis to ensure statistical accuracy. For examination with transmission electron microscope (TEM, JEM-2100F, JEOL, Tokyo, Japan) observation, the samples were mechanically polished to a thickness < 100 μm and then punched to form thin disks that had a diameter of 3 mm. They were electro-polished using a jet polisher (TenuPol-5, STRUERS, Denmark) at 22 V in the same solution used for the preparation of the EBSD samples.

3. Results and Discussion

3.1. Deformation Behavior at Room Temperature

The microstructures of the step-rolled alloy (i.e., STEP-0) and the subsequently annealed alloys (i.e., STEP-5 and STEP-6) were compared by EBSD analysis (Figure 2). The annealing had little effect on grain size: STEP-0 had a UFG structure with α-grain size d = 0.70 μm, STEP-5 had d = 0.72 μm, and STEP-6 had d = 0.74 μm. This small change implies that annealing induced recovery rather than recrystallization. In contrast, annealing significantly decreased the volume fraction of the β phase: STEP-0, 6.2%; STEP-5, 3.0%; STEP-6, 3.6%.

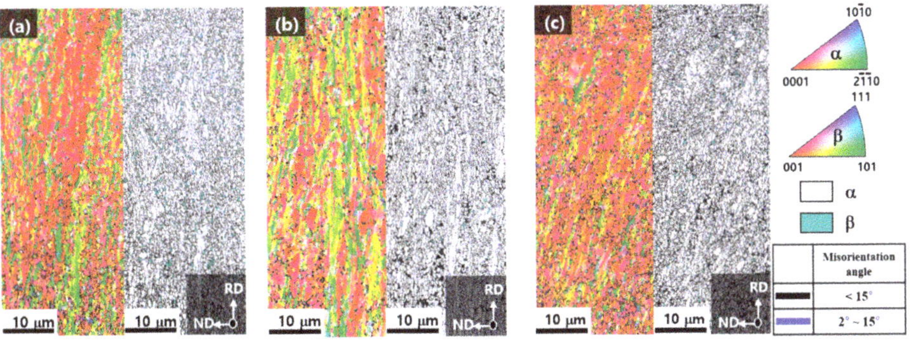

Figure 2. Electron backscatter diffraction (EBSD) inverse pole figure map and corresponding phase map of the samples: (**a**) STEP-0, (**b**) STEP-5, (**c**) STEP-6.

The subsequent annealing caused a marked difference in the RT tensile properties of the samples (Figure 3, Table 1). The STEP-0 sample had an increased yield strength (YS) and ultimate tensile strength (UTS) which were both 20% higher than those of conventional mill-annealed Ti-6Al-4V alloy, which has d = 10 μm [14]. However, STEP-0 showed a decrease in elongation to failure (EL) compared to the conventional sample [14]. The subsequent annealing of STEP-0 recovered the low EL without loss of mechanical strength.

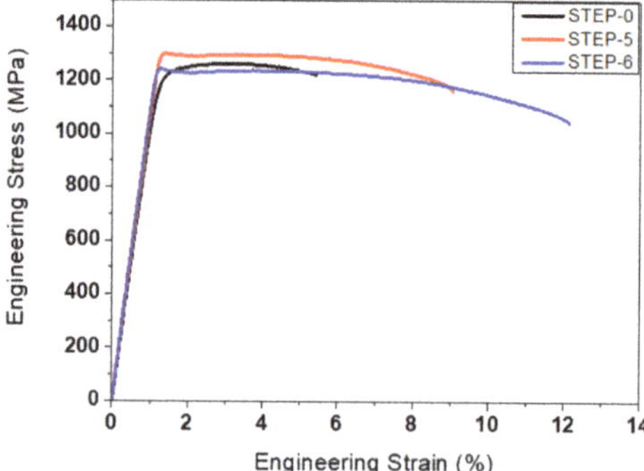

Figure 3. Engineering stress–strain curves of the investigated Ti-6Al-4V alloys at room temperature.

Table 1. Room-temperature tensile properties of the investigated Ti-6Al-4V alloys.

Sample	Yield Strength (YS) (MPa)	Ultimate Tensile Strength (UTS) (MPa)	Elongation to Failure (EL) (%)
STEP-0	1201	1263	5.4
STEP-5	1298	1311	9.0
STEP-6	1243	1243	12.1

Annealing temperature affected the change in strength and ductility differently in each sample. STEP-5 showed an increase in both whereas STEP-6 did not. TEM observation provided a rationalization for the simultaneous improvement in strength and ductility in STEP-5. The TEM micrograph (Figure 4) of STEP-0 showed the formation of dislocation tangles at the grain boundaries (Figure 4a, arrows). Annealing at 500 °C significantly diminished the frequency of dislocation tangles and decreased the dislocation density, resulting in the increased EL. STEP-5 showed fine (i.e., 100–200 nm in diameter) particles; a uniform distribution of fine particles effectively strengthens the matrix [15,16], so they may be the source of the increases in YS and EL in this sample. The selected-area diffraction pattern (Figure 4b, inset) characterizes these particles as α phase. In contrast, the STEP-6 sample showed no evidence of such fine particles. The α particles may have transformed to β phase at this annealing temperature. The absence of fine particles led to the loss of strengthening in STEP-6, compared to STEP-5.

Subsequent annealing affected the substructure of the UFG grains. STEP-0 showed only distributed dislocations (Figure 4c), whereas STEP-5 presented numerous subgrains with diameters of 20–50 nm (Figure 4d). At 500 °C, dislocations at grain boundaries were recovered and organized into subgrains inside a UFG α grain [17]. The consequent increase in the fraction of low-angle grain boundaries (LABs) would also contribute to the increased mechanical strength [18].

3.2. Deformation Behavior at Elevated Temperatures

EL at elevated temperatures was affected by strain rate, deformation temperature, and type of sample (Figure 5). ELs measured at a deformation temperature of 750 °C and 2×10^{-4} s^{-1} are not presented, because severe oxidation at the sample surface caused large variability in the measurements. An EL of 400% is regarded as the minimum requirement for superplasticity in metals [19]. The investigated alloys exhibited a significant EL of over 400% under most conditions, except at the lowest deformation temperature (650 °C) and the highest strain rate (10^{-2} s^{-1}). The subsequent annealing had an ambivalent effect on the difference between the mechanical properties at RT and those at elevated temperatures:

it rarely increased EL in the step-rolled alloys at elevated temperatures, and it even degraded the ductility at 650 °C.

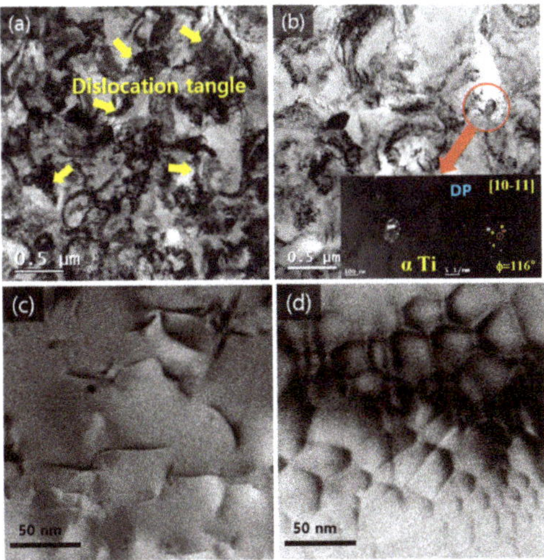

Figure 4. Bright-field transmission electron microscope (TEM) micrographs of (**a**,**c**) STEP-0 and (**b**,**d**) STEP-5. Insets in (**b**): dark-field image of α particle and corresponding selected-area diffraction pattern. Figure 4c,d are high-magnification images.

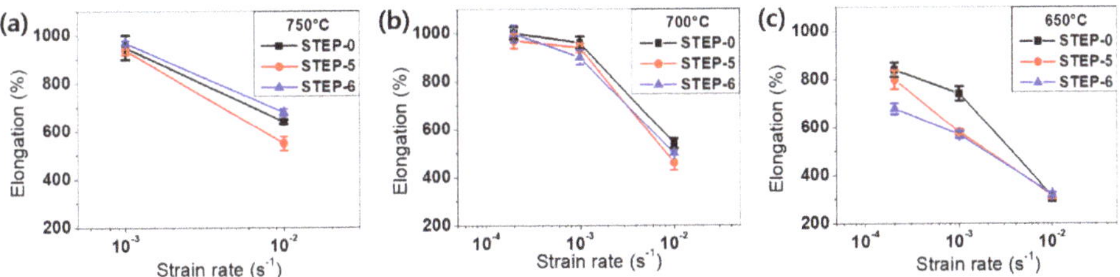

Figure 5. Variation of EL depending on the strain rate for the investigated alloys at (**a**) 750 °C, (**b**) 700 °C, and (**c**) 650 °C.

The stress–strain curves of STEP-0, STEP-5, and STEP-6 were affected by the deformation temperature and strain rate (Figure 6). At 750 °C and 10^{-3} s^{-1}, three specimens showed weak flow hardening behavior. This suggests that continuous dynamic recrystallization (CDRX) could not suppress grain growth, thereby diminishing the superplasticity [20]. The plateau region was confirmed during the deformation at 700 °C, indicating that CDRX strongly suppressed grain growth [21]. To obtain the highest and most efficient superplasticity, this plateau behavior should be achieved. Usually, EL is higher at 750 °C than at 700 °C, but in our results the ELs were similar. In this case, severe grain growth and oxidation degraded the superplasticity at 750 °C. At a deformation temperature of 650 °C, the specimens presented a flow softening followed by the plateau region. Despite the decreasing deformation temperature, they still exhibited significant EL values, suggestive of a low-temperature superplasticity.

The highest superplastic EL of STEP-0 among the investigated step-rolled alloys can be understood in terms of slope m (i.e., strain-rate sensitivity) in a log-scale plot of stress and strain rate (Figure 7). In general, UFG Ti-6Al-4V alloy demonstrates three regions with different slopes in this plot [6,22]. Among them, region II is characterized by (i) an intermediate 10^{-4} s$^{-1} \leq \dot{\varepsilon} \leq 2 \times 10^{-3}$ s^{-1}, (ii) high m, and (iii) significant superplastic EL. The high m of the STEP-0 sample indicates inhibition of the transition from region II to region III at a high strain rate of 10^{-2} s^{-1} [23], supporting a superplasticity obtained by the step rolling. Accordingly, the STEP-0 sample had a uniformly high m in the entire range of investigation employed in this work: $m = 0.34$ at 650 °C, 0.40 at 700 °C, and 0.55 at 750 °C. These m are comparable to or higher than those of UFG Ti-6Al-4V alloys fabricated by SPD processes: $0.34 \le \dot{\varepsilon} \le 0.43$ [4,6,7].

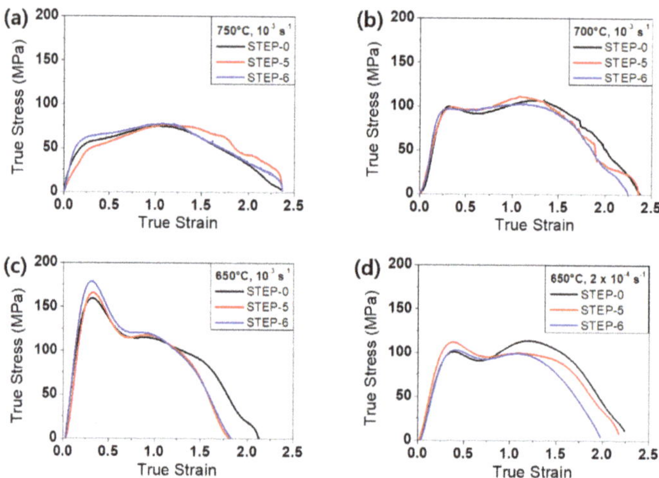

Figure 6. True stress–strain curves of the investigated Ti-6Al-4V alloys at (**a**) 750 °C and strain rate of 10^{-3} s^{-1}, (**b**) 700 °C and 10^{-3} s^{-1}, (**c**) 650 °C and 10^{-3} s^{-1}, and (**d**) 650 °C and 2×10^{-4} s^{-1}.

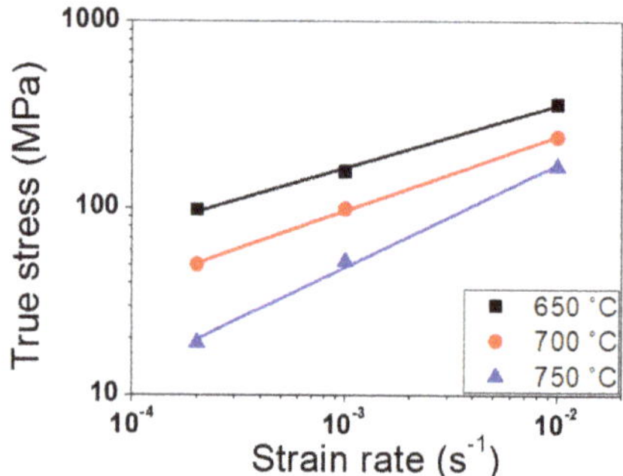

Figure 7. Variation of flow stress as a function of strain rate for STEP-0 specimens.

Step-rolled UFG Ti-6Al-4V yielded excellent superplasticity (i.e., EL up to 960%) that is comparable to the EL obtained using other fabrication methods. ECAP yielded UFG Ti-6Al-4V that had $d = 0.3$ μm, EL = 356% at 700 °C, and 10^{-3} s^{-1} [6]. HPT produced Ti-6Al-4V alloy that had d ~0.03 μm and EL = 820% under the same conditions [7]. Another HPT study produced $d = 0.3$ μm and the sample had EL = 676% at 725 °C and 10^{-3} s^{-1} [24]. Another UFG Ti-6Al-4V alloy ($d = 0.4$ μm) produced by hot rolling had EL = 400% at 700 °C and 10^{-3} s^{-1} [20]. A combination of forging and warm rolling refined the grains to $d = 0.3$ μm; the sample had EL = 550–900% at 700 °C [25].

Superplasticity is highly dependent on microstructural features. Grain refinement is regarded as the most important factor to activate superplastic behavior. Grain refinement expands the superplastic regime to lower temperatures and higher $\dot{\varepsilon}$, because the increasing fraction of grain boundaries provides increasing sources for GBS [5,9]. Grain morphology also affects superplasticity. Equiaxed grains yield a higher superplastic EL than elongated grains [26,27]. Dislocations distributed at grain boundaries also contribute to superplasticity by providing a fast diffusion path at high temperatures and by increasing the grain-boundary diffusion coefficient [28]. All of these microstructural features that are beneficial to superplasticity appeared in STEP-0, so its significant EL at 650–750 °C is understandable.

Comparison of the microstructural features at strains of ε = 0%, 100%, and 300% provided insight into the microstructural evolution during superplastic deformation at elevated temperatures (Figure 8). STEP-0 and STEP-6 were selected for this comparison, because STEP-6 was expected to show more distinct changes in microstructure than STEP-5 due to the higher annealing temperature of STEP-6. In the EBSD map, the STEP-0 and STEP-6 samples showed similar grain sizes before the deformation, whereas the former exhibited a smaller grain size after applying strains of 100% and 300%.

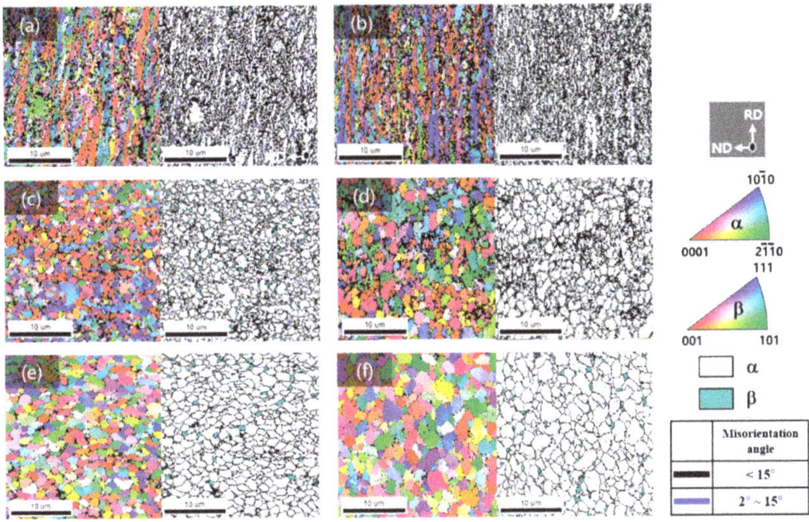

Figure 8. EBSD inverse pole figure maps and phase maps for (**a,c,e**) STEP-0 and (**b,d,f**) STEP-6. The micrographs were obtained after a strain of (**a,b**) 0%, (**c,d**) 100%, and (**e,f**) 300%.

Application of ε affected LAB fraction and grain size differently in the two samples (Figure 9). During the early stage of deformation, up to a 100% strain, the LAB fraction was considerably reduced, by 12.2%, in the STEP-0 sample but only by 6.1% in the STEP-6 sample. During this stage, grain size in the STEP-0 sample also increased by only 0.16 µm, whereas in the STEP-6 sample grain size increased by 0.68 µm. Moving from a 100% to a 300% strain, STEP-0 and STEP-6 exhibited a similar reduction in LAB fraction (5.6%) as well as a similar increase in grain size (0.46 µm).

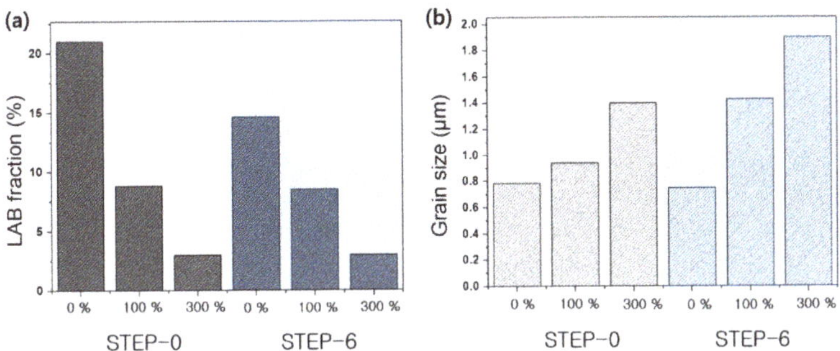

Figure 9. Microstructural features of STEP-0 and STEP-6 samples after strains of 0%, 100%, and 300% at 650 °C and 10^{-3} s^{-1}: (**a**) LAB fraction, (**b**) average grain size.

The distinct difference between the microstructural features of STEP-0 and those of STEP-6 arose from the active occurrence of CDRX in the STEP-0 sample. A continuous accumulation of dislocations during high-temperature deformation drove the formation of subgrains surrounded by LAB. Further straining generated recrystallized grains by increasing the misorientation of the subgrain boundaries [17,29]. The higher amount of pre-existing LAB in STEP-0 than in STEP-6 provided a higher driving force for CDRX in comparison with the annealed samples. This result suggests that CDRX of the STEP-0 sample occurred actively during the early stage of deformation, up to a 100% strain.

The hypothesis of the active occurrence of CDRX in STEP-0 during superplastic deformation is supported by the distribution of grain sizes (Figure 10). The STEP-0 sample showed an increasing fraction of grains with a size of ~1 μm in exchange for a decreasing fraction of coarse grains. In contrast, in the STEP-6 sample, the peak of the distribution curve shifted towards a higher grain size after a 100% strain at 650 °C and 10^{-3} s^{-1}. This is the typical trend that results from dominant grain-growth behavior at an elevated temperature. Such results suggest that the conversion of coarse grains into several fine grains (i.e., CDRX), occurred more actively in STEP-0 than in STEP-6. Consequently, active CDRX during the early stage of deformation in STEP-0 suppressed grain growth during elevated-temperature deformation and thereby contributed to its superior superplasticity.

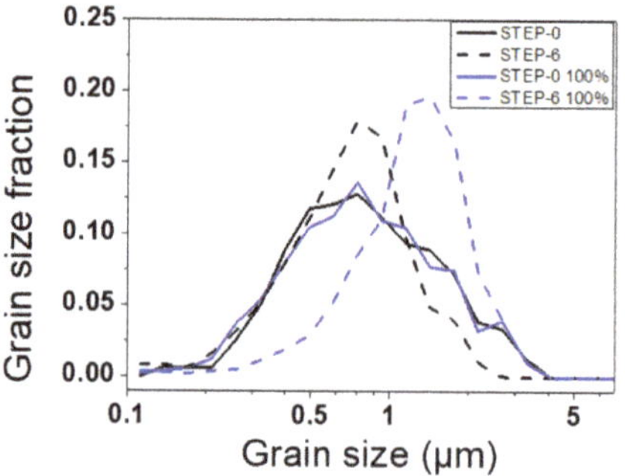

Figure 10. Distribution of grain sizes in STEP-0 and STEP-6 samples before and after applying a strain of 100% at 650 °C and 10^{-3} s^{-1}.

β particles also assisted in the increase in superplasticity in the STEP-0 sample. The sample possessed twice as much β phase as the STEP-5 and STEP-6 samples. Due to the difference in their crystallographic structures, grain-boundary diffusivity is over 100 times higher in the β phase than in the α phase [30]. Such a high diffusivity reduces the resistance to GBS activation. Moreover, resistance to GBS is lower at α/β interfaces than at α/α and α/β interfaces [31]. Therefore, the high β fraction was the secondary factor that increased the superplasticity of the STEP-0 sample.

4. Conclusions

This study investigated the RT and elevated-temperature deformation behaviors of UFG Ti-6Al-4V alloys fabricated by step rolling and subsequent annealing. Step rolling attained a significant grain refinement to a size of 0.70 μm, which was not much changed by subsequent annealing. The STEP-5 sample had significantly increased strength and ductility at RT. The strengthening was a result of the formation of fine α particles and (sub)grains, whereas the increased ductility was the consequence of a decreased frequency of dislocation tangles. The STEP-6 sample lost its strengthening effect because the fine particles disappeared. The STEP-0 sample had the highest superplasticity among the investigated samples; this result suggests that subsequent annealing has a variable influence that depends on the deformation temperature. The superplasticity arose from the equiaxed UFG structure and high dislocation density. Compared to the annealed samples, STEP-0 showed a rapid decrease in LAB fraction during the early stage of deformation. This result suggests that CDRX occurred actively in the STEP-0 sample at this stage and effectively suppressed grain growth during elevated-temperature deformation. The high β fraction of the STEP-0 sample also contributed to the GBS activation due to the considerable grain-boundary diffusivity and the presence of the α/β interface. Overall, active CDRX and a high content of β particles increased the superplasticity of the STEP-0 sample.

Author Contributions: Conceptualization, G.K. and C.S.L.; Writing—original draft, G.K., T.L., and Y.L.; Writing—review and editing, T.L. and C.S.L.; Supervision, C.S.L.; Investigation, G.K., Y.L., and J.N.K.; Resources, S.W.C. and J.K.H. All authors have read and agreed to the published version of the manuscript.

Funding: This research was funded by a grain (16-CM-MA-10) from Civil-Military Technology Cooperation Program funded by the Ministry of Trade, Industry and Energy, Korea.

Acknowledgments: The authors gratefully acknowledge the financial supports of the Ministry of Trade, Industry and Energy, Korea for this research.

Conflicts of Interest: The authors declare no conflict of interest.

References

1. Welsch, G. *Materials Properties Handbook: Titanium Alloys*; ASM International: Russell Township, OH, USA, 1993; ISBN 9780871704818.

2. Weisert, E. *Proc. AIME Conference on Superplastic Forming of Structural Alloys*; Paton, N.E., Hamilton, C.H., Eds.; TMS-AIME Publications: Warrendale, PA, USA, 1982.

3. Mishra, R.S.; Stolyarov, V.V.; Echer, C.; Valiev, R.Z.; Mukherjee, A.K. Mechanical behavior and superplasticity of a severe plastic deformation processed nanocrystalline Ti–6Al–4V alloy. *Mater. Sci. Eng. A* **2001**, *298*, 44–50. [CrossRef]

4. Zherebtsov, S.V.; Kudryavtsev, E.A.; Salishchev, G.A.; Straumal, B.B.; Semiatin, S.L. Microstructure evolution and mechanical behavior of ultrafine Ti6Al4V during low-temperature superplastic deformation. *Acta Mater.* **2016**, *121*, 152–163. [CrossRef]

5. Edington, J.W.; Melton, K.N.; Cutler, C.P. Superplasticity. *Prog. Mater. Sci.* **1976**, *21*, 61–170. [CrossRef]

6. Ko, Y.G.; Lee, C.S.; Shin, D.H.; Semiatin, S.L. Low-temperature superplasticity of ultra-fine-grained Ti-6Al-4V processed by equal-channel angular pressing. *Metall. Mater. Trans. A* **2006**, *37*, 381–391. [CrossRef]

7. Shahmir, H.; Naghdi, F.; Pereira, P.H.R.; Huang, Y.; Langdon, T.G. Factors influencing superplasticity in the Ti-6Al-4V alloy processed by high-pressure torsion. *Mater. Sci. Eng. A* **2018**, *718*, 198–206. [CrossRef]

8. Zherebtsov, S.; Kudryavtsev, E.; Kostjuchenko, S.; Malysheva, S.; Salishchev, G. Strength and ductility-related properties of ultrafine grained two-phase titanium alloy produced by warm multiaxial forging. *Mater. Sci. Eng. A* **2012**, *536*, 190–196. [CrossRef]

9. Lee, T.; Shih, D.S.; Lee, Y.; Lee, C.S. Manufacturing Ultrafine-Grained Ti-6Al-4V Bulk Rod Using Multi-Pass Caliber-Rolling. *Metals* **2015**, *5*, 777–789. [CrossRef]

10. Lee, T.; Park, K.-T.T.; Lee, D.J.; Jeong, J.; Oh, S.H.; Kim, H.S.; Park, C.H.; Lee, C.S. Microstructural evolution and strain-hardening behavior of multi-pass caliber-rolled Ti-13Nb-13Zr. *Mater. Sci. Eng. A* **2015**, *648*, 359–366. [CrossRef]

11. Lee, T.; Lee, S.; Kim, I.-S.; Moon, Y.H.; Kim, H.S.; Park, C.H. Breaking the limit of Young's modulus in low-cost Ti–Nb–Zr alloy for biomedical implant applications. *J. Alloy. Compd.* **2020**, *828*, 154401. [CrossRef]

12. Park, C.H.; Kim, J.H.; Yeom, J.-T.T.; Oh, C.-S.S.; Semiatin, S.L.; Lee, C.S. Formation of a submicrocrystalline structure in a two-phase titanium alloy without severe plastic deformation. *Scr. Mater.* **2013**, *68*, 996–999. [CrossRef]

13. Kim, D.; Won, J.W.; Park, C.H.; Hong, J.K.; Lee, T.; Lee, C.S. Enhancing Superplasticity of Ultrafine-Grained Ti–6Al–4V without Imposing Severe Plastic Deformation. *Adv. Eng. Mater.* **2019**, *21*, 1800115. [CrossRef]

14. Jeong, D.; Kwon, Y.; Goto, M.; Kim, S. High cycle fatigue and fatigue crack propagation behaviors of β-annealed Ti-6Al-4V alloy. *Int. J. Mech. Mater. Eng.* **2017**, *12*, 1. [CrossRef]

15. Morita, T.; Hatsuoka, K.; Iizuka, T.; Kawasaki, K. Strengthening of Ti–6Al–4V Alloy by Short-Time Duplex Heat Treatment. *Mater. Trans.* **2005**, *46*, 1681–1686. [CrossRef]

16. Valiev, R.Z.; Alexandrov, I.V.; Enikeev, N.A.; Murashkin, M.Y.; Semenova, I.P. Towards enhancement of properties of UFG metals and alloys by grain boundary engineering using SPD processing. *Rev. Adv. Mater. Sci.* **2010**, *25*, 1–10.

17. Sakai, T.; Belyakov, A.; Kaibyshev, R.; Miura, H.; Jonas, J.J. Dynamic and post-dynamic recrystallization under hot, cold and severe plastic deformation conditions. *Prog. Mater. Sci.* **2014**, *60*, 130–207. [CrossRef]

18. Muga, C.O.; Zhang, Z.W. Strengthening Mechanisms of Magnesium-Lithium Based Alloys and Composites. *Adv. Mater. Sci. Eng.* **2016**, *2016*, 1078187. [CrossRef]

19. Langdon, T.G. Seventy-five years of superplasticity: Historic developments and new opportunities. *J. Mater. Sci.* **2009**, *44*, 5998. [CrossRef]

20. Matsumoto, H.; Velay, V.; Chiba, A. Flow behavior and microstructure in Ti–6Al–4V alloy with an ultrafine-grained α-single phase microstructure during low-temperature-high-strain-rate superplasticity. *Mater. Des.* **2015**, *66*, 611–617. [CrossRef]

21. Semiatin, S.L.; Fagin, P.N.; Betten, J.F.; Zane, A.P.; Ghosh, A.K.; Sargent, G.A. Plastic Flow and Microstructure Evolution during Low-Temperature Superplasticity of Ultrafine Ti-6Al-4V Sheet Material. *Metall. Mater. Trans. A* **2010**, *41*, 499–512. [CrossRef]

22. Alabort, E.; Kontis, P.; Barba, D.; Dragnevski, K.; Reed, R.C. On the mechanisms of superplasticity in Ti–6Al–4V. *Acta Mater.* **2016**, *105*, 449–463. [CrossRef]

23. Lee, T.; Yamasaki, M.; Kawamura, Y.; Lee, Y.; Lee, C.S. High strain-rate superplasticity of AZ91 alloy achieved by rapidly solidified flaky powder metallurgy. *Mater. Lett.* **2019**, *234*, 245–248. [CrossRef]

24. Sergueeva, A.V.; Stolyarov, V.V.; Valiev, R.Z.; Mukherjee, A.K. Superplastic behaviour of ultrafine-grained Ti–6A1–4V alloys. *Mater. Sci. Eng. A* **2002**, *323*, 318–325. [CrossRef]

25. Salishchev, G.A.; Galeyev, R.M.; Valiakhmetov, O.R.; Safiullin, R.V.; Lutfullin, R.Y.; Senkov, O.N.; Froes, F.H.; Kaibyshev, O.A. Development of Ti–6Al–4V sheet with low temperature superplastic properties. *J. Mater. Process. Technol.* **2001**, *116*, 265–268. [CrossRef]

26. Paton, N.E.; Hamilton, C.H. Microstructural influences on superplasticity in Ti-6AI-4V. *Metall. Trans. A* **1979**, *10*, 241–250. [CrossRef]

27. Park, C.H.; Ko, Y.G.; Park, J.-W.W.; Lee, C.S. Enhanced superplasticity utilizing dynamic globularization of Ti-6Al-4V alloy. *Mater. Sci. Eng. A* **2008**, *496*, 150–158. [CrossRef]

28. Ovid'ko, I.A.; Sheinerman, A.G. Grain-boundary dislocations and enhanced diffusion in nanocrystalline bulk materials and films. *Philos. Mag.* **2003**, *83*, 1551–1563. [CrossRef]

29. Huang, K.; Logé, R.E. A review of dynamic recrystallization phenomena in metallic materials. *Mater. Des.* **2016**, *111*, 548–574. [CrossRef]

30. Wert, J.A.; Paton, N.E. Enhanced superplasticity and strength in modified Ti-6AI-4V alloys. *Metall. Mater. Trans. A* **1983**, *14*, 2535–2544. [CrossRef]

31. Kim, J.S.; Chang, Y.W.; Lee, C.S. Quantitative analysis on boundary sliding and its accommodation mode during superplastic deformation of two-phase Ti-6Al-4V alloy. *Metall. Mater. Trans. A* **1998**, *29*, 217–226. [CrossRef]

Article

Microstructural Characterization of Friction-Stir Processed Ti-6Al-4V

Sergey Mironov [1],*, Yutaka S. Sato [2], Hiroyuki Kokawa [2,3], Satoshi Hirano [4], Adam L. Pilchak [5] and Sheldon Lee Semiatin [5]

[1] Laboratory of Mechanical Properties of Nanoscale Materials and Superalloys, Belgorod National Research University, Pobeda 85, 308015 Belgorod, Russia

[2] Department of Materials Processing, Tohoku University, 6-6-02 Aramaki-aza-Aoba, Sendai 980-8579, Japan; ytksato@material.tohoku.ac.jp (Y.S.S.); kokawa@tohoku.ac.jp (H.K.)

[3] Shanghai Key Laboratory of Materials Laser Processing and Modification, School of Materials Science and Engineering, Shanghai Jiao Tong University, 800 Dongchuan Road, Minhang District, Shanghai 200240, China

[4] Hitachi Research Laboratory, Hitachi Ltd., 7-1-1 Omika-cho, Hitachi 319-1291, Japan; satoshi.hirano.xv@hitachi.com

[5] Air Force Research Laboratory, Materials and Manufacturing Directorate, Wright-Patterson AFB, OH 45433-7817, USA; adam.pilchak.1@us.af.mil or adam.pilchak.1@afresearchlab.com (A.L.P.); sheldon.semiatin.1@us.af.mil or slsemiatin@gmail.com (S.L.S.)

* Correspondence: mironov@bsu.edu.ru; Tel.: +7-4722-585455

Received: 2 July 2020; Accepted: 17 July 2020; Published: 20 July 2020

Abstract: The present work was undertaken to shed additional light on the globular-α microstructure produced during FSP of Ti-6Al-4V. To this end, the electron backscatter diffraction (EBSD) technique was employed to characterize the crystallographic aspects of such microstructure. In contrast to the previous reports in the literature, neither the texture nor the misorientation distribution in the α phase were random. Although the texture was weak, it showed a clear prevalence of the P_1 and C-fiber simple-shear orientations, thus providing evidence for an increased activity of the prism-<a> and pyramidal <c+a> slip systems. In addition, the misorientation distribution exhibited a crystallographic preference of 60° and 90° boundaries. This observation was attributed to a partial $\alpha \rightarrow \beta \rightarrow \alpha$ phase transformation during/following high-temperature deformation and the possible activation of mechanical twinning.

Keywords: Ti-6Al-4V; friction stir processing; electron backscatter diffraction; microstructure; texture

---

1. Introduction

Ti-6Al-4V is an important structural material which is widely used in aerospace, marine, and a number of other industries. The optimal balance of service properties in this alloy is often achieved by the formation of a "globular" microstructure. To this end, the material is subjected to a complex thermo-mechanical treatment, the final step of which involves processing below the β-transus temperature (at which $\alpha + \beta \rightarrow \beta$). Unfortunately, conventional industrial techniques (e.g., rolling, extrusion, forging) do not impose sufficient deformation to achieve the desired fully globular microstructure. Accordingly, the development of novel techniques is necessary to improve the performance of Ti-6Al-4V. One of these approaches is friction-stir processing (FSP) [1].

FSP involves a combination of very large strains and high temperatures and thus often leads to a substantial refinement in microstructure. Because of relatively high technological requirements (e.g., tool materials, working loads, etc.,), FSP of titanium alloys is relatively difficult, and until recently, microstructural observations of FSPed Ti-6Al-4V were limited. However, rapid progress in this technology during the past few years has promoted growing interest.

The first investigations of the FSP of Ti-6Al-4V revealed a relatively narrow processing window for this technique [2–18]. This is believed to be due to the relatively low ductility of Ti-6Al-4V even at elevated temperatures as well as its low thermal conductivity which gives rise to considerable temperature gradients during FSP. Moreover, a significant portion of the processing window often lies above β transus. FSP in such regimes does not provide the globular microstructure thus being not practical for material processing. Another important challenge is extensive wear of the processing tool [5,9,19–22]. This effect shortens tool life (thereby increasing processing cost) and results in a substantial contamination of the processed material by tool debris. Despite these shortcomings, careful control of process parameters may provide a desirable, fully globular microstructure which consists of fine-grain (or even ultrafine-grain) α phase [6,10,23–35]. In such instances, the processed material exhibits excellent superplastic properties [36–41].

Until recently, relatively little attention was given to examination of the crystallographic characteristics of such microstructures. It has sometimes been reported that the texture and misorientation distribution after FSP of Ti-6Al-4V are nearly random [12,24,27,28,34,35]. Despite valuable prior efforts to describe the underlying microstructural mechanisms [42], it is still not clear whether such results are isolated cases or represent reproducible, general trends. Hence, the present work was undertaken to shed additional light on this issue. To this end, the electron backscatter diffraction (EBSD) technique was employed to characterize the globular-α microstructure produced during FSP of Ti-6Al-4V.

2. Materials and Methods

The material used in the present effort was Ti-6Al-4V, which was supplied as 7-mm-thick plate in the mill-annealed condition. Microstructural details of the base material are shown in supplementary Figures S1–S3. This material was friction-stir processed using a cobalt-based FSP tool. Based on previous experience, a relatively-low spindle (rotation) rate of 120 rpm was used to keep the processing temperature below the β transus. To investigate two points in a possible processing window, trials using feed rates of 15 and 30 mm/min were conducted. The welding tool had a diameter of 17.5 mm and a probe of 5.8 mm in length. Additional processing details were proprietary to Hitachi Ltd. To maintain consistency with the terminology in the literature, the principal directions of the process geometry are denoted herein as the welding direction (WD), transverse direction (TD), and normal direction (ND).

Microstructural observations were performed on a transverse cross-section (TD × ND plane) using optical microscopy (OM), scanning electron microscopy (SEM), electron probe microanalysis (EPMA), and EBSD. For the OM, SEM, and EPMA examinations, sectioned samples were first ground with silicon carbide abrasive papers, mechanically polished with 1-μm diamond, and then chemically etched with Kroll's reagent (3% HNO_3, 2% HF, and 95% H_2O). A suitable surface finish for EBSD was obtained by mechanical polishing in a similar fashion followed by long-term (24 h) vibratory polishing with colloidal silica.

SEM and EBSD analysis were conducted using a Hitachi S-4300SE field-emission-gun SEM (FEG-SEM) (Hitachi, Tokyo, Japan) equipped with a TSL OIM^{TM} EBSD system (EDAX, Mahwah, NJ, USA) and operated at an accelerating voltage of 25 kV. All SEM examinations were made in the secondary-electron mode. To reveal the microstructure at different length scales, EBSD maps were acquired with scan step sizes of either 0.5 or 0.2 μm. For each diffraction pattern, nine Kikuchi bands were used to minimize mis-indexing errors. To ensure reliability of the EBSD data, small grains comprising one or two pixels were automatically removed from the maps by applying the grain-dilation feature of the EBSD software. To eliminate spurious boundaries caused by orientation noise, a lower-limit boundary misorientation cut-off of 2° was employed. Furthermore, a 15° criterion was applied to differentiate low-angle boundaries (LABs) and high-angle boundaries (HABs). The grain size was quantified from the EBSD data by applying either the grain-reconstruction approach [43] (i.e., converting each grain to a circle with equivalent area and calculating the associated circle-equivalent diameter) or the conventional grain-boundary intercept method. EPMA measurements were conducted with a JEOL XM-85300FBU FEG-SEM (JEOL, Tokyo, Japan) operated at an accelerating voltage of 15 kV.

3. Results and Discussion

3.1. Preliminary Analysis

Low-magnification optical images of the FSP'ed samples (Figure 1) revealed a distinct stir zone for both feed rates. In the sample friction-stirred at the higher rate, a tunnel-type defect was found (indicated by the arrow in Figure 1b). Considering the relatively small change in process variables for the two cases, this observation suggested a rather-narrow processing window, a finding consistent with prior work [12–18]. To avoid the possible influence of this defect on the interpretation of microstructure evolution and material flow, subsequent examination was thus focused on the defect-free material produced using a feed rate of 15 mm/min.

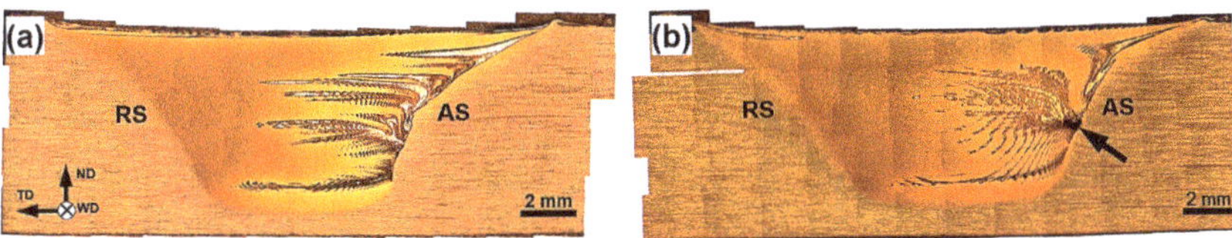

Figure 1. Low-magnification optical images showing a transverse cross section of samples processed using a feed rate of (**a**) 15 mm/min or (**b**) 30 mm/min. RS and AS denote the retreating side and the advancing side, respectively. The reference frame for both cases is shown in the bottom left corner of (**a**). In (**b**), the arrow indicates a tunnel-type defect.

One of the most striking features of the processed material was the very specific structural pattern which evolved at the advancing side and (Figure 1), to a lesser extent, at the upper surface of the stir zone (not shown). Such structures are often observed in friction-stirred Ti-6Al-4V and are usually attributed to tool wear [9,19–22]. SEM and EPMA of the material at the advancing side revealed the origin of this macrostructure (Figure 2). In the SEM image (Figure 2a), the α phase is dark, and the β phase is light. From this micrograph, it is appeared that the lighter features were β (or a β-rich) phase. The conclusion that some regions had transformed to β was rationalized by EPMA chemical maps (Figure 2b–e). These maps did indeed indicate a relatively-low concentration of titanium (Figure 2b) and an enrichment in cobalt, tungsten, and nickel (Figure 2c–e, respectively). Each of these elements are included in the specification for the processing-tool material. Therefore, it was likely that the observed clustering of these elements in the stir zone was associated with tool wear during FSP. The appropriate binary phase diagrams (not shown) indicated that additions of these elements to titanium *decrease* the β transus. Accordingly, a local inter-alloying of the stir zone material by these elements should result in stabilization of the β phase, as was observed (Figure 2a). Moreover, these observations are in good agreement with prior findings [9,19–22].

3.2. Microstructure Distribution within Stir Zone

In addition to the tool wear discussed in the previous section, noticeable variations in microstructure were found in the thickness direction of the stir zone (Figure 3). As indicated by the SEM micrographs, all of the microstructures were dominated by a relatively-fine globular-α phase. This indicates that the processing temperature was indeed below the β transus. In the upper and mid-thickness sections of the stir zone, in particular, a significant amount of secondary α (α_s) was also noted (Figure 3a,b). These observations suggested that the temperature in these areas likely exceeded ~900 °C, and that the material had experienced a *partial* $\alpha \rightarrow \beta \rightarrow \alpha_s$ phase transformation sequence.

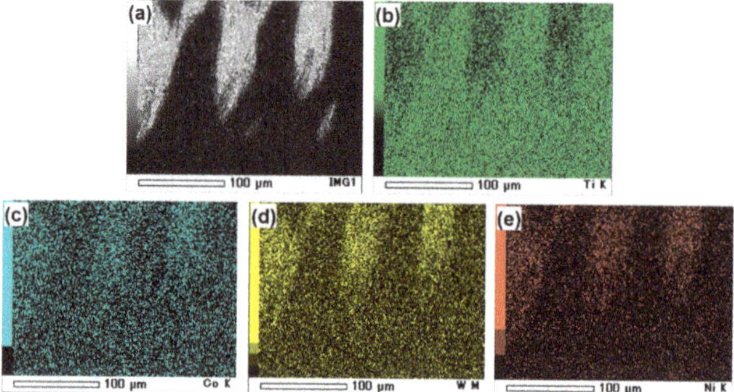

Figure 2. (**a**) SEM image and electron probe microanalysis (EPMA) (composition) maps for (**b**) titanium, (**c**) cobalt, (**d**) tungsten, and (**e**) nickel taken from the advancing side of the stir zone.

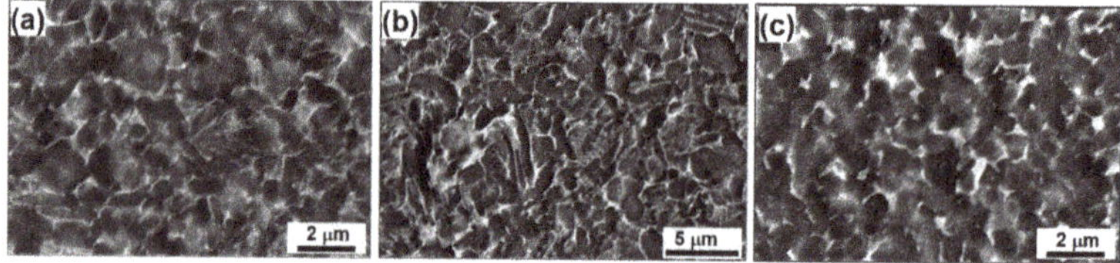

Figure 3. Typical SEM micrographs showing the microstructure distribution in the thickness direction of the stir zone: (**a**) Upper section, (**b**) mid-thickness, and (**c**) bottom section. Note the difference in magnification.

As expected, the finest microstructure was observed at the stir-zone root (Figure 3c). This is consistent with previous work, and is usually explained in terms of the relatively-low thermal conductivity of titanium alloys [13,15,18,24,44–46] and the fact that the main source of heating during FSP is friction between the workpiece surface and tool shoulder [1]. Accordingly, the local temperature is typically believed to decrease in the downward direction, thereby enhancing the microstructure-refinement effect at the stir zone root. Nevertheless, it was surprising to find a fine grain structure in the *upper* region of the stir zone as well (Figure 3a). The source of this effect is less clear, but is sometimes attributed to the relatively-large strain induced by the tool shoulder during FSP. Thus, the coarsest microstructure was observed at the *mid-thickness* of the stir zone (Figure 3b).

3.3. Microstructure Morphology and Grain Size

EBSD maps taken from the central section of the stir zone (Figure 4) provided deeper insight into microstructure evolution. The corresponding microstructure statistics derived from the EBSD maps are summarized in Figures 5–7. Due to difficulties in indexing the β phase, however, only limited data on this microstructural constituent could be obtained. Thus, the analysis in the present investigation was focused on the α phase, and the fine β particles (Figure 3b) were neglected during the evaluation of grain size and misorientation distribution for the α phase.

The *low-resolution* EBSD maps (Figure 4a) showed that the microstructure was very uniform. Equally important, there was no pronounced clustering of α grains sharing a common crystallographic axis (i.e., so-called "microtexture") in contrast to conventional processing of Ti-6Al-4V. This observation has been also highlighted by Pilchak, et al. [42]. In that work, the effect was attributed to the specific nature of deformation during FSP, which is close to simple shear and therefore has no true "end"/stable crystallographic orientation(s). The effect may also be enhanced by the relatively-complex strain path inherent to FSP in general.

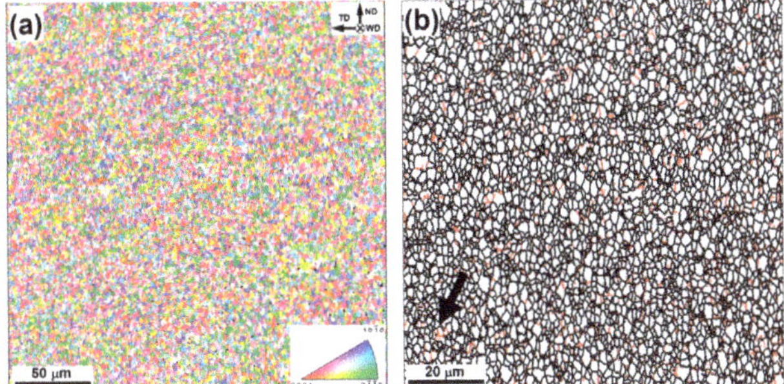

Figure 4. (a) Low-resolution EBSD inverse-pole-figure map and (b) higher-resolution EBSD grain-boundary map for the α phase taken from the central region of the stir zone. The reference frame for both maps is given in the top right corner of (a). In (b), low-angle boundaries (LABs) and high-angle boundaries (HABs) are depicted by red and black lines, respectively. Note: The black clusters denote the β phase.

As expected, the microstructure was dominated by the fine grains (Figure 4b) with a mean circle-equivalent diameter of 1.4 μm (Figure 5a) and mean HAB intercept length of 1 μm (Figure 5b). The HAB fraction was 86%. Not surprisingly, the remnant LABs were clustered primarily within relatively-coarse α grains which tended to subdivide them into smaller-scale fragments; an example is indicated by the arrow in Figure 4b. These observations suggested that microstructure evolution within the α phase during FSP occurred by a process of *continuous* dynamic recrystallization (CDRX), thus being consistent with similar reports in the literature [25,35,42]. Also in agreement with previous work, the α grains exhibited a nearly-globular shape, thus confirming the formation of a globular structure (Figure 4b).

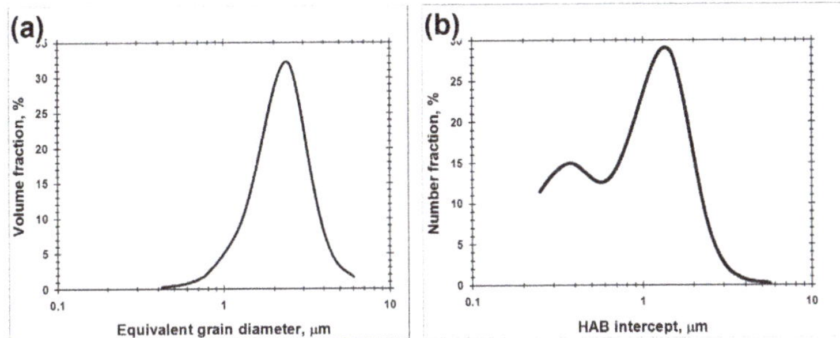

Figure 5. Grain-size distributions for the α phase measured by (a) the grain-reconstruction approach and (b) the grain-boundary intercept method from the EBSD map in Figure 4b.

3.4. Texture

It is commonly accepted that the deformation imposed during FSP is close to a mode of simple-shear [47,48]. However, according to Pilchak et al. [42], the strain path is simple shear only at limiting points around the tool; elsewhere, it is a mixture of simple and pure shear. The ideal orientations expected during simple shear of hcp crystals (such as α titanium) are given in Figure 6a and Table 1. For FSP, the shearing direction is tangential to the tool-rotation direction [48]. However, the orientation of the shear plane is usually less evident. Often, it is thought to be tangential to the tool-workpiece interface [49] or oriented nearly-parallel to the boundary between the stir zone and the thermo-mechanically affected zone (TMAZ) as in the truncated-cone model [50].

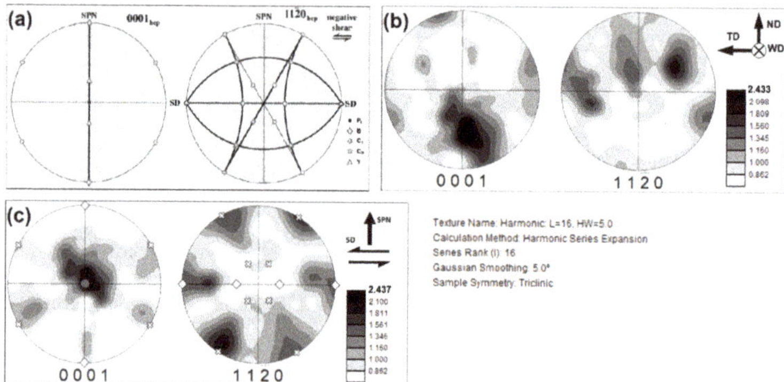

Figure 6. {0001} and $\{11\bar{2}0\}$ pole figures showing (**a**) ideal simple-shear textures expected for hexagonal close packed metals (after Fonda, et al. [48]), (**b**) those derived from Figure 4a, and (**c**) those derived from Figure 4a which were rotated 50° about the TD and then 15° about the ND to align with the presumed geometry of simple shear. For comparison purposes, several ideal simple-shear orientations are indicated in (**c**).

The textures developed during FSP were interpreted in terms of {0001} and $\{11\bar{2}0\}$ pole figures (Figure 6b) derived from the low-resolution EBSD map (Figure 4a). Assuming that the shear plane in the present work was parallel to the stir zone-TMAZ boundary, the measured pole figures were rotated 50° about the TD to align them with the presumed geometry of simple shear. They were also given an additional rotation of 15° about the ND in order to adjust the experimental data with the ideal textures. The rotated pole figures are shown in Figure 6c.

From the comparison of the rotated pole figures with the ideal textures in Figure 6a, it appeared that the FSP material was characterized by a preference for the P_1 $\{\bar{1}100\} < \bar{2}110 >$ and *C*-fiber components. Based on literature findings (Table 1), such results implied a prevalence of prism- <*a*> and pyramidal-<*c+a*> slip during FSP. In addition, there was some evidence for the development of a B-fiber component (Figure 6c), thus implying the possible activation of basal <*a*> slip as well (Table 1). However, these observations were not clear-cut, and thus further verification may be helpful. While the dominance of prism slip (with its low critical resolved shear stress) was expected, the activation of the pyramidal slip appeared somewhat surprising. It was likely necessitated by strain-compatibility requirements and perhaps the presence of the β phase. Despite these observations, the measured texture was weak with a maximum peak intensity of only ~2.5 times random (Figure 6c). This observation was in the line with previous texture studies [24,27,28,34,35].

It is also worth noting that the final microstructure was produced partially by a local $\alpha \rightarrow \beta \rightarrow \alpha_s$ phase transformation, as mentioned in Section 3.2. This process could influence texture evolution, and thus the analysis given in the present section may be oversimplified. Therefore, orientation measurements within the prior-β phase (and associated the secondary α) are needed to provide further insight into the evolution of texture during FSP.

Table 1. Ideal simple-shear textures in hexagonal close-packed metals (after Li [51], and Beausir et al. [52]).

Notation	Euler Angles (φ1, Φ, φ2)	Miller-Bravais Indices {*hkil*}<*uvtw*>	Primary Slip Mode
P-fiber	(0;0–90;0)	{*hkil*} $< \bar{2}110 >$	Prism <*a*> slip
P_1	(0;0;0)	$\{\bar{1}100\} < \bar{2}110 >$	Prism <*a*> slip
B-fiber	(0;90;0–60)	{0001} < *uvtw* >	Basal <*a*> slip
Y-fiber	(0;30;30–60)	-	Pyramidal <*a*> slip
C-fiber	(60;90;0–60)	-	Pyramidal <*c+a*> slip

3.5. Misorientation Distribution

Misorientation data extracted from the *high-resolution* EBSD map (Figure 4b) were arranged as distributions of misorientation angles and misorientation (rotation) axes (Figure 7). To assist in the interpretation of these experimental results, a so-called uncorrelated (or texture-derived) misorientation distribution was also calculated. For this approach, 1000 pixels were arbitrarily selected from the EBSD map and all possible misorientations between them were determined using one of the standard options in the EBSD software. Results from this texture-derived distribution were broadly similar to the distribution for a texture comprising randomly oriented grains (Figure 7a). This observation was likely due to the very weak texture that was developed within the stir zone, as discussed in the previous section. On the other hand, the measured grain-boundary misorientation distribution was noticeably different from both the texture-derived and random ones. Specifically, it was characterized by a pronounced low-angle maximum, additional misorientation peaks near ~60° and ~90° (Figure 7a), and the clustering of misorientation axes around several preferred crystallographic directions (Figure 7b).

The low-angle peak was likely attributable to the very large strain experienced by the material during FSP and the continuous nature of CDRX. Surprisingly, the rotation axes of the LABs were distributed in a near-random fashion (Figure 7b). This behavior contrasted with the preferential clustering of LAB axes near the <0001> pole often observed in heavily deformed α titanium [53–55]. The present finding may thus be a result of the complex character of slip involving the activation of both prism- <*a*> and pyramidal <*c+a*> modes, as discussed previously.

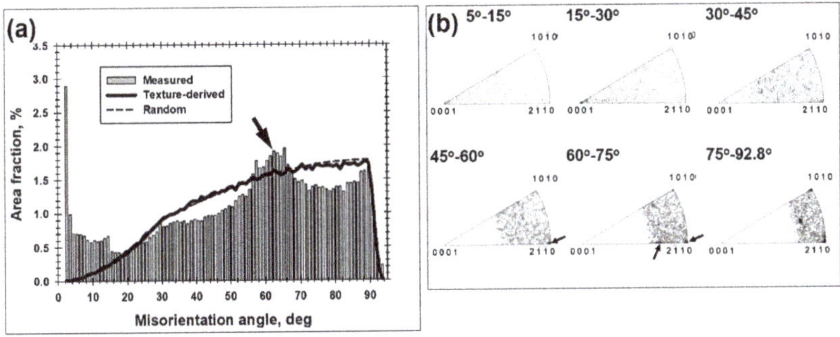

Figure 7. Distributions of (**a**) misorientation angle and (**b**) rotation axis for the α phase derived from the EBSD data in Figure 4b. The arrows show the misorientations which presumably originated from an α→β→α$_s$ phase transformation.

The noticeable proportion of ~60° boundaries was likely associated with the partial α→β→α$_s$ phase transformation, as discussed in Section 3.2. In titanium alloys, this transformation is normally governed by the Burgers orientation relationship, viz., $\{0001\}_\alpha // \{110\}_\beta$ and $< 11\bar{2}0 >_\alpha // < 111 >_\beta$. Because of the crystallographic symmetry of the α and β phases, there are 12 crystallographic variants of this relationship. The possible misorientations between the variants inherited from the same prior-β grain are shown in Table 2. From the table, it is seen that 3 of 5 such variants provide a peak near 60° in the misorientation-angle distribution and a clustering of rotation axes near the $< 2\bar{1}\bar{1}0 >$ pole, in broad agreement with the experimental data (indicated by the arrows in Figure 7). This provides an explanation for the origin of the 60° peak therefore. In addition, it is worth noting that the measured fraction of "inter-variant" boundaries constituted only ~6.5% of the total grain-boundary area. The relatively low fraction of such misorientations was probably associated with very fine-grain nature of the microstructure (Figure 3b) which typically results in the nucleation of only a single α variant within some prior β grains. On the other hand, an example of the prior-β grain structure containing several secondary-α colonies (which are responsible for the inter-variant misorientations) is shown in supplementary Figure S4.

Table 2. Predicted misorientations between α variants inherited from the same parent β grain (after Gey et al. [56] and Wang et al. [57]).

Misorientation (Angle-Axis Pair)	Symbol	Location of Misorientation Axes on Stereographic Triangle
$10.5° < 0001 >$	▲	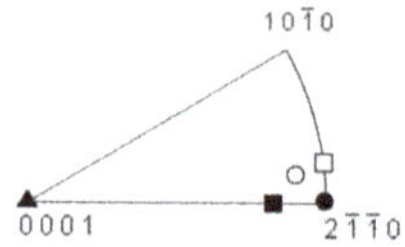
$60° < 11\bar{2}0 >$	●	
$60.8° < \overline{1.377};\bar{1};2.377;0.359 >$	○	
$63.3° < \overline{10};5;5;\bar{3} >$	■	
$90° < 1;\overline{2.38};1.38;0 >$	□	

The crystallographic preference for ~90°< $2\bar{1}\bar{1}0$ >, ~90°< $10\bar{1}0$ >, and ~90°< $16;\bar{4};\overline{12};3$ > misorientations (Figure 7b) is less clear. From a broad perspective, such boundaries in α titanium can be produced by mechanical twinning involving $\{10\bar{1}2\}$, $\{11\bar{2}3\}$, or $\{10\bar{1}2\} \rightarrow \{11\bar{2}3\}$ modes, respectively e.g., [58]. Indeed, evidence of twinning of the α phase has been found during uniaxial compression of Ti-6Al-4V at room temperature [59] and, very recently, during the hot compression of the single-phase-α alloy Ti-7Al [60].

It should be noted that lowering the FSP temperature (and the concomitant development of ultrafine microstructures similar to that in Figure 3c) suppresses the phase transformation, but may enhance mechanical twinning. Such changes would likely result in a different misorientation distribution. On the other hand, considering the relatively narrow processing window as well as the substantial temperature gradient within the stir zone, phase transformation and twinning may always exert an influence on microstructural evolution to some degree.

4. Conclusions

The present work was undertaken to provide insight into the globular microstructure developed during FSP of Ti-6Al-4V. To this end, the advanced capabilities of the EBSD technique were employed. The main results derived from this study were as follows.

The microstructure developed in the stir zone under nominally subtransus processing conditions results from a complex superposition of several processes. In addition to the strain-induced refinement (common to FSP), it is also influenced noticeably by the partial $\alpha \rightarrow \beta \rightarrow \alpha_s$ phase transformation (induced by the FSP thermal cycle) and inter-alloying due to the wear of the FSP tool. The markedly inhomogeneous microstructure distribution within the stir zone shows evidence of considerable variations in thermo-mechanical conditions.

The microstructure produced in the central section of stir zone is dominated by a fully globular α phase with a mean grain size of ~1 μm and HAB fraction of 86%, and an absence of microtexture. The LAB substructure within remnant, relatively coarse α grains suggest a continuous dynamic recrystallization mechanism of α phase refinement.

Although the texture in the α phase is very weak, there is a crystallographic preference for the formation of P_1 $\{\bar{1}100\}$ < $\bar{2}110$ > and *C*-fiber simple-shear components. This observation may be attributed to the dominance of prism- <*a*> and pyramidal <*a*+*c*> slip activity during FSP.

The misorientation distribution in the α phase is characterized by a noticeable proportion of 60° and 90° boundaries. The former boundaries are likely associated with a partial $\alpha \rightarrow \beta \rightarrow \alpha_s$ phase transformation, whereas the latter may indicate the possible activation of $\{10\bar{1}2\}$ and/or $\{11\bar{2}3\}$ twinning.

Supplementary Materials: The following are available online at http://www.mdpi.com/2075-4701/10/7/976/s1, Figure S1: SEM images of the microstructure of the base material at: (a) low magnification and (b) high magnification. Figure S2: EBSD characterization of the base material: (a) low-resolution orientation (inverse-pole-figure) map, and (b) (0001) and {11-20} pole figures showing the texture. In (a), LABs and HABs are depicted as white and black lines, respectively. Figure S3: EBSD characterization of the base material: (a) high-resolution orientation (inverse-pole-figure) map, and (b) misorientation distribution. In (a), LABs and HABs are depicted as white and black lines, respectively. In (b), the insert in the top right corner shows misorientation-axis distribution. Figure S4: SEM micrograph taken from the stir zone that exemplifies several secondary-alpha colonies within a prior-β grain.

Author Contributions: Conceptualization, S.H., Y.S.S., S.M.; methodology, S.H., Y.S.S., S.M.; formal analysis, S.M.; investigation, S.M.; resources, S.H., Y.S.S., H.K.; data curation, S.H., Y.S.S., S.M.; writing—original draft preparation, S.M.; writing—review and editing, Y.S.S., H.K., S.H., A.L.P., S.L.S.; visualization, S.M.; supervision, S.H., Y.S.S., H.K.; project administration, Y.S.S.; funding acquisition, S.H. All authors have read and agreed to the published version of the manuscript.

Funding: This research was based on results obtained from a project commissioned by the New Energy and Industrial Technology Development Organization (NEDO).

Conflicts of Interest: The authors declare no conflict of interest.

References

1. Mishra, R.S.; Ma, Z.Y. Friction stir welding and processing. *Mater. Sci. Eng. R* **2005**, *50*, 1–78. [CrossRef]
2. Juhas, M.C.; Viswanathan, G.B.; Fraser, H.L. Microstructural evolution in Ti alloy friction stir welds. In Proceedings of the Second Symposium on Friction Stir Welding, Gothenburg, Sweden, 26–28 June 2000.
3. Juhas, M.C.; Viswanathan, G.B.; Fraser, H.L. Characterization of microstructural evolution in a Ti-6Al-4V friction stir weld. In Proceedings of the Lightweight Alloys for Aerospace Application, TMS, Warrendale, PA, USA, 12–14 February 2001; Jata, K., Lee, E.W., Frazier, W., Kim, N.J., Eds.; TMS: Warrendale, PA, USA, 2001; pp. 209–217.
4. Ramirez, A.J.; Juhas, M.C. Microstructural evolution in Ti-6Al-4V friction stir welds. In *Material Science Forum*; Trans Tech Publications Ltd.: Zurich-Uetikon, Switzerland, 2003; Volume 426, pp. 2999–3004.
5. Pavka, P.A. Microstructural Evolution of Friction Stir Processed Ti-6Al-4V. Ph.D. Thesis, Ohio State University, Columbus, OH, USA, 2006.
6. Pilchak, A.L.; Juhas, M.C.; Williams, J.C. Microstructural changes due to friction stir processing of investment-cast Ti-6Al-4V. *Metall. Mater. Trans. A* **2007**, *38*, 401–408. [CrossRef]
7. Pilchak, A.L.; Li, Z.T.; Fisher, J.J.; Reynolds, A.P.; Juhas, M.C.; Williams, J.C. The relationship between friction stir processing (FSP) parameters and microstructure in investment cast Ti-6Al-4V. In *Friction Stir Welding and Processing IV*; Mishra, R.S., Mahoney, M.W., Lienert, T.J., Jata, K.V., Eds.; TMS: Warrendale, PA, USA, 2007; pp. 419–427.
8. Pilchak, A.L.; Juhas, M.C.; Williams, J.C. The effect of friction stir processing on microstructure and properties of investment cast Ti-6Al-4V. In Proceedings of the 11th World Titanium 2007 Conference, Kyoto, Japan, 3–7 June 2007; Ninomi, M., Akiyama, S., Ikeda, M., Hagiwara, M., Maruyama, K., Eds.; Japan Institute of Metals: Sendai, Japan, 2007; pp. 1723–1726.
9. Pilchak, A.L.; Juhas, M.C.; Williams, J.C. Observations of tool-workpiece interactions during friction stir processing of Ti-6Al-4V. *Metall. Mater. Trans. A* **2007**, *38*, 435–437. [CrossRef]
10. Pilchak, A.L.; Norfleet, D.M.; Juhas, M.C.; Williams, J.C. Friction stir processing of investment-cast Ti-6Al-4V: Microstructure and properties. *Metall. Mater. Trans. A* **2008**, *39*, 1519–1524. [CrossRef]
11. Pilchak, A.L.; Juhas, M.C.; Williams, J.C. A comparison of friction stir processing of investment cast and mill-annealed Ti-6Al-4V. *Weld. World* **2008**, *52*, 60–68. [CrossRef]
12. Lauro, A. Friction stir welding of titanium alloys. *Weld. Int.* **2012**, *26*, 8–21. [CrossRef]
13. Su, J.; Wang, J.; Mishra, R.S.; Xu, R.; Baumann, J.A. Microstructure and mechanical properties of a friction stir processed Ti-6Al-4V alloy. *Mater. Sci. Eng. A* **2013**, *573*, 67–74. [CrossRef]
14. Lippold, J.C.; Livingston, J.J. Microstructure evolution during friction stir processing and hot torsion simulation of Ti-6Al-4V. *Metall. Mater. Trans. A* **2013**, *44*, 3815–3825. [CrossRef]
15. Edwards, P.; Ramulu, M. Identification of process parameters for friction stir welding Ti-6Al-4V. *J. Eng. Mater. Technol.* **2010**, *132*, 031006–1. [CrossRef]
16. Edwards, P.; Ramulu, M. Effect of process conditions on superplastic forming behavior in Ti-6Al-4V friction stir welds. *Sci. Technol. Weld. Join.* **2009**, *14*, 669–680. [CrossRef]
17. Sanders, D.G.; Ramulu, M.; Edwards, P.D.; Cantrell, A. Effect on the surface texture, superplastic forming, and fatigue performance of Titanium 6Al-4V friction stir welds. *J. Mater. Eng. Perform.* **2010**, *19*, 503–509. [CrossRef]
18. Edwards, P.D.; Ramulu, M. Investigation of microstructure, surface and subsurface characteristics in titanium alloy friction stir welds of varied thicknesses. *Sci. Technol. Weld. Join.* **2009**, *14*, 476–483. [CrossRef]
19. Wang, J.; Su, J.; Mishra, R.S.; Xu, R.; Baumann, J.A. Tool wear mechanisms in friction stir welding of Ti-6Al-4V. *Wear* **2014**, *321*, 25–32. [CrossRef]

20. Pilchak, A.L.; Tang, W.; Sahiner, H.; Reynolds, A.P.; Williams, J.C. Microstructure evolution during friction stir welding of mill-annealed Ti-6Al-4V. *Metall. Mater. Trans. A* **2011**, *42*, 745–762. [CrossRef]
21. Fall, A.; Fesharaki, M.H.; Khodabandeh, A.R.; Jahazi, M. Tool wear characteristics and effect on microstructure in Ti-6Al-4V friction stir welded joints. *Metals* **2016**, *6*, 275. [CrossRef]
22. Wu, L.H.; Wang, D.; Xiao, B.L.; Ma, Z.Y. Tool wear and its effect on microstructure and properties of friction stir processed Ti-6Al-4V. *Mater. Chem. Phys.* **2014**, *146*, 512–522. [CrossRef]
23. Kitamura, K.; Fujii, H.; Iwata, Y.; Sun, Y.S.; Morisada, Y. Flexible control of the microstructure and mechanical properties of friction stir welded joints. *Mater. Design* **2013**, *46*, 348–354. [CrossRef]
24. Yoon, S.; Ueji, R.; Fujii, H. Effect of rotation rate on microstructure and texture evolution during friction stir welding of Ti-6Al-4V plates. *Mater. Character.* **2015**, *106*, 352–358. [CrossRef]
25. Zhou, L.; Liu, H.J.; Liu, P.; Liu, Q.W. The stir zone microstructure and its formation mechanism in Ti-6Al-4V friction stir welds. *Scripta Mater.* **2009**, *61*, 596–599. [CrossRef]
26. Liu, H.J.; Zhou, L.; Liu, Q.W. Microstructural characteristics and mechanical properties of friction stir welded joints of Ti-6Al-4V titanium alloy. *Mater. Design* **2010**, *31*, 1650–1655. [CrossRef]
27. Liu, H.J.; Zhou, L. Microstructural zones and tensile characteristics of friction stir welded joint of TC4 titanium alloy. *Trans. Nonferrous. Met. Soc. China* **2010**, *20*, 1873–1878. [CrossRef]
28. Zhou, L.; Liu, H.-J.; Wu, L.-Z. Texture of friction stir welded Ti-6Al-4V alloy. *Trans. Nonferrous Met. Soc. China* **2014**, *24*, 368–372. [CrossRef]
29. Farnoush, H.; Bastami, A.A.; Sadeghi, A.; Mohandesi, J.A.; Moztarzadeh, F. Tribological and corrosion behavior of friction stir processed Ti-CaP nanocomposites in simulated body fluid solution. *J. Mech. Beh. Biomed. Mater.* **2013**, *20*, 90–97. [CrossRef] [PubMed]
30. Esmaily, M.; Mortazavi, S.N.; Todehfalah, P.; Rashidi, M. Microstructural characterization and formation of α' martensite phase in Ti-6Al-4V alloy butt joints produced by friction stir and gas tungsten arc welding processes. *Mater. Design* **2013**, *47*, 143–150. [CrossRef]
31. Pasta, S.; Reynolds, A.P. Residual stress effects on fatigue crack growth in a Ti-6Al-4V friction stir welds. *Fatigue Fract. Eng. Mater. Struct.* **2008**, *31*, 569–580. [CrossRef]
32. Steuwer, A.; Hattingh, D.G.; James, M.N.; Singh, U.; Buslaps, T. Residual stress, microstructure and tensile properties in Ti-6Al-4V friction stir welds. *Sci. Technol. Weld. Join.* **2012**, *17*, 525–533. [CrossRef]
33. Muzvidziwa, M.; Okazaki, M.; Suzuki, K.; Hirano, S. Role of microstructure on the fatigue crack propagation behavior of a friction stir welded Ti-6Al-4V. *Mater. Sci. Eng. A* **2016**, *652*, 59–68. [CrossRef]
34. Yoon, S.; Ueji, R.; Fujii, H. Microstructure and texture distribution of Ti-6Al-4V alloy joints friction stir welded below β-transus temperature. *J. Mater. Process. Technol.* **2016**, *229*, 390–397. [CrossRef]
35. Ma, Z.Y.; Pilchak, A.L.; Juhas, M.C.; Williams, J.C. Microstructural refinement and property enhancement of cast light alloys via friction stir processing. *Scripta Mater.* **2008**, *58*, 361–366. [CrossRef]
36. Zhang, W.; Liu, H.; Ding, H.; Fujii, H. Superplastic deformation mechanism of the friction stir processed fully lamellar *Ti-6Al-4V alloy*. *Mater. Sci. Eng. A.* **2020**, *785*, 139390. [CrossRef]
37. Ramulu, M.; Edwards, P.D.; Sanders, D.G.; Reynolds, A.P.; Trapp, T. Tensile properties of friction stir welded and friction stir welded-superplastically formed Ti-6Al-4V butt joints. *Mater. Design* **2010**, *31*, 3056–3061. [CrossRef]
38. Sanders, D.G.; Ramulu, M.; Edwards, P.D. Superplastic forming of friction stir welds in Titanium alloy 6Al-4V: Preliminary results. *Mater. Sci. Eng. Technol.* **2008**, *39*, 353–357. [CrossRef]
39. Edwards, P.D.; Sanders, D.G.; Ramulu, M. Simulation of tensile behavior in friction stir welded and superplastically formed Titanium 6Al-4V alloy. *J. Mater. Eng. Perform.* **2010**, *19*, 510–514. [CrossRef]
40. Sanders, D.G.; Ramulu, M.; Klock-McCook, E.J.; Edwards, P.D.; Reynolds, A.P.; Trapp, T. Characterization of superplastically formed friction stir weld in titanium 6Al-4V: Preliminary results. *J. Mater. Eng. Perform.* **2008**, *17*, 187–192. [CrossRef]
41. Edwards, P.D.; Sanders, D.G.; Ramulu, M.; Grant, G.; Trapp, T.; Comley, P. Thinning behavior simulations in superplastic forming of friction stir processed titanium 6Al-4V. *J. Mater. Eng. Perform.* **2010**, *19*, 481–487. [CrossRef]
42. Pilchak, A.L.; Williams, J.C. Microstructure and texture evolution during friction stir processing of fully lamellar Ti-6Al-4V. *Metall. Mater. Trans. A* **2011**, *42*, 773–794. [CrossRef]
43. Humphreys, F.J. Quantitative metallography by electron backscattered diffraction. *J. Microsc.* **1999**, *195*, 170–185. [CrossRef]

44. Ji, S.; Li, Z.; Wang, Y.; Ma, L. Joint formation and mechanical properties of back heating assisted friction stir welded Ti-6Al-4V alloy. *Mater. Design* **2017**, *113*, 37–46. [CrossRef]

45. Buffa, G.; Fratini, L.; Schneider, M.; Merklein, M. Micro and macro mechanical characterization of friction stir welded Ti-6Al-4V lap joints through experiments and numerical simulation. *J. Mater. Process. Technol.* **2013**, *213*, 2312–2322. [CrossRef]

46. Yoon, S.; Ueji, R.; Fujii, H. Effect of initial microstructure on Ti-6Al-4V joint by friction stir welding. *Mater. Design* **2015**, *88*, 1269–1276. [CrossRef]

47. Fonda, R.W.; Bingert, J.F.; Colligan, K.J. Development of grain structure during friction stir welding. *Scripta Mater.* **2004**, *51*, 243–248. [CrossRef]

48. Fonda, R.W.; Knipling, K.E. Texture development in friction stir welds. *Sci. Technol. Weld. Join.* **2011**, *16*, 288–294. [CrossRef]

49. Park, S.H.C.; Sato, Y.S.; Kokawa, H. Basal plane texture and flow pattern in friction stir weld of a magnesium alloy. *Metall. Mater. Trans. A* **2003**, *34*, 987–994. [CrossRef]

50. Reynolds, A.P.; Hood, E.; Tang, W. Texture in friction stir welds of Timetal 21S. *Scripta Mater.* **2005**, *52*, 491–494. [CrossRef]

51. Li, S. Orientation stability in equal channel angular extrusion. Part. II: Hexagonal close-packed material. *Acta Mater.* **2008**, *56*, 1031–1043. [CrossRef]

52. Beausir, B.; Toth, L.S.; Neale, K.W. Ideal orientations and persistence characteristics of hexagonal close packed crystals in simple shear. *Acta Mater.* **2007**, *55*, 2696–2705. [CrossRef]

53. Mironov, S.Y.; Salischev, G.A.; Myshlayaev, M.M.; Pippan, R. Evolution of misorientation distribution during warm 'abc' forging of commercial-purity titanium. *Mater. Sci. Eng. A* **2006**, *418*, 257–267. [CrossRef]

54. Dyakonov, G.S.; Mironov, S.; Semenova, I.P.; Valiev, R.Z.; Semiatin, S.L. Microstructure evolution and strengthening mechanisms in commercial purity titanium subjected to equal-channel angular pressing. *Mater. Sci. Eng. A* **2017**, *701*, 289–301. [CrossRef]

55. Mironov, S.; Sato, Y.S.; Kokawa, H. Development of grain structure during friction stir welding of pure titanium. *Acta Mater.* **2009**, *57*, 4519–4528. [CrossRef]

56. Gey, N.; Hubert, M. Characterization of the variant selection occurring during the $\alpha \rightarrow \beta \rightarrow \alpha$ phase transformations of a cold rolled titanium sheet. *Acta Mater.* **2002**, *50*, 277–287. [CrossRef]

57. Wang, S.C.; Aindow, M.; Starink, M.J. Effect of self-accommodation on α/α boundary populations in pure titanium. *Acta Mater.* **2003**, *51*, 2485–2503. [CrossRef]

58. Dyakonov, G.S.; Mironov, S.; Semenova, I.P.; Valiev, R.Z.; Semiatin, S.L. EBSD analysis of grain-refinement mechanisms operating during equal-channel angular pressing of coppercial-purity titanium. *Acta Mater.* **2019**, *173*, 174–183. [CrossRef]

59. Prakash, D.G.L.; Ding, R.; Morat, R.J.; Jones, I.; Withers, P.J.; Quinta da Fonseca, J.; Preuss, M. Deformation twinning in Ti-6Al-4V during low strain rate deformation to moderate strains at room temperature. *Mater. Sci. Eng. A* **2010**, *527*, 5734–5744. [CrossRef]

60. Semiatin, S.L.; Levkulich, N.C.; Salem, A.A.; Pilchak, A.L. Plastic flow during hot working of Ti-7Al. *Metall. Mater. Trans A* **2020**, 1–16, in press. [CrossRef]

Technical Note

An Intermetallic NiTi-Based Shape Memory Coil Spring for Actuator Technologies

Ganesh Shimoga *, Tae-Hoon Kim and Sang-Youn Kim *

Interaction Laboratory, Future Convergence Engineering, Advanced Technology Research Center, Korea University of Technology and Education, Cheonan-si 31253, Chungcheongnam-do, Korea; taehoon@koreatech.ac.kr
* Correspondence: shimoga@koreatech.ac.kr (G.S.); sykim@koreatech.ac.kr (S.-Y.K.); Tel.: +82-041-560-1484 (S.-Y.K.)

Abstract: Amongst various intermetallic shape memory alloys (SMAs), nickel–titanium-based SMAs (NiTi) are known for their unique elastocaloric property. This widely used shape remembering material demonstrates excellent mechanical and electrical properties with superior corrosion resistance and super-long fatigue life. The straight-drawn wire form of NiTi has a maximum restorable strain limit of ~4%. However, a maximum linear strain of ~20% can be attained in its coil spring structure. Various material/mechanical engineers have widely exploited this superior mechanic characteristic and stress-triggered heating/cooling efficiency of NiTi to design smart engineering structures, especially in actuator technologies. This short technical note reflects the characteristics of the NiTi coil spring structure with its phase transformations and thermal transformation properties. The micro-actuators based on NiTi have been found to be possible, suggesting uses from biomedical to advanced high-tech applications. In recent years, the technical advancements in modular robotic systems involving NiTi-based SMAs have gained speculative commercial interest.

Keywords: actuators; mechatronics; NiTi coil spring; shape memory alloy; soft robotics

Citation: Shimoga, G.; Kim, T.-H.; Kim, S.-Y. An Intermetallic NiTi-Based Shape Memory Coil Spring for Actuator Technologies. *Metals* **2021**, *11*, 1212. https://doi.org/10.3390/met11081212

Academic Editors: Maciej Motyka and João Pedro Oliveira

Received: 11 June 2021
Accepted: 27 July 2021
Published: 29 July 2021

Publisher's Note: MDPI stays neutral with regard to jurisdictional claims in published maps and institutional affiliations.

1. Introduction

Tetsuro Mori, an eminent senior engineer of Yaskawa Electric Corporation, in 1969, coined the word "mechatronics". Technically speaking, mechatronics is an interdisciplinary branch of intensive engineering, which involves the unification of electronic and mechanical systems, with interlinking aspects of computer system engineering and control system engineering (see Figure 1). The best example of a mechatronic system is an industrial robot, which operates at the intersection of electronics, mechanics, and computing to perform its routine tasks [1–4]. In the past few decades, the advancement in mechatronics has implicated its use from biomedical fields to the technology arena [5,6].

Electromechanical actuators and electromagnetic sensors are the two most imperative components of mechatronic systems. Designing various electromechanical and electromagnetic device modules is crucial in the automotive and robotic industries. Utilization of the material properties and material selection criteria plays an important role in constructing electromechanical actuator modules [7,8]. Shape memory alloys (SMAs) are broadly used in the development of thermostatic and electromechanical actuators, to perform mechanical actions assisted by temperature change phenomena via electric currents. To improve the robustness, the compact and straightforward design strategy was usually adopted [9–11]. Amongst various type of SMAs, nickel–titanium (NiTi)-based SMAs (commonly known as NiTi) are preferably used in most smart engineering structures and actuator technologies due to their exceptional thermo-mechanical performances, practicability, shape memory characteristics, and pseudoelasticity [12–15].

The thermal shape memory effect of SMAs during the actuation cycle involves two stable phases, namely austenite and martensite. Under the influence of temperature and stress, three dissimilar crystal structures (viz. twinned martensite, detwinned martensite, and austenite) are possible with transformation in the atomic arrangements (see Figure 2). Due to alteration in the microstructural array arrangement during phase transition, the discrepancy in the mechanical properties (yield strength and Young's modulus) can be noticed [16–19]. In particular, NiTi can be deformed in its martensite phase and, upon heating, it can return to its original shape. Briefly, the austenite phase is cooled down to the twinned martensite phase. This martensite phase can be deformed by applied external stress. Then, when thermally stimulated, the material returns to its original shape. This anomaly can be referred as the "shape memory effect" [20]. NiTi exhibits Lüders-like deformation under a variety of testing conditions and, for easy understanding, three testing environments are provided here: (I) the tensile distortion related to the stress-induced martensitic transformation from the austenite phase, (II) the reverse transformation of the stress-induced martensite to austenite in "pseudoelasticity", and (III) the deformation in the martensitic state via a martensite variant reorientation process [21]. A stress plateau and a stress drop at the beginning of the process for the forward transformation upon loading and a stress minimum for the reverse transformation on unloading characterize the Lüders-like deformation behavior [21,22]. As we know, Schmid's law describes the slip plane and the slip direction of a stressed material well, and the martensite transformation shows good agreement with Schmid's law [23]. Indeed, martensite transformations are unique due to their reversible characteristics and self-accommodating nature of martensite plates [24].

NiTi can withstand large deformations, that is, austenite can, under high stress, transform into stress-induced martensite. Since SMAs are unidirectional actuators, external stress must be applied to strain it to its detwinned state, and to relapse its highly organized (austenite) crystal structure, thermal acceleration is essential [19]. Despite electroactive polymers and piezoelectric materials, NiTi SMAs can develop high stresses (~560 MPa in austenitic phase and ~100 MPa in martensitic phase) in wires of ~0.5 mm in thickness [25].

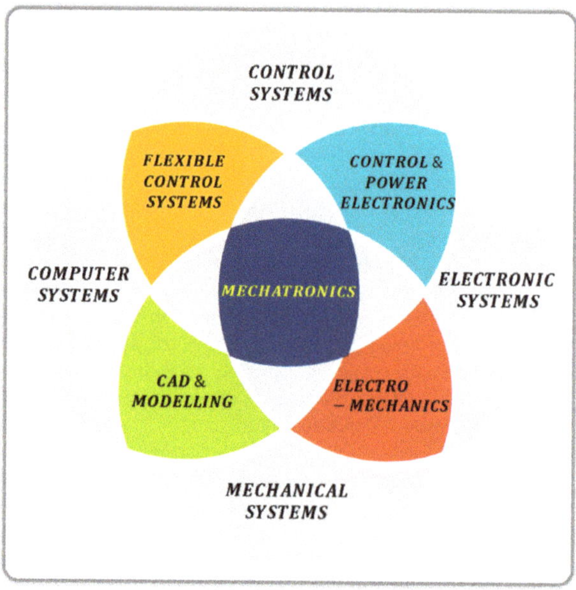

Figure 1. Euler diagram representation of mechatronics and its interdisciplinary branches of engineering.

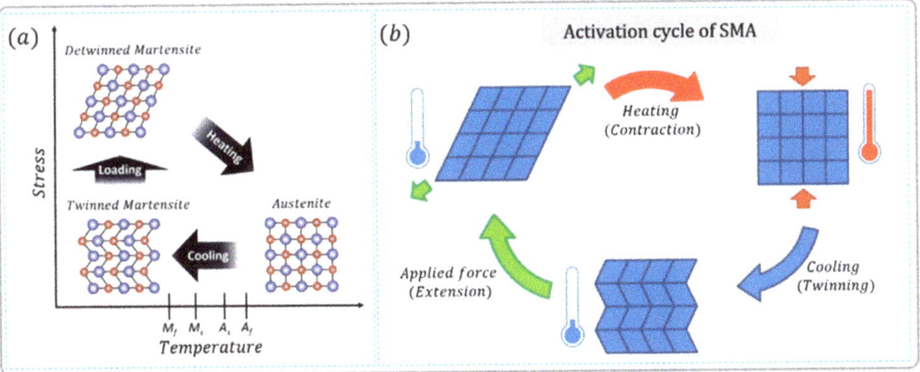

Figure 2. (**a**) Phase transition behavior of SMAs, adapted with permission from [18], Elsevier, 2016. (**b**) Activation cycle of SMAs under the influence of temperature, adapted with permission from [19], IEEE, 2014. (Note: A_f = Austenite finish temperature, A_s = Austenite start temperature, M_s = Martensite start temperature, and M_f = Martensite finish temperature).

Divergent applications of NiTi have been developed for actuator technologies, especially in robotic science and engineering [26]. The widespread use of NiTi is due to its fatigue behavior and sturdiness. Depending on the customized applications, NiTi is commercially available in various forms (see Figure 3). The corrosion resistance and biocompatibility of NiTi caused it to gain immense interest for biomedical applications by allowing it to be used in invasive surgical instruments and medical implants [27].

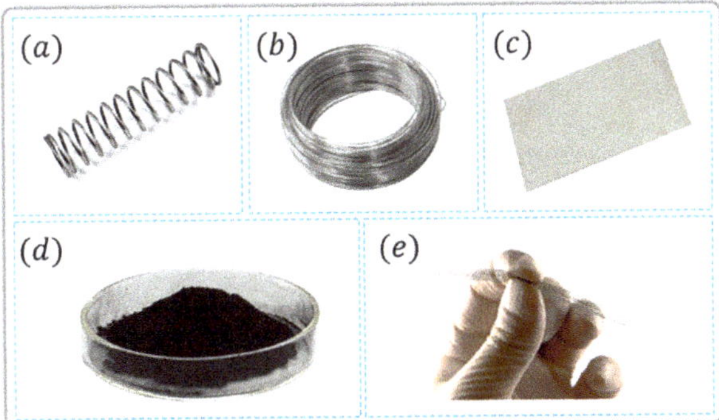

Figure 3. Various forms of commercially available NiTi: (**a**) spring, (**b**) wire, (**c**) thin film, (**d**) nanoparticles, (**e**) stent (demonstrating squeezing of a self-expanding braided NiTi stent used for endovascular surgery) [12].

There are three operational phases in NiTi, namely (I) an austenitic phase, (II) a martensitic phase, and (III) an intermediate R-phase. A well-ordered and highly symmetric body-centered cubic (BCC) crystal lattice structure denoted as a B2 structure was encountered in the austenitic phase, resembling cesium chloride (cubic crystal lattice structure), in which body-centered and corner atoms do not have similar neighborhood atoms [28,29]. In the case of the martensitic phase, a complex twinned B19′ (monoclinic) crystal structure can be observed with low symmetry, stabilized by residual stresses (see Figure 4). The arrangement of atoms in the martensitic phase can be compared to herringbone-patterned needle-like crystal arrays. The martensite is more ductile with a softer entity [30]. Compared to the martensitic phase, the austenitic phase is harder and more rigid. In some NiTi grades, primarily when additionally alloyed with a ferrous element, an intermediate R-phase can be encountered with a rhombohedral crystal structure displaying low-temperature hysteresis (1–10 °C) and low transformation strain [31–33].

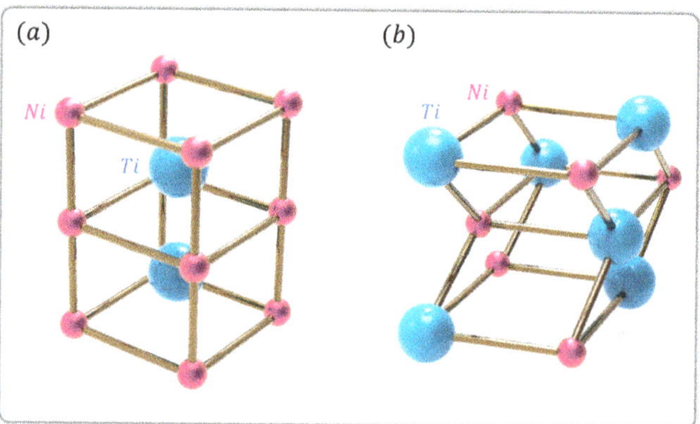

Figure 4. Crystal structure phases of NiTi displaying (**a**) B2 austenitic phase, and (**b**) B19′ martensitic phase (Designed by T.-H.K.).

Superelasticity is the characteristic property which is displayed by NiTi-based SMA wires. The superelasticity refers to the reversible transformation of the martensitic phase, which is triggered due to the variation in stress state. As we can see from Figure 5, an adequate tensile load activates the transformation of phases from austenite into detwinned martensite. While unloading, the stress-induced martensitic phase reverts to the austenitic phase via a partially detwinned martensitic phase. In general, the stress-induced martensitic phase is one of the martensitic phases, which forms from austenite in the existence of stress. There are many thermomechanical loading paths that can result in the formation of stress-induced martensitic phases. In general, both the 'superelastic' and 'rubber-like' behaviors are labeled as 'pseudoelasticity'. The reversible phase transformation caused by a thermomechanical loading path is called the superelastic behavior [34,35]. The rubber-like effect is an exclusive behavior of the martensite phase and occurs due to the reversible reorientation of martensite (the martensite crystals may be more pliable, equally stiff, or stiffer than austenite crystals, depending on the orientation of the loading direction with respect to the unit cells) [36].

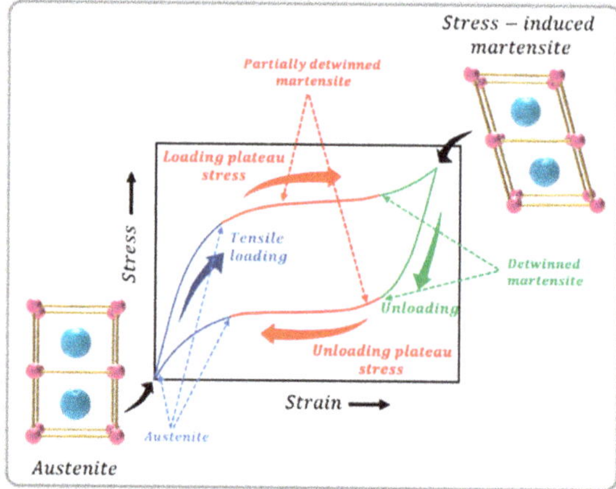

Figure 5. Typical stress–strain curve for the superelasticity of NiTi-based SMAs (Designed by T.-H.K. and G.S.).

2. Characteristics of NiTi Coils

In recent years, various mechatronic system module and robotic system studies have focused on utilizing NiTi-based materials as actuators. This proliferation of research activities in actuator technologies involving NiTi-based materials, especially wire and/or spring units to construct linear and rotary actuators, has been consistently documented [37–40]. The typical straight-drawn wire form of commercial NiTi has a maximum restorable strain limit of ~4%. A preeminent strain can be achieved by coiling the wire to resolve the applied force for increased strain. A maximum linear strain of ~20% can be achieved using coiled NiTi wires by successively tuning its shape memory properties in the linear coil model [41–43]. An example of a NiTi coil is schematically represented in Figure 6.

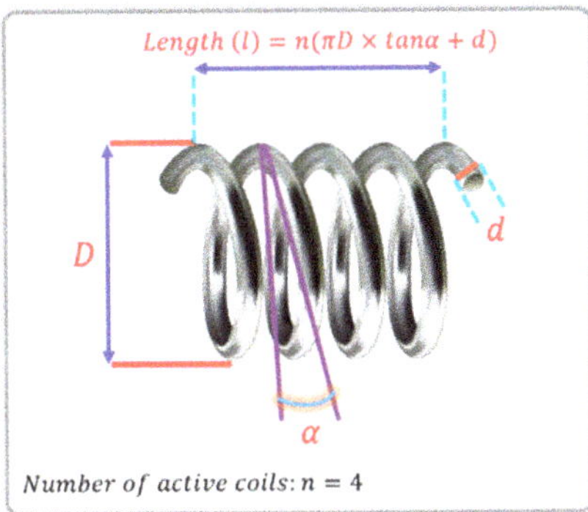

Figure 6. Geometrical design of a NiTi coil spring, representing thickness (*d*), outer diameter (*D*), length (*l*), number of turns in the coil (*n*), and initial pitch angle (*α*); (Designed by G.S.).

Depending on the crystal phases, we can notice two different shear moduli, G_A and G_M, respectively, for complete austenitic and martensitic crystal phases. The effective spring constant (*K*) can be expressed by the following Equation (1) and is a function of the shear modulus (*G*) of the respective NiTi phases; therefore, it is obvious that tuning of the stiffness and contractile properties can be possible by changing '*d*', '*D*', and '*n*' [43].

$$K = \frac{Gd^4}{8nD^3} \tag{1}$$

Considering the actuation cycle of a NiTi spring, it generally commences from the martensite phase with an external load of *F*. Later, the spring is stimulated by inducing a suitable temperature until the transition temperature, resulting in the contraction of spring length under the applied load. The effective displacement (δ_{eff}) created by this phenomenon can be determined by Equation (2) (please refer Figure 7 for better understanding of indexes *H*, *L*, and *M*).

$$\delta_{eff} = \delta_M + \delta_L - \delta_H \tag{2}$$

With the displacement of the coil δ_i under applied load *F*, where $i \in \{H, L, M\}$, using the general coiled spring deflection equation, Equation (1) can be written as Equation (3):

$$\delta_i = \frac{8FnD^3}{Gd^4}. \tag{3}$$

The shear strain γ of the spring can be written as

$$\gamma = \frac{\tau}{G} \tag{4}$$

where shear stress τ can be expressed as

$$\tau = \frac{8FD\kappa}{\pi d^3}.$$

(5)

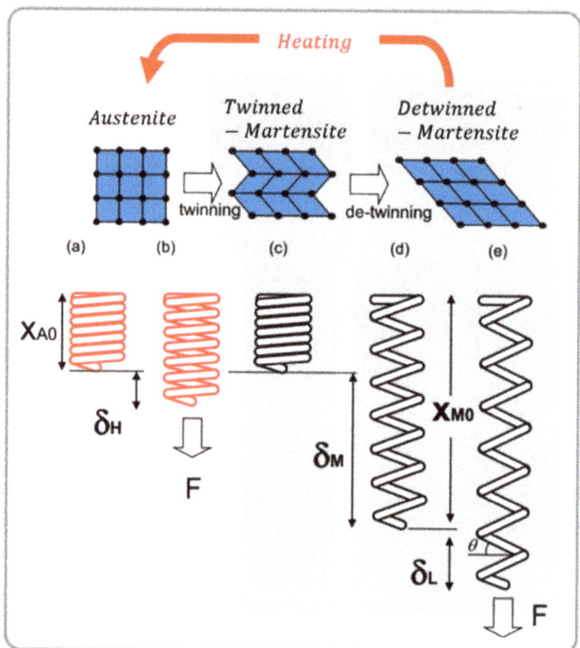

Figure 7. Five characteristic representative states of NiTi coil spring actuator. (**a**) Complete austenite phase without load, (**b**) complete austenite phase with load, (**c**) twinned martensite phase without load, (**d**) complete detwinned martensite phase without load, and (**e**) complete detwinned martensite phase with load. Adapted with permission from [43], IEEE, 2013.

From Wahl's formula [44], the stress correction factor κ can be written as

$$\kappa = \frac{4C - 1}{4C - 4} + \frac{0.615}{C}$$

(6)

where C is the spring index, and is defined as follows

$$C = \frac{D}{d}.$$

(7)

From Equations (3)–(5), the free length difference between the austenitic phase and martensitic phase can be calculated as follows:

$$\delta_M = \frac{\pi \gamma n D^2}{d\kappa}.$$

(8)

Consequently, the effective displacement from Equation (2) can be written as follows:

$$\delta_{eff} = \frac{\pi \gamma n D^2}{d\kappa} + \frac{8FnD^3_{eff}}{G_M d^4} - \frac{8FnD^3}{G_A d^4}.$$

(9)

The effective change in coil diameter of the spring (D_{eff}), due to the elongation under the applied load F in the martensite phase, can be written as below, where θ is the angle between the horizontal plane and the spring string (see Figure 7e).

$$D_{eff} = D \, Cos \, \theta \tag{10}$$

Figure 7 clearly demonstrates the actuation characteristics of a NiTi coil spring under different states. Under the thermal stimulation, when the temperature is applied to the NiTi spring, due to the contraction phenomenon, it converts to the austenitic phase (as shown in Figure 7a). The spring will have a displacement δ_H when a load F is applied in this state (see Figure 7b). No noticeable shape change can be observed, beyond the transition temperature and the phase conversion from martensitic phase to austenitic phase can be seen (see Figure 7c). The detwinning of the crystal structure can be seen if the load is affixed below the transition temperature. Figure 7e,d, respectively, represent the detwinned martensitic phase with a load F and detwinned martensitic phase as its free length after removing the detwinning load. The free length difference between the austenite and detwinned martensite phases is denoted by δ_M, where δ_L signifies the length difference between the loaded and unloaded martensitic phase under complete detwinned mode [45,46].

3. Characteristic Properties and Features of NiTi

As mentioned earlier, the drawbacks of SMA wires are limited stroke and the prerequisite of high recovery force. This can be minimized by shaping them into a spring structure. The developed actuators made of coiled spring structures can withstand the maximum force and stroke. Another peculiar advantage of SMAs is that they can work effortlessly in liquid surroundings without effectively losing their mechanical properties [47]. The key advantage is the propulsion mechanism of bioinspired robots that can be constructed logically using SMAs, which can work favorably in both air and underwater environments. With increasing propensity to develop lightweight and tiny actuators with a high power/weight ratio, the actuators made of SMAs show compelling advantages over the conventional hydraulic, pneumatic, and electric actuators, as a convincing substitute for conventional technologies [47–50]. The high work density (10^4–10^5 KJ m^{-3}), high force to weight ratio (~100), and quick actuation response (<1 s) of SMAs have inspired scientists to construct bioinspired soft actuation modules [51]. The SMAs suffer from a relatively lower actuation bandwidth (0.5–5 Hz) and energy efficiency of ~3% [52]; this is because of passive cooling at the terminal of the respective actuation cycles and wasting of energy due to heat loss in both cooling and activation cycles. To minimize these effects, SMA wires can be embedded inside various thermally conductive elastomeric materials while constructing soft robotic modules for morphing aircraft designs [53].

The metallic composition of nickel and titanium plays a crucial role in deciding the shape memory characteristics of NiTi. Normally, the nickel content varies from 49% to 57% (atomic % of Ni). The titanium composition in NiTi should be within 38–50% (atomic % of Ti) to acquire effective shape memory features. The characteristic physical, mechanical, and thermal transformation properties of NiTi are provided in Table 1 [16,52–62].

The NiTi-based SMAs with coil spring, wire, and foil structures are usually embedded into elastomeric materials to develop soft robotic actuators. When NiTi structures are assimilated into robotic mechanical joints [63–65], they can customarily provide large rotations with high torque because of their unique ability to withstand high strain (~200 to 1600%) and high force output [51,66]. Electrically driven tiny/micro-sized smart robotic actuators made with NiTi-based SMAs can perform natural sensitive biomimicking gesticulations like walking, jumping, climbing, whirling, and swimming. The advantages of using NiTi as one of the base materials is its corrosion resistibility and lightweight characteristics, especially when used in marine environments [67–69]. It is not surprising that SMAs ubiquitously reached technological maturity with promising industrial engineering applications in the mechatronics field amongst various smart materials. The unique phase transition (martensitic–austenitic) behavior anticipated their actuation properties and allowed the development of significantly lightweight, compact, and soundless industrial actuators [70].

Table 1. Characteristic physical, mechanical, and thermal transformation properties of NiTi-based SMAs [16,52–62].

Physical Properties	
Density (kg m^{-3})	6450−6500
Melting point (°C)	ca. 1250−1310
Thermal conductivity of the austenite phase (W m^{-1} K^{-1})	ca. 18
Thermal conductivity of the martensite phase (W m^{-1} K^{-1})	ca. 9
Resistivity of the austenite phase (µΩ cm)	~100
Resistivity of the martensite phase (µΩ cm)	~70
Corrosion resistance	Excellent
Biocompatibility	Excellent
Magnetic susceptibility of the austenite phase (emu g^{-1})	3.7×10^{-6}
Magnetic susceptibility of the martensite phase (emu g^{-1})	2.4×10^{-6}
Magnetic permeability	<1.002
Specific heat (cal g^{-1} °C^{-1})	0.2
Damping capacity (SDC %)	15−20
Heat capacity (J kg^{-1} K^{-1})	390
Mechanical properties	
Young's modulus of the austenite phase (GPa)	ca. 83
Young's modulus of the martensite phase (GPa)	ca. 23−41
Poisson's ratio	0.33
Elongation to fracture of the austenite phase	1−2%
Elongation to fracture of the martensite phase	Up to 60%
Plateau stress austenite phase (MPa)	200−650
Plateau stress martensite phase (MPa)	70−200
Yield strength austenite phase (MPa)	195−700
Yield strength martensite phase (MPa)	~70−140
Tensile strength (MPa)	895
Tensile strain (fully annealed form)	20−60
Normal working stress (GPa)	0.5−0.9
Ultimate tensile strength (GPa)	0.9
Fracture toughness (MPa $\sqrt{m}$)	895
Transformation temperature and strain properties	
Transformation temperature range (°C)	−50 [a]−110
Transformation enthalpy (kJ kg^{-1} K^{-1})	0.47−0.62
Transformation strains: up to 1 cycle	Up to 8%
Transformation strains: up to 100 cycles	Up to 5%
Transformation strains: up to 100,000 cycles	Up to 3%
Transformation strains: above 100,000 cycles	ca. 2%
[b] Overall thermal hysteresis (°C)	30−80
Hysteresis of martensite phase (°C)	30−40
Hysteresis of R-phase (°C)	2−5
Heat of transformation in martensite phase (J mole^{-1})	295
Heat of transformation in R-phase (J mole^{-1})	55
Latent heat of transformation (cal g^{-1}-atom)	40

[a] Low transformation temperature can be achieved by increasing nickel metal content; however, the nature of the material becomes highly brittle. [b] The hysteresis is calculated for complete load–unload cycles; however, the value is reduced for partial cycles.

The blended elemental powder metallurgy (BEPM) technique is one of the most common and cost-effective methods to produce intermetallic NiTi-based SMAs. As indicated from the binary phase diagram of NiTi [71], the eutectic composition transpires at 942 °C (at metallic compositions of 24.5 at. % of Ni, between the β-Ti and NiTi$_2$ intermetallic phase). It is believed that a stable Ni$_3$Ti phase was evidently formed from the sequential decomposition of metastable phases (Ni$_4$Ti$_3$ and Ni$_3$Ti$_2$ phases), above the eutectic point for a longer duration [72].

The microstructure image displayed in Figure 8a represents the 80 µm thick intermetallic phase of NiTi annealed at 900 °C for 100 h. As seen from Figure 8a, the gradual cooling process has generated two-phase regions in the NiTi phase. Hence, this unperturbed binary reactive diffusion process between Ni and Ti can be expressed as NiTi $\rightarrow$ Ni$_3$Ti $\rightarrow$ NiTi$_2$. Bertheville et al. [73] interpreted the needle-shaped particles observed in Figure 8b as martensitic NiTi, which can be easily discriminated from the austenitic phase using SEM and microstructure analysis and can be attributed to Ni$_4$Ti$_3$ particles [74]. As speculated by the authors, at elevated temperatures, the Ni$_4$Ti$_3$ phase gradually decomposed to the most stable and prominent Ni$_3$Ti phase via a precipitation sequence expressed as Ni$_4$Ti$_3$ $\rightarrow$ Ni$_3$Ti$_2$ $\rightarrow$ Ni$_3$Ti. Novák et al. [75] concluded that the self-propagating high-temperature synthesis (SHS) conditions can affect the microstructure of NiTi phases. Rapid heating (heating rate >300 °C min^{-1}), producing dominant NiTi and Ti$_2$Ni phases, can be perceived easily from microstructural analysis [75].

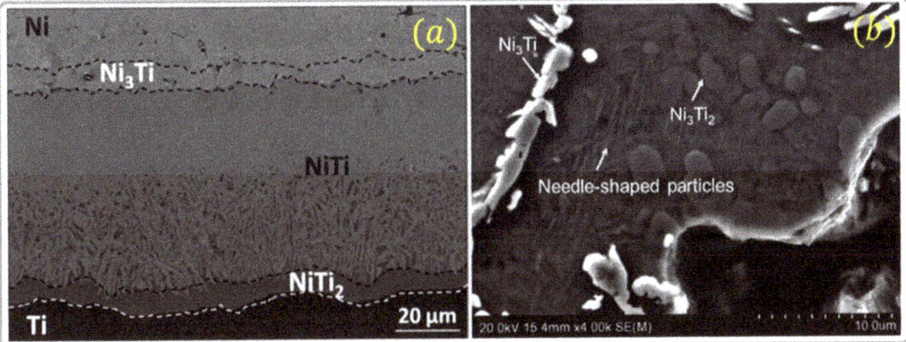

Figure 8. (**a**) The microstructure of interdiffusion zone in NiTi binary diffusion couple after annealing at 900 °C for 100 h. Adapted with permission from [71], Walter de Gruyter GmbH, Berlin/Boston, 2018. (**b**) SEM image of as-sintered samples prepared under processing conditions (pressing pressure: 400.0 MPa, sintering final temperature: 1050 °C, from ambient to 700 °C at heating rate 10 °C min^{-1}, 700–900 °C at heating rate 2 °C min^{-1}, 900–1050 °C at heating rate 10 °C min^{-1}, sintering holding time: 4 h). Adapted with permission from [74], Elsevier, 2018.

Aging of Ni-rich alloys at 400 °C can easily form noticeable lenticular Ni$_4$Ti$_3$ precipitates. The stress fields due to the precipitates formed can result in the formation of an intermediate R-phase (the phase in between the austenite and martensite phases). This rhombohedral structure of the crystal phase (R-phase) generally disappears with heat treatment at higher temperatures [76,77]. Recent studies on 55 at. % of Ni in NiTi SMAs revealed that the composition exhibits transformation temperatures in the range of −10 °C to 60 °C. This composition alloy is referred to as a chemically multi-phased alloy, so it exhibits low transformation strains and shows excellent corrosion resistance behavior as compared to stainless steel even in harsh saline environments [78]. Since these composition alloys shows exceptional thermomechanical stability, they do not require cold working and can be hot formed into various complex shapes [79,80].

Nam et al. [81] preferentially replaced Ni in NiTi SMAs with copper (Cu) metal to obtain NiTiCu alloys. It was observed that the addition of Cu reduces the hysteresis of the shape memory response, and we can also notice a decrease in the transformation strain. In Ni$_{40}$Ti$_{50}$Cu$_{10}$ (i.e., 10 at. % of Cu), the transformation hysteresis is much reduced to approximately 4%. In addition, the pseudoelastic hysteresis is less than 100 MPa, when compared to binary alloy (Ni$_{50}$Ti$_{50}$). The addition of Cu greatly reduces the sensitivity of the martensitic phase start temperature to composition, influencing the material behavior associated with the change in phase transition [82]. Amongst diverse compositions of Cu, ≤5–10 at. % of Cu in NiTi is preeminent because it is associated with small hysteresis and transformation, which is ideal for actuators. An at. % of Cu above 10% in NiTi influences the brittleness and the applications are constrained for actuator technologies [83].

The demand for high-temperature shape memory alloys (HTSMAs) has increased in the past few decades for aircraft engines and/or down-hole applications in oil-based industries [84]. The prime material criteria of HTSMAs is to have transformation temperatures > 100 °C and to have good actuating ability under high-temperature conditions [84]. The alloys produced by adding ternary elements, such as palladium (Pd), platinum (Pt), and gold (Au), to NiTi-based SMAs can widen the transformation temperature range to 100–800 °C [85,86]. Nicholson et al. [87] studied the thermomechanical behavior of NiTiPdPt HTSMA springs. The authors observed that the transformation strains are generally smaller compared to conventional NiTi alloys. The spring actuators made of $Ni_{19.5}Ti_{50.5}Pd_{25}Pt_5$ (at. %) were subjected both monotonic axial loading and thermomechanical cycling. Larger strokes were obtained for the case of the rotationally unconstrained spring and maximum strokes were obtained at an intermediate load [87].

Due to the high cost associated with Pd, Pt, and Au, as alloying metals to NiTi, metals like hafnium (Hf) and zirconium (Zr) have been investigated to extend the commercial viability [88,89]. The composition $Ni_{50.3}Ti_{29.7}Hf_{20}$ (at.%) is the most widely studied because of its desirable actuation properties. The composition displays austenite finish temperatures (A_f) of 150–200 °C (depending on slight alterations in the composition), and transformation strains above 3.5% in tension, 6% in torsion, and 2% in compression. In addition, it shows reasonable cyclic stability at high stresses, and high temperature superelasticity with more than 3% recoverable strain at > 200 °C [90]. The early works on niobium (Nb) as an alloying metal to NiTi for the composition $Ni_{47}Ti_{44}Nb_9$ (i.e., 9 at. % of Nb) showed that the insoluble globular precipitates of nearly pure Nb influence the deformation stress, equivalent to the detwinning stress of martensite [91]. The large thermal hysteresis of the composition is associated with the partitioning of the strain into a recoverable part (due to the NiTi phase) and an irrecoverable part (due to the Nb precipitates). The irregularities in the composition influence the mechanical properties and the material does not display complete recovery during deformation. Therefore, a Nb concentration around 3 at.%. has shown promising shape memory effects [92,93].

Miniature NiTi spring actuators have conceivable applications in technological sectors. The performance of the actuator is highly dependent on the diameter of the coil (D), the diameter of the wire (d), and the number of active rings (n) involved in the construction of spring. Many researchers have demonstrated and discussed thermodynamic models of NiTi springs for use as spring actuators [94–96]. The previously developed models do not explain the change in the spring's free length caused by the phase change phenomena in the NiTi microstructure. In the case of a complete martensite phase, the newly changed length X_{Mo} due to the elongation of the spring can be easily seen as well as its diminutive free length X_{Ao} in the complete austenite phase (see Figure 6). Additionally, the spring constant in the austenitic phase is about two to three times better than in its martensitic phase [60].

By adopting the techniques developed previously by Seok et al. [43] and Kim et al. [97], low spring index actuators were documented by Holschuh et al. [98]. The authors tuned the force-displacement characteristics of the spring structure by altering the geometry. In brief, winding a 305 µm NiTi wire (ommercially known as 'Flexinnol' by Dynalloy Inc.) was carried out over a 635 µm stainless steel core, to achieve a spring index (C) ~3.08. The winding was accomplished as shown in Figure 9a, so as to attain a consistent pitch angle and tight packing density. After winding the wire at room temperature, the annealing of the wires was carried out at 450 °C for a short duration (~10 min) to fix the austenitic memory state of the spring before quenching in cold water. These annealing parameters are crucial [43], and decide the permanent plastic deformation after each actuation cycle of the spring (see Figure 9b). The annealing temperature decides various parameters of the spring actuator, specifically, (I) large annealing temperatures can affect the spring's ability to return to its original shape, (II) increasing the annealing temperature decreases the detwinning force. The actuated NiTi coil in the twinned martensite state returns to its detwinned martensite state via martensite transformation. The external work should be necessary to perform this detwinning process and the extension of the coil results due to tensile force (pull force). This pull force is highly reliant on the annealing temperature, and is a mechanical characteristic of the actuators [43].

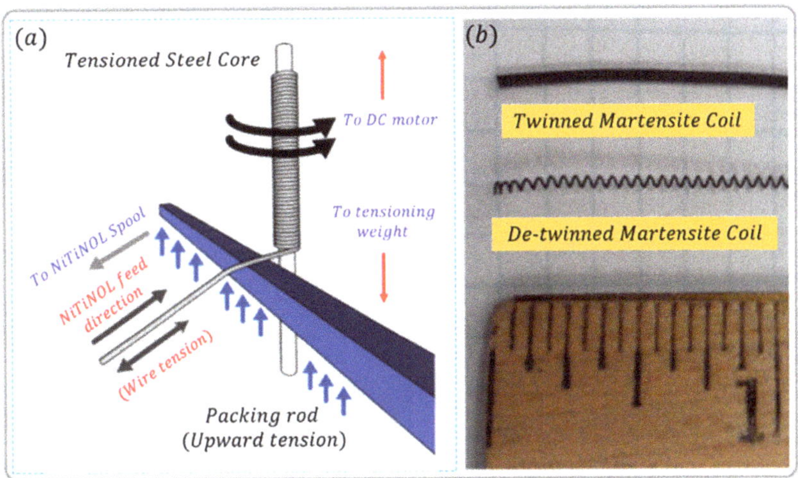

Figure 9. (**a**) Demonstration sketch of the winding technique developed by Seok et al. [43], Kim et al. [97], and Holschuh et al. [98], to achieve low spring index NiTi coil actuators. Adapted with permission from [43,97,98], IEEE, 2013, 2009, and 2015, respectively. (**b**) Pictograph representing finished twinned and detwinned NiTi coil spring actuators. Adapted with permission from [98], IEEE, 2015.

The as-prepared low spring index (C ~ 3.08) NiTi coils by Holschuh et al. [98] demonstrated the increase in the actuation force upon applied voltage and extensional strain (up to 7.24 N). The authors firstly developed the compression tourniquet system embedded with these low index NiTi coils. During the activation cycle of the actuator, the system enables over a 70% increase in the applied pressure. This strategy can be effectively used to develop wearable smart fabrics, which dynamically tune their shape and size, for advanced applications ranging from healthcare sectors to designer smart shrink-wrapping spacesuits [99]. Tourniquet systems in association with sensors can expand the progressive impact in the event of any battlefield injury/bleeding or accidents, by impulsive compression of smart suits, and save lives.

Conceptual miniature robotic designs inspired by nature have advantageous technical implications in the human–robot interaction field [100,101]. Inspired by the locomotion patterns of an inchworm, Lobontiu et al. [102] and Lin et al. [103] have developed piezo-based and polymer-based tiny robots, respectively. Koh et al. [104] designed a sliding mini-robot, which bends its body in an omega (Ω) shape during crawling using NiTi coil springs. The authors named it 'Omegabot'. To perform crawling locomotion, the total degrees of freedom were minimized by segmenting the robot with rigid links entailing four bar linkages and a steering mechanism, as shown in Figure 10a,b. The Z-shaped bending arises each time, generating one degree of freedom every time. The steering mechanism has three degrees of freedom (pitch, yaw, and roll) controlled by spherical six-bar linkages. The Omegabot prototype pictograph developed by Koh et al. [104] is presented in Figure 10c. These miniature robots can perform effortlessly in extremely harsh environments to execute emergency tasks [105,106].

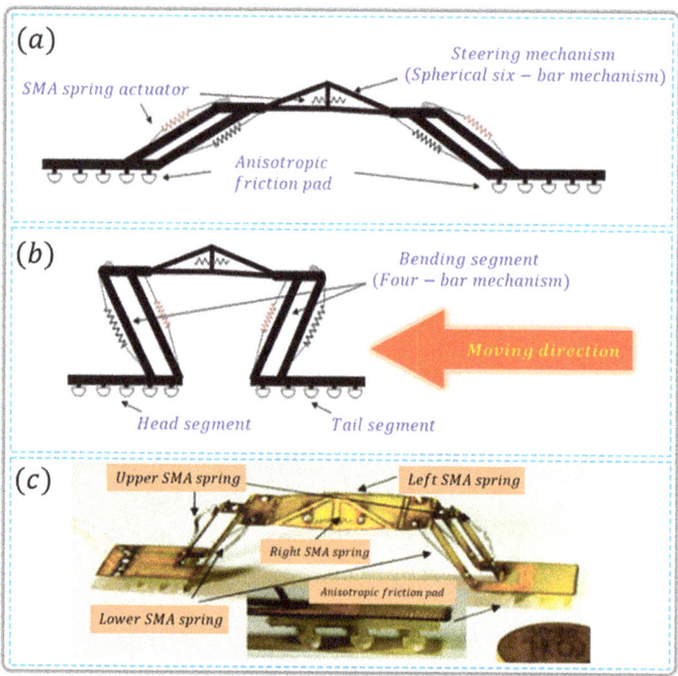

Figure 10. (a,b) Schematic design of the Omegabot developed by Koh et al. [104], (c) final prototype of an Omegabot. Note: red coils show the activated actuator. Adapted with permission from [104], IEEE, 2013.

High-cycle fatigue behavior of designed engineering materials is crucial for uninterrupted performances of actuators [107]. The pioneering works of Melton and Mercier [108] documented some fascinating trends. By varying A_f (austenite finish temperature) values between 10 and 110 °C, the binary compositions (NiTi) were studied at room temperature to obtain stimulating results. Interestingly, superelastic NiTi provides longer sustainability with superior fatigue limits than thermal martensite subjected to stress-control fatigue. The authors related this examination to the difference in plateau strength between austenite and martensite. The 10^7-cycle stress fatigue limit is approximately 80% of the respective stress plateau (superelastic) or detwinning plateau (martensite) [108]. Recently, Jaureguizahar et al. [109] performed fatigue cycling of NiTi wire (diameter = 0.5 mm, 50.9 at. % Ni) using servo-hydraulic testing machines (MTS 810 and INSTRON 8800) equipped with respective environmental chambers. The results revealed that without pseudoelastic transformation, for the austenite and martensite phase, fatigue lives of 10^7 and 9.33×10^6 were observed, respectively [109].

4. Conclusions

This short technical note is a concise summary of the understanding of the properties of NiTi coil springs. Due to their unique thermo-mechanical performances, shape memory behavior, and high fatigue characteristics, NiTi-based smart and compact engineering structures have exceptional importance in actuator technologies. Their great elasticity under stress, corrosion resistivity, biocompatibility, and original shape remembering proficiency in thermal conditions have created excessive applications in various technological fields, including biomedical sectors. The NiTi coil springs can be attained by coiling straight-drawn NiTi wires; this can maximize the restorable linear strain limit up to 20%. The linear actuation stroke of NiTi wire actuators can be improved significantly by converting them into coil spring structures. The specific inevitabilities and applicability of developed miniature robots activated by these coil spring structures could affect the final shape of the developed actuator. The inventive bioinspired designs and novel architecture of mini-robotic systems made of NiTi could comprehensively stimulate the commercial perspective [110,111].

Author Contributions: Conceptualization, data curation/formal analysis, writing original draft, G.S.; data curation/formal analysis, T.-H.K.; manuscript review, final editing and supervision, S.-Y.K. All authors have read and agreed to the published version of the manuscript.

Funding: This work was supported by the Priority Research Centers Program through the National Research Foundation of Korea (NRF) funded by the Ministry of Education (NRF-2018R1A6A1A03025526). This work was supported by a National Research Foundation of Korea (NRF) grant funded by the Ministry of Education (NRF-2020R1I1A3065371). This work was also supported by the Technology Innovation Program (10077367, Development of a film-type transparent/stretchable 3D touch sensor/haptic actuator combined module and advanced UI/UX) funded by the Ministry of Trade, Industry & Energy (MOTIE, Korea).

Institutional Review Board Statement: Not applicable.

Informed Consent Statement: Not applicable.

Data Availability Statement: Not applicable.

Acknowledgments: The authors acknowledge Cooperative Equipment Center at KoreaTech for formal discussions.

Conflicts of Interest: The authors declare no conflict of interest.

References

1. Wikiwand. Mechatronics. Available online: https://www.wikiwand.com/en/Mechatronics (accessed on 16 April 2021).
2. Cetinkunt, S. *Mechatronics with Experiments*, 2nd ed.; John Wiley & Sons Ltd.: Chichester, UK, 2015; ISBN 9781118802465.
3. Moheimani, S.O.R. Mechatronics: 2021 and beyond. *Mechatronics* **2020**, *71*, 102446. [CrossRef]
4. Dixit, U.S.; Hazarika, M.; Davim, J.P. History of Mechatronics. In *A Brief History of Mechanical Engineering. Materials Forming, Machining and Tribology*; Springer: Cham, Switzerland, 2017; ISBN 978-3-319-42916-8. [CrossRef]
5. Liu, Y.-H. Medical Mechatronics for Healthcare. *J. Healthc. Eng.* **2018**, *2018*. [CrossRef]
6. Sharma, R.; Dhiman, B. Mechatronics Around the World—At a Glance. *J. Mechatron. Robot.* **2021**, *5*, 1–7. [CrossRef]
7. Pawlak, A.M. *Sensors and Actuators in Mechatronics: Design and Applications*, 1st ed; CRC Press, Taylor & Francis Group: Boca Raton, FL, USA, 2007; ISBN 9780849390135.
8. Gorodetskiy, A.E. Smart Electromechanical Systems Modules. In *Smart Electromechanical Systems*; Gorodetskiy, A., Ed.; Studies in Systems, Decision and Control; Springer: Cham, Switzerland, 2016; ISBN 978-3-319-27547-5. [CrossRef]
9. Krishnan, R.V.; Bhaumik, S.K. Shape memory alloys: Properties and engineering applications. In *Smart Materials, Structures, and Systems*; Proc. SPIE 5062; International Society for Optics and Photonics: Bangalore, India, 2003. [CrossRef]
10. Zareie, S.; Issa, A.S.; Seethaler, R.J.; Zabihollah, A. Recent advances in the applications of shape memory alloys in civil infrastructures: A review. *Structures* **2020**, *27*, 1535–1550. [CrossRef]
11. Lobo, P.S.; Almeida, J.; Guerreiro, L. Shape Memory Alloys Behaviour: A Review. *Proc. Eng.* **2015**, *114*, 776–783. [CrossRef]
12. Crawford, M. Metal That Remembers Its Shape, The American Society of Mechanical Engineers. (1 July 2017). Available online: https://www.asme.org/topics-resources/content/metal-that-remembers-its-shape (accessed on 16 April 2021).
13. Wadood, A. Brief Overview on Nitinol as Biomaterial. *Adv. Mater. Sci. Eng.* **2016**, *2016*. [CrossRef]
14. Abubakar, R.A.; Wang, F.; Wang, L. A review on Nitinol shape memory alloy heat engines. *Smart Mater. Struct.* **2020**, *30*, 013001. [CrossRef]
15. Wang, Z.; Everaerts, J.; Salvati, E.; Korsunsky, A.M. Evolution of thermal and mechanical properties of Nitinol wire as a function of ageing treatment conditions. *J. Alloy. Compd.* **2020**, *819*, 153024. [CrossRef]
16. Huang, W. On the selection of shape memory alloys for actuators. *Mater Des.* **2002**, *23*, 11–19. [CrossRef]
17. Kim, D.J.; Kim, H.A.; Chung, Y.-S.; Choi, E. Pullout resistance of straight NiTi shape memory alloy fibers in cement mortar after cold drawing and heat treatment. *Compos. Part B Eng.* **2014**, *67*, 588–594. [CrossRef]
18. Han, M.-W.; Rodrigue, H.; Cho, S.; Song, S.-H.; Wang, W.; Chu, W.-S.; Ahn, S.-H. Woven type smart soft composite for soft morphing car spoiler. *Compos. Part B Eng.* **2016**, *86*, 285–298. [CrossRef]
19. Swensen, J.P.; Doller, A.M. Optimization of Parallel Spring Antagonists for Nitinol Shape Memory Alloy Actuators. In Proceedings of the IEEE International Conference on Robotics and Automation (ICRA), Hong Kong, China, 31 May–7 June 2014. [CrossRef]

20. Stachiv, I.; Alarcon, E.; Lamac, M. Shape Memory Alloys and Polymers for MEMS/NEMS Applications: Review on Recent Findings and Challenges in Design, Preparation, and Characterization. *Metals* **2021**, *11*, 415. [CrossRef]
21. Liu, Y.; Liu, Y.; Humbeeck, V.J. Luders-like deformation associated with martensite reorientation in NiTi. *Scr. Mater.* **1998**, *39*, 665213. [CrossRef]
22. Alkan, S.; Sehitoglu, H. Prediction of transformation stresses in NiTi shape memory alloy. *Acta Mater.* **2019**, *175*, 182–195. [CrossRef]
23. Polatidis, E.; Šmíd, M.; Kuběna, I.; Hsu, W.-N.; Laplanche, G.; Swygenhoven, H.V. Deformation mechanisms in a superelastic NiTi alloy: An in-situ high resolution digital image correlation study. *Mater. Des.* **2020**, *191*, 108622. [CrossRef]
24. Gall, K.A. The effect of Stress State and Precipitation on Stress-Induced Martensitic Transformations in Polycrystalline and Single Crystal Shape Memory Alloys: Experiments and Micro-Mechanical Modeling. Ph.D. Thesis, University of Illinois, Urbana, IL, USA, 1998. Available online: https://www.proquest.com/docview/304447772?pq-origsite=gscholar&fromopenview=true (accessed on 16 July 2021).
25. Dynalloy, Inc. Flexinol®Actuator Wire Technical and Design Data. Available online: https://www.dynalloy.com/tech_data_wire.php (accessed on 16 April 2021).
26. Kheirikhah, M.M.; Rabiee, S.; Edalat, M.E. A Review of Shape Memory Alloy Actuators in Robotics. In *Lecture Notes in Computer Science*; Ruiz-del-Solar, J., Chown, E., Plöger, P.G., Eds.; RoboCup 2010: Robot Soccer World Cup XIV. RoboCup 2010; Springer: Berlin/Heidelberg, Germany, 2011; Volume 6556. [CrossRef]
27. Kapoor, D. Nitinol for Medical Applications: A Brief Introduction to the Properties and Processing of Nickel Titanium Shape Memory Alloys and their Use in Stents. *Johns. Matthey Technol. Rev.* **2017**, *61*, 66–76. [CrossRef]
28. Otsuka, K.; Ren, X. Physical metallurgy of Ti–Ni-based shape memory alloys. *Prog. Mater. Sci.* **2005**, *50*, 511–678. [CrossRef]
29. Contardo, L.; Guenin, G. Training and two-way memory effect in Cu-Zn-Al alloy. *Acta Metall. Mater.* **1990**, *38*, 1267–1272. [CrossRef]
30. Chekotu, J.C.C.; Groarke, R.; O'Toole, K.; Brabazon, D. Advances in Selective Laser Melting of Nitinol Shape Memory Alloy Part Production. *Materials* **2019**, *12*, 809. [CrossRef] [PubMed]
31. Saedi, S. Shape Memory Behavior of Dense and Porous NiTi Alloys Fabricated by Selective Laser Melting. Ph.D. Thesis, University of Kentucky, Lexington, KT, USA, 2017. Available online: https://uknowledge.uky.edu/me_etds/90/ (accessed on 19 April 2021).
32. Huang, X.; Ackland, G.J.; Rabe, K.M. Crystal structures and shape-memory behaviour of NiTi. *Nat. Mater* **2003**, *2*, 307–311. [CrossRef] [PubMed]
33. Parlinski, K.; Parlinska-Wojtan, M. Lattice dynamics of NiTi austenite, martensite and R-phase. *Phys. Rev. B* **2002**, *66*, 064307. [CrossRef]
34. Proffit, W.; Fields, H. *Contemporary Orthodontics*, 5th ed.; Elsevier: St. Louis, MO, USA, 2012; pp. 359–394. ISBN 9780323291521.
35. Lagoudas, D.C. *Shape Memory Alloys, Modeling and Engineering Applications*; Springer: Austion, TX, USA, 2008; ISBN 9780387476858.
36. Bucsek, A.N.; Paranjape, H.M.; Stebner, A.P. Myths and Truths of Nitinol Mechanics: Elasticity and Tension–Compression Asymmetry. *Shap. Mem. Superelasticity* **2016**, *2*, 264–271. [CrossRef]
37. Yuan, H.; Fauroux, J.-C.; Chapelle, F. A review of rotary actuators based on shape memory alloys. *J. Intell. Mater. Syst. Struct.* **2017**, *28*, 1863–1885. [CrossRef]
38. Jani, J.M.; Leary, M.; Subic, A. Designing shape memory alloy linear actuators: A review. *J. Intell. Mater. Syst. Struct.* **2016**, *28*, 1699–1718. [CrossRef]
39. Song, G. Design and control of a Nitinol wire actuated rotary servo. *Smart Mater. Struct.* **2007**, *16*, 1796. [CrossRef]
40. Jani, J.M. Design Optimisation of Shape Memory Alloy Linear Actuator Applications. Ph.D. Thesis, RMIT University, Melbourne, Australia, 2016. Available online: https://core.ac.uk/download/pdf/43495833.pdf (accessed on 19 April 2021).
41. Kónya, A.; Maxin, M.; Wright, K.C. New Embolization Coil Containing a Nitinol Wire Core: Preliminary In Vitro and In Vivo Experiences. *J. Vasc. Interv. Radiol.* **2001**, *12*, 869–877. [CrossRef]
42. Koh, J.-S. Design of Shape Memory Alloy Coil Spring Actuator for Improving Performance in Cyclic Actuation. *Materials* **2018**, *11*, 2324. [CrossRef]
43. Seok, S.; Onal, C.D.; Cho, K.-J.; Wood, R.J.; Rus, D.; Kim, S. Meshworm: A Peristaltic Soft Robot with Antagonistic Nickel Titanium Coil Actuators. *IEEE/ASME Trans. Mechatron.* **2013**, *18*, 1485–1497. [CrossRef]
44. Tan, P.S.; Farid, A.A.; Karimzadeh, A.; Koloor, S.S.R.; Petrů, M. Investigation on the Curvature Correction Factor of Extension Spring. *Materials* **2020**, *13*, 4199. [CrossRef]
45. Liu, Y. Detwinning process and its anisotropy in shape memory alloys. *Proc. SPIE Smart Mater.* **2001**, *4234*, 82–93. [CrossRef]
46. Liu, Y. On the Detwinning Mechanism in Shape Memory Alloys. In Proceedings of the IUTAM Symposium on Mechanics of Martensitic Phase Transformation in Solids. Solid Mechanics and Its Applications, Hong Kong, China, 11–15 June 2001; Sun, Q.P., Ed.; Springer: Dordrecht, The Netherlands, 2002; Volume 101. ISBN 978-94-017-0069-6. [CrossRef]
47. Mehrpouya, M.; Bidsorkhi, H.C. MEMS Applications of NiTi Based Shape Memory Alloys: A Review. *Micro Nanosyst.* **2016**, *8*, 79–91. [CrossRef]
48. Costanza, G.; Tata, M.E. Shape Memory Alloys for Aerospace, Recent Developments, and New Applications: A Short Review. *Materials* **2020**, *13*, 1856. [CrossRef]
49. Soother, D.K.; Daudpoto, J.; Chowdhry, B.S. Challenges for practical applications of shape memory alloy actuators. *Mater. Res. Express* **2020**, *7*, 073001. [CrossRef]

50.   Huang, X.; Ford, M.; Patterson, Z.J.; Zarepoor, M.; Pan, C.; Majidi, C. Shape memory materials for electrically-powered soft machines. *J. Mater. Chem. B* **2020**, *8*, 4539–4551. [CrossRef] [PubMed]

51.   Rich, S.I.; Wood, R.J.; Majidi, C. Untethered soft robotics. *Nat. Electron.* **2018**, *1*, 102–112. [CrossRef]

52.   Jani, J.M.; Leary, M.; Subic, A.; Gibson, M.A. A review of shape memory alloy research, applications and opportunities. *Mater. Des.* **2014**, *56*, 1078–1113. [CrossRef]

53.   Barbarino, S.; Saavedra Flores, E.L.; Ajaj, R.M.; Dayyani, I.; Friswell, M.I. A review on shape memory alloys with applications to morphing aircraft. *Smart Mater. Struct.* **2014**, *23*, 063001. [CrossRef]

54.   Johnson Matthey Inc. Available online: http://www.jmmedical.com/ (accessed on 20 April 2021).

55.   Kumar, G. Modeling and Design of one Dimensional Shape Memory Alloy Actuators. Master's Thesis, The Ohio State University, Columbus, OH, USA, 2000. Available online: https://etd.ohiolink.edu/apexprod/rws_olink/r/1501/10?p10_etd_subid=63836&clear=10 (accessed on 20 April 2021).

56.   Andani, M.T. Constitutive Modeling of Superelastic Shape Memory Alloys Considering Rate Dependent Non-Mises Tension-torsion Behavior. Master's Thesis, The University of Toledo, Toledo, OH, USA, 2013. Available online: https://etd.ohiolink.edu/apexprod/rws_etd/send_file/send?accession=toledo1371577037&disposition=inline (accessed on 20 April 2021).

57.   Fugazza, D. Use of Shape-Memory Alloy Devices in Earthquake Engineering: Mechanical Properties, Advanced Constitutive Modelling and Structural Applications. Ph.D. Thesis, University of Pavia, Pavia, Italy, 2005. Available online: https://citeseerx.ist.psu.edu/viewdoc/download?doi=10.1.1.623.2545&rep=rep1&type=pdf (accessed on 20 April 2021).

58.   SMA Group. Available online: http://smagroup.blogspot.com/2009/04/shape-memory-alloy-sma-also-known-as.html (accessed on 20 April 2021).

59.   Srinivasan, A.V.; McFarland, D.M. *Smart Structures: Analysis and Design*; Cambridge University Press: Cambridge, UK, 2001; ISBN 0521650267.

60.   Otsuka, K.; Wayman, C.M. *Shape Memory Materials*; Cambridge University Press: Cambridge, UK, 1999; ISBN 9780521663847.

61.   Lexcellent, C. *Shape-Memory Alloys Handbook*; Wiley-ISTE Publisher Ltd.: London, UK, 2013; ISBN 9781848214347.

62.   Tanzi, M.-C.; Fare, S.; Candiani, G. Part II Biomaterials and Biocompatibility, Chapter 4. Biomaterials and Applications. In *Foundations of Biomaterials Engineering*, 1st ed.; Acedemic Press—Elsevier: Cambridge, MA, USA, 2019.

63.   Taylor, A.J.; Slutzky, T.; Feuerman, L.; Ren, H.; Tokuda, J.; Nilsson, K.; Tse, Z.T.H. MR-Conditional SMA-Based Origami Joint. *IEEE/ASME Trans. Mechatronics* **2019**, *24*, 883–888. [CrossRef]

64.   Cheng, S.S.; Kim, Y.; Desai, J.P. Modeling and characterization of shape memory alloy springs with water cooling strategy in a neurosurgical robot. *J. Intell. Mater. Syst. Struct.* **2017**, *28*, 2167–2183. [CrossRef]

65.   Yang, H.; Xu, M.; Li, W.; Zhang, S. Design and Implementation of a Soft Robotic Arm Driven by SMA Coils. *IEEE Trans. Ind. Electron.* **2019**, *66*, 6108–6116. [CrossRef]

66.   An, S.M.; Ryu, J.; Cho, M.; Cho, K.J. Engineering design framework for a shape memory alloy coil spring actuator using a static two-state model. *Smart Mater. Struct.* **2012**, *21*, 055009. [CrossRef]

67.   Sreekumar, M.; Nagarajan, T.; Singaperumal, M.; Zoppi, M.; Molfino, R.M. Critical review of current trends in shape memory alloy actuators for intelligent robots. *Ind. Robot* **2007**, *34*, 285–294. [CrossRef]

68.   Cuéllar, W.H.C. BR3: A Biologically Inspired Fish-Like Robot Actuated by SMA-Based Artificial Muscles. Ph.D. Thesis, Universidad Politecnica de Madrid, Madrid, Spain, 2015. Available online: http://oa.upm.es/36254/1/WILLIAM_HERNAN_CORAL_CUELLAR.pdf (accessed on 20 April 2021).

69.   Ulloa, C.C.; Terrile, S.; Barrientos, A. Soft Underwater Robot Actuated by Shape-Memory Alloys "JellyRobcib" for Path Tracking through Fuzzy Visual Control. *Appl. Sci.* **2020**, *10*, 7160. [CrossRef]

70.   Spaggiari, A.; Castagnetti, D.; Golinelli, N.; Dragoni, E.; Mammano, G.S. Smart materials: Properties, design and mechatronic applications. *Proc. Inst. Mech. Eng. Part L J. Mater. Des. Appl.* **2019**, *233*, 734–762. [CrossRef]

71.   Wierzba, B.; Nowak, W.J.; Serafin, N. The Interface Reaction between Titanium and Iron-Nickel alloys. *High Temp. Mater. Proc* **2018**, *37*. [CrossRef]

72.   Hornbuckle, B.C.; Xiao, X.Y.; Noebe, R.D.; Martens, R.; Weaver, M.L.; Thompson, G.B. Hardening behavior and phase decomposition in very Ni-rich Nitinol alloys. *Mater. Sci. Eng.* **2015**, *639*, 336–344. [CrossRef]

73.   Bertheville, B.; Neudenberger, M.; Bidaux, J.-E. Powder sintering and shape-memory behaviour of NiTi compacts synthesized from Ni and TiH$_2$. *Mater. Sci. Eng.* **2004**, *384*, 143–150. [CrossRef]

74.   Khanlari, K.; Ramezani, M.; Kelly, P.; Cao, P.; Neitzert, T. Mechanical and microstructural characteristics of as-sintered and solutionized porous 60NiTi. *Intermetallics* **2018**, *100*, 32–43. [CrossRef]

75.   Novák, P.; Mejzlíková, L.; Michalcová, A.; Čapek, J.; Beran, P. Effect of SHS conditions on microstructure of NiTi shape memory alloy. *Intermetallics* **2013**, *42*, 85–91. [CrossRef]

76.   Buehler, W.J.; Wiley, R.C. Nickel-Base Alloys. U.S. Patent US3174851A, 23 March 1965.

77.   Miyazaki, S.; Otsuka, K. Deformation and transformation behavior associated with the r-phase in Ti-Ni alloys. *Metall. Trans. A* **1986**, *17*, 53–63. [CrossRef]

78.   Dębska, A.; Gwóździewicz, P.; Seruga, A.; Balandraud, X.; Destrebecq, J.-F. The Application of Ni–Ti SMA Wires in the External Prestressing of Concrete Hollow Cylinders. *Materials* **2021**, *14*, 1354. [CrossRef]

79. Clingman, D.J.; Calkins, F.T.; Smith, J.P. Thermomechanical properties of Ni60%weightTi40%weight. In *Proceedings of the SPIE, Smart Structures and Materials: Active Materials: Behavior and Mechanics*; International Society for Optics and Photonics: Bellingham, WA, USA, 2003; Volume 5053, pp. 219–229. [CrossRef]

80. Mabe, J.; Ruggeri, R.; Calkins, F.T. Characterization of nickel-rich nitinol alloys for actuator development. In Proceedings of the International Conference on Shape Memory and Superelasticity Technologies, Pacific Grove, CA, USA, 7–11 May 2006. [CrossRef]

81. Nam, T.H.; Saburi, T.; Shimizu, K. Cu-Content Dependence of Shape memory Characteristics in Ti-Ni-Cu Alloys. *Mater. Trans. JIM* **1990**, *31*, 959–967. [CrossRef]

82. Liu, Y. Mechanical and thermomechanical properties of a $Ti_{0.50}Ni_{0.25}Cu_{0.25}$ melt spun ribbon. *Mater. Sci. Eng. A* **2003**, *354*, 286–291. [CrossRef]

83. Xie, Z.L.; Van Humbeeck, J.; Liu, Y.; Delaey, L. TEM study of $Ti_{50}Ni_{25}Cu_{25}$ melt spun ribbons. *Scripta Materialia* **1997**, *37*, 363–371. [CrossRef]

84. Ma, J.; Karaman, I.; Noebe, R.D. High temperature shape memory alloys. *Int. Mater. Rev.* **2013**, *55*, 257–315. [CrossRef]

85. Benafan, O.; Garg, A.; Noebe, R.D.; Bigelow, G.S.; Padula, S.A., II; Gaydosh, D.J.; Vaidyanathan, R.; Clausen, B.; Vogel, S. Thermomechanical behavior and microstructural evolution of a Ni-rich $Ni_{24.3}Ti_{49.7}Pd_{26}$ high temperature shape memory alloy. *J. Alloy. Compd.* **2015**, *643*, 275–289. [CrossRef]

86. Yamabe-Mitarai, Y.; Arockiakumar, R.; Wadood, A.; Suresh, K.S.; Kitashima, T.; Hara, T.; Shimojo, M.; Tasaki, W.; Takahashi, M.; Takahashi, S.; et al. Ti(Pt, Pd, Au) based High Temperature Shape Memory Alloys. *Mater. Today Proc.* **2015**, *2*, S517–S522. [CrossRef]

87. Nicholson, D.E.; Padula, S.A., II; Noebe, R.D.; Benafan, O.; Vaidyanathan, R. Thermomechanical behavior of NiTiPdPt high temperature shape memory alloy springs. *Smart Mater. Struct.* **2014**, *23*, 125009. [CrossRef]

88. Oliveira, J.P.; Schell, N.; Zhou, N.; Wood, L.; Benafan, O. Laser welding of precipitation strengthened Ni-rich NiTiHf high temperature shape memory alloys: Microstructure and mechanical properties. *Mater. Des.* **2019**, *162*, 229–234. [CrossRef]

89. Ramos, A.P.; de Castro, W.B.; Costa, J.D.; de Santana, R.A. Influence of Zirconium Percentage on Microhardness and Corrosion Resistance of $Ti_{50}Ni_{50-x}Zr_x$ Shape Memory Alloys. *Mater. Res.* **2019**, *22*, e20180604. [CrossRef]

90. Bigelow, G.S.; Benafan, O.; Garg, A.; Noebe, R.D. Effect of Hf/Zr Ratio on Shape Memory Properties of High Temperature Ni $_{50.3}Ti_{29.7}(Hf,Zr)_{20}$ Alloys. *Scr. Mater.* **2021**, *194*, 113623. [CrossRef]

91. Zhang, C.S.; Zhao, L.C.; Duerig, T.W.; Wayman, C.M. Effects of deformation on the transformation hysteresis and shape memory effect in a $Ni_{47}Ti_{44}Nb_9$ alloy. *Scr. Metall. Mater.* **1990**, *24*, 1807–1812. [CrossRef]

92. Zhao, L.C.; Duerig, T.W.; Justi, S.; Melton, K.N.; Proft, J.L.; Yu, W.; Wayman, C.M. The study of niobium-rich precipitates in a Ni-Ti-Nb shape memory alloy. *Scr. Metall. Mater.* **1990**, *24*, 221–226. [CrossRef]

93. He, X.M.; Rong, L.J.; Yan, D.S.; Li, Y.Y. TiNiNb wide hysteresis shape memory alloy with low niobium content. *Mater. Sci. Eng. A* **2004**, *371*, 193–197. [CrossRef]

94. Dong, Y.; Boming, Z.; Jun, L. A changeable aerofoil actuated by shape memory alloy springs. *Mater. Sci. Eng. A* **2008**, *485*, 243–250. [CrossRef]

95. Kim, B.; Lee, M.G.; Lee, Y.P.; Kim, Y.; Lee, G. An earthworm-like micro robot using shape memory alloy actuator. *Sens. Actuators A Phys.* **2006**, *125*, 429–437. [CrossRef]

96. Lee, H.J.; Lee, J.J. Evaluation of the characteristics of a shape memory alloy spring actuator. *Smart Mater. Struct.* **2000**, *9*, 817. [CrossRef]

97. Kim, S.; Hawkes, E.; Cho, K.; Jolda, M.; Foley, J.; Wood, R. Micro artificial muscle fiber using NiTi spring for soft robotics. In Proceedings of the 2009 IEEE/RSJ International Conference on Intelligent Robots and Systems, St. Louis, MO, USA, 10–15 October 2009; pp. 2228–2234. [CrossRef]

98. Holschuh, B.; Obropta, E.; Newman, D. Low Spring Index NiTi Coil Actuators for Use in Active Compression Garments. *IEEE/ASME Trans. Mechatron.* **2015**, *20*, 1264–1277. [CrossRef]

99. Chu, J. Shrink-wrapping spacesuits, spacesuits of the future may resemble a streamlined second skin. *MIT News* Spetember-2014. Available online: https://news.mit.edu/2014/second-skin-spacesuits-0918 (accessed on 22 April 2021).

100. Shimoga, G.; Choi, D.-S.; Kim, S.-Y. Bio-Inspired Soft Robotics: Tunable Photo-Actuation Behavior of Azo Chromophore Containing Liquid Crystalline Elastomers. *Appl. Sci.* **2021**, *11*, 1233. [CrossRef]

101. Clabaugh, C.; Matarić, M. Robots for the people, by the people: Personalizing human-machine interaction. *Sci. Robot.* **2018**, *3*, eaat7451. [CrossRef]

102. Lobontiu, N.; Goldfarb, M.; Garcia, E. A piezoelectric-driven inchworm locomotion device. *Mech. Mach. Theory* **2001**, *36*, 425–443. [CrossRef]

103. Lin, H.; Leisk, G.; Trimmer, B. GoQBot: A caterpillar-inspired softbodied rolling robot. *Bioinspiration Biomim.* **2011**, *6*, 026007. [CrossRef]

104. Koh, J.-S.; Cho, K.-J. Omega-Shaped Inchworm-Inspired Crawling Robot with Large-Index-and-Pitch (LIP) SMA Spring Actuators. *IEEE/ASME Transactions Mechatron.* **2013**, *18*, 419–429. [CrossRef]

105. Wong, C.; Yang, E.; Yan, X.-T.; Gu, D. Autonomous robots for harsh environments: A holistic overview of current solutions and ongoing challenges. *Syst. Sci. Control Eng.* **2018**, *6*, 213–219. [CrossRef]

106. Tsitsimpelis, I.; Taylor, C.J.; Lennox, B.; Joyce, M.J. A review of ground-based robotic systems for the characterization of nuclear environments. *Prog. Nucl. Energy* **2019**, *111*, 109–124. [CrossRef]

107. Wheeler, R.W. Actuation Fatigue Characterization Methods and Lifetime Predictions of Shape Memory Alloy Actuators. Ph.D. Thesis, Texas A&M University, College Station, TX, USA, 2017. Available online: https://core.ac.uk/download/pdf/147255901.pdf (accessed on 8 June 2021).
108. Melton, K.N.; Mercier, O. Fatigue of NITI thermoelastic martensites. *Acta Metall.* **1979**, *27*, 137–144. [CrossRef]
109. Jaureguizahar, S.M.; Chapetti, M.D.; Yawny, A.A. Fatigue of NiTi shape memory wires. *Procedia Struct. Integr.* **2016**, *2*, 1427–1434. [CrossRef]
110. Nespoli, A.; Besseghini, S.; Pittaccio, S.; Villa, E.; Viscuso, S. The high potential of shape memory alloys in developing miniature mechanical devices: A review on shape memory alloy mini-actuators. *Sens. Actuators A Phys.* **2010**, *158*, 149–160. [CrossRef]
111. Zainal, M.A.; Sahlan, S.; Ali, M.S.M. Micromachined Shape-Memory-Alloy Microactuators and Their Application in Biomedical Devices. *Micromachines* **2015**, *6*, 879–901. [CrossRef]

MDPI
St. Alban-Anlage 66
4052 Basel
Switzerland
Tel. +41 61 683 77 34
Fax +41 61 302 89 18
www.mdpi.com

Metals Editorial Office
E-mail: metals@mdpi.com
www.mdpi.com/journal/metals